Integration of Biomaterials for Gene Therapy

Scrivener Publishing
100 Cummings Center, Suite 541J
Beverly, MA 01915-6106

Publishers at Scrivener
Martin Scrivener (martin@scrivenerpublishing.com)
Phillip Carmical (pcarmical@scrivenerpublishing.com)

Integration of Biomaterials for Gene Therapy

Edited by
Rishabha Malviya
Sonali Sundram
and
Neelam Jain

Scrivener Publishing

WILEY

This edition first published 2023 by John Wiley & Sons, Inc., 111 River Street, Hoboken, NJ 07030, USA
and Scrivener Publishing LLC, 100 Cummings Center, Suite 541J, Beverly, MA 01915, USA

For more information about Scrivener publications please visit www.scrivenerpublishing.com.

Wiley Global Headquarters
111 River Street, Hoboken, NJ 07030, USA

For details of our global editorial offices, customer services, and more information about Wiley products visit us at www.wiley.com.

Library of Congress Cataloging-in-Publication Data

ISBN 978-1-394-17473-7

Cover image: Pixabay.Com
Cover design by Russell Richardson

Set in size of 11pt and Minion Pro by Manila Typesetting Company, Makati, Philippines

Printed in the USA

10 9 8 7 6 5 4 3 2 1

Contents

Foreword

It gives me immense pleasure to write the foreword for this book edited by Dr. Rishabha Malviya. He is well-known and dignified in the pharmacy field, and his research has an excellent reputation. The quality of his work is always impressive, and this time is no exception.

Gene and cell therapies are seeing a resurgence, with an acceleration in scientific innovation and renewed interest in how they should benefit society. The book *Integration of Biomaterials for Gene Therapy* will enhance our understanding of biomaterial's functionality in gene therapy, which is a promising treatment option for many diseases such as cancer, inherited disorders, and certain viral infections. Scientists who are involved in drug delivery targeting will find this book to be a useful resource, because it provides essential information about a key biomaterial-integrated gene therapy, discusses how biomaterials can assist in targeted gene treatment, and brings together professionals in the field to explain the optimization, development, and integration of biomaterials for gene therapy. Moreover, this book emphasizes bio-based materials that have been used already in potential applications of biotherapy, including gene delivery, immunotherapy, stem-cell therapy, and tissue engineering.

In this book, various international professionals present a broad view of the subject, and the subsequent efforts of its readers will enhance the effectiveness of biomaterial-based delivery. I truly appreciate the time and effort that the authors have invested in this project. I believe that each one has taken extra pains to complete this book, paying great attention to its quality and timeliness.

Dr. Ramchander Khatri
Asst. Controller of Examination
Delhi Pharmaceutical Sciences and Research University,
(Govt. of NCT of Delhi), New Delhi, India

Preface

Many organic and inorganic, synthetic, or naturally sourced materials are considered usable in biomedical and pharmaceutical sciences because they are not toxic to humans. These materials are commonly referred to as biomaterials because they can be introduced into living human tissues for prostheses, as well as for drug delivery, diagnosis, treatments, tissue regeneration, and a variety of other clinical applications. For a while now, nanoparticles and bioabsorbable polymers have substantially expanded the fields of application of biomaterials, gaining much greater attention from the biomedical world. The development of biomaterials and their production are not new concepts in science and research. More recently, biomaterials have been employed to deliver genes in a more programmable, controlled, and targeted manner. Unfortunately, despite recent developments, there has been little success in transferring innovative medication and gene delivery systems to the clinical practice. Consequently, the need to close gaps in the translational application of biomaterial-based drug and gene delivery systems is urgent. Results can be improved and made more efficient by having a thorough understanding of the nature, characteristics, properties (physical, chemical, and mechanical), and biological performance of the various biomaterials. To address the need, this book covers biomaterials-mediated gene delivery utilizing gene vectors and the fundamental facts about biomaterials, including various biological and engineering challenges.

This book comprises research and review articles from leading researchers with multidisciplinary experience. It discusses many broad topics, including nanoparticle-enabled gene therapy, inorganic nanocarrier-based gene delivery, non-viral delivery of nucleic acid, biocompatible hydrogels, silk, and polysaccharides-based gene delivery. Other gene delivery topics discussed include the use of smart and engineered biomaterials, combined therapy with growth factors and cell transportation, and the prospects and challenges in the treatment of different diseases, including cancer.

The book bridges the knowledge of pharmaceutics, engineering, basic sciences, and clinical research fields in a way that will help the research community to expedite the clinical application of these therapies for various diseases and conditions.

The Editors

Acknowledgment

"Trust in the lord with all your heart and lean not on your understanding; in all your ways submit to him, and he will make your path straight"

First, I would like to thank God Almighty for giving me the life, opportunity, and guidance to achieve my goals and be successful.

I am deeply indebted to Galgotias University which has allowed me to rise as a researcher and provided continuous motivation and support. Thank you for believing in me.

A few people have helped to make this book successful, and I am very grateful to all. I extend very special thanks to my friends who deserve special acknowledgment and thanks.

I could not have undertaken this journey without the support of my family. Their endless love and encouragement make possible the completion of all my tasks. Words cannot express my gratitude towards them.

As the principal editor of this book, I wish to express my sincere gratitude to the prestigious Wiley-Scrivener Publishing imprint for their excellent services. I have especially enjoyed my scholarly work with Scrivener as they have been timely, creative, and very helpful on my previous books with them.

Rishabha Malviya

1

Biocompatible Hydrogels for Gene Therapy: Advancement and Applications

Ankita Gupta[1] and Swatantra K.S. Kushwaha[2*]

[1]Department of Pharmaceutics, Bombay College of Pharmacy, Mumbai, India
[2]Krishna Institute of Pharmacy and Sciences, Kanpur, India

Abstract

Gene therapy is gaining researchers interest to be a beneficial armamentarium in the treatment of various acquired and inherited diseases. Hydrogels carry the good potential to modify present gene therapy due to their superior biocompatibility, capability to mimic an environment that supports new cells and tissue growth, and regulate gene delivery to promote differentiation and improve therapeutic after effects. Hydrogel can also be integrated with genes to initiate differentiation for tissue regeneration. An emerging promising strategy is to design hydrogel for dual cell and genetic material distribution utilized to engineer the bones and tissue to stem cell differentiation down the chondrogenic lineage to produce cartilage. Hydrogel-based delivery of genes was also investigated for the repair of cardiac tissue and coding of Plasmid DNA. Hydrogel bio-materials can overcome barriers to gene therapy by meticulously regulating the expression of transgene vectors. The limitations include preventing genotoxicity arising from integrated vectors for gene delivery, regulating genome off targeting, increasing transfer efficacy of genes to a level effective for cure of ailments, reducing the body's defenses generated by vectors administered *in-vivo*, moving to common consent for dealing with concerns like justified germline editing. The emerging scope of hydrogel for gene therapy, and its multi-variant aids for the well-being of living beings, as studied via technical innovations along with scientific accomplishments across earlier numerous ages, supports continual hopefulness along with escalating hard work for formulating the delivery system best amongst treatment of serious diseases.

Keywords: Hydrogel, gene therapy, natural polymer, tissue engineering, vectors, PEG, VEGF, plasmid DNA

**Corresponding author*: swatantrakushwaha15@gmail.com

Rishabha Malviya, Sonali Sundram and Neelam Jain (eds.) Integration of Biomaterials for Gene Therapy, (1–12) © 2023 Scrivener Publishing LLC

1.1 Introduction

The material of engrossment in this chapter is on hydrogel for gene therapy. Hydrogels are a 3D network formed by cross-linking of hydrophilic polymers that can imbibe huge amounts of water and biological fluids thereby maintaining its structure. Its implicit biocompatibility and physical properties resembling natural tissue make it apt for a myriad of *in vivo* applications. Recent applications for hydrogel are in the areas of gene therapy, therapeutic delivery, tissue repair & engineering, wound repair, bio-sensors elements, pH sensitivity and as diagnostics tools.

Gene delivery through hydrogel is investigated to be an interesting tool for multiple clinical applications; a few are regenerative medicine, inherited disorders treatment via gene therapy and drug delivery. The large water imbibing capability and least reactive synthetic features of hydrogel makes it apt for delivery of gene by preserving lentiviral vectors activity and protect it from host immune response. Hydrogels are also designed for use in restoration of tissues as three dimensional bio-printed structures [1], injectable [2], stimuli-responsive, or adhesive systems [3].

Gene delivery can be utilized for short-term as well as extended transgene presentation at particular location, resulting in local or systemic effects. It can be also utilized by altering gene sequence or for multiple gene delivery escaping re-fabrication of delivery system. This flexibility aids its increase utilization for biomaterial- gene delivery technique yield the most therapeutic benefit. The impeding factors to be addressed are virus activity, dose, immune response generated by the host. Biomaterial platforms could be one of the methods to address above limitation encouraging stability of vector and also protecting form host immune system. Hydrogel also investigated by researchers for tissue engineering, it encourages permeation of cells, also drive to encapsulate vectors that code for elements that aids development of tissues. Sustained and localized expression of lentiviral vector at implantation site can be achieved by the hydrogel delivery tool.

The tunable structure of hydrogel assists to regulate it swelling and degradation which is leveraged to drive gene therapies as well as regulate behavior of transplanted cells. Entrapment and modification of vectors within hydrogels restricts its elimination through defense mechanism of humans. Hydrogels can also be fabricated for onsite delivery or get biologically stimulated, drive penetration of stem cells, cartilage cells, etc. and to develop tissues and bones. Hydrogel can be synthesized by either cross-linking as well as self-arrangement of hydrophilic man made or organic polymer like chitosan, dextran, cellulose, poly (N-isopropyl acryl amide), polyurethane respectively.

It's a challenging task, though, to control the properties of hydrogel in the complex tissue environment of healthy and diseased cells. Moreover, it is also adorned using bio-similar characteristics to assist binding and transfer of specialized cells, resulting in increased ability of cells to undergo transduction. In contrast to other biomaterial delivery systems, hydrogels provide a hydrated, less reactive, aqueous synthetic environment that assist in entrapment of gene vectors. Hydrogel might be utilized for several therapies. For example, it might be used to deliver vectors for gene technologies, and may improvise functions of cells being transplanted, hydrogels intended for VEGF is also investigated to cause neo-vessel development in addition to inhibit thrombosis in stent.

Hydrogel are also adorned to initiate interactions among chemicals and lentiviral vectors resulting in successful retention or release of vector in gel assisting distribution of transduced cells. Hydrogel cell infiltration rate is also found to influence location as well as the extent of transgene expression. Significant research has been conducted in the design, optimization, and applications of hydrogels for therapeutic purpose. But still its large-scale manufacturing has not been achieved due to lack of novel materials, low efficiency in technical terms for bulk production of hydrogels, and scope to formulate scaffolds containing biologically active macromolecules. The goal for this chapter is to publish study report on cutting-edge research work on development and applications of hydrogels with major emphasis on its use as gene delivery.

1.2 Hydrogels Classification

1. Depending on source of polymer: It can be synthesized by either natural/synthetic polymers.
2. Depending on composition of polymer:
   (a) Homo-polymeric: Single species of monomer is used to develop polymer network.
   (b) Co-polymeric: Polymeric network chains consist of alternate arrangement of different monomers with hydrophilic component.
   (c) Multiplier: Twining of cross-linked and non-cross-linked polymer in a network.
3. Depending on configuration
   Amorphous, Crystalline, and Semi crystalline Polymers

4. Depending on physical appearance
   Depending on polymerization technique hydrogels can be classified to occur in form of Films, Matrix or spheres.
5. Depending on E-charge they posses
   Depending on charge be present on polymer chains it can classified as Nonionic, Ionic, Amphoteric or Zwitter ion.

1.3 Fabrication of Hydrogels and Its Desirable Technical Features

Hydrogels can be developed by self-assembly process or by cross-linking of polymers of hydrophilic nature to form a web like system [4]. Its spongy as well as hydrous morphology result in formation of gels and aids in swelling in living micro-environment, that eases it localized administration through injection without any invasive surgery. A hydrogel synthesized by following methods:

Physical approach: It includes hydrogen bonding/ionic/hydrophobic interaction.
Chemical approach: It includes photo-polymerization or cross-linking pathway.

The core technical features of an ideal hydrogel include:

- The high capacity of absorption in saline.
- Type of application influences expected rate of absorption (particle size and porosity).
- High under load absorbency.
- Low content of soluble matter and residue matter.
- Good stability during storage.
- Biodegradable.
- Maintain normal pH after swelling also.
- Photo stability, Colorless, odorless, and inert.

1.4 Factors to be Tuned for Gene Encapsulation in Hydrogels

Diffusion of vector hydrogel is depended on various parameters such as amount of pores present, hydrophobic or hydrophilic, degradation mechanism and interaction rate of material with vector. Virus retention ability and the capability of cells to infiltrate hydrogels are major factors influencing transgene expression. Polymers of natural origin are preferred owing to good bio-compatibility as well as its intrinsic ability of natural connections with cells that influence addition with cells along with its differentiation. Whereas polymers of synthetic origin have ability to fabricate hydrogels with precisely controlled physical characteristics such as its degradation, influencing development of tissues along with transfer of genes. It also can modify duration and extent of presentation of transgene at grafted location as compared to bolus injection.

The delivery efficiency of gene is dependent on combined effects of activity of vector and efficiency of hydrogel on host response. Sustained release is obtained by vectors entrapped within hydrogels. It aids in keeping high vector concentrations at local site of administration resulting in cellular infiltration. Hydrogel synthesized using natural or synthetic polymers can be prepared under mild conditions and therefore are better for entrapping gene therapy vectors without influencing activity of vectors.

Mesh size of hydrogels are one of the important factor influencing release or retention of vector. If the mesh size is more than the diameter of vector then the release is rapid with diffusion whereas if vector diameter is more than mesh size then the release rate depends on hydrogel degradation rate. Smart hydrogels that are formed by stimuli sensitive polymers, variation in there pore size occurs on stimulation and are excellent in controlling vector release.

Interactions among hydrogels and vector also prevent vector diffusion away from hydrogel resulting in increased localized concentration of vector and thereby aids effective gene transfer. Linkage of vectors with biological substrate also aids attachment of cells of and can also retain virus and cell for efficient gene transfer. Hydrogel prepared from cationic polymers, poly ethyelene glycol adorned with peptides with cationic and chitosans are utilized to enhance retaining of vectors. Lentiviral capsid proteins are modified to Co-bind with components of fibrin at time of development of polymers and deliver vector via degradation of hydrogels.

Optimization of interaction strength between vector and hydrogels must be done such that it must be not too strong to prevent release of vector to prevent its internalization and interaction by nearby cells nor too weak to inhibit vector retention. Rate of distribution is adjusted to maintain 3D retention of gene vectors. Maximum infection level has been observed with lentivirus encapsulated in low-medium fibrin concentration, whereas vector diffusion rate was found to be inefficient at high fibrin concentration.

While above searches reveal the budding efficacy of hydrogels centered on natural polymers for localized gene delivery, synthetic hydrogels are utilized to meticulously control mode and dynamics of cell infiltration as well as degradation rate of hydrogels. Such as, PEG hydrogels cross-linked by protease-sensitive peptides. On movement of cells via hydrogels, degradation of crosslinks permit synchronized permeation of cells along with delivery of gene. Utilization of manmade gels similarly permit addendum of specialized and protease-susceptible peptides that favorably influence infection of specific cell type.

For facilitating robust cell-infiltration, synthetic hydrogels are integrated with interconnected macropores utilizing diverse approaches. Amending size of macropores along with density can regulate cell infiltration in parallel of degradation as well as vector release of hydrogels. Hydrogels also fabricated to foster permeation and transfer of genetic material of specific cell types through integrating short biologically active peptide obtained by extra-cellular medium like RGD peptides to scaffoldings. Regarding gels based on synthetic polymers, the meticulous regulation of adhesive sites densities is utilized to modify potency of biomaterial supporting cell growth and attachment and types of receptor in extra cellular cells which facilitate above connections.

Speed of matrix degradation which transforms movement of cell via matrix impacts combination with tissue of the host. Incorporation with bio compatible material designed towards current precise signs like spatial, mechanical, and chemical as well as gene therapy vectors embodies an encouraging approach to foster development and restoration of tissues.

1.5 Recent Advances on Hydrogels for Gene Delivery

Hydrogels are utilized as regenerative medicine because of its capability to develop an environment that aids transplanted or endogenic permeating cells that facilitate them to recondition the purpose of tissue vanished because of ailment. Moreover, it can be used like transportation tool for medicinal genes that guide and augment the role of relocated or endogenic cells. Some of the latest advancements are discussed in this section.

Lee *et al.* [5] fabricated a chitosan based thermo-responsive hydrogels for gene therapy to enrich the local transgene expression. Photo-irradiation approach was utilized for chemical cross-linking. The duration of photo-irradiation as well as content of chitosan was found to regulate rate of release of plasmid DNA incorporated in hydrogel.

Gene therapy could be used to aid dermal recurring injuries exhibiting transformed expression of genetic material and de-regulation of mRNA.

Though, it is challenging to carry extended and controlled delivery of therapeutics due to its irregular topology.

Jose and Researchers designed a model for dressing of lesion centered upon viscous and elastic hydrogels prepared of polyβ-amino esters and polymers made of polyethylene glycol. It was attested to be decomposable, cell-compatible, and first and foremost, injectable. The later features endorsed it to be used *in-situ* gel form that can be easily applied to irregular, deep wounds like chronic ulcers [6].

Ravishankar *et al.* utilized chitosan and lignin to develop biocompatible hydrogel. High firmness as well as viscous nature of chitosan was found to be enhanced on integration of lignin. *In vitro* studies reported hydrogels to be benign to mesenchymal stem cells and also *in-vivo* assay it was found to be nontoxic to zebra fish at concentration of 10 mg/mL [7].

Lei *et al.* [8] displayed utilization of fibrin hydrogels for competent, cell-controlled, and 3D retained gene transfer to targeted cells. Researchers studied fibrin-facilitated gene transfer via incorporating plasmid DNA inside hydrogels matrix through polymerization as well as utilized dual approaches for gene transfection and cells, retained outside or inside hydrogel. Retention of target cells along with lipoplexes inside hydrogels granted better gene transfer efficacy compared to cells seeded on outside surface of gel. Fibrin hydrogels amended transport of genes via reducing toxicity effects on cells of transaction mediators like lipofectamine.

In conventional gene transfer approach target cells were exposed to large content of lipoplexes promptly, although in case of fibrin related hydrogels, cells for gene delivery were bared with exposed to lipoplexes steadily degrading the hydrogel. Its steady contact with cell damaging mediators mitigated stimulation of pro-apoptotic pathways thus reduced necrosis and apoptosis of cells.

Cyclodextrin depended supramolecular hydrogels comprising cationic polymers also displayed reduced cell toxicity along with effectual transfection in genes in cell cultivations. Prime characteristics in developed matrix are that polyplexes linked to DNA and any polymer can be reformed via formation of complex because of high proportion of cyclodextrin [9, 10]. Pegylation of PEO-adamantane via arrangement of complex connecting adamantane to β-cyclodextrinon complex of polymers with DNA can be observed for stabilizing polyplex nanoparticles, resulting in better transfection in genes.

Polyethylene glycol related photo-cross-linked gels for delivery of gene in sustained mode, got explored with the experiment of Anseth researchers. Polymer matrix was developed by poly ethylene glycols, which were core fastened to poly lactic acid and poly caprolactone block to encourage degradation, and adorned using meth-acrylate moieties that promote its fast photo-polymerization. Release from pDNA was modified for several days, having straight arrangement devoid of the preliminary sudden release through amending variety and configuration of polyester.

Nikolaev *et al.* [11] fabricated PCL along with fibrin containing system of hydrogel providing on site release of plasmids holding genes coding for vascular endothelial and fibroblast growth factor, rooted inside sciatic nerve of rat having nerve diastasis. On site jab of plasmid at place of traumatic wound promoted restoration at sciatic nerve of rat along with functions of motor and sensory nerves retrieved.

In spite of great potential of hydrogels, its usage is at initial stage for delivering gene in a controlled manner to repair cartilage. Osteochondral and chondral components can be particularly hopeful to deliver the polymeric gene because of limited flow of blood at the site which worsens transfer of DNA. Gonzalez-Fernandez *et al.* [12] displayed spatial porous bioinks which offer a momentary 3D transfer of gene via moderating the porous features. Hydrogels derived from MC-Alginate were stacked by osteogenic, chondrogenic and SOX9 genes to yield automatically supported, scaffolding by activation of genes. Resultant mode of gene delivery, *in vitro* stimulated osteogenesis and chondrogenesis in mesenchymal stem cells. On implantation inside cells, 3D printed bioinks fabricates reinforced advancement of bony tissues with new blood vessels laden with cartilage layer.

Wegman and group [13] experimented efficacy for development of bones via means of hydrogel derived from alginates. GAMs compared to plasmid DNA-GAMs, resulted in improved production of cartilage and muscle packed with plasmid DNAs encoding for osteogenic factors. It was merged to specialized markers *in vitro*. On transfection of cells with above system resulted in expression of genes for more than 2 weeks.

Spinal cord injuries frequently directed to dysfunction of nervous system because of malfunctioning in re-development of axon. Fabrication of spatial nanofibrous hydrogels scaffoldings as biostimulated and biocompatible media that offer extended delivery of genes and contact guidance, demonstrated to be appealing methodology aimed at healing of neurological wounds. Nguyen *et al.*, [14] developed spatial hydrogels derived from collagen embedded with Poly-e-caprolactone polymers co-polymerized to ethyene phosphate via electron spinning techniques to release protein like neutrotrophins and micro RNA *in vivo*.

1.6 Conclusion

Hydrogels carry vast prospects to improvise existing technology for gene delivery and have a firm effect on gene therapy technologies as well as regenerative medicine. Contrary to injecting a bare lentivirus, its entrapment inside gels or synthetic amendment using polymers enhance *in vivo* retention as well as inhibit deactivation of a vector through the body's defense mechanism. In addition to this, it can retain localized, elevated concentration of vector that surmounts bulk movement restrictions for gene delivery.

Furthermore, it also displays peptide motifs which encourage adhesion of cells and retention of vectors. This approach increases co-localization of target cells and vectors, thus stimulating gene transfer. It also has immense possibility for various gene as well as reformative therapies. It increases capacity for lentiviral vector therapy via restricting it to localized tissues, protecting scavenging the immune system along with increased efficacy of transduction. In case of restoration therapy medicine as well as gene engineering purposes, hydrogels are also modified for rendering cell adhesion, bio-degradable characteristics which imitate relevant tissues *in vivo*.

Hydrogel possesses extended prospect to regulate transgene expression at various levels, time span and specific targets, which majorly blocks successful delivery of gene. Entrapment of gene vectors in gels or hydrogel reformation protects deactivation of a vector via the host defense mechanism. Moreover, it is tailored by bio imitating properties that support connection and movement of specialized cells, to enhance transduction in cells. Structure of gel dependent gene therapy for reformation treatment must maintain equilibrium between the reduced immune response and control vector clearance during the increment of cell permeation.

Gene vectors entrapped inside hydrogels which formulate for controlling activation of immune cells possess restricted delivery of gene, that arise by less recruitment of cells because of reduced cell-template connections specific for them which regulate stimulation of cells in the immune system. In reformative treatment, this is developed for promotion of cellular processes related to development of tissues, and release of vectors results in enhanced bioactivation.

Effectual transfer of genes exhibit an equilibrium in vector retaining against its release from hydrogels, the vector steadiness and speed of permeation in cell inside system. Several prospects persist for more enhancing hydrogel characteristics, emerging tools to enhance cell transduction along with offering multidimensional and sequential restriction on the expression of genes which simulate complicated models of gene expression existing in tissue growth. It offers an exclusive means which is engineered to coordinate with necessities for effective gene delivery and fostering tissue formation.

To conclude, hydrogels for gene transfer correspond to a hopefully budding scope for exploration of forthcoming efficient as well as secure medication of a plethora of ailments. Above reviewed approach assists by-pass existing restraints in gene remedy, they are also modifiable as bespoke for restoration of tissues.

References

1. Singh, S., Choudhury, D., Yu, F., Mironov, V., Naing, M.W., *In situ* bioprinting—Bioprinting from benchside to bedside. *Acta Biomater.*, 101, 14–25, 2020.
2. Peña, B., Laughter, M., Jett, S., Rowland, T.J., Taylor, M.R.G., Mestroni, L., Park, D., Injectable hydrogels for cardiac tissue engineering. *Macromol. Biosci.*, 18, e1800079, 2018.
3. Sood, N., Bhardwaj, A., Mehta, S., Mehta, A., Stimuli-responsive hydrogels in drug delivery and engineering. *Drug Delivery*, 23, 748–770, 2016.
4. Valle, D., Hydrogels for biomedical applications: Cellulose, chitosan, and protein/peptide derivatives. *Gels*, 3, 27, 2017.
5. Lee, J.I., Kim, H.S., Yoo, H.S., DNA nanogels composed of chitosan and Pluronic thermo-sensitive and photo-cross-linking properties. *Int. J. Pharm.*, 373, 93–99, 2009.
6. Duran-Mota, J.A., Quintanas, J., Yani, B.D., Almquist, Borrós, S., Oliva, N., Polyplex-loaded hydrogels for local gene delivery to human dermal fibroblasts. *ACS Biomater. Sci. Eng.*, 7, 9, 4347–436, 2021.
7. Ravishankar, K., Venkatesan, M., Desingh, R.P. *et al.*, Biocompatible hydrogels of chitosan-alkali lignin for potential wound healing applications. *Mater. Sci. Eng. C*, 102, 447–457, 2019.
8. Lei, P., Padmashali, R.M., Andreadis, S.T., Cell-controlled and spatially arrayed gene delivery from fibrin hydrogels. *Biomaterials*, 30, 22, 3790–9, Aug 2009.
9. Pack, D.W., Hoffman, A.S., Pun, S., Stayton, P.S., Design and development of polymers for gene delivery. *Nat. Rev. Drug Discovery*, 4, 7, 581–593, 2005.
10. Davis, M.E. and Brewster, M.E., Cyclodextrin-based pharmaceutics: Past, present and future. *Nat. Rev. Drug Discovery*, 3, 12, 1023–1035, 2004.
11. Nikolaev, S.I., Gallyamov, A.R., Mamin, G.V., Chelyshev, Y.A., Poly(ε-Caprolactone) nerve conduit and local delivery of vegf and fgf2 genes stimulate neuroregeneration. *Bull. Exp. Biol. Med.*, 157, 155–158, 2014.

12. Gonzalez-Fernandez, T., Rathan, S., Hobbs, C., Pitacco, P., Freeman, F.E., Cunniffe, G.M., Dunne, N.J., McCarthy, H.O., Nicolosi, V., O'Brien, F.J. *et al.*, Pore-forming bioinks to enable spatio-temporally defined gene delivery in bioprinted tissues. *J. Control. Release*, 301, 13–27, 2019.
13. Wegman, F., Bijenhof, A., Schuijff, L., Öner, F.C., Dhert, W.J.A., Alblas, J., Osteogenic differentiation as a result of BMP-2 plasmid DNA based gene therapy *in vitro* and *in vivo*. *Eur. Cells Mater.*, 21, 230–242, 2011.
14. Nguyen, L.H., Gao, M., Lin, J., Wu, W., Wang, J., Chew, S.Y., Three-dimensional aligned nanofibers-hydrogel scaffold for controlled non-viral drug/gene delivery to direct axon regeneration in spinal cord injury treatment. *Sci. Rep.*, 7, 1–12, 2017.

2

Use of Polysaccharides: Novel Delivery System for Genetic Material

Prashant Kumar[1], Swatantra K.S. Kushwaha[2*], Neelottama Kushwaha[2], Abhishek Singh[1] and Surya Nath Pandey[1]

[1]Teerthanker Mahaveer College of Pharmacy, Teerthanker Mahaveer University, Moradabad, Uttar Pradesh, India
[2]Krishna Institute of Pharmacy and Science, Kanpur, Uttar Pradesh, India

Abstract

Due to their biocompatibility, biodegradability, and distinctive bioactive features, polysaccharides have been identified as attractive candidates for a variety of biological applications in gene therapy and tissue engineering. Gene delivery is a method to relocate external DNA into the presenter cells for genetic research or gene remedy of human hereditary disorders. Gene therapy is one of the promising therapies for the management of various diseases like cancers; Alzheimer's viral infection as well as inherited diseases polysaccharides (carbohydrates) is usually isolated from plants and animal sources abundant in nature. For the delivery of genes, many polysaccharide-based biomaterials are being researched, including starch, cellulose, chitosan, dextran, and alginates. Further, these polymers are widely accepted by the body due to their biochemical similarity with human extracellular matrix components regulating cell signaling and tissue growth. Further, polysaccharides-based biomaterials are designed as graphene, metal oxide, and carbon nanotubes utilized in diverse fields of research. Recently, the major focus of research has been concentrating on the identification of the efficient and safe substitutes for the viral vehicles required in gene delivery. Future research can be expected to face significant hurdles related to understanding the molecular interactions of interacting compounds, the destiny of polysaccharide nanoparticles *in vivo*, and the molecular interactions between polysaccharides and loaded drugs/genes. The current chapter is primarily concerned with an overview of polysaccharide-based biomaterials for a variety of biomedical applications, including gene transfer. Additionally, the focus will be on structure, cross-linking

**Corresponding author*: swatantrakushwaha15@gmail.com

Rishabha Malviya, Sonali Sundram and Neelam Jain (eds.) Integration of Biomaterials for Gene Therapy, (13–46) © 2023 Scrivener Publishing LLC

mechanisms, biosynthesis, chemical characteristics, biological properties, current problems, clinical usage, recent advancements, and methods to build polysaccharide-derived biomaterials. Finally, numerous future prospects for the design and fabrication of polysaccharides-based materials to accomplish their extensive commercial usage are discussed.

Keywords: Polysaccharides, bioactive, gene delivery, clinical trials, biodegradable, tissue engineering

2.1 Introduction

Polysaccharides are an essential class of biopolymers with decent biocompatibility, easy accessibility, biological acceptability, and prototypes for structural modifications. Polysaccharides derived from natural sources offer antibacterial, antiviral, and anti-inflammatory potential, and they own outstanding biocompatibility. As a result, they have received a lot of attention as efficient biomaterials for use in biomedical applications. As regenerative therapies and drug delivery systems, engineered carbohydrate-based biomaterials offer an alternative. These include cellulose, dextran, hyaluronic acid, chitosan, and other polysaccharides originating from plants or animals. These are the most widely used sources of biopolymers because they can be used to create composites, films, sheets, hydrogels, fibers and composites for the features including tissue engineering, medication delivery systems, and wound healing [1, 2]. Polysaccharides have a wide range of uses, including food, pharmaceutical carriers, biomaterials, hydrogels, food stabilizers/thickeners, biosorbents, and enclosing agents. Monosaccharides naturally polymerize in specific patterns or sequences that are appropriate for each one's function or involvement in the production of polysaccharides. By adding additional functional group(s), certain polysaccharides are further modified, giving rise to unique characteristics [3, 4].

Due to their advantageous properties of low immunogenicity, biodegradability, biocompatibility, and natural abundance, natural polysaccharides have drawn considerable interest in the development of microcarriers and nanocarriers. In this chapter, the most recent developments in gene delivery using natural polysaccharide-based micro- and nanocarriers are explored. Further, on the micro- or nanoscale, polysaccharides are used to create biomaterials and formulations [5].

Gene delivery has been hailed as a potent method for treating both acquired diseases such as multigenetic disorders and those brought on by invading virus genes as well as many inherited diseases [6, 7].

A number of techniques are used by gene delivery systems to aid in the absorption of the chosen gene by the cell. Building a gene delivery system necessitates a thorough study of the relationship between the delivery method and the target cell. It's important to comprehend intercellular traffic and targeting mechanisms in order to develop more effective gene delivery techniques. It is the most popular gene via endocytosis, particularly facilitating cellular uptake of non-viral gene delivery [8, 9].

Cellular release occurs to start DNA transcription and translation as well as the production of the associated protein after the delivery system has been taken up by the cell through endocytosis. In order to enhance gene activity, a successful gene delivery process must reduce the possibility of an inhibiting the inflammatory reaction while also getting past certain obstacles at each stage. Viral gene delivery systems are made up of viruses that have underwent engineering towards becoming replication-deficient. These viruses are able to carry the genes to cells for expression. For the delivery of viral genes, lentiviruses, retroviruses, and adenoviruses are utilized [10–12].

The mechanism for delivering genes is a crucial technological element that is required for both fundamental study and therapeutic gene therapy. The use of polymer-based carriers have received a lot of interest since they are simpler to prepare and less harmful than viral carriers [13, 14]. However, there are still issues like toxicity spurred on by excessive contact between cationic gene transporters and the cytoplasmic membrane and reduced gene expression than viral carriers. For efficient gene expression, it is important to regulate nuclear entrance, endosome disruption, and cellular uptake of the polycation/pDNA complex. Moreover, gene delivery through polysaccharides methods such cationic dextran, cationic pullulan, chitosan, and chizophyllan18 have been created thus far due of their low toxicity and great biocompatibility [15].

2.1.1 Commonly Available Polysaccharide-Based Natural Biomaterials with Their Physicochemical Properties

- Chitosan

  Chitosan is thought to be the second most common natural polysaccharide on the market, behind cellulose. For 50,000–800,000 Daltons, chitin, and chitosan have been the main subject of study for several researchers [16].

  Natural sources of chitosan include certain fungus like Mucor, Rhizopus, and Absidia. Chitosan is primarily a deacetylated derivative of chitin. Generally speaking, chemical and enzymatic methods are used to extract chitosan [17].

The natural polysaccharide chitin has a significant derivative called chitosan (CS). It is a naturally occurring, biocompatible cationic polysaccharide that contains amino acids. Because it contains amino groups, chitosan is simple to chemically alter. The capacity of CS-based DDSs to cling to negatively charged cell membranes, which is favorable to cell endocytosis, results in minimal immunogenicity and immunogenicity [18, 19].

The arrangement of various sugar molecules is evident in chitosan's structure. Due to its special attributes to be biocompatible, biodegradable, and nontoxic, chitosan has become crucially significant in biomedical applications [20]. Numerous factors, comprising crystallinity, molecular weight, degree of deacetylation, etc., have a substantial impact on the physicochemical properties of chitosan. It is soluble in weak acids but insoluble in water and in alkaline conditions [21].

It has been established that chitosan-based biomaterials possess analgesic, antibacterial, antifungal, and mucoadhesive characteristics. These features are crucial and extremely valuable in tissue engineering and medication delivery applications. A major drawback to the use of chitosan in biological applications has been its solubility issue. The issue has so far been consistently addressed by numerous researchers through chemical modifications [22]. Additionally, the functional groups present in it might be chemically changed due to the development of carrier systems with good potential for transdermal, ocular, oral and nasal administrations [23]. Particularly, the production of chitosan (Figure 2.1) nanoparticles and their physicochemical conformation (such as concentration of protein and chitosan, deacetylation level

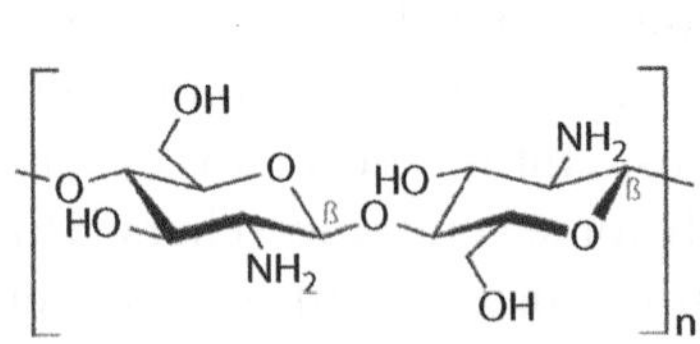

Figure 2.1 Chitosan.

and molecular weight) influence their potential to improve mucoadhesive property and facilitate drug/gene administration [24].

- Alginate

The main source of alginic acid is found inside the cell walls of common brown algae. Because of their peculiar egg-box shape and other chemical properties, gels can easily be created in aqueous media. Alginate can produce hydrogel through intramolecular and intermolecular cross-linking with divalent metal ions (Ca2+, Cu2+, Zn2+, Pb2+, etc.) in aqueous solution [25, 26]. Alginate (C6H8O6) n is an anionic based block co-copolymer composed of mannuronic and L-guluronic acids, with the M/G ratios vary depending on the sources. It is a linear, unbranched polymer [27].

Alginate is a polysaccharide that is called as frequently experience bioerosion, an extraordinarily gradual development that occurs in the microenvironment of a body and is also enzyme mediated, like many other polysaccharides. This is crucial to keep in mind when creating scaffolds as it facilitates tissue engineering and provides cells with a simpler way to get nutrients, the disposal of metabolic waste, appropriate oxygen, and more room to develop. Alginate solutions are insoluble in organic solvents and slightly more viscous in acidic environments. Their solubility in water depends on the solvent's pH, the presence of free metal cations, and the medium's ionic strength. Since alginate has so many different applications, numerous research teams have thoroughly examined the chemical changes applied to it. Using this polysaccharide, it is feasible to successfully produce enzymes, alkyl chains, hydrophobic polymers, amino acids and other functions by utilizing the carboxylic groups (-COOH) and cis diols contained in the alginate monomers [28–30].

Because of their biocompatibility, cross-linking activity, and gelling properties, alginate-based biomaterials and nanofibers are employed in the tissue engineering such wound dressing and drug delivery systems. The four most popular procedures for creating alginate nanoparticles include reverse microemulsion, decarboxylation, cross-linking, and evaporation [31–34]. Alginate salts that are water soluble are changed into water-insoluble salts by incorporating divalent groups (calcium, strontium as well barium), which cause the ionic cross-linkage between residues [35]. The cross-linked gel is created by electrostatic contact involving carboxyl groups and divalent ions. This phenomenon has a considerable impact on medication release. Owing to the anionic nature of alginate it can interact with cationic components. Because it contains positively charged medicines and molecules, it can be used to prepare delivery systems. The most often employed alginate based matrix for the development of drug- and gene-delivery systems is calcium alginate [36, 37].

Chitosan-coated alginate gel beads can be stored and circulated *in vivo* to extend the residence period. Additionally, compared to alginate or chitosan separately, alginate-chitosan polymeric complex was demonstrated to be a promising approach for the regulated release of integrated compounds. Additionally, isoniazid, pyrazinamide, ethambutol, and rifampicin's bioavailability has been improved with the use of alginate (Figure 2.2) nanoparticles [38–40].

- Dextran

Dextran, commonly referred to as glucan, is a starch-like water-soluble polysaccharide complex. It is frequently used as an adjunctive therapy for conditions like hemorrhagic shock and thrombus. Dextran is innocuous to living things and has excellent hydrophilicity and water solubility. Dextran's molecular chain has several hydroxyl groups, making chemical modification easy [41].

By using such carriers, medications can be released under regulated conditions without being inactivated by bodily fluids or enzymes. In the presence of dextranse, the poly(DL-lactide co-glycolide), modified dextran with PLGA grafts core-shell-structured carrier demonstrated a rapid drug release rate [42, 43].

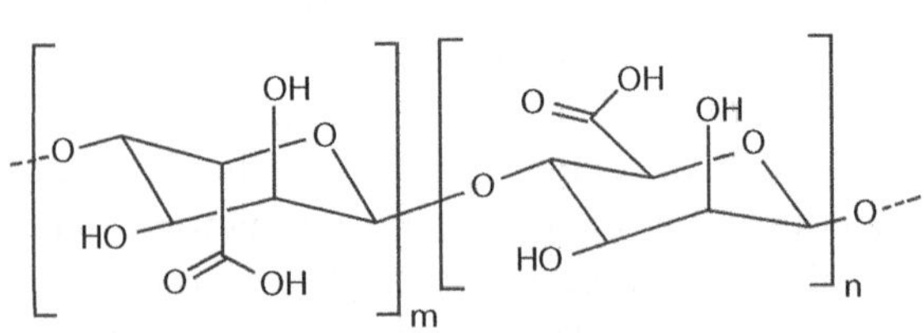

Figure 2.2 Alginate.

Lactic acid bacteria use sugar as a source to create the extracellular bacterial polymer dextran. Dextran, a homopolymer polysaccharide with the unit formula C18H32O16, is composed of glycosidic links with lengths ranging from 3 to 2000 kDa, which combined form linear chains and branches, respectively. The microbial strains and extraction technique have an impact on the physicochemical characteristics of dextran. Dextran is easily dissolved in distilled water and a few solvents, including formamide, glycerol, methyl sulphide, and ethylene glycol [44]. Dextran-based polymers have a variety of applications in the food business, cosmetics industry, wastewater treatment industry, and biomedical applications because of their biocompatibility, water holding capacity, low toxicity, and biodegradability [45]. The structure of dextran, which is similar to many other polysaccharides, is essential to understanding how it performs various tasks. Dextran's characteristics are significantly influenced by both its molecular mass and degree of branching. A close look at the dextran structure reveals that it has hydroxyl groups and aldehyde groups (CHO) at terminal point, two conventional sites for chemical functionalization (Figure 2.3) [46].

BSA dextran conjugates have also been demonstrated to successfully encapsulate hydrophobic medications like ibuprofen. The decoration of dextran nanocarriers for medicinal applications has received particular attention due to its biocompatibility, biodegradability, nonantigenicity, and

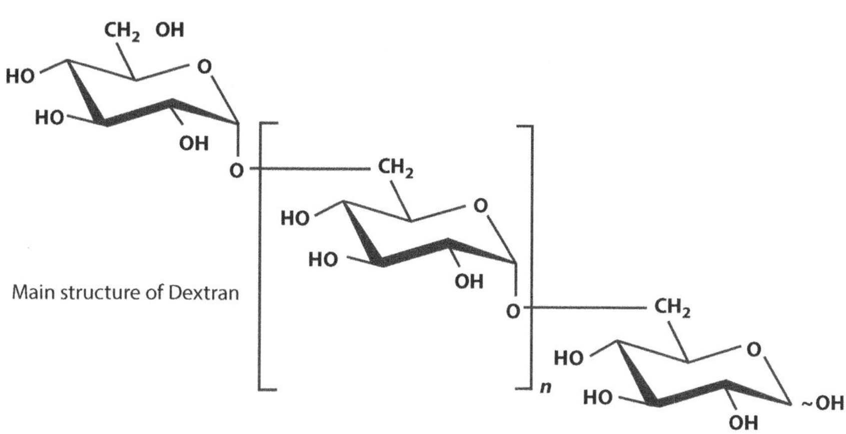

Figure 2.3 Dextran.

non-immunogenicity, dextran is a desirable material to use in the production of nanoparticles that will be used to deliver therapeutic targets. For example, doxorubicin has been delivered to the cancer cell's nuclei using advanced drug delivery systems called dextran nanoparticles in chemotherapy [47–49].

- Cellulose

  Cellulose is the most typical and prevalent biopolymer known to exist on Earth. In 1838, Anselma Payen found cellulose in plant tissues and isolated it; he then identified its molecular formula as (C6H10O5) n. The first thermoplastic polymer made from cellulose was celluloid, created by Hyatt Manufacturing Company. Hermann Staudinger's discovery of cellulose's polymeric structure is merited [50]. The sources of cellulose include the plant kingdom (including jute, wheat straws, wood, hemp and cotton), as well as marine life (including tunicates), an algae (including valonia), bacteria (including acanthamoeba, acetobacter, and achromobacter), mineral and fungi sources. Cellulose is typically hydrophobic by nature and insoluble in the majority of organic solvents due to its large molecular weight and crystalline structure. On how much of it has been polymerized, most of the features of cellulose depend. With the use of processes like esterification and oxidation of cellulose, among others, many cellulose derivatives have been produced with better physicochemical qualities than the original form, such as higher water solubility and decent solubility in other organic solvents. Cellulose (Figure 2.4) and its derivatives are useful materials for their utilization in medicine administration, wound dressing, and various associated biomedical applications due to their low toxicity, desired biodegradability, significant biocompatibility, abundance in nature, and remarkable mechanical qualities [51, 52].

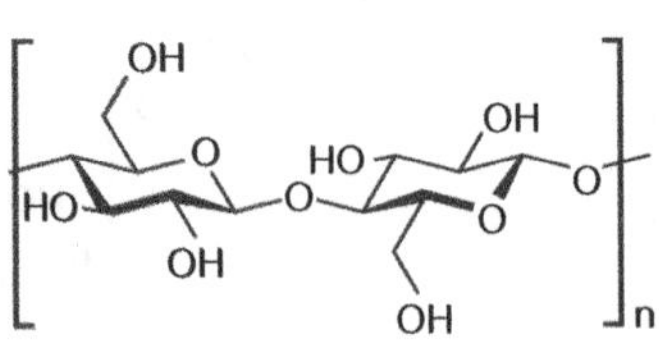

Figure 2.4 Cellulose.

- Chondroitin Sulfate

  The extracellular matrix (ECM) and cell surfaces contain a family of polysaccharides known as chondroitin sulphate (CS), which is often found coupled to proteins as proteoglycans. They are 4)-GlcA-(13)-GalNAc-alternating linear polysaccharides. 1-disaccharide units, whose hydroxyl (-OH) groups can be sulfated in different amounts and locations to create different CSs (GlcA = glucuronic acid, GalNAc = N-acetylgalactosamine) [53]. Chondroitin sulphates are hybrid structures made primarily of 4- or 6-O-GalNAc monosulfated disaccharides with only a small amount of non-sulfated disaccharides (CS-0). However, marine species have been shown to have disulfated (CS-B, CS-D, CS-E, CS-K, and CS-L) along with trisulfated (CS-M, CS-S, and CS-T) disaccharides [54]. An animal cartilage component called chondroitin sulphate is a glycosaminoglycan. Chondroitin sulphate can detect and connect with the CD44 receptor when it interacts with cancer cells, which significantly enhances the uptake of chondro sulfate-based DDS by cancer cells. Additionally, chondroitin sulphate can dissolve quickly in the presence of degrading enzymes (such hyaluronidase), this results in the drug release of DDS based on chondroitin sulphate [55]. Despite having high water solubility, the direct use of unprocessed chondroitin sulfate as a potent drug carrier will only lead to minimal drug usage. Studies have shown that cross-linking with chondroitin sulphate can generate a conjugate system that significantly increases the effectiveness of hydrophobic active compounds against tumors [56].

  The cationic polymers and cationic liposomes needed to bind to pDNA in order to produce stable cationic polyplexes and lipoplexes. In B16-F10 cells, these cationic compounds showed excellent transgene effectiveness, but they also showed significant cytotoxicity and potent erythrocyte agglutination. The polyplexes and lipoplexes might be encapsulated by CS to produce stable anionic particles without changing their structural makeup. The ternary complexes surrounded by CS in B16-F10 cells showed good transgene efficacy with minimal cytotoxicity and agglutination. When given intravenously to mice, polyplexes exhibited a limited amount of transgenic effectiveness, but those that were encased in CS had far higher transgene effectiveness. The CS capsulation, however, reduced the high lipoplex transgenic effectiveness [57, 58]. Tetsuro Tomiyama *et al.* investigated the potential of adding CS to PPC, which contains 5 cationic amino acids, to enhance gene expression in tumor cells. The plasmid DNA (pDNA)/PPC/CS combination has a size and zeta potential of 200 nm and -10 to -15 mV, respectively. The pDNA/PPC/CS combination showed weaker stability and gene regulation than the pDNA/PPC complex in experiments on tumor cells. Gene expression did not alter noticeably between positive and negative polymers, either. The cause of these outcomes was the rapid disintegration of the pDNA/PPC/CS complexes in the presence of serum. Consequently, we propose that adding negatively charged CS to polymers with a low charge density may result in low stability and difficult gene control [59].

2.1.2 Advantages of Polysaccharides-Based Potent Biomaterials

Conversely, natural-based biomaterials provide a number of advantages above synthetic biomaterials. In addition to possessing biomimetic characteristics and being environmentally safe, they may be made from a variety

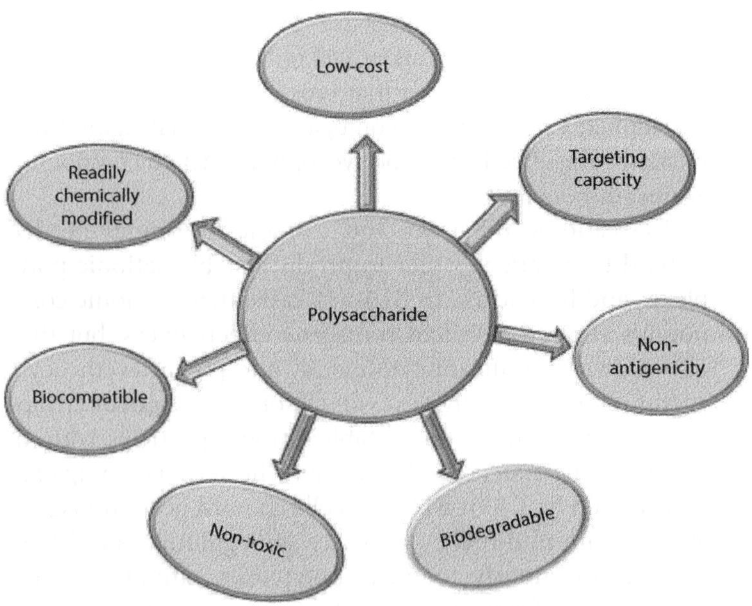

Figure 2.5 Advantages of polysaccharides-based biomaterials.

of biodegradable waste materials and abundant agricultural food supplies, which lowers the cost of their synthesis and processing. Natural biomaterials demonstrate high biocompatibility and biodegradability, and they may be transformed into a variety of enzymatically modified derivatives that promote cell adhesion and proliferation with little risk of tissue rejection (Figure 2.5) [60–62].

2.1.3 Limitations

Although there are a number of limitations on how these biopolymers can be used in clinical settings, ongoing research efforts have shown promising outcomes. For instance, clinical application alginate and chitosan wound dressings/hemostats have just been granted FDA approval [63].

2.2 Cross-Linking Techniques for Engineering Polysaccharides-Based Biomaterials

2.2.1 Physical Cross-Linking

The actual advantage of physically cross-linking polysaccharide hydrogels for biomedical applications is that there is no need of extra cross-linkers during the synthesis process, reducing any potential systemic toxicity with the generated biomaterials. Physical cross-linking hydrogel production has gained popularity due of how straightforward it is to make them. It is promising to modify and control the end stage gel formation's characteristics and macro- and micro-structures by altering the polysaccharide selection, the type of additives, and the temperature. Ionic interface and self-assembly are used to create biomaterials based on polysaccharides that are physically cross-linked [64–66].

Ionic polysaccharides like alginate, chitosan, and pectin have potent functional entities like carboxylate, amine, and sulphates which take part in ionic cross-linking with a range of divalent and trivalent cationic (Ca2+, Sr2+, Zn2+, Fe3+, Sc3+, Al3+, Ga3+) or anionic (tripolyphosphate (TPP)) counterions. Low methoxy pectin and sodium alginate were cast together to make continuous, homogenous and clear edible composite films, according to Nei *et al.* (LMP) [67–70].

As a result of ionic interactions, TPP has also been used by various research groups to create CS NPs for a variety of medicinal applications. Long-lasting drug release has been accomplished by physically cross-linking anionic cross-linkers such polyaspartic acid anxanthanum gum to cationic chitosan oligomers. The cross-linking of alginate has also been investigated for ionic interactions to produce films, fibers, hydrogels, nanoparticles, microspheres, and 3D printed frameworks that are efficient for drug delivery and tissue regeneration [71, 72].

2.2.2 Chemical Cross-Linking

By adding particular functional groups, it is possible to create polysaccharides that are chemically cross-linked. Chemically cross-linked hydrogels are more stable, strong, and have better mechanical qualities including bending, shear stress, and tensile strength etc. than hydrogels created through physical cross-linking. The usage of chemicals, however, could produce toxic biomaterials, therefore choosing the right cross-linking agent is essential for biomedical applications. The literature uses a range of methods rely on different chemical processes, including Michael addition, click chemistry, the formation of Schiff bases, and radical polymerization. The synthesis of polysaccharides containing both electrophilic (such as acrylate, and vinyl sulfone groups) and nucleophilic (such as SH, phosphine) groups has been the subject of numerous experiments to date. To create polysaccharides-based biomaterials, the Michael addition method has been used to cross-link thiolate dextran conjugates, vinyl sulfone comprising dextran acrylic acid cross-linked chitosan, tri(2-carboxyethyl) phosphine cross-linked chitosan, dopamine-bearing HA and SH functional groups like pluronic F127 and methacrylate polymers. Further, Michael addition is considered a straightforward process, but it has significant limitations when used to biological molecules since it may cause negative effects by competing with nucleophiles. An alternative to chemically cross-linking hydrogels to create anticipated biomaterials is radical polymerization of vinyl groups. An initiator degrades to release free radicals in radical polymerization, which then interacts with double bond to produce polymerization [73–77]. In this scenario, accurate control of reaction constraints, such as temperature, polymer concentration, and initiator concentration, is essential to avoid possible termination reactions and produce the desired outcome. There have been numerous reports of chemically cross-linked hydrogels based on polysaccharides for biomedical uses in the literature.

Biomaterials based on polysaccharides benefited from chemical cross-linking because it increases the mechanical properties and system stability. Because the chemical cross-linking sites based on hydrazone, disulfide, ketals, etc. are stimuli-responsive, their usage in the biomedical field is increasing. But aside from the toxicity concern, chemical cross-linkers may also provide polysaccharides-based biomaterials a specific color, which could limit their applicability in tissue engineering, especially in skin regeneration or replacement [78–81].

2.2.3 Ionically Cross-Linked

Ionic cross-linking requires a softer experimental procedure and more adaptable ways for reversible ionic cross-linking in contrast to covalent cross-linking, which provides an easy and biocompatible replacement for charged polysaccharide. Anionic substances like tripolyphosphate and multivalent counterions like Ca2+, Mg2+, and Ba2+ are examples of substitutes for polysaccharides and their derivatives that can produce ionic cross-linking (such as polyvinyl alcohol, polyacrylic acid, polyacrylamide) [82, 83]. Using the common polycation and polyanionic polysaccharides, chitosan and alginate, Mizrahy *et al.* established an ionic cross-linking process. The cross-linked network in chitosan is predominantly formed by ionic interactions between negatively charged cross-linkers and positively charged groups on the chitosan molecules. There are numerous cross-linkers available for the cross-linking of chitosan. For instance, the polyanion cross-linker tripolyphosphate (TPP), which is frequently used to ionic cross-link chitosan, has been extensively studied for usage as a drug delivery vector [84–86]. Table 2.1 describes the patents granted on the polysaccharides based biomaterials for gene therapy.

Table 2.1 List of patents on polysaccharides-based biomaterials for gene delivery.

S.N.	Patents no. and year	Title	Type of patent	Reference
1	US7914819B1 Mar. 29, 2011	Polysaccharide-based biomaterials	US PATENTS	[87]
2	US9655844B2 May 23, 2017	Biocompatible polysaccharide-based hydrogels	US PATENTS	[88]
3	CN106426729A, 2022	Microneedle-based natural polysaccharide for drug delivery systems (DDS): progress and challenges. Pharmaceuticals,	China	[89]
4	US10058620B2, 2015	Dextran-peptide hybrid for efficient gene delivery	US PATENTS	[90]

(Continued)

Table 2.1 List of patents on polysaccharides-based biomaterials for gene delivery. (*Continued*)

S.N.	Patents no. and year	Title	Type of patent	Reference
5	US7914819B1, 2007	Polysaccharide-based biomaterials	US PATENTS	[91]
6	EP2509644A2, 2009	Dextrin hydrogel for biomedical applications	European Patent Office	[92]
7	US20110076332A1, 2010	Dextran-chitosan-based *in-situ* gelling hydrogels for biomedical applications	US PATENTS	[93]
8	DE102014016901A1, 2014	New dextran derivatives as drug carrier system	German patent	[94]
9	US3766165A, 1972	Polysaccharides and their preparation	US PATENTS	[95]
10	US20080139441A1, 2007	Antimicrobial and bacteriostatic-modified polymers for cellulose fibers	US PATENTS	[96]
11	WO1995024929A3, 1995	Polymeric gene delivery system	PCT	[97]
12	US20120220749A1, 2012	Combinatorial polymeric compositions for drug delivery	US PATENTS	[98]

2.3 Approaches to Design Polysaccharide-Derived Biomaterials

2.3.1 Etherification of Polysaccharides

By attaching an alkylating agent to the saccharide alcohol when a base is present, the etherification process produces substances such hydroxyl propylated polysaccharides. Hydropyrrolation of polysaccharides adds hydroxypropyl groups via an etherification process with propylene oxide. This might be

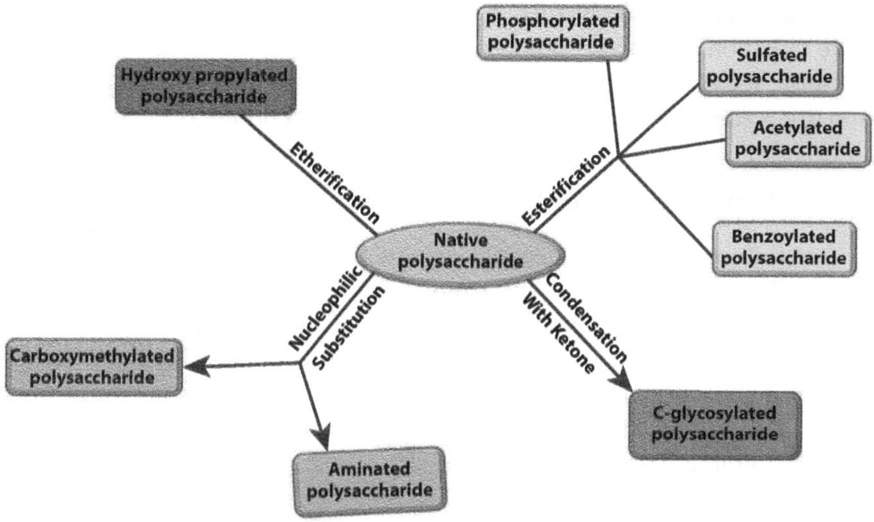

Figure 2.6 Reactions of polysaccharides.

achieved via combining isopropanol, polysaccharide, and sodium hydroxide. After mixing for 30 minutes, propylene oxide should be added and kept at 40°C for 8 hours. Despite the fact that hydroxypropylation is easy, affordable, non-toxic, and very effective for increasing polysaccharide solubility in water, it has received less attention when applied to enhance the biological characteristics of polysaccharides (Figure 2.6) [99, 100].

2.3.2 Esterification of Polysaccharide

In the process of esterification, the saccharide alcohol combines with acylating substances to make acetylated, benzoylated, sulfated, and phosphorylated polysaccharides, among other compounds [101].

2.3.3 Sulfation of Polysaccharides

Sulfation comprises substituting one of the polysaccharide's hydroxyl groups with a sulphate group through the esterification reaction. Usually, SO3 Pyridine is utilized as the sulfation agent in sulfation reactions, and pyridine is utilized as an aprotic to achieve a homogenous reaction mixture. However, it has also apparently been done with the use of other solvents, such as formamide, dimethylformamide, and dimethylsulfoxide [102]. Sulfation processes are typically conducted at 45°C for 6 h or 60°C for 15 minutes, with the excess amount of HCl generated during the synthesis of the sulfating agent then being neutralized using NaOH. A recent study found that the use of microwave radiation produced homogenous reaction products, a higher degree of sulfation, fewer side reactions, uniform heat distribution, and shorter reaction times subsequently leading to uniform reaction products [103, 104].

2.3.4 Benzoylations of Polysaccharides

The technique of adding benzoyl groups to polysaccharides in place of hydroxyl is referred as benzoyolation. Benzyolated polysaccharides are synthesized through a formamide-based mechanism for solubilization of 1, 2-benzene dicarboxylic acid, 4, 8-dimethylaminopyridine, and toluenesulfonyl chloride anhydride. The incorporation of benzoyl groups may modify the polysaccharide's polarity and have an impact on its high nucleophilic characteristics [105, 106].

2.3.5 Acetylation of Polysaccharides

In order to acetylate polysaccharides, hydroxyl group(s) must be substituted with an acetyl group(s). Acetylation makes polysaccharides more hydrophobic, which strengthens their capacity to interact with non-polar molecules. A ubiquitous derivatization agent is acetic anhydride, notwithstanding variations in the methods used to achieve acetylation of polysaccharides. Use of acetylation procedures because it has extra benefits including no side product generation, less cost, availability, improved reaction efficiency, and gentle reaction conditions [107, 108].

2.3.6 Phosphorylation of Polysaccharides

Phosphate groups are substituted for hydroxyl groups in the process of phosphorylating polysaccharides. The generation of phosphorylated polysaccharides can occur when the polysaccharide based solution reacts with the derivatization reagent (usually polyphosphoric acid). Native polysaccharides exhibited flexible chains with stretched chain sequences and became more soluble after being phosphorylated. Using solubilizing substances like formamide, native polysaccharides have been broken down [109, 110].

2.3.7 Amination of Polysaccharides

Polysaccharide hydroxyl groups are transformed to amino groups during amination. Fucoidan was aminated in an alkaline medium utilizing water as the solvent, epichlorohydrin as the activating ingredient, and ammonium hydroxide as a key derivatizing agent. DMSO was used as a solvent in a different system for amination of oat-glucan, along with paraformaldehyde for oxidation, ammonium acetate for derivatization, and sodium cyanoborohydride for imine reduction. In comparison to a polysaccharide that has not been changed, amination causes the production of a cationic-polysaccharide derivative [111].

2.3.8 Carboxymethylation of Polysaccharides

A polysaccharide's hydroxyl group is converted to a carboxymethyl group via carboxymethylation. Three steps make up the reaction, which would be carried out in an alkaline conditions: derivatization, product recovery, and solubilization of polysaccharides.

Native polysaccharides have been solubilized using a variety of chemicals, such as 2-propanol with 80% ethanol. Native polysaccharides are depolymerized as a result of carboxymethylation, which also causes a rise in solubility, a reduction of the molecular weight distribution, a complex hyperbranched conformation, and a reduction in intrinsic viscosity [112–114].

2.3.9 Condensation with Ketones

C-glycosylated polysaccharides are produced when polysaccharides and ketones are conjugated. In order to replace the polysaccharide's hydroxyl group with a glycosyl group, condensation process with ketone is used to C-glycosylate polysaccharides. In accordance with the reaction protocol, polysaccharides are combined with aqueous NaHCO3, pentane-2, 4-dione (a derivatization agent), and the mixture is agitated while being heated to 100 °C for 20 hours.

The production of a C-C bond results from this process, known as Knoevenagel condensation. After polysaccharides are C-glycosylated, 1 H-NMR signals indicate a shift in spectra indicating the - and -anomers, respectively. The functionality of sugars has improved significantly due to c-glycosylation, yet research on the Knoevenagel reactions of carbohydrates is scarce [115–117].

2.4 Biomedical Applications of Polysaccharide-Derived Biomaterials

Polysaccharides, which are naturally occurring, renewable molecules with a wide range of medicinal applications, have gained relevance. Polysaccharides have a wide range of uses, including medication delivery, tissue engineering, and 3D bioprinting. But the areas that receive the most attention in detail are:

2.4.1 Polymer-Drug Conjugates

Even the most basic polysaccharide molecules, which consist of repeating monosaccharide units with hydroxyl groups as the sole functional group, offer the possibility of conjugation. Despite the fact that only a small number of polysaccharide conjugates have received approval or are presently undergoing clinical trials, they have been well-known for a long time. There have been many polysaccharides used in the conjugations that have amine, carboxylic acid, and aldehyde functionality. "Click chemistry" or standard reaction conditions make it easier to conjugate the majority of bioactive compounds (drugs, proteins, and nucleic acids) [118, 119]. Clinical trials for polysaccharide conjugates are being shown.

- Delivery and Imaging
- Tissue Engineering and Support
- Bio adhesives Based on Polysaccharide Conjugate
- Conjugated Vaccine

2.5 Advanced Biomaterials for Wound Dressings

Due to their practicality, low cost, and extended wound management, wound dressings have emerge1d as a critical component of wound healing treatments. Marine polysaccharides are a subset of the majority of marine organisms and are emerging biomaterials. It has a variety of bioactivities that enable it to be converted into a variety of wound dressings. As a result, for the purposes of studying wound therapy, it is crucial to have a thorough understanding of how marine polysaccharides are used in wound dressings [120].

2.6 Scaffolds for Tissue Engineering and Development of Bioinks for 3D Bioprinting

Scientists have developed bioprinting techniques to produce constructs that can solve clinical issues as a result of limitations in wound treatment. The three-dimensional (3D) positioning of cells, molecules, and biomaterials can be spatially controlled thanks to the bioprinting method. These features offer fresh opportunities to improve functional outcomes and homology to native skin. However, the creation of hydrogel bioink with improved printability and bioactive properties is necessary for the clinical value. In this study, we created a bioink for the 3D bioprinting of a dermis layer by fusing the exceptional viscoelastic behavior of nanofibrillated cellulose (NFC) with the rapid cross-linking properties of alginate (ALG), carboxymethyl cellulose (CMC), and encapsulated human skin cells (hSF) [121, 122].

2.7 Recent Utilization of Polysaccharides

2.7.1 Bone/Cartilage Regeneration

The action of joints largely depends on articular cartilage. The fundamental for the joint's capacity to move adequately is complete articular cartilage. Articular cartilage is a single tissue that lacks a lymphatic or blood supply. The ability of cartilage to mend itself is limited, and it generally cannot be restored, as cartilage cells divide relatively slowly. Defects in articular cartilage are therefore a significant problem in orthopedic surgery. The feeling of joint pain will be brought on by primary osteoarthritis or trauma. Since cartilage cannot heal itself, injuries are sometimes left untreated for years, which can cause severe degeneration. Artificial options for biological repair include synthetic scaffolds. Their significance over natural scaffolds is that they can sustain weight-bearing forces by controlling their mechanical performance. Furthermore, it has been documented that several synthetic polymers can effectively repair cartilage deformities [123, 124].

2.7.2 Cardiac Regeneration

Loss of cardiac function arises from physical damage to the heart brought by cardiovascular disease. Despite the fact that drugs can alleviate symptoms, it is better to treat the underlying cause by repairing damaged tissues because this has a positive impact on patients' prognoses. Cardiothoracic surgeons can repair the heart valves and ventricular septal wall utilizing a range of surgical procedures in addition to heart transplants. Different biomaterials are employed to replace and repair damaged heart tissues [125].

2.7.3 Neural Regeneration

An multidisciplinary discipline known as "regenerative medicine" uses a combination of cells, bioactive substances, and biomaterial scaffolds to replace or repair diseased or damaged tissues. Polysaccharides have gained increased attention recently as potential biomaterials for the building of scaffolds due to their natural origin, homology to the extracellular matrix, biocompatibility, and lack of toxicity [126].

2.7.4 Skin Regeneration

The re-epithelialization and tissue regeneration of wounds can be accelerated by polysaccharides. They may also prevent the development of skin-infecting germs or cause them to die. Polysaccharides may shield the skin from UV rays, decrease oxidative stress, and guard against cancer [127].

2.7.5 As a Hemostatic Agent

However, without the assistance of hemostatic materials, their own clotting procedure cannot be completed in time for efficient hemostasis. Rapid hemostasis and the development of stable blood clots are crucial for preventing large blood loss from extreme bleeding in a living body. Traditional wound closure techniques comprising suturing and stapling are typically more likely to result in additional tissue damage, activated inflammatory responses, brief usage durations, and required follow-up treatments in clinical settings. These common methods of wound closure are by no means adequate, particularly for large wounds that need the quick closure of fluids or gases. It is widely anticipated that numerous tissue adhesives, sealants, and hemostatic substances will help with these problems [128].

2.7.6 Stimulation of Angiogenesis

Recent research has shown that naturally occurring plant polysaccharides have potent anti-properties. Therefore, previous studies have shown that dandelion polysaccharide inhibits the proliferation of hepatocellular carcinoma (HCC) cells both *in vitro* and *in vivo* (DP). In this study, we investigated the effects of DP on HCC angiogenesis and putative molecular mechanisms by which DP regulates angiogenesis. Experiments on wound healing and transwell invasion showed that DP prevented HUVECs from migrating and invading *in vitro*. Tube formation tests conducted *in vitro* and *in vivo*, as well as tests using the chick chorioallantoic membrane (CAM) and immunohistochemistry (IHC), all revealed that DP impeded vasculogenesis. By using immunofluorescence labeling and Western blotting, it was also shown that DP therapy decreased the protein levels of some significant elements involved in the angiogenesis of HCC. Additionally, these polysaccharides have a variety of biological uses, including the distribution of growth factors, the delivery of medications, the transfer of gene materials, and the encapsulation of proteins [129–131]. According to a separate study, a collagen/alginate scaffold produced an ideal root canal healing environment that was appropriate for regenerative endodontics [132].

Biocompatibility with human cells, the systematic repair of injured tissue, immune-stimulating action, and rapid induction of hemostasis, free radical scavenging potential, and antibacterial activity are among them. Thus according recent studies, chitin and chitosan are the most adaptable materials for use in surgical aids including coated sterile stents and anti-adhesion gels as well as elaborate diagnostics, devices for specifically recognizing malignant cells, and medication and gene administration. Biomaterials made of polysaccharides work well as tablet binders and natural disintegrates and present an enticing alternative for sustained-release systems. Biomaterials made from polysaccharides are frequently utilized as crude material for the paper industry to make products like paper and cardboard. Biomaterials produced from polysaccharides have a variety of applications in the pharmaceutical, agricultural, food, and biomedical industries. A credible alternative for sustained-release systems, polysaccharide is a natural disintegrant and tablet binder.

2.8 Toxicity Concerns of Polysaccharide-Derived Biomaterials

There should be less reason to worry about their toxicity because of their claimed nontoxicity, biodegradability, and biocompatibility. Natural polysaccharides that interconnect and cross-link with other synthetic polymeric chains have been used to create SPHs of higher generation. Exposures to the hazardous polysaccharide-iron combination, as per reports from poison control centers. No substantial negative effects related to PIC exposure were reported to the AAPCC.

2.9 Preclinical and Clinical Studies on Gene Delivery Using Polysaccharide-Based Biomaterials

According to studies, nucleic acids might be efficiently delivered to brain tumors *in vitro* and *in vivo* using chitosan-based drug delivery systems.

Braeckmans K *et al.* 2010 used fluorescence fluctuation spectroscopy and single particle tracking (SPT) to investigate the stability of siRNA-chitosan carriers in human plasma.

Kang *et al.* 2017 Applied an RNAi-based therapeutic strategy to silence RANK signaling using thermosensitive chitosan hydrogel as siRNA reservoir and vector. They reported that RAW264.7 cells treated with hydrogel loaded with siR-RANK demonstrated a significant prolonged RANK mRNA knockdown effect.

Mahsa *et al.* 2023 explored Cellulose-based composite scaffolds for bone tissue engineering and localized drug delivery and concluded that Cellulosic biomaterials are mechanically strong because of their crystallinity. Further, the utilization of cellulose for 3D printing bio-nanocomposite expressed major advantages in drug delivery and tissue engineering applications for the following reasons: easy processability, sustainability, wide range of availability and biochemical functionalization with other materials.

Qiong *et al.* 2014 had designed and evaluated dextran–peptide hybrid for gene delivery and concluded that dex–peptide-hybrid with the low molecular weight of dextran and the high degree of substitution is a very promising material for safe and efficient gene therapy [133–138].

2.10 Challenges and Future Directions

The problem still lies in creatively combining the various capabilities into a single complex vector, despite the fact that research has so far created a strong arsenal of polymer tools and design methodologies. Several tactics have surfaced to present fresh and intriguing viewpoints on overcoming this difficulty. Using more organized macromolecular structures, such as interlocking supramolecules with many components or 2- and 3-dimensional scaffolds, is one method for developing new gene delivery systems. Another approach improves current vector platforms (covalent functionalization of polymer chains, for instance) using combinatorial approaches and high throughput technologies in order to more quickly discover new vectors and analyze the enormous parameter space of vector design. The advancements of increasingly sophisticated analytical methods to quantify the many aspects of gene delivery will also be evident in studies on the subject. Research will greatly benefit from approaches that can more precisely describe the physicochemical characteristics of multifaceted vectors, evaluate the precise delivery-enhancing effects of different physicochemical vector features, and quantify the behavior of polyplexes within the cellular environment.

The ongoing work toward addressing the biological environment that gene delivery systems must navigate is a significant field of gene delivery research. The complexity of the cellular environment, which is still unknown, is highlighted by the discovery recently by several groups of the participation of hitherto undetected intracellular entities and pathways in gene transport. It will be crucial to better vector designs to resolve the biological landscape and its implications for gene delivery [139].

2.11 Future Prospects

In empirical investigations, biodegradable polymers for gene transfer and therapeutics have garnered a lot of interest. Although biodegradable polymer vectors have superior biodegradability, biocompatibility, and non-toxicity, their advancement from *in vitro* to *in vivo* and ultimately clinical trials still faces a significant hurdle. Future research on biodegradable polymer vectors should concentrate on building biodegradable polymer vectors with a clear structure-function link in order to achieve better *in vivo* stability in order to get around these obstacles. Next-generation gene-delivery vector development also depends on the creation of effective vectors that can release their loaded gene payloads in a timely and spatially-controlled way.

2.12 Conclusion

Gene Delivery Materials Based on Polysaccharides gives logically and clearly quantitative background on novel polysaccharide materials, highlighting the most intriguing gene delivery possibilities. The chapter is primarily concerned with the many kinds of polysaccharide nanocomposites for gene delivery. The main goal of recent research has been to find secure and effective replacements for the viral vehicles needed for gene delivery. Future research will face several obstacles, nevertheless, including understanding the chemical interactions of interacting compounds, the destiny of polysaccharide nanoparticles *in vivo*, and the molecular interactions between polysaccharides and loaded drugs/genes. The current chapter is primarily concerned with an overview of polysaccharide-based biomaterials for a variety of biomedical applications, including gene transfer. Additionally, the focus will be on structure, cross-linking mechanisms, biosynthesis, chemical characteristics, biological properties, current problems, clinical usage, recent advancements, and methods to build polysaccharide-derived biomaterials.

References

1. Tekale, S.U., Kanagare, A.B., Dhirbassi, A.V., Domb, A.J., Pawar, R.P., Polysaccharide-based biomaterials: Overview, 1–26, 2022, https://doi.org/10.1039/9781839166235-00001.
2. Mohammed, A.S., Naveed, M., Jost, N., Polysaccharides; classification, chemical properties, and future perspective applications in fields of pharmacology and biological medicine (a review of current applications and upcoming potentialities). *J. Polym. Environ.*, 29, 8, 2359–71, 2021 Aug.
3. Prajapati, V.D., Jani, G.K., Moradiya, N.G., Randeria, N.P., Pharmaceutical applications of various natural gums, mucilages and their modified forms. *Carbohydr. Polym.*, 92, 2, 1685–99, 2013 Feb 15.
4. Chen, F. and Huang, G., Preparation and immunological activity of polysaccharides and their derivatives. *Int. J. Biol. Macromol.*, 112, 211–6, 2018 Jun 1.
5. Liu, Z., Jiao, Y., Wang, Y., Zhou, C., Zhang, Z., Polysaccharides-based nanoparticles as drug delivery systems. *Adv. Drug Delivery Rev.*, 60, 15, 1650–62, 2008 Dec 14.
6. Nayerossadat, N., Maedeh, T., Ali, P.A., Viral and nonviral delivery systems for gene delivery. *Adv. Biomed. Res.*, 1, 1, 27, 2012 Jan 1.
7. Miller, A.D., Human gene therapy comes of age. *Nature*, 357, 6378, 455–60, 1992 Jun 11.
8. Cevher, E., Sezer, A.D., Çağlar, E., Gene delivery systems: Recent progress in viral and non-viral therapy, in: *Recent Advances in Novel Drug Carrier Systems*, vol. 31, pp. 437–70, 2012 Oct.
9. Gantenbein, B., Tang, S., Guerrero, J., Higuita-Castro, N., Salazar-Puerta, A.I., Croft, A.S., Gazdhar, A., Purmessur, D., Non-viral gene delivery methods for bone and joints. *Front. Bioeng. Biotechnol.*, 8, 598466, 2020 Nov 19.
10. Zhang, X. and Godbey, W.T., Viral vectors for gene delivery in tissue engineering. *Adv. Drug Delivery Rev.*, 58, 4, 515–34, 2006 Jul 7.
11. Conwell, C.C. and Huang, L., Recent advances in non-viral gene delivery. *Adv. Genet.*, 53, 1–8, 2005 Jan 1.
12. Cevher, E., Sezer, A.D., Çağlar, E., Gene delivery systems: Recent progress in viral and non-viral therapy, in: *Recent Advances in Novel Drug Carrier Systems*, vol. 31, pp. 437–70, 2012 Oct.
13. Liu, Z.H., Jiao, Y.P., Wang, Y.F., Zhou, C.R., Zhang, Z.Y., Self assembly of protein with polymer. *Adv. Drug Delivery Rev.*, 60, 1650–1662, 2008.
14. Nguyen, D.N., Green, J.J., Chan, J.M., Langer, R., Anderson, D.G., Polymeric materials for gene delivery and DNA vaccination. *Adv. Mater.*, 21, 8, 847–67, 2009 Feb 23.
15. Toita, S., Morimoto, N., Akiyoshi, K., Functional cycloamylose as a polysaccharide-based biomaterial: Application in a gene delivery system. *Biomacromolecules*, 11, 2, 397–401, 2010 Feb 8.
16. Ramya, R., Venkatesan, J., Kim, S.K., Sudha, P.N., Biomedical applications of chitosan: An overview. *J. Biomater. Tissue Eng.*, 2, 2, 100–11, 2012 Jun 1.

17. Younes, I. and Rinaudo, M., Chitin and chitosan preparation from marine sources. Structure, properties and applications. *Mar. Drugs*, 13, 3, 1133–74, 2015 Mar 2.
18. Chirio, D., Peira, E., Sapino, S., Dianzani, C., Barge, A., Muntoni, E., Morel, S., Gallarate, M., Stearoyl-chitosan coated nanoparticles obtained by microemulsion cold dilution technique. *Int. J. Mol. Sci.*, 19, 12, 3833, 2018 Nov 30.
19. Feng, C., Li, J., Kong, M., Liu, Y., Cheng, X.J., Li, Y., Park, H.J., Chen, X.G., Surface charge effect on mucoadhesion of chitosan based nanogels for local anti-colorectal cancer drug delivery. *Colloids Surf. B: Biointerfaces*, 128, 439–47, 2015 Apr 1.
20. Debele, T.A., Peng, S., Tsai, H.C., Drug carrier for photodynamic cancer therapy. *Int. J. Mol. Sci.*, 16, 9, 22094–136, 2015 Sep 14.
21. Bakshi, P.S., Selvakumar, D., Kadirvelu, K., Kumar, N.S., Chitosan as an environment friendly biomaterial–a review on recent modifications and applications. *Int. J. Biol. Macromol.*, 150, 1072–83, 2020 May 1.
22. Ambhore, N.P., Dandagi, P.M., Gadad, A.P., Formulation and comparative evaluation of HPMC and water soluble chitosan-based sparfloxacin nanosuspension for ophthalmic delivery. *Drug Deliv. Transl. Res.*, 6, 1, 48–56, 2016 Feb.
23. Amidi, M., Mastrobattista, E., Jiskoot, W., Hennink, W.E., Chitosan-based delivery systems for protein therapeutics and antigens. *Adv. Drug Delivery Rev.*, 62, 1, 59–82, 2010 Jan 31.
24. Xu, Y. and Du, Y., Effect of molecular structure of chitosan on protein delivery properties of chitosan nanoparticles. *Int. J. Pharm.*, 250, 1, 215–26, 2003 Jan 2.
25. Lee, C., Shin, J., Lee, J.S., Byun, E., Ryu, J.H., Um, S.H., Kim, D.I., Lee, H., Cho, S.W., Bioinspired, calcium-free alginate hydrogels with tunable physical and mechanical properties and improved biocompatibility. *Biomacromolecules*, 14, 6, 2004–13, 2013 Jun 10.
26. Chen, L., Shen, R., Komasa, S., Xue, Y., Jin, B., Hou, Y., Okazaki, J., Gao, J., Drug-loadable calcium alginate hydrogel system for use in oral bone tissue repair. *Int. J. Mol. Sci.*, 18, 5, 989, 2017 May 6.
27. Augst, A.D., Kong, H.J., Mooney, D.J., Alginate hydrogels as biomaterials. *Macromol. Biosci.*, 6, 8, 623–33, 2006 Aug 7.
28. Kamaly, N., Yameen, B., Wu, J., Farokhzad, O.C., Degradable controlled-release polymers and polymeric nanoparticles: Mechanisms of controlling drug release. *Chem. Rev.*, 116, 4, 2602–63, 2016 Feb 24.
29. Szabó, L., Gerber-Lemaire, S., Wandrey, C., Strategies to functionalize the anionic biopolymer Na-alginate without restricting its polyelectrolyte properties. *Polymers*, 12, 4, 919, 2020 Apr.
30. Hou, J., Li, C., Guan, Y., Zhang, Y., Zhu, X.X., Enzymatically cross-linked alginate hydrogels with improved adhesion properties. *Polym. Chem.*, 6, 12, 2204–13, 2015.

31. Taemeh, M.A., Shiravandi, A., Korayem, M.A., Daemi, H., Fabrication challenges and trends in biomedical applications of alginate electrospun nanofibers. *Carbohydr. Polym.*, 228, 115419, 2020 Jan 15.
32. Salatin, S. and Jelvehgari, M., Natural polysaccharide based nanoparticles for drug/gene delivery. *Pharm. Sci.*, 23, 2, 84–94, 2017 Feb 9.
33. Nesamony, J., Singh, P.R., Nada, S.E., Shah, Z.A., Kolling, W.M., Calcium alginate nanoparticles synthesized through a novel interfacial cross-linking method as a potential protein drug delivery system. *J. Pharm. Sci.*, 101, 6, 2177–84, 2012 Jun 1.
34. Teng, Z., Luo, Y., Wang, Q., Nanoparticles synthesized from soy protein: Preparation, characterization, and application for nutraceutical encapsulation. *J. Agric. Food Chem.*, 60, 10, 2712–20, 2012 Mar 14.
35. De, S. and Robinson, D., Polymer relationships during preparation of chitosan–alginate and poly-l-lysine–alginate nanospheres. *J. Controlled Release*, 89, 1, 101–12, 2003 Apr 14.
36. Sun, J. and Tan, H., Alginate-based biomaterials for regenerative medicine applications. *Materials*, 6, 4, 1285–309, 2013 Mar 26.
37. Zhao, Y., Carvajal, M.T., Won, Y.Y., Harris, M.T., Preparation of calcium alginate microgel beads in an electrodispersion reactor using an internal source of calcium carbonate nanoparticles. *Langmuir*, 23, 25, 12489–96, 2007 Dec 4.
38. Kemp, K., Griffiths, J., Campbell, S., Lovell, K., An exploration of the follow-up needs of patients with inflammatory bowel disease. *J. Crohn's Colitis*, 7, 9, e386–95, 2013 Oct 1.
39. Zahoor, A., Sharma, S., Khuller, G., Inhalable alginate nanoparticles as antitubercular drug carriers against experimental tuberculosis. *Int. J. Antimicrob. Agents*, 26, 4, 298–303, 2005 Oct 1.
40. Ahmad, Z., Pandey, R., Sharma, S., Khuller, G.K., Pharmacokinetic and pharmacodynamic behaviour of antitubercular drugs encapsulated in alginate nanoparticles at two doses. *Int. J. Antimicrob. Agents*, 27, 5, 409–16, 2006 May 1.
41. Hennink, W.E., De Jong, S.J., Bos, G.W., Veldhuis, T.F., Van Nostrum, C.F., Biodegradable dextran hydrogels cross-linked by stereocomplex formation for the controlled release of pharmaceutical proteins. *Int. J. Pharm.*, 277, 1-2, 99–104, 2004 Jun 11.
42. Jung, S.W., Jeong, Y.I., Kim, Y.H., Choi, K.C., Kim, S.H., Drug release from core-shell type nanoparticles of poly (DL-lactide-co-glycolide)-grafted dextran. *J. Microencapsulation*, 22, 8, 901–11, 2005 Jan 1.
43. Debele, T.A., Mekuria, S.L., Tsai, H.C., Polysaccharide based nanogels in the drug delivery system: Application as the carrier of pharmaceutical agents. *Mater. Sci. Eng.: C*, 68, 964–81, 2016 Nov 1.
44. Kothari, D., Tingirikari, J.M., Goyal, A., *In vitro* analysis of dextran from Leuconostoc mesenteroides NRRL B-1426 for functional food application. *Bioact. Carbohydr. Dietary Fibre*, 6, 2, 55–61, 2015 Oct 1.

45. Baruah, R., Maina, N.H., Katina, K., Juvonen, R., Goyal, A., Functional food applications of dextran from Weissella cibaria RBA12 from pummelo (Citrus maxima). *Int. J. Food Microbiol.*, 242, 124–31, 2017 Feb 2.
46. Vocelle, D., Chesniak, O.M., Malefyt, A.P., Comiskey, G., Adu-Berchie, K., Smith III, M.R., Chan, C., Walton, S.P., Dextran functionalization enhances nanoparticle-mediated siRNA delivery and silencing. *Technology*, 4, 01, 42–54, 2016 Mar 31.
47. Li, J. and Yao, P., Self-assembly of ibuprofen and bovine serum albumin–dextran conjugates leading to effective loading of the drug. *Langmuir*, 25, 11, 6385–91, 2009 Jun 2.
48. Salmaso, S. and Caliceti, P., Stealth properties to improve therapeutic efficacy of drug nanocarriers. *J. Drug Delivery*, 2013, 1–19, 2013.
49. Li, Y.L., Zhu, L., Liu, Z., Cheng, R., Meng, F., Cui, J.H., Ji, S.J., Zhong, Z., Reversibly stabilized multifunctional dextran nanoparticles efficiently deliver doxorubicin into the nuclei of cancer cells. *Angew. Chem.*, 121, 52, 10098–102, 2009 Dec 21.
50. Klemm, D., Heublein, B., Fink, H.P., Bohn, A., Cellulose: Fascinating biopolymer and sustainable raw material. *Angew. Chem. Int. Ed.*, 44, 22, 3358–93, 2005 May 30.
51. Seddiqi, H., Oliaei, E., Honarkar, H., Jin, J., Geonzon, L.C., Bacabac, R.G., Klein-Nulend, J., Cellulose and its derivatives: Towards biomedical applications. *Cellulose*, 28, 4, 1893–931, 2021 Mar.
52. Javanbakht, S., Hemmati, A., Namazi, H., Heydari, A., Carboxymethylcellulose-coated 5-fluorouracil@MOF-5 nano-hybrid as a bio-nanocomposite carrier for the anticancer oral delivery. *Int. J. Biol. Macromol.*, 155, 876–82, 2020 Jul 15.
53. Bedini, E., Laezza, A., Iadonisi, A., Chemical derivatization of sulfated glycosaminoglycans. *Eur. J. Org. Chem.*, 2016, 18, 3018–42, 2016 Jun.
54. López-Álvarez, M., López-Senra, E., Valcárcel, J., Vázquez, J.A., Serra, J., González, P., Quantitative evaluation of sulfation position prevalence in chondroitin sulphate by Raman spectroscopy. *J. Raman Spectrosc.*, 50, 5, 656–64, 2019 May.
55. Shi, Y.G., Meng, Y.C., Li, J.R., Chen, J., Liu, Y.H., Bai, X., Chondroitin sulfate: Extraction, purification, microbial and chemical synthesis. *J. Chem. Technol. Biotechnol.*, 89, 10, 1445–65, 2014 Oct.
56. Radhakrishnan, K., Tripathy, J., Datey, A., Chakravortty, D., Raichur, A.M., Mesoporous silica–chondroitin sulphate hybrid nanoparticles for targeted and bio-responsive drug delivery. *New J. Chem.*, 39, 3, 1754–60, 2015.
57. Hamada, E., Kurosaki, T., Hashizume, J., Harasawa, H., Nakagawa, H., Nakamura, T., Kodama, Y., Sasaki, H., Anionic complex with efficient expression and good safety profile for mRNA delivery. *Pharmaceutics*, 13, 1, 126, 2021 Jan 19.
58. Kurosaki, T., Kitahara, T., Fumoto, S., Nishida, K., Yamamoto, K., Nakagawa, H., Kodama, Y., Higuchi, N., Nakamura, T., Sasaki, H., Chondroitin sulfate

capsule system for efficient and secure gene delivery. *J. Pharm. Pharm. Sci.*, 13, 3, 351–61, 2010 Sep 6.

59. Tomiyama, T., Toita, R., Kang, J.H., Koga, H., Shiosaki, S., Mori, T., Niidome, T., Katayama, Y., Effect of introduction of chondroitin sulfate into polymer-peptide conjugate responding to intracellular signals. *Nanoscale Res. Lett.*, 6, 1, 1–7, 2011 Dec.
60. Alaribe, F.N., Manoto, S.L., Motaung, S.C., Scaffolds from biomaterials: Advantages and limitations in bone and tissue engineering. *Biologia*, 71, 4, 353–66, 2016 May 1.
61. Wan, M.C., Qin, W., Lei, C., Li, Q.H., Meng, M., Fang, M., Song, W., Chen, J.H., Tay, F., Niu, L.N., Biomaterials from the sea: Future building blocks for biomedical applications. *Bioact. Mater.*, 6, 12, 4255–85, 2021 Dec 1.
62. Luckachan, G.E. and Pillai, C.K., Biodegradable polymers-a review on recent trends and emerging perspectives. *J. Polym. Environ.*, 19, 3, 637–76, 2011 Sep.
63. Gardikiotis, I., Cojocaru, F.D., Mihai, C.T., Balan, V., Dodi, G., Borrowing the features of biopolymers for emerging wound healing dressings: A review. *Int. J. Mol. Sci.*, 23, 15, 8778, 2022 Aug 7.
64. Sood, A., Gupta, A., Agrawal, G., Recent advances in polysaccharides based biomaterials for drug delivery and tissue engineering applications. *Carbohydr. Polym. Technol. Appl.*, 2, 100067, 2021 Dec 25.
65. Pushpamalar, J., Meganathan, P., Tan, H.L., Dahlan, N.A., Ooi, L.T., Neerooa, B.N.H.M., Essa, R.Z., Shameli, K., Teow, S.Y.. Development of a polysaccharide-based hydrogel Drug Delivery System (DDS): An update. *Gels*. 7, 4, 153, 2021 Sep 27.
66. Severino, P., da Silva, C.F., Andrade, L.N., de Lima Oliveira, D., Campos, J., Souto, E.B., Alginate nanoparticles for drug delivery and targeting. *Curr. Pharm. Des.*, 25, 11, 1312–34, 2019 Mar 1.
67. Taemeh, M.A., Shiravandi, A., Korayem, M.A., Daemi, H., Fabrication challenges and trends in biomedical applications of alginate electrospun nanofibers. *Carbohydr. Polym.*, 228, 115419, 2020 Jan 15.
68. Nešić, A., Onjia, A., Davidović, S., Dimitrijević, S., Errico, M.E., Santagata, G., Malinconico, M., Design of pectin-sodium alginate based films for potential healthcare application: Study of chemico-physical interactions between the components of films and assessment of their antimicrobial activity. *Carbohydr. Polym.*, 157, 981–90, 2017 Feb 10.
69. Lei, Y., Yao, Q., Jin, Z., Wang, Y.C., Intelligent films based on pectin, sodium alginate, cellulose nanocrystals, and anthocyanins for monitoring food freshness. *Food Chem.*, 10, 134528, 2022 Oct.
70. Kiadeh, S.Z., Ghaee, A., Pishbin, F., Nourmohammadi, J., Farokhi, M., Nanocomposite pectin fibers incorporating folic acid-decorated carbon quantum dots. *Int. J. Biol. Macromol.*, 216, 605–17, 2022 Sep 1.
71. Hu, H. and Xu, F.J., Rational design and latest advances of polysaccharide-based hydrogels for wound healing. *Biomater. Sci.*, 8, 8, 2084–101, 2020.

72. Cui, R., Zhang, L., Ou, R., Xu, Y., Xu, L., Zhan, X.Y., Li, D., Polysaccharide-based hydrogels for wound dressing: Design considerations and clinical applications. *Front. Bioeng. Biotechnol.*, 10, 845735, 2022.
73. Azeredo, H.M. and Waldron, K.W., Cross-linking in polysaccharide and protein films and coatings for food contact–A review. *Trends Food Sci. Technol.*, 52, 109–22, 2016 Jun 1.
74. Tang, X., Gu, X., Wang, Y., Chen, X., Ling, J., Yang, Y., Stable antibacterial polysaccharide-based hydrogels as tissue adhesives for wound healing. *RSC Adv.*, 10, 29, 17280–7, 2020.
75. Zhang, R., Yu, B., Tian, Y., Pang, L., Xu, T., Cong, H., Shen, Y., Diversified antibacterial modification and latest applications of polysaccharide-based hydrogels for wound healthcare. *Appl. Mater. Today*, 26, 101396, 2022 Mar 1.
76. Ranganathan, N., Bensingh, R.J., Kader, M.A., Nayak, S.K., Synthesis and properties of hydrogels prepared by various polymerization reaction systems. Published online, Springer International Publishing, New York City, USA, pp. 487–511, 2019.
77. James, S.P., Zhang, M., Schwartz, H., UHMWPE/hyaluronan microcomposite biomaterials, in: *UHMWPE Biomaterials Handbook*, pp. 259–276, Academic Press, New York City, USA, 2009 Jan 1.
78. Gao, T., Gillispie, G.J., Copus, J.S., Pr, A.K., Seol, Y.J., Atala, A., Yoo, J.J., Lee, S.J.. Optimization of gelatin-alginate composite bioink printability using rheological parameters: A systematic approach. *Biofabrication*. 10, 3, 034106, 2018 Jun 29;
79. Miao, T., Wang, J., Zeng, Y., Liu, G., Chen, X., Polysaccharide-based controlled release systems for therapeutics delivery and tissue engineering: From bench to bedside. *Adv. Sci.*, 5, 4, 1700513, 2018 Apr.
80. Sood, A., Gupta, A., Agrawal, G., Recent advances in polysaccharides based biomaterials for drug delivery and tissue engineering applications. *Carbohydr. Polym. Technol. Appl.*, 2, 100067, 2021 Dec 25.
81. Fu, S., Dong, H., Deng, X., Zhuo, R., Zhong, Z., Injectable hyaluronic acid/poly(ethylene glycol) hydrogels cross-linked via strain-promoted azide-alkyne cycloaddition click reaction. *Carbohydr. Polym.*, 169, 332–40, 2017 Aug 1.
82. Miao, T., Wang, J., Zeng, Y., Liu, G., Chen, X., Polysaccharide-based controlled release systems for therapeutics delivery and tissue engineering: From bench to bedside. *Adv. Sci.*, 5, 4, 1700513, 2018 Apr.
83. Basu, A., Kunduru, K.R., Abtew, E., Domb, A.J., Polysaccharide-based conjugates for biomedical applications. *Bioconjugate Chem.*, 26, 8, 1396–412, 2015 Aug 19.
84. Mizrahy, S. and Peer, D., Polysaccharides as building blocks for nanotherapeutics. *Chem. Soc. Rev.*, 41, 7, 2623–40, 2012.
85. Liu, Z., Jiao, Y., Wang, Y., Zhou, C., Zhang, Z., Polysaccharides-based nanoparticles as drug delivery systems. *Adv. Drug Delivery Rev.*, 60, 15, 1650–62, 2008 Dec 14.

86. Rana, V., Rai, P., Tiwary, A.K., Singh, R.S., Kennedy, J.F., Knill, C.J., Modified gums: Approaches and applications in drug delivery. *Carbohydr. Polym.*, 83, 3, 1031–47, 2011 Jan 30.
87. X. Wen and Y. Qiu, Polysaccharide-based biomaterials. US Patents, 2011.
88. S. Gerecht, Y.-I. Shen, C.C. Ho, G. Sun, Biocompatible polysaccharde-based hydrogels. US Patents, 2017.
89. Damiri, F., Kommineni, N., Ebhodaghe, S.O., Bulusu, R., Jyothi, V.G., Sayed, A.A., Awaji, A.A., Germoush, M.O., Al-Malky, H.S., Nasrullah, M.Z., Rahman, M.H., Microneedle-based natural polysaccharide for drug delivery systems (DDS): Progress and challenges. *Pharmaceuticals*, 15, 2, 190, 2022 Feb 3.
90. G. Cheng, Q. Tang, B. Cao, inventors; University of Akron, assignee, Dextran-peptide hybrid for efficient gene delivery. United States Patent US 10,058,620, 2018 Aug 28.
91. X. Wen and Y. Qiu, Clemson University Research Foundation. US Patents, 2007.
92. F. Miguel, P. Da Gama, M. Cabral, M. Molinos, Universidade do Minho, Dextrin hydrogel for biomedical applications. European Patent Office, 2009.
93. X. Yu, C.M. Valmikinathan, Y. Cheng, Stevens Institute of Technology. Dextran-chitosan based *in-situ* gelling hydrogels for biomedical applications. US Patents, 2010.
94. D. Fischer, T.H. Matthias, Z.K. Hotzel, Friedrich Schiller Universtaet Jena FSU, New dextran derivatives as drug carrier system. German Patent, 2014.
95. H. Rennhard, Pfizer Inc, Polysaccharides and their preparation. US Patents, 1972.
96. H. Xiao, Y. Guan, A. Zheng, University Of New Brunswick, US Patents, 2007.
97. E. Mathiowitz, Y. Shik, J.K. Boekelheide, Brown University, Polymeric gene delivery system. PCT, 1995.
98. F. Dasgupta, Innovotech LLC, Combinatiorial polymeric compositions for drug delivery. US Patents, 2012.
99. Xia, S., Zhang, L., Davletshin, A., Li, Z., You, J., Tan, S., Application of polysaccharide biopolymer in petroleum recovery. *Polymers*, 12, 9, 1860, 2020 Aug 19.
100. Vickers, N.J., Animal communication: When i'm calling you, will you answer too? *Curr. Biol.*, 27, 14, R713–5, 2017 Jul 24.
101. Wong, T., Brault, L., Gasparotto, E., Vallée, R., Morvan, P.Y., Ferrières, V., Nugier-Chauvin, C., Formation of amphiphilic molecules from the most common marine polysaccharides, toward a sustainable alternative? *Molecules*, 26, 15, 4445, 2021 Jul 23.
102. Palma, V., Barba, D., Cortese, M., Martino, M., Renda, S., Meloni, E., Microwaves and heterogeneous catalysis: A review on selected catalytic processes. *Catalysts*, 10, 2, 246, 2020 Feb 18.
103. Heinze, T., Liebert, T., Koschella, A., *Esterification of polysaccharides*, Springer Science & Business Media, Springer International Publishing, New York City, USA, 2006 Aug 2.

104. Feng, Y., Li, W., Wu, X., He, L., Ma, S., Rapid and efficient microwave-assisted sulfate modification of lentinan and its antioxidant and antiproliferative activities *in vitro*. *Carbohydr. Polym.*, 82, 3, 605–12, 2010 Oct 15.
105. Murthy, K.C., Jayaprakasha, G.K., Patil, B.S., Obacunone and obacunone glucoside inhibit human colon cancer (SW480) cells by the induction of apoptosis. *Food Chem. Toxicol.*, 49, 7, 1616–25, 2011 Jul 1.
106. Zhang, Z., Zhang, Q., Wang, J., Zhang, H., Niu, X., Li, P., Preparation of the different derivatives of the low-molecular-weight porphyran from Porphyra haitanensis and their antioxidant activities *in vitro*. *Int. J. Biol. Macromol.*, 45, 1, 22–6, 2009 Jul 1.
107. Wang, J., Liu, L., Zhang, Q., Zhang, Z., Qi, H., Li, P., Synthesized oversulphated, acetylated and benzoylated derivatives of fucoidan extracted from Laminaria japonica and their potential antioxidant activity *in vitro*. *Food Chem.*, 114, 4, 1285–90, 2009 Jun 15.
108. Pawar, P.M., Koutaniemi, S., Tenkanen, M., Mellerowicz, E.J., Acetylation of woody lignocellulose: Significance and regulation. *Front. Plant Sci.*, 4, 118, 2013 May 21.
109. Huang, Q. and Zhang, L., Preparation, chain conformation and anti-tumor activities of water-soluble phosphated (1→ 3)-α-d-glucan from Poria cocos mycelia. *Carbohydr. Polym.*, 83, 3, 1363–9, 2011 Jan 30.
110. Wang, J., Zhang, Q., Zhang, Z., Zhang, J., Li, P., Synthesized phosphorylated and aminated derivatives of fucoidan and their potential antioxidant activity *in vitro*. *Int. J. Biol. Macromol.*, 44, 2, 170–4, 2009 Mar 1.
111. Wang, J., Wang, F., Zhang, Q., Zhang, Z., Shi, X., Li, P., Synthesized different derivatives of low molecular fucoidan extracted from Laminaria japonica and their potential antioxidant activity *in vitro*. *Int. J. Biol. Macromol.*, 44, 5, 379–84, 2009 Jun 1.
112. Huang, S., Chen, F., Cheng, H., Huang, G., Modification and application of polysaccharide from traditional Chinese medicine such as Dendrobium officinale. *Int. J. Biol. Macromol.*, 157, 385–93, 2020 Aug 15.
113. Vickers, N.J., Animal communication: When i'm calling you, will you answer too? *Curr. Biol.*, 27, 14, R713–5, 2017 Jul 24.
114. Wang, Y., Yu, Y., Mao, J., Carboxymethylated β-glucan derived from Poria cocos with biological activities. *J. Agric. Food Chem.*, 57, 22, 10913–5, 2009 Nov 25.
115. Wojaczyńska, E., Steppeler, F., Iwan, D., Scherrmann, M.C., Marra, A., Synthesis and applications of carbohydrate-based organocatalysts. *Molecules*, 26, 23, 7291, 2021 Nov 30.
116. Schoeneberg, C., Schilling, M., Burggraf, M., Fochtmann, U., Lendemans, S., Reduction in mortality in severely injured patients following the introduction of the "treatment of patients with severe and multiple injuries" guideline of the German society of trauma surgery–a retrospective analysis of a level 1 trauma center (2010–2012). *Injury*, 45, 3, 635–8, 2014 Mar 1.
117. Queneau, Y., Pinel, C., Scherrmann, M.C., Some chemical transformations of carbohydrates in aqueous medium. *C. R. Chim.*, 14, 7-8, 688–99, 2011 Jul 1.

118. Goodarzi, N., Varshochian, R., Kamalinia, G., Atyabi, F., Dinarvand, R., A review of polysaccharide cytotoxic drug conjugates for cancer therapy. *Carbohydr. Polym.*, 92, 2, 1280–93, 2013 Feb 15.
119. Basu, A., Kunduru, K.R., Abtew, E., Domb, A.J., Polysaccharide-based conjugates for biomedical applications. *Bioconjugate Chem.*, 26, 8, 1396–412, 2015 Aug 19.
120. Shen, S., Chen, X., Shen, Z., Chen, H., Marine polysaccharides for wound dressings application: An overview. *Pharmaceutics*, 2021, 13, 1666.
121. Zhao, M., Wang, J., Zhang, J., Huang, J., Luo, L., Yang, Y., Shen, K., Jiao, T., Jia, Y., Lian, W., Li, J., Functionalizing multi-component bioink with platelet-rich plasma for customized *in-situ* bilayer bioprinting for wound healing. *Mater. Today Bio.*, 16, 100334, 2022 Dec 1.
122. Zidarič, T., Milojević, M., Gradišnik, L., Stana Kleinschek, K., Maver, U., Maver, T., Polysaccharide-based bioink formulation for 3D bioprinting of an *in vitro* model of the human dermis. *Nanomaterials*, 10, 4, 733, 2020 Apr 11.
123. Buckwalter, J.A. and Mankin, H.J., Articular cartilage: Degeneration and osteoarthritis, repair, regeneration, and transplantation. *Instr. Course Lect.*, 47, 487–5, 1998.
124. Bujia, J. and Sittinger, M., *In-vitro*-Züchtung von Knorpelgewebe für die rekonstruktive Chirurgie: Einfluss von L (+)-Laktat und Glykolat auf kultivierte humane Chondrozyten. *Laryngo-Rhino-Otologie*, 74, 03, 183–7, 1995 Mar.
125. Lam, M.T. and Wu, J.C., Biomaterial applications in cardiovascular tissue repair and regeneration. *Expert Rev. Cardiovasc. Ther.*, 10, 8, 1039–49, 2012 Aug 1.
126. Psarrou, M., Mitraki, A., Vamvakaki, M., Kokotidou, C., Stimuli-responsive polysaccharide hydrogels and their composites for wound healing applications, *Polymers*, 15, 986, 2023.
127. Albuquerque, P.B., de Oliveira, W.F., dos Santos Silva, P.M., dos Santos Correia, M.T., Kennedy, J.F., Coelho, L.C., Skincare application of medicinal plant polysaccharides—A review. *Carbohydr. Polym.*, 277, 118824, 2022 Feb 1.
128. Li, D., Chen, J., Wang, X., Zhang, M., Li, C., Zhou, J., Recent advances on synthetic and polysaccharide adhesives for biological hemostatic applications. *Front. Bioeng. Biotechnol.*, 8, 926, 2020 Aug 14.
129. Ren, F., Wu, K., Yang, Y., Yang, Y., Wang, Y., Li, J., Dandelion polysaccharide exerts anti-angiogenesis effect on hepatocellular carcinoma by regulating VEGF/HIF-1α expression. *Front. Pharmacol.*, 11, 460, 2020 Apr 8.
130. Volova, T.G., Vinnik, Y.S., Shishatskaya, E.I., Markelova, N.M., Zaikov, G.E., *Natural-based polymers for biomedical applications*, Apple Academic Press, Taylor and Francis Group, UK, 2017 Apr 21.
131. Shelke, N.B., James, R., Laurencin, C.T., Kumbar, S.G., Polysaccharide biomaterials for drug delivery and regenerative engineering. *Polym. Adv. Technol.*, 25, 5, 448–60, 2014 May.

132. Devillard, R., Rémy, M., Kalisky, J., Bourget, J.M., Kérourédan, O., Siadous, R., Bareille, R., Amédée-Vilamitjana, J., Chassande, O., Fricain, J.C., *In vitro* assessment of a collagen/alginate composite scaffold for regenerative endodontics. *Int. Endod. J.*, 50, 1, 48–57, 2017 Jan.
133. Posadas, I., Monteagudo, S., Ceña, V., Nanoparticles for brain-specific drug and genetic material delivery, imaging and diagnosis. *Nanomedicine*, 11, 7, 833–49, 2016 Apr.
134. Islam, S.U., Shehzad, A., Ahmed, M.B., Lee, Y.S., Intranasal delivery of nanoformulations: A potential way of treatment for neurological disorders. *Molecules*, 25, 8, 1929, 2020 Jan.
135. Braeckmans, K., Buyens, K., Bouquet, W., Vervaet, C., Joye, P., Vos, F.D., Plawinski, L., Doeuvre, L., Angles-Cano, E., Sanders, N.N., Demeester, J., Sizing nanomatter in biological fluids by fluorescence single particle tracking. *Nano Lett.*, 10, 11, 4435–42, 2010 Nov 10.
136. Kang, S.H., Revuri, V., Lee, S.J., Cho, S., Park, I.K., Cho, K.J., Bae, W.K., Lee, Y.K., Oral siRNA delivery to treat colorectal liver metastases. *ACS Nano*, 11, 10, 10417–29, 2017 Oct 24.
137. Janmohammadi, M., Nazemi, Z., Salehi, A.O., Seyfoori, A., John, J.V., Nourbakhsh, M.S., Akbari, M., Cellulose-based composite scaffolds for bone tissue engineering and localized drug delivery. *Bioact. Mater.*, 20, 137–63, 2023 Feb 1.
138. Tang, Q., Cao, B., Lei, X., Sun, B., Zhang, Y., Cheng, G., Dextran–peptide hybrid for efficient gene delivery. *Langmuir*, 30, 18, 5202–8, 2014 May 13.
139. Wong, S.Y., Pelet, J.M., Putnam, D., Polymer systems for gene delivery—Past, present, and future. *Prog. Polym. Sci.*, 32, 8-9, 799–837, 2007 Aug 1.

3

Polysaccharide-Based Biomaterials for Gene Delivery

Ankita Moharana[1], Abhitav Tiwari[2*], Shalini Perada[2], Shivlal Yadav[3] and Om Prakash Kumar[4]

[1]*Department of Pharmacy, Arka Jain University, Mohanpur, Tata Jamshedpur, Jharkhand, India*
[2]*R & D, Viatris, Hyderabad, Telangana, India*
[2]*RITEE College of Pharmacy, Chhatauna Mandir Hasaud, Raipur, Chhattisgarh, India*
[3]*Department of Pharmacy, Central University of Rajasthan, Rajasthan, India*
[4]*Sachchidanand Sinha College, Aurangabad, Bihar, India*

Abstract

There are various physical and chemical cross-linking strategies for designing biomaterials derived from polysaccharides is presented. There has been a significant need in the field of biological research for innovative composite materials based on polysaccharides. In order to accomplish a challenging job in a complicated biological context, polysaccharides-based biomaterials are being intensively researched due to their inherent biocompatible, cheap, naturally accessible, and biodegradable qualities. Accordingly, this article details the most up-to-date methods for using polysaccharides produced functional biomaterials, with a focus on their use in drug administration, gene therapy, and tissue engineering. Polysaccharides are a possible alternative to nonbiodegradable chemical polymers owing to their biodegradability and sustainability in the environment. They consist of a long chain of monosaccharides that are joined together by glycosidic linkages. Composite nanomaterials based on natural polysaccharides, metal oxides, and carbon-based nanomaterials have been created. To comprehend the structural property of polysaccharide-based nanocomposite, the surface area/volume ratio is essential. Polysaccharides-based formulations have recently been investigated for their use in the creation of innovative drug delivery systems that may release numerous medicines concurrently or sequentially, hence improving the efficacy of targeted gene therapy. The research discussed here inspires the development

**Corresponding author*: abhitavggu@gmail.com

Rishabha Malviya, Sonali Sundram and Neelam Jain (eds.) Integration of Biomaterials for Gene Therapy, (47–76) © 2023 Scrivener Publishing LLC

of new biomaterials based on polysaccharides that exhibit superior characteristics and functionalities. Finally, future prospects for expanding their commercial application are discussed. In this article, we focus on biomaterials based on polysaccharides, which have recently emerged as a flexible platform for a wide range of biomedical applications, such as drug delivery, gene therapy, and tissue regeneration. Different polysaccharides' structures and characteristics are discussed, as are methods for functionalizing them. The development and use of polysaccharide-based nanocomposites in modeling are topics covered in this book, making it a valuable resource for those studying these topics. This is a much-needed volume on a promising new area of research; polysaccharide-based nanocomposites have promising future uses in areas such as gene delivery and tissue engineering, putting them at the forefront of contemporary medicine.

***Keywords*:** Polysaccharide, biopolymer, nanocomposite, gene therapy, tissue engineering

3.1 Background

Gene therapy is a novel approach to healthcare that utilizes genetic information for illness prevention or treatment. Gene therapy has the potential to replace invasive procedures like surgery and harmful medications by simply putting a therapeutic gene into a patient's cells [1]. Several methods of gene therapy are now under investigation, including (i) the substitution of a healthy gene for a disease-causing mutation; (ii) a 'knockout' or inactivation of a mutant protein-coding gene not working correctly; & (iii) the introduction of protective genes into cells.

Gene therapy has great promise as a treatment for cancer, hereditary illnesses, and certain forms of viral infection, but it is still in its infancy and requires more study to ensure its safety before it can be widely used. So far, gene therapy has only been researched for incurable disorders. Gene expression may be activated by genetic molecules that enter the nucleus of host cells. Gene therapy is a method of treating genetic problems by directing a patient's somatic cells to produce therapeutically active copies of a protein [2]. Designing an effective gene delivery system requires careful consideration of how the system will interact with the cell being targeted. Gene delivery systems comprise a plasmid-based gene expression system, a gene encoding a specific therapeutic protein, and a gene delivery system that controls the distribution of the plasmid carrying the gene expression instructions to the desired location inside the body. The foreign genetic material must be stable inside the host cells for the gene delivery mechanism to be effective [3]. The use of viral-based vectors for transgenic expression was pioneered in the 1980s. In 1984, the vaccinia virus was employed as a vector in vaccination against Hepatitis B in chimpanzees. However, cellular phenotypic change due to DNA exposures in the body was first revealed to have occurred via a non-viral gene delivery mechanism.

3.2 Introduction

Glycosidic linkages are the structural backbone of polysaccharides, which are made up of monosaccharides. Polysaccharides can be obtained from a variety of resources, including algae (alginate and carrageenan) and plants (cellulose, pectin, guar gum, chitosan, hyaluronan, and chondroitin). Polysaccharide-based biomaterials have been used in the past for drug, protein/peptide, and nucleic acid delivery systems, as described in several previous publications [4]. They have a broad variety of uses, and their physicochemical characteristics are very structure-dependent. This is because of the existence of hydroxyl, carboxyl, and amino groups, which provide a functional purpose many different types of polysaccharides may be readily changed for certain uses [5]. Determining the efficacy of polysaccharide-based drug carriers is crucial since the change in the pharmacokinetic and pharmacological properties of a polysaccharide polymer are significantly impacted by the structure of its polysaccharide backbone. Due to their advantageous properties of low immunogenicity, biodegradability, biocompatibility, and natural abundance, natural polysaccharides have gained extensive attention in the creation of microcarriers and nanocarriers.

Tissue regeneration for wound healing relies heavily on biomaterials, which facilitate host cell adhesion, migration, and integration by providing a site that promotes tissue growth by facilitating cell adhesion and migration. These materials' regeneration potential may be increased by the targeted administration of substances (such as cytokines or growth hormones) that can stimulate tissue growth [6]. Transmission of Genes is a flexible method for sustaining protein synthesis in the wound, since transfected (for non-viral vectors) or transduced (for viral vectors) cells carrying the transgene may operate in the role of regional bioreactors for the target protein. The physical characteristics of vectors used in gene therapy are often unaffected by the nucleic acid sequence being transported, making it possible to use the same vector to transport a wide variety of sequences or combinations of sequences. Vectors can be plasmids, which are stable circular strands of DNA that can be supplied directly from the source, or they can be nucleic acids packed into viral or nonviral systems, which are often more successful than naked plasmids [7]. Increasing the likelihood of cellular internalization may be achieved by loading these vectors onto biomaterial scaffolds and engineering them for controlled release of their payload.

Biomaterials, both synthetic and biologically generated, have become standard in clinical applications of drug delivery and diagnostics [8]. Typically, biomaterials encapsulate the agents in a way that allows them to be transported to the sites of action without being eliminated by the body's natural detoxification processes. Moreover, they may integrate features that may react to physiological cues to improve the functioning of given agents, and they may offer resistance to substances in the body's natural setting. The ability to integrate therapeutic and diagnostic agents is particularly important in the context of attempts to use treatments based on nucleic acids, sometimes known as gene therapies [9]. Nucleic acids offer the unbounded potential for molecular treatment to rectify defective physiology with the knowledge of patient-specific supplementary information. Clinicians are keen to find alternatives to conventional small-molecule medicines due to their wide-ranging impacts on several cellular systems and the subsequent development of drug resistance when rogue cells initiate natural physiological mechanisms to counteract the drug's effects. Different possibilities with gene treatments have attracted doctors since the development of recombinant technology to manufacture natural and designed protein mass [10]. Recent molecular understanding of the role of DNA and RNA in the control of cell physiology has resulted in a surge of interest in the commercialization of both traditional nucleic acids (such as antisense oligonucleotides, plasmid DNA, etc.) and novel agents such as short interfering RNA (siRNA), non-coding microRNAs (miRs), and messenger RNA (mRNA). As an example of the latter, mRNA therapies developer Moderna Therapeutics just completed the biggest biotech IPO in the United States in 2018.

The goal of gene delivery systems is to direct where a gene is expressed in the body by influencing its identification and subsequent intracellular trafficking and nuclear translocation by way of a cell-surface receptor [11]. To prevent a gene expression system from being degraded too quickly in the extracellular environment, gene delivery systems should be able to transport genes to their intended cells in either a nonspecific or cell-specific manner. Other parts of a gene delivery system may facilitate the intracellular trafficking of a gene expression system. The development of a superior synthetic gene carrier is a major challenge for non-viral gene therapy. Multiple systems based on lipids, peptides, and polymers are being studied for their potential to facilitate gene transport. Methods such as electroporation, microinjection, and the "gene cannon" are studied as methods of gene transfer; however, these technologies may have trouble gaining patient consent since they require surgical exposure to target regions. An effective high-voltage discharge is necessary for Plasmid DNA to be deposited on gold particles in the gene cannon to penetrate their intended organs of effect.

There are two types of polysaccharide materials: polyelectrolytes and non-polyelectrolytes. Anionic polyelectrolytes include chitosan, alginate, heparin, pectin, and hyaluronic acid, while cationic polyelectrolytes include alginate, heparin, pectin, and hyaluronic acid [12]. In addition to PEGylation, a newer method that involves coating nanoparticles with polysaccharides and oligosaccharides has been suggested. Coated materials may actively target by interacting with a specific class of receptors on the surface of cells and tissues 16. Polyelectrolyte complexation, which relies on making nanoparticles out of polysaccharides, is a well-established, uncomplicated procedure that relies on the electrostatic interaction of polysaccharides with various oppositely charged polymers [13].

3.3 Gene Therapy

The experimental practice of altering genes or the expression of those genes to cure or prevent illness is known as gene therapy. It may be utilized to treat monogenic illnesses by restoring cell function or to confer novel capacities on cells [14]. With gene therapy, genes are edited, replaced, or altered in expression rather than pharmaceuticals being used.

3.3.1 Why Does Gene Therapy Need a Carrier?

Many biological hurdles must be overcome before gene treatments may reach their target. Their primary objective must be to avoid degradation on route to the intended organ, tissue, or cell. Because of the existence of endo- and exonucleases in physiological media, oligonucleotides, for example, have a relatively limited half-life [15].

Gene therapies must overcome many biological obstacles to reach their intended organ or tissue. The intravenous approach is often used (IV). To reach their intended destination, gene therapies in circulation must first pass through the endothelium. The characteristics of this barrier vary from organ to organ and tissue to tissue. Fenestrated vasculature in organs including the spleen, liver, and tumor makes endothelium crossing more manageable.

Numerous research efforts are focused on improving gene therapy methods for treating neurological disorders [16]. The blood-brain barrier (BBB) is an obstacle to gene treatments that are intended for central nervous system administration. The brain's vascular system has this biological barrier in the form of a continuous endothelial membrane. Tight connections are formed between adjacent endothelial cells to prevent the passage of harmful substances like poisons and pathogenic germs.

About two-thirds (66.4%) of ongoing gene therapy clinical studies focus on cancer. Gene therapies in solid tumors must penetrate an extracellular matrix rich in collagen released by cancer-associated fibroblasts before they can reach the cancer cells [17]. High interstitial fluid pressure (IFP) in solid tumors, caused by leaky vasculature, poor lymphatic drainage, and the fast proliferation of cancer cells, is a barrier to the absorption of medications, particularly those with a high molecular weight, such as gene therapies.

Furthermore, the molecular structure of the cellular membrane must be traversed for gene therapies to be effective. The molecular weights of nucleic acids are rather high. Due to the presence of a phosphate group in their backbone, they are negatively charged and so readily dissolve in water [18]. Because of these features, they are unable to pass through biological membranes. For internalization inside cells, endocytosis is the primary pathway. Endosomes, intracellular vesicles that contain multiple digestive enzymes that may break down nucleic acids, are the target for the delivery of gene therapies [19]. However, nucleic acids, particularly DNA, are not stable at such a low pH, and the endosomal pH decreases from early to late endosomes and then to lysosomes. The low pH and catalytic enzymes in the early stages of the endosomal process necessitate the early release of gene therapies.

In conclusion, the cytoplasm is where RNA-based gene therapies like silencing RNAs exert their therapeutic effects [20]. In contrast, DNA-based gene therapy methods are required for the nucleus by passing through either the double nuclear membrane or a nuclear pore.

Therefore, it is clear that a delivery mechanism is required to transport gene therapies to the nucleus, where they can have the desired effect, can avoid degradation, and can cross cellular membranes.

3.3.2 Translational Relevance

To induce transgenic expression, the vector must first be transported to the outer membrane, inside the cell, and finally the nucleus [21]. Although the primary focus of biomaterial scaffolds that distribute vectors is on overcoming mass transport restrictions, the scaffold design may also affect processes like internalization routes. Regeneration is affected by the environment in which the transgene is expressed by cells both inside and around the material [22]. The effectiveness with which the transgene is expressed is affected by many factors, including the scaffold's design (rate of degradation, porosity, and architecture), all of which influence how quickly cells infiltrate and what kinds of cells invade. More efficient and regulated delivery methods are required to drive functional tissue engineering for various uses, and the criteria for creating the vector and the scaffold, have begun to be established and refined in the papers reviewed below.

Material, vector, and implant site factors include (i) the duration and intensity of transgene expression, (ii) the type of cells that infiltrate the scaffold and express the transgene, (iii) the density of transgene-expressing cells within and around the implant, (iv) the effect of material and vector on the foreign body response, and (v) the factors that should be expressed to optimize tissue formation [23]. Fundamental research at the interface of gene therapy and tissue engineering may help solve these issues and improve the technique for potential therapeutic applications.

3.4 Gene Delivery Systems Based on Polysaccharides

In this paper, we disclose a family of biodegradable cationic polymers prepared by grafting oligoamine residues onto natural polysaccharides [24]. The grafting concept was created, whereby side chain oligomers are connected to linear or branched hydrophilic polysaccharides, to interact with the anionic surfaces of double-stranded DNA in a three-dimensional fashion. The polysaccharide-based gene delivery approach demonstrates the importance of the cationic polysaccharide structure on transfection activity. Mucosal transport of complexes, including peptides, proteins, oligonucleotides, and plasmids was studied using colloidal polysaccharide particles. The use of nano-sized natural polysaccharide particles to transport genes and drugs has also been investigated. Hydrophobic polysaccharide self-assembly, electrolyte complexation, and covalent/ionic cross-linking were all suggested as mechanisms for creating polysaccharide-based nanoparticles [25]. Researchers have looked at the efficacy of natural polysaccharides like chitosan and pectin as vectors for transporting genes and drugs. Poly-N-acetyl-D-glucosamine, a polymer with a positive charge and significant mucoadhesion, is a substance that has been highly suggested by the chitosan family. Different structural segments, such as Pectin composed of homo-galacturonan (HG) and rhamnogalacturonan (RG-I), respectively represent smooth and hairy regions.

3.4.1 Polysaccharide-Based Nanocarriers

Benefits from polysaccharides include their charge, solubility, and other properties in the creation of nanoparticles using several different methods [26]. Several writers have surveyed the literature on nanoparticle production and its uses with a particular polysaccharide in mind. Amphiphilic polysaccharide derivatives modified with hydrophobic moieties self-assemble, polysaccharides can be cross-linked with small molecule cross-linkers, and charged polysaccharides can easily bind with oppositely charged macromolecules are the main methods used to prepare polysaccharide-based nanoparticles.

3.4.2 Polysaccharide Nanoparticles for Gene Delivery

To cure or prevent genetic disorders or dysfunctions, gene therapy often involves inserting foreign genes into host cells. Viral vectors include, but are not limited to, retroviruses, adenoviruses, adeno-associated viruses (AAVs), and herpes simplex viruses that may transfect both actively proliferating and quiescent cells. Researchers' focus has shifted in recent years to the discovery of secure and effective substitutes for viral vehicles [27]. When opposed to viral vectors, non-viral vectors offer several benefits, including ease of preparation, minimal immunogenicity, the ability to carry genes of varying sizes, and fewer clinical hazards. There is promising evidence that nanoparticles, a novel type of non-viral vector, may transfer DNA from the outside of the cell to the nucleus. The most often utilized polymers for gene delivery are polysaccharides and other cationic polymers.

If you want highly effective gene vectors, chitosan is a good option. Because of its cationic composition, chitosan facilitates the ionic contact between the DNA's negative charge and the amine groups on the strand's positive charge [28]. Anionic alginate coating of chitosan decreases its interaction with DNA, mediating its dissociation upon entrance into the cell. Thus, p-DNA and siRNA that have been encapsulated may be protected against nuclease destruction. It has also been suggested that chitosan nanoparticles, rather than liposomes, may be superior in transporting siRNA cargo into their intended targets. Non-viral vectors for gene transfection have been explored, specifically, the injection of cationic chitosan/carboxymethyl cellulose nanoparticles coated with plasmid DNA directly into the skin of mice 81. However, chitosan's poor water solubility may restrict its binding capabilities with nucleic acid. Grafting PEG onto chitosan nanoparticles is one way to boost their solubility and stability [29]. Another strategy for increasing water solubility and facilitating access to interior binding sites is acidic hydrolysis.

To introduce human papillomavirus type 16 into CHO cells, chitosan nanoparticles were previously created. SDS-PAGE and western blotting studies confirmed the efficient expression of E7 proteins. Finally, chitosan nanoparticles were proposed as a safe and effective non-viral approach for delivering nucleotides into CHO cells in this work. Interleukin-12 is a cytokine with immunomodulatory properties that is essential for fostering potent anticancer activity [30]. To transfect the interleukin-12 gene into dendritic cells via the mannose receptor-mediated endocytosis method, mannosylated chitosan nanoparticles have been created. There is some evidence that this approach might be effective in eliciting a durable cell-mediated immune response. Given that v3 integrin is overexpressed in many types of cancer tissues, this trait may be exploited for administration to malignant cells. Recently, a method was disclosed for the three integrins receptor-specific delivery of siRNA to tumor cells using a cyclic Arg-Gly-Asp (RGD) peptide conjugated to chitosan nanoparticles [31]. Chitosan and a negatively charged polymer, like alginate, make a formidable pair, has been proposed, however, to decrease the affinity between chitosan and the encapsulated gene, and improve transfection efficiency. It has been shown that alginate chitosan complexes are superior to either alginate or chitosan alone in protecting entrapped compounds [32]. In addition, the synthesis can see how they affect the performance of alginate/chitosan nanoparticles as carriers for antisense oligonucleotides. Gene escapes from the endosome and is localized inside the cell following alginate nanoparticle endocytosis due to fast erosion and osmotic swelling. To improve transfection efficiency in NIH 3T3 cells, gene escape was facilitated by covering chitosan nanoparticles with alginate, which occurred before the vesicles entered the digesting endolysosomes. The NIH 3T3 cells were able to transfect plasmid-loaded calcium-alginate nanoparticles, as reported by Jin-Oh You *et al.*

These nano complexes were more effective than polyethyleneimine in compacting plasmid DNA (PEI) [33]. Transfection of the interleukin-10 gene into macrophages using tuftsin-labeled alginate nanoparticles has been described as part of anti-inflammatory gene therapy. Since HA is highly concentrated in the tissues of the eye, researchers are now focusing on developing methods to employ HA as a mediator for the delivery of genes to the eye. To target gene delivery into the cornea and conjunctiva, researchers created a synthetic gene nanocarrier built from the mucoadhesive polysaccharides chitosan and HA. To better treat joint problems, researchers looked at the possibility of using successful gene transfer into chondrocytes using HA/chitosan nanoparticles. HA's anionic nature prevents it from encasing negatively charged molecules like siRNA. But this issue may be easily solved by modifying nanoparticles including components such as fatty amines and cationic polyamines, which lessen the negative charge density in the polymer framework and increase the loading capacity of siRNA [34]. It was also demonstrated that CD44 receptors, which are overexpressed in cisplatin-resistant tumors, may be efficiently targeted by HA nanoparticles loaded with siRNA and cisplatin. High levels of -gal expression may be achieved in COS-7 cells by transfection with the pBUDLacZ plasmid using hydrogel pullulan nanoparticles. These nanoparticles were just as effective as the Lipofectamine 2000 formulation in inducing protein expression. Pectin nanoparticles were prepared for use as gene delivery vectors via a divalent cation (Ca, Mg, or Mn) combines with pectin to form an isotropic gel. One may say that there was a significant improvement in viral transduction efficiency with this batch of particles, and cytotoxicity was kept to a minimum [35].

3.5 Practical Application of Gene Delivery Systems

Only the optimal method of gene delivery will ensure the success of gene therapy. Gene delivery systems in therapeutic applications typically employ cell-selective, non-immunogenic vectors to achieve therapeutically relevant levels of transgene expression. Analytical methods can be used to study patterns in gene expression and DNA or RNA microarrays used in various types of gene sequencing, potentially leading to the identification of thousands of genes [36]. The gene delivery system has been used to develop a hybrid biosynthetic vector for use as a vaccine delivery mechanism. By fusing E. coli with synthetic polymers, the vector can circumvent the common challenges associated with delivering genes. The gene therapy system offers promising new options for treating many medical issues, such as cancer, infectious diseases, and genetic abnormalities. The non-viral vector's low immunogenicity and relatively cheap cost to reproduce are two of its best features. The DNA size for packing into the non-viral vector and modifying with ligands for precise cell targeting is unrestricted [37].

It has been investigated if synthetic polymeric materials could serve as vectors for the transfer of genes across animals. One such candidate is plasmid DNA, which can deliver the gene to the nucleus of the desired cells without harm. Exciting novel therapeutic approaches based on the delivery of genes via mechanisms other than viruses have also emerged [38].

3.6 Polysaccharide-Based Nanoparticles

3.6.1 Chitosan

Chitosan is a glycosidic (1,4)-linked polysaccharide made up of D-glucosamine and N-acetyl-D-glucosamine. Chitosan, a cationic compound, binds to the DNA's negatively charged phosphate groups. It has been established that the preparation media, the plasmid-to-chitosan ratio, and the molecular weight of the chitosan, all have an impact on the chitosan/plasmid complex's colloidal and surface characteristics. Transfection of several cell lines, including COS-1 cells, is possible using a plasmid/chitosan combination [39]. High molecular weight (above 100 kDa) chitosan dissolves in acetic acid-treated water, yielding low molecular weight (22 kDa) chitosan that is extremely soluble under physiological conditions and efficiently compacts DNA. Although PLL/DNA complexes were more effective, low-molecular-weight chitosan/DNA complexes also showed promising results in terms of transfection effectiveness, they are safer for the cells.

By synthesizing hydrophobically modified chitosan with deoxycholic acid, the research team was able to transfect chitosan/DNA complexes into COS7 cell lines [40]. To specifically target cells that express a galactose-binding membrane lectin, researchers partly replaced a-lactose with varying quantities using sodium cyanoborohydride and reductive amination of lactosyl residues. Expression vector delivery into BNL CL2 and Hep G2 cells was not improved by using lactosylated chitosans, which were replaced with 3, 10, or 20% lactose residues. In contrast, lactosylated PEI increased *in vitro* transfection effectiveness by a factor of 1000 (Up to 108 RLU/mg of protein). That there is a deficiency in receptors on the cells is unlikely to be the cause of lactosylated chitosan's ineffectiveness. Lactosylation of chitosan, on the other hand, was shown to decrease the complexes' surface charge, which causes them to cluster. Lactosylation and reductive amination may also play a role in this process greatly decreasing chitosan's affinity for DNA. Transfection effectiveness was not increased by the addition of an endosmotic drug like chloroquine, however, in the presence of 10% serum, lactosylated chitosan/DNA complexes were able to efficiently transfect HeLa cells [41]. There was a linear increase in gene expression from 24 to 96 hours in cells transfected with lactosylated chitosan, but the efficiency of PEI-mediated transfection dropped two orders of magnitude under the same conditions (after 96 hours, chitosan was shown to be 10 times more efficient in transfection than PEI) [42].

To transfect mucosal cells lining the intestines and the lymphoid tissue that surrounds the intestines, chitosan/DNA complexes may be transferred across the mucosal borders, since chitosan is a mucoadhesive polymer. The administration of chitosan/pCMV-Arah2 complexes orally results in transgene expression in the intestinal epithelium has been observed by other researchers. Immunized animals with chitosan/pDNA complexes had much lower levels of immunoglobulin E, histamine in the blood, and vascular permeability in response to allergen than unimmunized mice or bare DNA-treated mice. These findings provide support for the hypothesis that chitosan/DNA nanoparticle oral allergen-gene immunization may successfully regulate the anaphylactic response in mice.

3.6.2 Dendrimers

Highly branching spherical polymers are known as starburst polyamidoamine (PAMAM) dendrimers [42]. Each generation of PAMAM dendrimers has a unique surface charge and diameter due to the increasing complexity of the synthesis process; for example, fifth-generation dendrimers are the result of five polymerization cycles. Both ammonia and ethylenediamine may serve as the starting point for the stepwise polymerization process that results in PAMAM dendrimers, but which ultimately determines the shape, density, and surface charge of the final product [43]. Researchers may find "intact" dendrimers, as well as "fractured" dendrimers. Dendrimers with no breaks have dendrimers typically have two arms per branch point, although dendrimers with breaks may have less than two arms per branch point.

Plasmid condensation results from electrostatic interactions between the terminal primary amines of dendrimers and the DNA phosphate groups. Researchers have found that particle size, surface charge, and gene transfer efficiency are all affected by the concentration of dendrimers in dendrimer/plasmid complexes generated utilizing the 5th generation of dendrimers. Therefore, the ability of the formulation to transfect into cells is affected by the size, shape, and number of primary amine groups on the surface of the polymer. Starburst dendrimers have mediated efficient but nonspecific *in vitro* gene transfer to eukaryotic cells. Therapeutic genes were tested for their ability to be delivered to cancer cells using a Plasmid vector based on the Epstein-Barr virus (EBV) that has been conjugated to the PAMAM dendrimer (EBV/polyplex) [44]. Dendrimer/EBV plasmid complexes are highly efficient at transferring suicide genes in both *in vitro* and *in vivo* studies (e.g., the complex can successfully transfer the herpes simplex virus type 1 thymidine kinase (HSV-1 TK) gene into Ewing's sarcoma A4573 cells *in vitro* and into subcutaneous tumors in immunodeficient mice).

PMMA, or poly(2-methylamino ethyl) methacrylate (pDMAEMA). To facilitate transfection into different cell types, pDMAEMA, a water-soluble cationic polymer, may attach to DNA, compressing it in the process. Poly(2-(trimethyl amino)ethyl methacrylate) (pDMAEMA) and its quaternary amine analog poly(2-(trimethyl amino)ethyl methacrylate) (pDMAEM) (48) were compared to determine whether or not the presence of tertiary amine groups (pKa 7.5) had any beneficial benefits (pTMAEMA). Transfection and interaction data suggested that pTMAEMA lacks the "proton sponge" function of pDMAEMA and PEI, leading the researchers to believe that it is incapable of escaping the endosomes [45].

Compared to pTMEAMA, pDMEAM has much greater transfection efficiency. Gel electrophoresis, fluorescence spectroscopy, and circular dichroism spectroscopy all show DNA association/dissociation qualities are similar to pDMAEMA and PLL, although DNA is tightly condensed by pTMEMA, perhaps explaining why it is less effective at transfecting cells than pDMAEMA.

The maximum transfection efficiency was achieved at a pDMAEMA/DNA ratio of 3:1 (w/w), where homogeneous complexes with a diameter of 150 nm could be formed. Contrary to expectations, the presence of serum proteins did not reduce the transfection efficacy of the complexes [46]. The stability of pDMAEMA/DNA was also tested in 20 mM Hepes (pH 7.4) buffer with 10% sucrose at 4, 20, and 400 C. (51). Transfection efficiency was significantly reduced when pDMAEMA/DNA complexes were stored at temperatures above 30C for more than 10 months.

Poly(ethyleneimine) (PEI) and other polymers containing tertiary amines have been found to have endosmotic activity. Similar to DMAEMA is N, N'-diethyl aminoethyl methacrylate (DEAEMA). Four-vinyl pyridine (5.45 pKa) and diethylamine (9.5 pKa) both contain tertiary amine groups. Therefore, a pH-sensitive polymer may be created by copolymerizing 4VP with DEAEMA, which may improve membrane fusing.

3.6.3 Alginate

Brown algae from the ocean (like *Laminaria hyperborea*) and bacteria from the ground are the sources of the naturally occurring hydrophilic polymer alginate (ex. Azobactervinelandii) [47]. Polymers of alginate are made up of b-D-mannuronic acid and a-L-glucuronic acid residues alternate in 1,40-linked blocks. As a result, the molecular weight, sequence of organization, and composition of alginate polymers are the primary factors in determining their physical qualities. Alginates are widely utilized in the pharmaceutical and food sectors due to their desirable properties, including solubility in water, innocuousness, and gelation. Alginate, for an instance, maybe gelled externally or internally with the help of materials with two positive charges (divalent cations) such as calcium and barium. Alginate gel's ability to release drugs slowly is tied to its ability to expand and erode, both of which are influenced by the gel formulation method chosen. Intriguingly, alginate gels shrink in acidic conditions yet rapidly disintegrate in alkaline ones. Alginate formulations are preferred for intestinal medication administration because of their pH-dependent swelling tendency. Alginate's mucoadhesive ability makes it a valuable synthetic extracellular matrix material that has several uses where it may entrap a wide variety of bioactive compounds, cells included. For instance, alginate hydrocolloid compositions provide prolonged drug release [48]. For instance, alginate hydrocolloid encapsulation may allow vascular endothelial growth factor (VEGF) to be safely administered for up to 14 days without compromising the protein's biological activity. Plasmid DNA (pDNA) was successfully encapsulated into gel-structured alginate nanoparticles with a loading efficiency of 60% thanks to alginate's cross-linking activity. Compared to other polysaccharides, alginate hydrogels have had fewer therapeutic applications despite reports of their usefulness in the local and sustained delivery of different growth factors. To put it simply, this is probably because of excessive clumping in the circulatory system.

3.6.4 Hyaluronan

Hyaluronic acid (HA), also known as hyaluronan, is present in the synovial fluids, vitreous humor, and connective tissues of the body. HA is a high molecular weight, non-sulfated glycosaminoglycan. Hydrophilic linear glycosaminoglycan HA is a naturally occurring polysaccharide made up of the disaccharide units (1, 4)-b-D-glucuronic acid and (1, 3)-b-N-acetyl-D-glucosamine in alternating order. The non-immunogenic, non-toxic, biodegradable, and immune-innocuous nature of HAs has made them a popular choice [49]. HA and its derivatives have been used effectively in the treatment of arthritis, as scaffolds in wound healing procedures, in ophthalmic surgery, and tissue augmentation throughout the course of many decades. These cell surface glycoproteins are overexpressed in many types of tumor cells, and their role in the HA receptor-ligand interaction has been extensively explored. As a result of this receptor-ligand interaction, HA nanoparticle absorption by tumor cells may be greatly improved.

As a result, researchers have been looking at different HA nanoparticles as potential new anti-cancer gene carriers.

3.6.5 Pullulan

Pullulan is a linear polysaccharide made up of a-(1-6) maltotriosyl units and is soluble in water. To a large extent, pullulan due to its adhesiveness and relatively low water viscosity, among other biological and physicochemical features is determined by its distinctive linking pattern. Hydrophobic modification of water-soluble pullulan enables the polymer to self-aggregate into stable nanoparticles, allowing for reliable drug loading. Caiti-modified pullulan has been studied as a potential pDNA encapsulant for use in both *in vitro* and *in vivo* settings because of its non-ionic nature in nucleic acid delivery [50]. Nanoparticles with pullulan modifications may be used to transport cancer drugs. In particular, because of their strong affinity for asialoglycoprotein receptors, which are overexpressed on the sinusoidal membranes of hepatocytes, pullulan, and its derivatives may be useful for liver targeting. As a result of its high adhesiveness, pullulan is a promising candidate for use in the development of artificial extracellular matrices.

3.6.6 Pectin

To put it simply, Polysaccharide pectin is a structural component of plants. The degree to which pectin's galacturonic acid residues are esterified is a key component in influencing its solubility, gelling, and film-forming capabilities [51]. Both the stomach's acidic environment and the small intestine's alkaline environment are safe for pectin, only to be broken down in the large intestine by pectinases released by bacteria. The qualities indicated make pectin-based delivery systems an excellent choice for shielding protein and polypeptide-based medications throughout their journey through the digestive tract and into the large intestine [52]. Pectin-based delivery systems have shown promise in protecting loaded drugs throughout the gastrointestinal tract, although their efficacy may be limited by pectin's very hydrophilic nature. Therefore, future research must focus on the identification of pectin derivatives with low water solubility. Inactivating proteases, applying cross-linking calcium ions, increasing coating polymer thickness, and including physical layers that separate drug-pectin are all viable options for avoiding this issue. Table 3.1 provides a summary of the several types of polysaccharide-based nanoparticles that may be employed as drug-delivery vehicles [53].

Table 3.1 Nanoparticle drug carriers based on polysaccharides and their classification.

Nanoparticle system	Nanoparticle system	Comments
Chitosan	Exhibits the capacity to relax the bonds that hold epithelial cells together and a mucoadhesive property	Its cationic properties make it an effective carrier for electronegative compounds like DNA.
Alginate	Shows mucoadhesive and gelling properties	Its anionic properties act as a carrier for cationic drugs.
Heparin	An anticoagulant polysaccharide that is anionic and strongly sulfated	This is the optimal growth factor delivery technique.
Hyaluronic acid	Affinity for absorbing water and gelling when exposed to it	Promotes CD44 receptor-mediated endocytosis, allowing for passive tumor targeting.
Dextran	A non-toxic neutral polysaccharide	Nanoparticles can be broken down by an enzyme called dextranase.
Pulluan	A neutral polymer generated by a particular fungus	Relative high cost of pullulan has hindered its applicability
Pectin	A gelling and film-forming polysaccharide that is anionic	Pectinase, an enzyme released by colonic bacteria, degrades nanoparticles.

3.6.7 Heparin

Heparin is a polysaccharide with anionic and highly sulfated structures that has potent anticoagulant effects. To interact with heparin, several protein components, including various growth factors, including heparin-binding domains. Nanoparticles based on heparin have the potential to serve as innovative methods for protection by strengthening the connection between the factors and their receptors, heparin-binding growth factors are protected from proteolytic and chemical degradations and their release is regulated. Heparin has shown antitumor activity, much like chitosan. Heparin's anticancer capability is connected to its capacity to decrease tumor angiogenesis and metastasis by preventing VEGF and bFGF from functioning normally [54]. However, low-molecular-weight tiny heparins are more effective in inhibiting tumor development.

3.6.8 Other Polysaccharides

There are a wide variety of different polysaccharides that have been investigated for potential use in biomedicine as vehicles for the transport of drugs and genes. Medicines, protein/peptides, pDNA, oligonucleotides, and siRNA are just some of the many substances that have been encapsulated and delivered using nanoformulations, and their templates include cyclodextrin, pectin, b-glucan, and guar gum.

3.7 DNA Delivery

In recent years, there has been a dramatic increase in the need for biopharmaceuticals. Gene therapy, in particular, has great promise for the future of medicine. Since gene therapy is being recognized as a possible contender for the treatment of many stubborn illnesses, over the past twenty years, there has been a lot of focus on the study of gene delivery. When used therapeutically, gene therapy requires the identification of specific target genes. To be effective, these therapeutic gene therapies require efficient targeted delivery to the intended cells. Small RNA molecules, oligonucleotides, and plasmid DNA are all examples of nucleic acid all examples of therapeutic genes that may be damaged by serum nucleases when they are exposed [55]. Furthermore, there are significant issues with the transfection efficiency and cellular absorption of naked therapeutic genes in gene therapy.

Over 1340 clinical studies using gene therapy have been completed or are now underway since the first experiment was conducted in 1989. However, a limited number of gene therapy drugs approved for commercial release thus far. The absence of effective, safe, and manageable means of delivering genes has been identified as the primary roadblock to the expansion of human gene therapy. Therefore, a reliable vector that can shield the genes from damage and guide them to the correct cells is essential for an effective gene delivery system. Both viral and non-viral vectors have been used extensively created for use in gene therapy during the last decade. Viral vectors are very effective, but they have important drawbacks such as immunogenic characteristics and the possibility of causing a mutagenic infection [56].

To transfer larger quantities of therapeutic genes, non-viral vectors are preferable to viral vectors, but their transfection efficiency is often subpar. The use of nanoparticle-based polysaccharides as alternative non-viral vectors for gene delivery is also being investigated.

Valid non-viral DNA delivery vectors will safeguard the DNA until it reaches its final destination. In addition, the gene-carrying vector must be small enough to be efficiently imported into the cell's cytoplasm. Cationic carriers are widely believed to be effective in condensing and delivering DNA *in vitro* and *in vivo*, and they are used to create complexes with negatively charged pDNA [57]. Non-viral vectors for gene delivery are under intense scrutiny, and cationic phospholipids and polymers are at the top of the list of potential options. Cationic polymers also have greater stability than cationic lipids. With their advantageous biocompatibility, biodegradability, and low toxicity, many kinds of polysaccharides are currently being utilized as man-made cationic polymers.

The application of chitosan analogs as a non-viral vector for gene transfer is gaining momentum. As was previously established, chitosan is unique among polysaccharides in that it contains free amine groups. To concentrate genes into nanoscale complexes, polysaccharides, which normally do not provide cationic charges, must be to a greater degree altered with cationic polymers. While pDNA is anionic, the major amine groups on chitosan's backbone give the polymer cationic charges that form electrolyte complexes with anionic pDNA. Further, these functional amine groups have the potential to operate as active sites in a variety of chemical reactions.

The use of modified polysaccharide nanoparticles has been shown to increase the efficiency of transfection and prolong the blood circulation of pDNA. Hydrophobized dextran-spermine and chitosan-graft-spermine copolymers are two examples of pDNA delivery systems that work well in both *vitro* and *in vivo*. In this case, chitosan-graft-spermine was synthesized using periodate-oxidized chitosan and spermine (CHI-gSPE). To better understand how chitosan-graft-spermine nanoparticles function as vectors for gene transfer, Jiang *et al.* take into account factors like particle size and zeta potential, which may have a significant impact on nanoparticle durability and gene delivery effectiveness [58]. To maximize their effectiveness, CHI-g-SPE/DNA complexes were designed as 55.7 nm-sized, generally homogeneous nanoparticles. After they joined together to form complexes, their surface charges quickly spiked to the positive side. Binding to anionic cell surfaces, a crucial step in the uptake process, is facilitated by this positive surface charge. There was also no detectable cytotoxicity of the CHI-g-SPE/DNA complexes *in vitro*. A minimum of 83% cell viability was observed after CHI-g-SPE copolymer treatment in A549, WI-38, and HepG2 cells at concentrations as high as 20 and 50 lg/mL. However, at the same dosages, cell viability significantly decreased in PEI 25 K-treated cells. Transfection effectiveness was monitored by using the green fluorescent protein (GFP)-expressing plasmid pcDNA3.1, and the transfection of DNA into mature cells using CHI-g-SPE/DNA complexes was shown to be more efficient than using chitosan/DNA complexes. The effectiveness of aerosol delivery of CHI-g-SPE was evaluated by exposing C57BL/6 mice to CHI-g-SPE/GFP-DNA complexes for 48 hours by the nose-only exposure strategy [59]. Inhalation of the CHI-g-SPE/GFP-DNA complexes resulted in a GFP signal in lung tissues, showing efficient delivery of GFP-DNA to the lung and high absorption efficiency by tumor cells. The CHI-g-SPE showed promise in safe and effective gene delivery for lung cancer treatment via aerosol administration. A wide range of chitosan nanoparticles is effective in the oral and intravenous delivery of certain target DNA molecules.

To increase receptor-mediated phagocytic uptake, the murine IL-10 gene was encased in a gel-like alginate nanoparticle containing a tuftsin targeting sequence (TKPR). In the grand scheme of things, this facilitates the direct delivery of genes to macrophages. When treating inflammation *in vivo*, it's important to target macrophages the encapsulated gene must be highly efficient at loading (*60%), stable, and intracellularly available. Future medical advancements using therapeutic genes may benefit from several methods employing non-viral polysaccharide nanoparticles for gene delivery.

3.7.1 siRNA Delivery

Gene silencing employs a double-stranded RNA molecule that is 21-25 bases in length; this molecule is known as small interfering RNA (siRNA). The siRNA molecules congregate into RNA-induced silencing complexes (RISCs) in the cytoplasm, where they direct the enzymatic cleavage of homologous mRNA, resulting in the selective destruction of the target mRNA and a subsequent decrease in protein output [60]. Because of its specificity in silencing target genes, siRNA has been the subject of substantial research in the therapeutic sector during the last decade. Multiple studies have shown that small interfering RNAs (siRNAs) can cure many different disorders, including AMD, infections, and cancer. Naked siRNA has a large molecular weight and an anionic charge, making it difficult for it to cross the cell membrane. Naked siRNA is also easily degraded by the nucleases present in blood and other bodily fluids. As a result, siRNA needs a delivery vector system to avoid destruction by enzymes and to access the cytoplasm, where it can exert its effects. The feasibility of using various vectors, such as viruses and phages, to transport tiny interfering RNAs has been explored [61]. Low loading capacity, carcinogenic potential, and immunogenic effects are problems with both viral and non-viral vectors used for DNA delivery however, viral vectors are more efficient. Many polysaccharide nanoparticles have been carefully explored as potential non-viral vectors for the creation of new siRNA delivery methods. For siRNA delivery, several polysaccharide vectors have been proposed, including chitosan and dextran-based nanoparticles.

Several types of polysaccharide-based nanoparticles have been identified in previous research as nonviral vectors for siRNA delivery. More than a decade has passed since chitosan was first utilized as a vehicle for transporting genetic material like plasmid DNA (pDNA) or oligo(dT)-nucleotides (ODN) [62]. Complexes of siRNA duplexes with the polymer chitosan were initially described in 2006 as a method for delivering siRNA through chitosan nanoparticles. Nanoparticles made from chitosan were able to suppress endogenous increased green fluorescent protein in human lung cancer cells (H1299) and mouse peritoneal macrophages (EGFP) through siRNA-mediated RNA interference. Recent research has shown that siRNA may be efficiently transfected into human colon cancer cells (HCT116) by being encapsulated in chitosan or PEGylated chitosan nanoparticles, with encapsulation efficiencies of up to 97%. Research into the benefits of chitosan-based nanoparticles for siRNA delivery, and this review summarizes these findings.

Cationic dextran nanogels were first developed by Raemdonck *et al.* for siRNA administration; they use photochemical internalization to trigger siRNA endosomal escape. Amphiphilic photosensitizers are used in a technique called photochemical internalization (PCI) to cause the breakdown of endosomal vesicles [63]. Cationic dextran nanogels are taken up by photosensitizers, which cause causes the release of harmful oxygen radicals (ROS). Since oxidative damage may render the endosomal barrier ineffective. Studies have shown that sequence-specific silencing of EGFP expression may be achieved using dextran nano gels containing siRNAs targeting EGFP. Time-dependent gene silencing tests revealed changes in the depth of gene silencing across dextran nanogels with varied starting cross-link densities, but no significant effects on the length of EGFP silence.

Amplification of the Expression of Green Fluorescent Protein The Alexa Fluor647-labeled dextran nanogels was rapidly taken up by Huh-7 cells, and the number of Alexa Fluor647-positive cells did not start to decrease until day 6 post-transfection. However, the average intracellular fluorescence decreased as the nanogel particles were disseminated throughout the daughter cells. One-time administration of siRNA can effectively silence genes and may be prolonged using dextran-siRNA nano gels that undergo photochemical internalization [64].

However, therapeutic applications of siRNA *in vivo* are already underway, despite the relative newness of RNAi methods. Since many cancer-related genes are significantly up-regulated in cancer patients, these genes have been identified as potential targets of siRNA therapies. One of the most important VEGF siRNA targets in cancer therapy is VEGF itself. Some tumor cells, including squamous carcinoma cells, rely heavily on the anti-apoptotic Bcl-2 protein involved in the differentiation of myeloid leukemia cells (Mcl-1). The team designed and tested self-assembling nanocarriers made from oligo-chitosan for delivering siMcl-1 *in vivo*. Hydrophobic a-tocopherol-succinate and hydrophilic oligo-chitosan were used to replenish cationic units and form nanoparticles in this work. The oligo-mersomes (TCOsomes) based on atocopherol-oligochitosan spontaneously formed siRNA/TCOsome complexes through a cationic interaction with the negatively charged siRNAs. Human epidermoid cancer cells responded to the siRNA/TCOsome complexes quite well (KB cells). Significant GFP silencing was also achieved *in vitro* using the siRNA/TCOsome complexes. To sum up, intratumoral injection of the development of xenografted tumors was significantly slowed by treatment with siMcl-1 using siRNA/TCOsome complexes. The siRNA/TCOsome complex formulation offers various benefits over liposomes or already-existing nanoparticles. A more even distribution of sizes was seen in siRNA/TCOsome complexes compared to liposomes. In particular, the TCOsome-based nanoparticles' size and surface charge can be finely tuned. This characteristic might be employed to transport siRNA to the appropriate target areas *in vivo* since nanoparticle size is directly correlated with renal clearance or localization *in vivo*. More research on TCOsomes' self-assembling and size-controllable capabilities is needed before they can be used to produce *in vivo* siRNA carriers for therapeutic applications.

MDR-1 is an additional promising siRNA target for cancer treatment. One of the biggest obstacles to effective cancer treatment is drug resistance. Overexpression of P-glycoprotein (P-gp) by the up-regulated MDR-1 gene is a well-established factor in the establishment of multidrug resistance. P-glycoprotein (P-gp) is a transmembrane glycoprotein that might promote the efflux of drugs and reduce the cytotoxicity of medications, especially anti-cancer drugs. To combat the issue of medication resistance in osteosarcoma, Susa *et al.* engineered dextran nanoparticles loaded with soluble multidrug resistance-associated protein 1. Thiolate lipid-modified dextran polymer may entrap siRNA molecules and disseminate siMDR-1 effectively in drug-resistant osteosarcoma cell lines (KHOSR2 and U-2OSR2 cells). Because of this, siMDR-1-carrying dextran nanoparticles effectively decreased P-gp at the protein level, rendering KHOSR2 and U-2OSR2 cells more sensitive to doxorubicin. Another example of siMDR-1 delivery is the use of HA-NPS for the delivery of siRNA and miRNA for gene silencing in a wide range of cancer cells with positive results. Together, DOX and siMDR-1 coloaded HA-NPs render drug-resistant tumor cells more amenable to therapy and suppress tumor growth *in vivo*.

Cancer is a life-threatening illness since it may spread to many organs and kill people there. Even while the intratumoral injection is a very efficient method of delivering medications and drug-loaded nanoparticles to a single tumor site, it is not feasible in situations involving many tumors. Systemic administration of nanoparticles carrying therapeutic siRNA is necessary for effective cancer therapy; however, there are several challenges associated with this approach, including the formation of aggregations in serum conditions and the high accumulation of nanoparticles at untargeted sites [65]. To systemically administer siRNA therapies, researchers created tumor-homing nanoparticles based on glycol chitosan (GC). Both the tumor-targeting ability and the efficiency with which the target gene was silenced *in vivo* were significantly improved by the GC nanoparticles. Excellent tumor-targeting ability *in vivo* has been proven using self-assembled GC nanoparticles. Due to the absence of a positive charge, nanoparticles containing negatively charged siRNA cannot be successfully created with GC polymer alone. Polyethyleneimine (PEI) was added to GC to make it more positively charged. Each polymer was also functionalized with hydrophobic 5b-cholanic acid, which fostered the formation of self-assembled nanoparticles via strong interactions between 5b-cholanic acids contained inside the polymers. The cytotoxicity of the GC/PEI hybrids was significantly lower than that of pure GC or PEI nanoparticles. Cy5.5-siRNA-GC/PEI nanoparticles were successfully used to target tumors when delivered intravenously, as confirmed by *in vivo* real-time near-infrared fluorescence (NIRF) imaging. The GC/PEI nanoparticles sustained strong NIRF signals at the tumor site for up to 2 days. Systemic therapy with siRFP-GC/PEI nanoparticles (50 lg of siRNA/animal) every 48 hours significantly reduced RFP expression in a mouse model of cancer in which RFP is expressed. GC and other polysaccharide-based nanoparticles showed great potential in this study for the systemic delivery of therapeutic siRNA, as seen by these results.

siRNA polyplex creation has been a problem, and an alternative route toward more stable siRNA polyplexes, polymeric siRNA, has been presented as a solution to the problem of low colloidal stability of siRNA polyplexes compared to Pdna [66]. Furthermore, between the thiolate GC and the thiol group of the self-assembled siRNA, a mild buffer disulfide cross-linking is favored. In addition to sub-therapeutic dosages of DOX, a new development of P-glycoprotein inhibition is the aim of siRNA delivery. DOX-loaded and siRNA-complexed Cy5.5-labeled GCNPs accumulated in the tumor microenvironment and tumor blood vessels, as shown by near-infrared fluorescence microscopy. At 51 days post-treatment, the aforesaid combination resulted in tumor weights 9 times lower than those in the control groups. Combination therapies effectively inhibited tumor development by suppressing mechanisms of Drug Resistance Involving P-Glycoprotein Targeting.

In another work, siRNA was delivered to the liver using a PEI-introduced pullulan. As was previously noted, pullulan is a polysaccharide that has a strong affinity for the hepatic asialoglycoprotein receptors. This quality of pullulan has led to its most common usage in the targeting of the liver by various biopharmaceutical drugs [67]. When comparing PEI/siRNA complexes to PEI-pullulan/siRNA complexes, the latter showed worse liver targeting capacity. Furthermore, substantial toxicity and fatality *in vivo* were not seen after intravenous injection of PEI-pullulan/siRNA complexes. According to these results, the PEI-pullulan polymer shows potential as a liver-targeted siRNA delivery method.

Many polysaccharide-based delivery vectors for siRNAs have been developed to facilitate the delivery of siRNAs against a wide range of gene targets. Antisense interference with vascular endothelial growth factor (siRNA targeting VEGF) has been studied extensively as a cancer treatment; recent years have seen a rise in interest in cancer therapy with siRNA that targets AMD and p-glycoprotein (MDR-1) [68].

Over the last several years, polysaccharide nanoparticles have also been the subject of intense research for their potential use in the transport of all these siRNAs. siRNAs are unique in that their target sequences vary depending on the gene being targeted, but otherwise, they are all quite similar in terms of electrical charge, molecular weight, and chemical composition. Because of this, once a nanoparticle is developed for siRNA delivery, it can be used for several siRNA applications across a wide range of disease types [69, 70]. Even though the issue of poor is still no consensus on how to maximize polysaccharide nanoparticle transfection efficiency remain strong contenders for siRNA delivery methods.

3.8 Conclusion

Recognizing and naming the interactions that occur between biomaterial-based gene delivery and the biological milieu of our body has become an absolute need in recent years. To improve this knowledge, more advanced experiments and technologies are required. Once we do that, we can start looking for ways to improve the engineering of gene carriers, leading to greater gene therapy effectiveness and safer delivery. Indeed, solving the current challenge in gene therapy will need close cooperation among scientists from a variety of disciplines. Nucleic acids, in general, are hampered by drawbacks such as poor stability, weak bioactivity, and harmful consequences. Natural and manmade polymers have been investigated for their potential as drug carrier devices. with the hopes of overcoming these shortcomings and gaining benefits in drug activity. However, among these polymers, polysaccharides were shown to offer particularly promising uses when it comes to using nanotechnology for gene transfer in the medical and pharmaceutical fields. As a result, polysaccharide-based nanoparticles have been aggressively researched due to their many desirable characteristics. Polysaccharide nanoparticles are an effective carrier that may be used to encapsulate genes and increase target specificity. Due to its biological basis, it is more likely to get FDA clearance for human therapeutic uses of nanoparticles. Because of these characteristics, we believe that many scientists will continue to use polysaccharides in their nanoparticle preparations. Future therapies will be less invasive thanks to breakthroughs in nanotechnology and the development of more adaptable and effective gene-polysaccharide nanoparticles.

References

1. Uthaman, S., Lee, S.J., Cherukula, K., Cho, C.S., Park, I.K., Polysaccharide-coated magnetic nanoparticles for imaging and gene therapy. *Biomed. Res. Int.*, 2015, 14, 2015.
2. Vandamme, T.F., Lenourry, A., Charrueau, C., Chaumeil, J., The use of polysaccharides to target drugs to the colon. *Carbohydr. Polym.*, 48, 219, 2002.
3. Ganguly, K., Chaturvedi, K., More, U.A., Nadagouda, M.N., Aminabhavi, T.M., Polysaccharide-based micro/nanohydrogels for delivering macromolecular therapeutics. *J. Control. Release*, 193, 162, 2014.

4. Aminabhavi, T.M. and Deshmukh, A.S., Polysaccharide-based hydrogels as biomaterials, in: *Polymeric Hydrogels as Smart Biomaterials*, p. 45, Springer, Cham, 2016.
5. Byeon, H., Choi, S., Choi, J., Kim, I., Shin, B., Lee, E., Four-arm PEG cross-linked hyaluronic acid hydrogels containing PEGylated apoptotic TRAIL protein for treating pancreatic cancer. *Acta Biomater.*, 10, 142–50, 2014.
6. Xu, K., Lee, F., Gao, S., Chung, J., Yano, H., Kurisawa, M., Injectable hyaluronic acid-tyramine hydrogels incorporating interferon-alpha2a for liver cancer therapy. *J. Control. Release*, 166, 203–10, 2013.
7. Khan, M. and Mujahid, M., A review on recent advances in chitosan-based composite for hemostatic dressings. *Int. J. Biol. Macromol.*, 124, 138–47, 2019.
8. Matica, M.A., Aachmann, F.L., Tøndervik, A., Sletta, H., Ostafe, V., Chitosan as a wound dressing starting material. Antimicrobial properties and mode of action. *Int. J. Mol. Sci.*, 20, 5889, 2019.
9. Wagner, E., Zenke, M., Cotton, M., Beug, H., Birnstiel, M.L., Transferrin–polycation conjugates as carriers for DNA uptake into cells. *Proc. Natl. Acad. Sci. U.S.A.*, 87, 3410–3414, 2019.
10. Gottschalk, S., Cristiano, R.J., Smith, L.C., Woo, S.L., Folate-receptormediated DNA delivery into tumor cells: Protosomal disruption results in enhanced gene expression. *Gene Ther.*, 1, 185–191, 1994.
11. Ochiya, T., Takahama, Y., Nagahara, S., Sumita, Y., New delivery system for plasmid DNA *in vivo* using atelocollagen as a carrier material: The Minipellet. *Nat. Med.*, 5, 707–710, 1999.
12. Remes, A. and Williams, D.F., Immune response in biocompatibility. *Biomaterials*, 13, 731–743, 1992.
13. Plank, C., Mechtler, K., Szoka, F.C., Jr., Wagner, E., Activation of the complement system by synthetic DNA complexes: A potential barrier for intravenous gene delivery. *Hum. Gene Ther.*, 7, 1437–1446, 1996.
14. Mohammad, S.F., Chuang, H.Y., Mason, R.G., Roles of polymer size and ε-amino groups in polylysine–platelet interaction. *Thromb. Res.*, 1, 193–20, 1997.
15. Nishikawa, M., Takamura, S., Yamashita, F., Meijer, D.K., Hashida, M., Swart, D.J., Pharmacokinetics and *in vivo* gene transfer of plasmid DNA complexed with mannosylate poly(L-lysine) in mice. *J. Drug Targeting*, 8, 29–38, 2000.
16. Diebold, S.S., Lehrmann, H., Kursa, M., Wagner, E., Cotton, M., Zenke, M., Efficient gene delivery into human dendritic cells by adenovirus polyethylenimine and mannose polyethylenimine transfection. *Hum. Gene Ther.*, 10, 775–786, 1999.
17. Mahato, R.I., Takemura, S., Akamatsu, K., Nishikawa, M., Takakura, Y., Hashida, M., Physicochemical and disposition characteristics of antisense oligonucleotides complexed with glycosylated poly(L-lysine). *Biochem. Pharmacol.*, 53, 887–895, 1997.

18. Reddy, J.A., Dean, D., Kennedy, M.D., Low, P.S., Optimization of folateconjugated liposomal vectors for folate receptor-mediated gene therapy. *J. Pharm. Sci.*, 88, 1112–1118, 1999.
19. Leamon, C.P., Weigl, D., Hendren, R.W., Folate copolymer-mediated transfection of cultured cells. *Bioconjug. Chem.*, 10, 947–957, 1999.
20. Kim, J.-S., Maruyama, A., Akaike, T., Kim, S.W., Terplex DNA delivery system as a gene carrier. *Pharm. Res.*, 15, 116–121, 1998.
21. Harbottle, R.P., An RGD-oligolysine peptide: A prototype construct for integrin-mediated gene delivery. *Hum. Gene Ther.*, 9, 1037–1047, 1999.
22. 103Nishikawa, M., Takemura, S., Takakura, Y., Hashida, M., Targeted delivery of plasmid DNA to hepatocytes *in vivo*: Optimization of the pharmacokinetics of plasmid DNA/galactosylated poly(L-lysine) complexes by controlling their physicochemical properties. *J. Pharmacol. Exp. Ther.*, 287, 408–415, 1998.
23. Ferkol, T., Perales, J., Mularo, F., Hanson, R.W., Receptor-mediated gene transfer into macrophages. *Proc. Natl. Acad. Sci. U.S.A.*, 93, 101–105, 1996.
24. Choi, Y.H., Liu, F., Park, J.-S., Kim, S.W., Lactose-poly(ethylene glycol)-grafted poly-L-lysine as hepatoma cell-targeted gene carrier. *Bioconjug. Chem.*, 9, 708–718, 1998.
25. Choi, Y.H., Liu, F., Choi, J.S., Kim, S.W., Park, J.-S., Characterization of a targeted gene carrier, lactose-polyethylene glycol-grafted poly-L-lysine, and its complex with plasmid DNA. *Hum. Gene Ther.*, 10, 2657–2665, 1999.
26. Kim, J.-S., Maruyama, A., Akaike, T., Kim, S.W., *In vitro* gene expression on smooth muscle cells using a terplex delivery system. *J. Control. Release*, 47, 51–59, 1999.
27. Yu, L., Nielsen, M., Han, S.-O., Kim, S.W., Terplex DNA gene carrier system targeting to artery wall cells. *J. Control. Release*, in press, 2, 302–317, 2000.
28. 109Benns, J.M. and Kim, S.W., Tailoring new gene delivery designs for specific targets. *J. Drug Targeting*, 8, 1–12, 2000.
29. Suh, W., Chung, J.-K., Park, S.-H., Kim, S.W., Anti-JL1 antibody conjugated poly(L-lysine) for targeted gene delivery to leukemia T cells. *J. Control. Release*, in press, 4, 581, 2000.
30. Simões, S., Slepushkin, V., Pires, P., Gaspar, R., Pedroso de Lima, M.C., Düzgünes, N., Mechanism of gene transfer mediated by lipoplexes associated with targeting ligands or pH-sensitive peptides. *Gene Ther.*, 6, 1798–1807, 2000.
31. Schoen, P., Chonn, A., Cullis, P.R., Wilschut, J., Scherrer, Gene transfer mediated by fusion protein hemagglutinin reconstituted in cationic lipid vesicles. *Gene Ther.*, 6, 823–832, 1999.
32. 113Mechtler, K. and Wagner, E., Gene transfer mediated by influenza virus peptides: The role of peptide sequences. *New J. Chem.*, 21, 105–111, 1997.
33. Ishiguro, R., Matsumoto, M., Takahashi, S., Interaction of fusogenic synthetic peptide with phospholipid bilayers: Orientation of the peptide α-helix and binding isotherm. *Biochemistry*, 35, 4976–4983, 1997.

34. Park, J.H., Saravanakumar, G., Kim, K., Kwon, I.C., Targeted delivery of low molecular drugs using chitosan and its derivatives, *Adv. Drug Deliv. Rev.*, 62, 28, 2010.
35. Dutta, P.K., Dutta, J., Tripathi, V.S., Chitin and chitosan: Chemistry, properties and applications, *J. Sci. Ind. Res.*, 63, 20, 15, 2004.
36. Berger, J., Reist, M., Mayer, J.M., Felt, O., Gurny, R., Structure and interactions in covalently and ionically cross-linked chitosan hydrogels for biomedical applications, *Eur. J. Pharm. Biopharm.*, 57, 35, 2004.
37. Amidi, M., Romeijn, S.G., Borchard, G., Junginger, H.E., Hennink, W.E., Jiskoot, W., Chitosan-based delivery systems for protein therapeutics and antigens, *J. Control. Release*, 111, 107, 2006.
38. Thanou, M., Florea, B.I., Geldof, M., Junginger, H.E., Borchard, G., Quaternized chitosan oligomers as novel gene delivery vectors in epithelial cell lines, *Biomaterials*, 23, 153, 2002.
39. Wang, Y.S., Jiang, Q., Li, R.S., Liu, L.L., Zhang, Q.Q., Wang, Y.M., Zhao, J., Cholesterol-modified amino-pullulan nanoparticles as a drug carrier: Comparative study of cholesterol-modified carboxyethyl pullulan and pullulan nanoparticles, *Nanotechnology*, 6, 165, 2008.
40. Park, K., Kim, J.H., Nam, Y.S., Lee, S., Nam, H.Y., Kim, K., Park, J.H., Kim, I.S., Choi, K., Kim, S.Y., Kwon, I.C., Effect of polymer molecular weight on the tumor targeting characteristics of self-assembled glycol chitosan nanoparticles, *J. Control. Release*, 122, 305, 2007.
41. Aminabhavi, T.M. and Dharupaneedi, S.P., *Chitosan based biomaterials*, vol. 1, p. 295, Woodhead Publishing, Cambridge, 2017.
42. Kim, K., Kwon, S., Park, J.H., Chung, H., Jeong, S.Y., Kwon, I.C., Design and synthesis of lactose, galactose and cholic acid related dual conjugated chitosan derivatives as potential anti liver cancer drug carriers. *Biomacromolecules*, 6, 1154, 2015.
43. Kim, J.H., Kim, Y.S., Park, K., Kang, E., Lee, S., Nam, H.Y., Kim, K., Park, J.H., Chi, D.Y., Park, R.W., Kim, I.S., Choi, K., Kwon, I.C., Self-assembled glycol chitosan nanoparticles for the sustained and prolonged delivery of antiangiogenic small peptide drugs in cancer therapy. *Biomaterials*, 29, 1920, 2008.
44. Mitra, S., Gaur, U., Ghosh, P.C., Maitra, A.N., Tumour targeted delivery of encapsulated dextran–doxorubicin conjugate using chitosan nanoparticles as carrier. *J. Control. Release*, 74, 317, 2001.
45. Kim, J.H., Kim, Y.S., Park, K., Lee, S., Nam, H.Y., Min, K.H., Jo, H.G., Park, J.H., Choi, K., Jeong, S.Y., Park, R.W., Kim, I.S., Kim, K., Kwon, I.C., Insight of key factors influencing tumor targeting characteristics of glycol chitosan-based nanoparticles and *in vivo* applications. *J. Control. Release*, 127, 41, 2008.
46. Ganguly, K., Kulkarni, A.R., Aminabhavi, T.M.,Polysaccharide-based hydrogels as biomaterials in drug delivery. *Drug Delivery*, 2, 1–2, 2015.

47. Agnihotri, S.A., Mallikarjuna, N.N., Aminabhavi, T.M., Recent advances on chitosan-based micro- and nanoparticles in drug delivery. *J. Control. Release*, 100, 5, 2004.
48. Mansouri, S., Lavigne, P., Corsi, K., Benderdour, M., Beaumont, E., Fernandes, J.C., Chitosan-DNA nanoparticles as non-viral vectors in gene therapy: Strategies to improve transfection efficacy. *Eur. J. Pharm. Biopharm.*, 57, 1, 2004.
49. Mao, H.Q., Roy, K., Troung-Le, V.L., Janes, K.A., Lin, K.Y., Wang, Y., August, J.T., Leong, K.W., Chitosan-DNA nanoparticles as gene carriers: Synthesis, characterization and transfection efficiency. *J. Control. Release*, 70, 399, 2001.
50. Xue, W.J., Feng, Y., Wang, F., Guo, Y.B., Li, P., Wang, L., Liu, Y.F., Wang, Z.W., Yang, Y.M., Mao, Q.S., Preparation and anti-tumor metastasis of carboxymethyl chitosan. *Sci. Rep.*, 6, 22149, 2016.
51. Lee, D.W., Yun, K.S., Ban, H.S., Choe, W., Lee, S.K., Lee, K.Y., Preparation and characterization of chitosan/polyguluronate nanoparticles for siRNA delivery. *J. Control. Release*, 139, 146, 2009.
52. Huh, M.S., Lee, S.Y., Park, S., Lee, S., Chung, H., Lee, S., Choi, Y., Oh, Y.K., Park, J.H., Jeong, S.Y., Choi, K., Kim, K., Kwon, I.C., Tumor-homing glycol chitosan/polyethylenimine nanoparticles for the systemic delivery of siRNA in tumor-bearing mice, *J. Control. Release*, 144, 134, 2010.
53. Jain, S., Tran, T.H., Amiji, M., Macrophage repolarization with targeted alginate nanoparticles containing IL-10 plasmid DNA for the treatment of experimental arthritis. *Biomaterials*, 61, 162, 2015.
54. Rouzes, C., Gref, R., Leonard, M., Delgado, A.D., Dellacherie, E., Dextran-poly-ε-caprolactone micro- and nanoparticles: Preparation, characterization and tamoxifen solubilization. *J. Biomed. Mater. Res.*, 50, 557, 2000.
55. Sant, S., Poulin, S., Hildgen, P., Effect of polymer architecture on surface properties, plasma protein adsorption, and cellular interactions of pegylated nanoparticles. *J. Biomed. Mater. Res. Part A*, 87a, 885, 2008.
56. Sugahara, S., Kajiki, M., Kuriyama, H., Kobayashi, T.R., Doxorubicin release from self-assembled nanoparticles of deoxycholic acid-conjugated dextran. *J. Control. Release*, 117, 40, 2007.
57. Naeye, B., Raemdonck, K., Remaut, K., Sproat, B., Demeester, J., De Smedt, S.C., PEGylation of biodegradable dextran nanogels for siRNA delivery. *Eur. J. Pharm. Sci.*, 40, 342, 2010.
58. Naeye, B., Deschout, H., Roding, M., Rudemo, M., Delanghe, J., Devreese, K., Demeester, J., Braeckmans, K., De Smedt, S.C., Raemdonck, K., Hemocompatibility of siRNA loaded dextran nanogels. *Biomaterials*, 32, 9120, 2011.
59. Tripathi, S.K., Goyal, R., Gupta, K.C., Polysaccharide-based nanoparticles for gene delivery. *Soft Matter*, 7, 11360, 2011, 2011.08. 015 56.
60. De Backer, L., Naessens, T., De Koker, S., Zagato, E., Demeester, J., Grooten, J., De Smedt, S.C., Raemdonck, K., Hybrid pulmonary surfactant-coated

nanogels mediate efficient *in vivo* delivery of siRNA to murine alveolar macrophages. *J. Control. Release*, 217, 53, 2015.
61. Massia, S.P. and Stark, J., Immobilized RGD peptides on surface-grafted dextran promote bio specific cell attachment. *J. Biomed. Mater. Res.*, 56, 390, 2001.
62. Lapcik, L., Lapcik, L., De Smedt, S., Demeester, J., Chabrecek, P., *Chem. Rev.*, 98, 2663, 1998. 60. Kuo, J.W., Polymeric gene delivery systems. *Practical aspects of hyaluronan based medical products*, p. 61, CRC/Taylor & Francis, Boca Raton, 2006.
63. Balazs, E.A. and Denlinger, J.L., J. *Rheumatol. Suppl.*, 39, 3, 1993. 62. Burd, D.A., Greco, R.M., Regauer, S., Longaker, M.T., Siebert, J.W., Garg, H.G., Polysaccharide-based nanoparticles for gene delivery. *Br. J. Plast. Surg.*, 44, 579, 1991.
64. Choi, K.Y., Silvestre, O.F., Huang, X.L., Min, K.H., Howard, G.P., Hida, N., Jin, A.J., Carvajal, N., Lee, S.W., Hong, J.I., Chen, X.Y., Application of nanocarriers in co-loading siRNA and chemotherapeutic drugs to reverse multidrug resistance of tumor. *ACS Nano*, 8, 4559, 2014.
65. Kim, J.H., Kim, Y.S., Kim, S., Park, J.H., Kim, K., Choi, K., Chung, H., Jeong, S.Y., Park, R.W., Kim, I.S., Kwon, I.C., Insight of key factors influencing tumor targeting characteristics of glycol chitosan-based nanoparticles and *in vivo* applications. *Macromol. Res.*, 20, 9, 111, 228, 2006.
66. Lee, S.J., Huh, M.S., Lee, S.Y., Min, S., Lee, S., Koo, H., Chu, J.U., Lee, K.E., Jeon, H., Choi, Y., Choi, K., Byun, Y., Jeong, S.Y., Park, K., Kim, K., Kwon, I.C., Tumor-homing poly-siRNA/glycol chitosan self-cross-linked nanoparticles for systemic siRNA delivery in cancer treatment. *Angew. Chem. Int. Ed. Engl.*, 51, 7203, 2012.
67. Yhee, J.Y., Song, S., Lee, S.J., Park, S.G., Kim, K.S., Kim, M.G., Son, S., Koo, H., Kwon, I.C., Jeong, J.H., Jeong, S.Y., Kim, S.H., Kim, K., Combination of nitric oxide and drug delivery systems: Tools for overcoming drug resistance in chemotherapy. *J. Control. Release*, 198, 1, 2015.
68. Yoon, H.Y., Son, S., Lee, S.J., You, D.G., Yhee, J.Y., Park, J.H., Swierczewska, M., Lee, S., Kwon, I.C., Kim, S.H., Kim, K., Pomper, M.G., Glycol chitosan nanoparticles as specialized cancer therapeutic vehicles: Sequential delivery of doxorubicin and Bcl-2 siRNA. *Sci. Rep.*, 4, 1–12, 2014.
69. Kang, J.H., Tachibana, Y., Kamata, W., Mahara, A., Harada-Shiba, M., Yamaoka, T., Liver-targeted siRNA delivery by polyethylenimine (PEI)-pullulan carrier. *Bioorg. Med. Chem.*, 18, 3946, 2010.
70. Kaneo, Y., Tanaka, T., Nakano, T., Yamaguchi, Y., Evidence of receptor-mediated hepatic uptake of pullulan in rats. *J. Control. Release*, 70, 365, 2001.

4

Hydrogel-Based Gene Therapy

Shweta Kumari[1*], Dipti Jena[2], Vedant Kumar Prajapati[3], Shashi Ranjan Singh[3] and Garima Tripathi[3]

[1]T. John College of Pharmacy, Basavanapura, Bengaluru, Karnataka, India
[2]Department of Pharmacy, Arka Jain University, Mohanpur, India
[3]Narayan Institute of Pharmacy, Gopal Narayan Singh University Jamuhar, Rohtas, Bihar, India

Abstract

Gene therapy (GT) relies heavily on the adoption of safe delivery vehicles that do not cause any adverse side effects or immunological reactions. Biodegradable and biocompatible polymeric nanocarriers are being used for gene delivery due to their unique properties such as excellent bio- incompatibility, prolonged gene circulation time, prevention of gene degradation, passive targeting, and the ability to modify polymer structure. Biomaterials are playing a pivotal role in medical technologies that aim to transport medicinal chemicals to patients in hospitals and other clinical settings. Recent advances in our understanding of the role biomaterials play in gene therapy have stoked excitement in the generation of several nucleic acid agents for the treatment of numerous diseases. This investigation focuses on the importance of interdisciplinary teams in the development of contemporary hydrogels for performing complex biological activities like cancer immunotherapy, tissue engineering, and immunization.

Hydrogels are biocompatible hydrogels that can be used for integrated sensing and therapeutics. They have potential applications in gene transfer, regenerative medicine, genetic diseases, and medication delivery. Hydrogels have high water content and mild gelation conditions, making them useful for gene transfer. They can also be used to promote transplanted or endogenously infiltrating cells, heal or restore the function of damaged or injured tissues, and act as vectors for therapeutic genes. New developments in hydrogel manufacture for distribution have resulted from a greater knowledge of the design aspects, which include

**Corresponding author*: shweta.ktr24@gmail.com

Rishabha Malviya, Sonali Sundram and Neelam Jain (eds.) Integration of Biomaterials for Gene Therapy, (77–116) © 2023 Scrivener Publishing LLC

physical and biological components, transgene expression, cell engraftment, and cell function.

Keywords: Gene therapy, hydrogels, biocompatible, medication immunotherapy, vectors

4.1 Introduction

Synthetic hydrogels have come a long way since their inception in the biological systems engineering field in the 1960s. Over the last 60 years, there has been a meteoric rise in the amount of literature on this issue, from a few thousand pieces in 1982 to over a hundred thousand publications by 2020 and approx. six hundreds newly publications have been published in web of science by 2021 along with the yearly pace of publication has been over 10,000 for the last three years [1, 2]. Hydrogels have made significant contributions to modern life, perhaps most notably via the development of soft contact lenses, which have ushered in a new category of optically adjustable soft materials and spawned a billion-dollar industry. As a new field of research, the local controlled release of bioactive chemicals has already shown promise, with preliminary studies showing the efficacy of customized hydrogels in delivering various medications. By the 1990s, hydrogels had established themselves as a crucial component of tissue regeneration, having first been used in reconstructive surgery in the 1970s [3]. There has been much research on hydrogel materials, and as technology develops, intriguing new applications for them are being found. It is well known that hydrogel possess a variety of applications , including coatings for electronics, sanitary engineering, in the field of robotics, and adoptive cell treatment [4]. Polymers may be dissolved in water to form hydrogels, which are cross-linked, 3D structures (hydrophilic polymers). Long-term medication delivery might be enhanced by using a hydrogel with a wide pore size and high water content. They can take in a lot of water without breaking down, and they also grow quite a bit. The hydrogel is an NDDS, or new drug delivery system [5]. The controlled release of a medicine is made possible by embedding the drug's active pharmaceutical ingredient within the polymeric system's structure. The polymer, in both its natural and synthetic forms, has found use as a medicine carrier. Synthetic polymers are artificial polymers made up of a few of separate cross-linking agents or a limited number of linked cross-linking agents [6]. The long-term growth and infection of animal tissues provided by animal collagen may have negative effects on humans. Hydrogel makes use of these polymers because they are all-around great choices: they are harmless, easily available, and biodegradable. Hydrogels based on sodium alginate are modified with a no. of monomers for controlled release of medicines. These monomers include acrylamide, methacrylamide, and N-isopropyl acrylamide. Some examples of synthetic polymers used to create hydrogels are poly (hydroxyl ethyl methacrylate), poly (ethylene glycol), poly (amides), and poly (vinyl alcohol) (acrylic acid). Sterile goods, sealing, novel drug delivery methods, biomedical tissue creation, regenerative medicine, diagnostics, proteins, and peptides, and drug administration in malignancy or focused therapy are just some of the many areas where hydrogel advances have been put to use [7]. Hydrogels have a variety of applications in the healthcare and pharmaceutical sectors. Hydrogels are synthetic biomaterials that mimic the appearance and functions of real tissue very closely. Because of their high water content and simple construction, they are very similar to ordinary tissue. The primary objectives of designing pharmaceutical transport frameworks are to prevent the premature degradation of an active therapeutic molecule, to improve medicinal efficacy, and to reduce undesirable consequences [8]. Strategies such as dosage caps, dose reduction regimens, and controlled release schedules may be utilized to maintain therapeutic drug levels over lengthy periods of time. Hydrogel frameworks that respond to external stimuli have been investigated for their potential utility in medication administration due to their responsiveness to environmental variations. Many different materials exist, each with its own set of unique natural and physiochemical qualities that suggest a broad variety of therapeutic applications. Antibodies or nucleic acids may be included into the hydrogel system [9]. Most "smart" hydrogels respond to external stimuli like heat, electricity, or light by undergoing reversible form changes. Hydrogel actuators (HAs) and shape memory hydrogels (SMHs) have grabbed the attention of scientists because of their potential uses in sectors as diverse as biomedicine, soft robotics, and artificial muscles. Since smart hydrogels have an inhomogeneous formation and manufacturing technologies, a broad variety of production techniques may be used to create hydrogel-based actuators [10]. One reason for this is the need for multi-stimuli-sensitive chemical structures to be integrated in a hydrogel matrix in order to create the kind of multi-stage actuation seen in living organisms. More in-depth study is required to learn the inner workings of the human mind that lead to the adoption of complex and new pharmacological interventions. The production of inhomogeneous hydrogels via the improvement of manufacturing procedures [11] is a key step toward the synthesis of multi-stimulative hydrogels with complex but designed self-folding, twisting, and bending capabilities.

During tissue regeneration in response to sickness or injury, endogenous healing programs and inhibiting processes need external inputs to ensure their full effectiveness. To control the local environment and drive tissue-specific function, the extracellular matrix (ECM) is necessary. Therefore, biomaterial scaffolds that imitate the ECM have been designed to aid in tissue regeneration, serve as carriers for cell transplantation, and enhance cell survival, differentiation, and engraftment [12]. Hydrogels are adaptable in mechanical characteristics and have viscoelastic properties similar to tissue, making them ideal for modeling the extracellular matrix (ECM) [13]. They are also biocompatible, permeable to oxygen, nutritional growth boosters, and metabolic waste. [14] (Figure 4.1). Origin, chemical composition, environmental stresses, cross-link density, features, spatial organization, and ionic charge are all ways to classify hydrogels (Figure 4.2) [15].

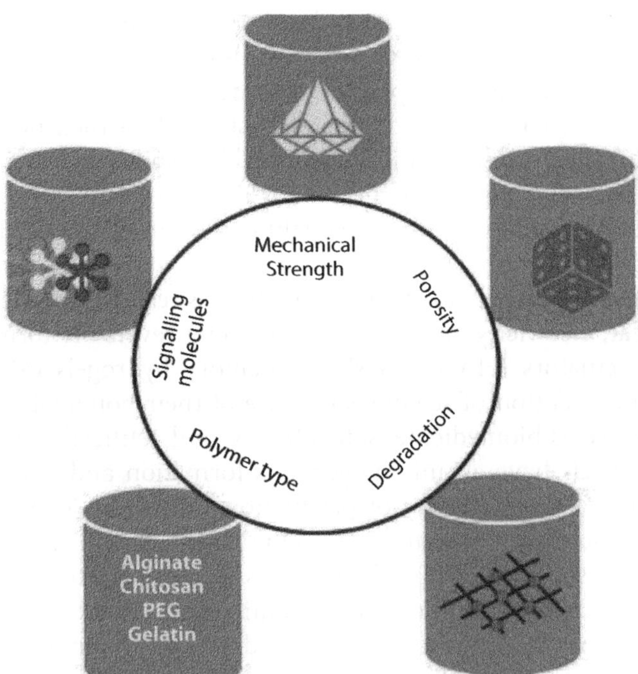

Figure 4.1 Properties of biocompatible hydrogel.

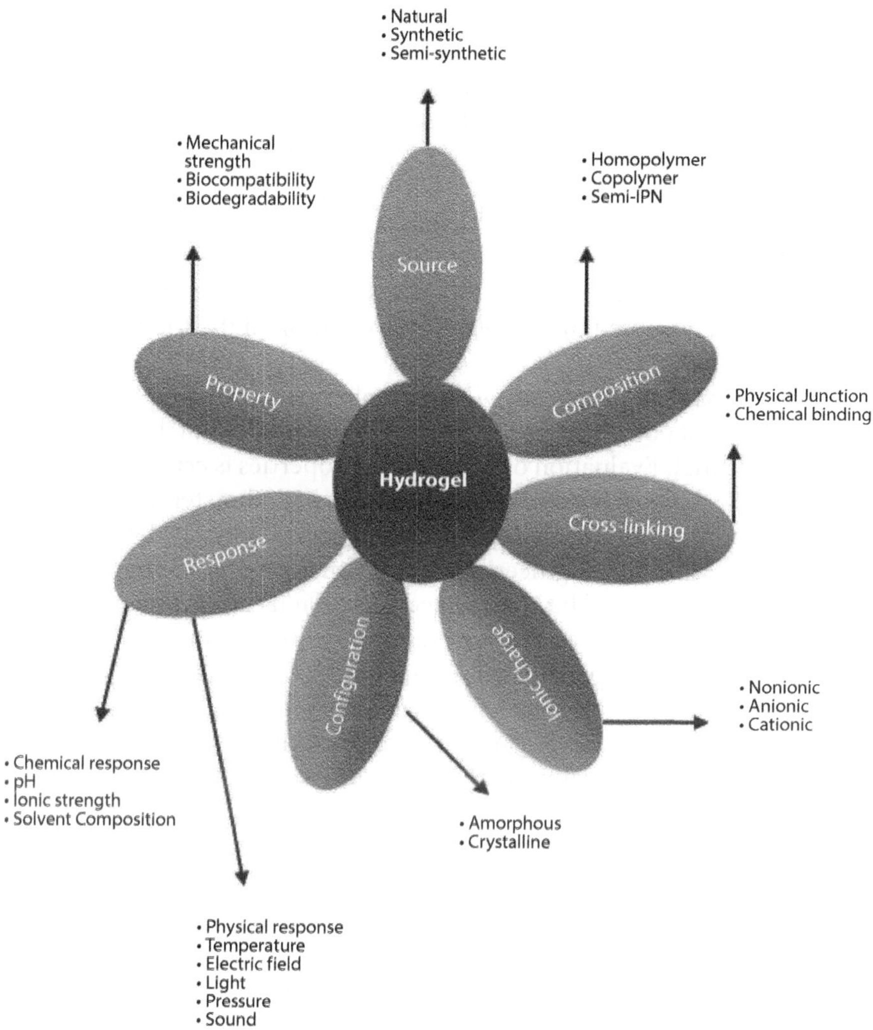

Figure 4.2 Classification of hydrogel.

Properties of Hydrogels

(a) Swelling properties – Cross-linking operators have a major role in the development of hydrogels. It's a determinant that influences volume of water is absorbed. A hydrogel may undergo rapid and reversible transformations in response to even a little alteration in atmospheric conditions. Hydrogel's physical action is affected by the pH, temperature, electric signal, and proximity to chemical or other ionic species in the surrounding As hydrogel forms, the precipitate grows and water content increases, which can be seen to the naked eye [16]. Gravimetric analysis may be used to calculate the amount of fluid that can be integrated into a hydrogel, and the swelling ratio can be used to illustrate the amount of fluid absorbed. Hydrogels may adapt to shifting environmental pH due to the presence of acidic or basic functional atoms. Its swelling profile and consequent volume variations are determined by the degree of ionization of the functional groups [17].

(b) Mechanical properties – From a pharmacological and biomedical point of view, the hydrogel's mechanical qualities are essential. Evaluation of mechanical properties is crucial for a wide range of biomedical applications, including tendon and ligament healing, wound dressing material, drug delivery lattices, tissue engineering, and ligament replacement. The hydrogel should be constructed such that it can maintain the necessary physical surface area while transporting the substance at the specified pace. The desired hydrogel mechanical characteristic may be achieved by adjusting the amount of cross-linking. Stronger hydrogels can be created at a higher concentration of the cross- linking agent; however, it decreases the hydrogel's elongation by a great amount, making the hydrogel more brittle. Altering the concentration of cross-linking agents may cause hydrogel to take on a variety of mechanical properties [18].

(c) Biocompatible properties – Biocompatibility and nontoxicity of hydrogels are essential for their use in biomedicine. When a substance is biocompatible, it may elicit the desired host response under controlled laboratory conditions [19]. The ability of a substance to function in a certain host environment is known as biocompatibility. There are two main parts ~~to~~ of biocompatibility: "biofunctionality" refers to a material's ability to perform the intended work. The absence of mutagenesis, cytotoxicity, or maybe carcinogenesis constitutes bio-security, also known as a suitable host response both globally and locally (the surrounding tissue) [20].

(d) Porosity and permeation – Hydrogels, or smaller holes inside the system, may form by stage partitioning process during the union. Hydrogels' normal pore estimate, the pore measure dispersion, and the pore interconnections are all crucial but difficult-to-manipulate facets of the matrix that are often lumped into a single quantity known as tortuosity. The following three factors effectively impact the pore-measure allocations of hydrogels: 1) The degree to which polymer chains are chemically connected, as shown by the ratio of cross-linker to monomer. The hydrogel's initial atom concentration is linked to the degree of a grouping of the physical entanglements of the polymer strands. The initial cationic or anionic concentration affects the net charge of the polyelectrolyte hydrogel [21].

4.2 Gene Therapy

An innovative kind of molecular medicine, gene therapy will have profound consequences for human health in the twenty-first century. The aim is the effective delivery of genes with long-term expression. Transgene expression can be maintained longer through gene transfer than through the direct administration of a therapeutic product with a short pharmacological half-life [22]. Gene transfer involves the introduction of one or more foreign genes or genetic sequences into a specific, identified target cell population using a chosen gene vehicle to manage a specific disease in patients. Gene editing and reprogramming used in gene therapy (inherited monogenic diseases, cancers, infectious diseases, degenerative conditions, etc.) include (direct, sequence-specific modification of endogenous genes in the cells using zinc-finger nucleases (ZNFs), transcription activator-like effector nucleases (TALENs), and clustered regularly interspaced short palindromic repeats-associated Cas9 (CRISPR/Cas9) The most often transferred sequences in human gene therapy studies include tumor suppressors, suicide genes and enzymes, antigens, cytokines, silencers, and replication inhibitors to battle cancer, and growth factors/receptors and signaling molecules for regenerative medicine [23].

Ex vivo gene therapy uses differentiated or progenitor cells, like adult mesenchymal stem cells (MSCs), embryonic stem cells (ESCs), and induced pluripotent stem cells (iPSCs), to produce "personalized" cells by reprogramming the patient's differentiated cells and then re-implanting them into the body. The two most popular methods for delivering genes are nonviral and viral vectors; each has advantages and limitations in terms of effectively and correctly altering the target cells and tissues in patients [24].

4.2.1 Nonviral Vectors

Direct DNA absorption (bare DNA), receptor-mediated DNA transfer (DNA-protein ligand conjugation), and physical and chemical approaches are all examples of nonviral gene delivery. Due to the lack of potentially damaging viral material being carried in the cells and the lack of immune responses induced in the recipient, nonviral vectors are regarded as a safe approach to genetically edit cells in patients [25]. Due to their ease of production on a large scale and suitability for tailored quality control, they make up around 24% of all patient-initiated clinical procedures. They are inefficient for *in vivo* settings; however, better suited for *ex vivo* gene therapy. [26]. These inefficiencies include low gene transfer (transfection) (20–40%), the need for cell division, the degradation of the material (at least some of it) by transport to the lysosomal compartment, and relatively short-term expression of the therapeutic gene being transferred (some days).

4.2.2 Viral Vectors

There are currently two main types of gene delivery strategies: physicochemical techniques and biological vectors. Although each of these strategies has its advantages, it is safe to argue that none is optimal for effective gene transport or the prolonged synthesis of a therapeutic gene product. Although many other viral vectors exist, we will be focusing on only four and highlighting their drawbacks and benefits [27].

Retroviral Vectors

The infected cell converts the retrovirus's RNA genome into DNA, making it a member of the family of viruses known as retroviruses. Genes for gag, pol, and env are located in the genome and are framed by the long terminal repeats (LTRs). These mark the beginning and end of the viral genome and are necessary for integration into the host genome. Expression of the viral genes is regulated by the LTRs, which also act as enhancer-promoter sequences [28]. The viral RNA may be recognized from other RNAs in the cell by the last sections of the genome, known as the packing sequence. Transgenes, like the factor IX gene, may be inserted into the viral genome by genetic engineering. The LTRs of viruses may regulate transgenic transcription, or artificial enhancer-promoter elements can be inserted to facilitate transgene expression.

The viral proteins (such as those produced by the gag, pol, and env genes) are then injected into a packaging cell together with the chimeric genome, but without the LTRs or packing sequence [29]. To create a retroviral vector, just the chimeric viral genomes need to be assembled. The transgene is transferred when the culture media in which the packing cells were grown is added to the target cells. One milliliter of the viral concoction may usually infect a million target cells in a growth plate. The inability of retroviral vectors to infect non-dividing cells such as those seen in muscle, brain, lung, and liver is a significant drawback of these vectors. To use a recombinant retroviral vector, it is necessary to harvest cells from the target tissue, cultivate them *in vitro*, and then infect them. The cells engineered to produce the alien protein are ultimately reintroduced into the recipient animal. We call this kind of gene therapy "*ex vivo*" [30].

Lentiviral Vectors

Lentiviruses are also members of the retrovirus family, however, unlike other members of this group, they may infect cells whether they are dividing or not. The human immunodeficiency virus (HIV) is the well-known lentivirus because of its use as a vector for *in vivo* gene transfer when it has been rendered inactive. In addition to the three core genes (gag, pol, and env) seen in basic retroviruses, HIV also contains genes for six auxiliary proteins (tat, rev, vpr, vpu, nef, and vif) [31]. Lentivirus vectors were designed using retrovirus vectors as a template; they include the transgene sandwiched between LTRs and a packaging sequence. Certain auxiliary proteins may be removed without diminishing vector generation or infection efficacy. HIV-based vectors are limited to infect CD4+ cells due to the presence of the env gene product; thus, different env sequences from RNA viruses are used in place of this gene (such as glycoprotein from the vesicular stomatitis virus). It is now possible to manufacture these vectors with >109 virus particles per milliliter. Injecting a lentivirus vector into a rodent's brain, liver, muscle, eye, or pancreatic-islet cells results in long-lasting expression throughout the animal's adult life. As a result, lentiviral vectors appearing to offer a significant opportunity for the long-term expression of a gene *in vivo* [32].

Adenoviral Vectors

Adenoviruses are a group of DNA viruses that cause mild respiratory infections in humans. They are capable of infecting both actively proliferating cells and quiescent ones. More than a dozen genes are included in their genomes, and they seldom integrate into the host genome. This is because they reproduce as nuclear episomal components inside the host cell. By exchanging the El gene, required for viral replication, with the gene of interest (such as factor IX) and an enhancer-promoter region, replication-deficient adenovirus vectors may be created [33].

Adeno-Associated Viral Vectors

Adeno-associated virus (AAV) is a safe, user-friendly virus since it only contains a single strand of DNA. Twisted terminal repetition demarcates the start and end of the virus and surround the viral cap and rep genes as well as the packaging sequence. The viral capsid (coat) proteins are encoded by the cap gene product, whereas the rep gene product is essential for viral replication and integration. These auxiliary genes for AAV replication are supplied by helper viruses (usually adenovirus or herpes simplex virus). Viral DNA preferentially integrates onto human chromosome 19 when the rep gene product is present. A variety of cell types are susceptible to infection by this virus. Substituting the transgene for the rep and cap genes yields an AAV vector [34].

4.3 *In Vivo* Gene Therapy Using Hydrogels

Traditional methods of cell administration have relied on injecting high-density cell suspensions directly into the site of illness or injury. Poor therapeutic response is common with direct cell injection approaches because of fast cell death, low or moderate engraftment of transplanted cells, and limited control over cell destiny [35]. These issues may be resolved by transplanting cells in a biocompatible scaffold, which may improve survival and extend retention at the location. In addition, hydrogels often share mechanical characteristics and water content with the original ECM. The morphology, viability, differentiation, and function of cells may all be impacted by the niche environment's mimicry [36]. There are two primary ways hydrogels can be used for stem cell transplantation: non-integrating methods, which encapsulate cells to keep them separate from the host tissue, and integrating methods, which allow the transplanted cells to make direct contact with the host tissue either immediately via a microporous design or overtime via biodegradation. Due to the sheer volume of systems studied in pre-clinical models, an entire list is impossible. Instead, we have chosen to highlight recent articles that are on the cusp of big animal or clinical translation.

Here, we'll look at how the survival of transplanted cells and their final role [37] impact the designs of these hydrogels.

4.4 Encapsulating Cells in Hydrogels for Gene Therapy Delivery

For transplanted cells to have a greater chance of surviving, an encapsulating device must provide a habitat conducive to normal cellular activity, while also serving as an immune-regulatory barrier by isolating or modulating the surrounding environment. The gelation procedure, hydrogel structure, and material composition all have a role in controlling its function. Thera Cyte is representative of a popular encapsulation method since it features a porosity vascularizing outside membrane that aids in tissue integration and an inside impermeable barrier that safeguards the transplanted allogeneic islets [38]. Transplanted neonatal pancreatic tissue survived and responded to glucose levels in non- obese diabetic mice for at least 50 days. Although the first version of this technology did not work well in clinical trials, the overall approach has been refined by many other businesses, such as Living Cell Technologies, Beta Logics, Viacyte, and Encaptra [39]. The Encaptra system uses a single immunoisolating membrane that also lets oxygen and nutrients in. Viacyte is testing the device's effectiveness and safety using stem-cell-derived cell sources in a phase I/II clinical study. Additional encapsulation tools that have progressed to human trials have recently been studied in depth. Hydrogels provide the same chance to overcome obstacles such as immune cell infiltration as these devices do, with the added benefits of improved transport and more modifiable characteristics [40]. Improved cell survival and therapeutic efficiency may be achieved by tailoring hydrogels' biomechanical characteristics and adhesion locations. To improve cell motility, viability, and function, researchers are creating hydrogels that use the Thera Cyte design's core delivery technique and give adjustable qualities to the device's outer and the interior. Alginate, a naturally occurring polymer derived from algae, has been widely studied for cell encapsulation due to its biocompatibility, low toxicity, comparative affordability, and moderate gelation by the addition of divalent cations, such as Ca2+. Cell adhesion and migration both may be enhanced by modifying alginate. To facilitate stem cell proliferation and migration after the inflammatory storm generated by surgery, a double-layered alginate hydrogel system was developed with matrix-metalloproteinases and Arg-Gly-Asp (RGD) peptide in the inner layer. Using a rat brain damage model, researchers found that a double-layered alginate hydrogel improved neural stem cell (NSC) survival and development after NSC transplantation [41]. Although this strategy was developed with NSCs in mind (because of their lower risk of teratoma development in comparison to hESCs and iPSCs), it is readily adaptable to other transplanted cell types. Success with alginate-based biomaterials has been shown in rodent models, however, translations to bigger animal models, such as monkeys and humans, had less success. Two clinical transplantations of human islets in barium- alginate and calcium and barium-alginate spheres had limited success, which has been ascribed in part to a foreign body reaction following transplantation [42]. However, no inflammatory responses detected It has been recently claimed that, depending on the spheres' diameter, fibrosis may be mitigated or even eradicated. Alginate, on the other hand, has been functionalized with a variety of chemical groups to search for chemistries that might prevent a fibrotic response [43]. Recently, hydrogels made from chemically modified alginates such as triazole-thiomorphiline dioxide (TMTD) were shown to be effective against fibrosis at the implant site in both rodents and non- human primates. C57BL/6J diabetic mice were treated with streptozotocin (STZ), and b cells generated from hESCs were transplanted into the animals using TMTD alginate hydrogel. There was no detectable foreign body reaction to the hydrogel, and the hESC-derived b cells were able to engraft and maintain euglycemia in immune-competent animals for 174 days. Hydrogel formulations that transcend the immunological barrier preventing long-term cell activity [44] are now a viable option for research in autoimmune diseases, both in animal models and in future human investigations. Hydrogels made from natural materials have a long history, but synthetic polymers are quickly replacing them because of their superior clinical transferability and reproducibility. Many people employ non-biodegradable polyethylene glycol (PEG) hydrogels for encapsulation for these reasons. Additionally, the tissue-like permeability and low inflammatory response provided by PEG's adaptable viscoelastic characteristics make it a promising candidate for regenerative medicine. Allogeneic islets transplanted into the epididymal fat pad of mice showed long-term remission of diabetes after receiving a conformal covering of PEG and Matrigel [45]. Immune cell penetration and T cell allogeneic priming were prevented by the addition of Matrigel into the PEG coating, which provided essential ECM interactions while maintaining a low permeability. PEG-based hydrogels are utilized for a variety of applications due to their biocompatibility, structural support, and simplicity of functionalization. Studies on the delivery of artificial ovarian tissue have demonstrated that encapsulating immature ovarian follicles in PEG-RGD hydrogels enhances primordial follicle growth and graft survival. Ovariectomized adult mice re-established their estrous cycle after 2 weeks of receiving a subcutaneous transplant of encapsulated ovarian tissue. Ovarian follicles, in contrast to islets, are avascular and resistant to hypoxia, meaning they make the most of immune-isolation techniques. To create an encapsulating gel with microchannels for vascularization, a PEG hydrogel was created utilizing lithographic methods for cells that are more reliant on graft vascularization. Furthermore, to create microchannels in a controlled area of the nano-porous hydrogel, this encapsulation technique combines encapsulation with printing technologies [46]. Nutrients and oxygen may be delivered more efficiently to encapsulated cells using the scaffold architecture described here. In addition, the network of bigger-sized channels might be used for pre-vascularization *in vitro* or to assist penetration of the host vasculature after implantation. Tissue and vascular ingrowth inside the microchannels of the gels were seen in transplanted islets, resulting in normoglycemia after transplantation, and maintained glucose control for the whole two months of the trial, until the device was removed. Multiple cell encapsulation technologies are now available to cure a wide range of illnesses. There are several obstacles to bring these cell-based therapeutics into clinical practice, and recent breakthroughs in material design, immunomodulation, and encapsulation methods are crucial to overcome them [47].

4.5 Hydrogels for Integrative Tissue Engineering and Cell Delivery

To aid in fusing with the host tissue, hydrogels might be microporous in structure or rendered biodegradable. Integration with the host tissue may aid in vascularization, which is necessary for survival, and efficient cell-cell communication, which may guide differentiation. Hydrogels with micropores are used for transcytosis. Cells may be seeded onto a scaffold with an interconnected microporous structure, and the host cells can penetrate the scaffold after transplantation to integrate with the transplanted cells [48]. The porous architecture provides a pathway for nutrition delivery and debris elimination while also facilitating cell attachment and tissue expansion. Perforated biomaterials encourage cell migration and cell-cell interactions that are essential for controlling cell division, maturation, and the structure of some tissue regeneration, in contrast to contained designs (e.g., cartilage and liver). When cells are encased within gel scaffolds, they get imprisoned in the 3D polymer networks, which may prevent or significantly delay the aforementioned biological functions [49]. As a result, adding cell-dwelling gaps larger than ten microns to gel structures may encourage cell dispersion, immigration, multiplication, and subsequently angiogenesis to create access to nutrients and other systemic stimuli. Sphere templating, lyophilization, and porogen leaching are the techniques used to create microporous hydrogels for *in vitro* and *in vivo* applications. The problem in using prefabricated microporous scaffolds is that it is difficult to distribute cells uniformly across the scaffold to allow effective regeneration of highly complex tissues [50]. It may be more challenging to achieve heterogeneous cell dispersion using porous structure hydrogel architectures with a passageway structure than encapsulation methods, which let cells be evenly dispersed during initial production. Although several seeding methods have been developed to provide more even cell dispersion, many of these approaches are time-consuming and have limited clinical relevance [51]. Static seeding, where a concentrated cell culture is passively put on a scaffold, is the most popular approach for seeding microporous scaffolds. The ineffective seeding efficiency and inadequate cell penetration [52] are the drawbacks of this method. By initially creating m-pore 3.5% hydraulic hydrogels and seeding them with mouse MSCs mixed into a thin 2.5% micro porous matrixins hydrogel precursor solution, one researcher was able to enhance the efficacy of cell seeding. The cells are evenly distributed throughout the hydrogel because the precursor solution is fluid at ambient temperature and flows through the hydrogel's pores [53]. In contrast to the conventional method of establishing cells on a hydrogel that has already been prepared, 3d printing may seed cells while the hydrogel is still being created, producing cell components that are extremely structured. This method is useful for overcoming problems caused by insufficient cell densities, unpredictable seeding locations, and a disorganized cell distribution in the scaffold. Researchers have recently developed a novel bioprinting approach based on the use of cell-laden GelMA inks as a bioprintable bulk matrix for the development of vascularized heterogeneous tissue constructions. The combination of a vascular network, various cell types, and ECM in a 3D microengineered system may provide a setting conducive to cell adhesion, remodeling, and migration. The microporous architecture has the drawback that the scaffolds must be shaped before implantation, making it difficult to target a particular area [54]. Injecting microscale hydrogel particles and allowing them to self-assemble into a bulk granular material is an interesting way to make injectable but porous materials. Microporous annealed particle (MAP) scaffolds were the first example of this strategy, and they work by injecting microscale hydrogel spherical beads into a cavity and then annealing them to each other to produce a bulk gel with the space between the beads functioning as pores in the gel. Unlike other porous hydrogels, which have a void percentage of up to 80%, MAP scaffolds contain a continuous micron-sized porosity structure that permits infiltration by transplanting cells and surrounding host tissue without degrading the scaffold [55]. Moreover, MAP gels may mold to the contour of a wound and encourage integration, leading to the formation of cellular structures. Lower inflammation at wound locations has been seen after injecting MAP hydrogel into skin and brain wounds [56]. Injectability is a key feature of the microporous injectable hydrogel design that may increase the potential uses of microporous hydrogels. Particle jamming is a recent innovation in granular hydrogels that eliminates the requirement for annealing. Jammed granular material acts as a solid, which is being utilized in this study [57].

4.6 Biocompatible Hydrogels for Transferring Cells

Cell transmission in encapsulated hydrogels, for example, is covered in that section's discussion of degradable encapsulating hydrogels, may act as an immune-isolation barrier at first, but their breakdown over time can enhance cell penetration and integration with the host. Cell behavior may be altered by hydrogel breakdown, which causes changes in mechanics and swelling. Hydrogels may be broken down by hydrolytic or enzymatic processes if fragile bonds are created by forming them with degradable reagents (such peptide, proteins, or proteolytic enzymes) [58]. It is generally accepted that the hydrogel's breakdown corresponds to the frequency of tissue regeneration alternatively, infiltration is sought. Using biodegradable alginate, The researchers recently created beads enclosing rat MSCs and fibrinogen and implanted the particles in a calcification bone defect to evaluate their potential for regeneration of bone [59]. It was possible to regulate the hydrogel's disintegration rate to promote rapid bone tissue development without sacrificing cell viability or allowing environmental contamination via the release of potentially dangerous chemicals. Although many naturally occurring polymers are biodegradable, the materials' inherent characteristics may limit their use in tissue engineering. In contrast, deterioration in synthetic polymers may be managed by predictable methods, such as by adjusting the cross-linking density of critical segments. Hydrogel cross-linking, for instance, may be achieved by combining the peptide sequence with the hydrogel's polymer backbone that is degradable and wound healing microenvironment specific cell secreted proteases such as MMPs [60].

Recent research has shown that by cross-linking PEG gels enclosing a peptide that really is MMP-degradable on chondrocytes and MSCs, the resorption of the scaffold can be controlled to correspond with the rate at which the matrix is produced by cells during cartilage healing. Cell-mediated degradable hydrogels exhibited considerably higher Collagen and GAG deposition are both crucial biomarkers of tissue regeneration, compared to non-degradable hydrogels. An artificial, two-component PEG hydrogel macro device for transferring is lets to extra hepatic transplantation location was assessed for its degradative qualities [61]. To improve regional blood flow, the hydrogel has a cross linked inner layer of non-degradable PEG dithiol and a cross linked outer layer of a proteolytically sensitive peptide. Islets that were encapsulated inside the device showed great survivability, and the use as hallucinogenic, biodegradable layer of hydrogel improved the density of the transplant's vessels area. until the findings show that additional tuning of parameters like islet load is necessary for achieving normoglycemia with the device, they do highlight the value of degradable interfaces for promoting engraftment. Biodegradable hydrogels may also be used as injectables, where they provide a unique benefit by gelling *in vivo* to deliver cells to precise areas through minimally invasive treatments. Injectable hydrogels can undergo polymerization in reaction to a change in temperature, pH, physical cross-linking, solvent exchange, crystallization, or even just thickening, following the moving of the injection shear [62]. Poly(N-isopropylacrylamide) is a popular polymer for various biomedical uses because its lower critical solution temperature is near to human body. Recently researchers examined the efficacy of hydrogels containing poly(N-isopropylacrylamide) (PNIPAM) as carriers for intramyocardial administration of brown adipose-derived stem cells in rats with myocardial infarction. Because of its quick and subphysiological phase transition temperatures, hydrogel allows for the noninvasive delivery of a liquid cell suspension that gels in place to create a cell-loaded scaffold, effectively isolating therapy to the injection site. Myocardial infarction engraftment was greatly improved, and treatment efficacies were boosted [63]. This study opens up new possibilities for treating disorders including myocardial ischemia, for which previous therapies have had little effectiveness due to inadequate targeting. An injectable and shear-thinning HA hydrogel are created by researchers to transport endothelium precursor cells to the myocardial infarction relieving this ailment. The hydrogel was made to disassemble at the injection site after being forced through a syringe by shear. This administration method provided several advantages that were demonstrated *in vivo*, including increased cell confinement and angiogenesis, less adverse refurbishing, and better heart activity. These biodisposible hydrogels may be modified with trophic factors to improve cell viability or function and then injected into a patient [64]. Degradable PEG hydrogels have been developed to encapsulate islets. Furthermore, vascular endothelial growth factor (VEGF) has been added to these hydrogels to stimulate regional angiogenesis. Using a similar strategy, we were able to improve MSC lifespan in a thermo-sensitive, bio disposable hydrogel by including a basic heparin-binding growth factor release mechanism [65]. This hydrogel was injected *in situ* at an extrahepatic location to promote engraftment and diabetes reversal. Following intramuscular injection of MSCs, hydrogels containing the pro-survival factor bFGF improved blood vessel density, limb perfusion, and muscle diameter in addition to increasing MSC survivability. Although these hydrogels can be functionalized to increase the delivery of oxygen and nutrients to the region of interest, research into their capacity to stimulate neovascularization is still in its infancy due to the lengthy time frame of angiogenesis (days to weeks). Recent research has shown the need of considering local degradation when designing tissue engineering constructions. Hydrogels are gaining popularity as a potential synthetic tissue scaffold owing to their high biocompatibility, intrinsic resemblance to the natural ECM, and their ability to provide a favorable environment for cellular proliferation and survival. These latest developments are mentioned above to highlight crucial concerns for engineering hydrogels for successful cell distribution [66]. Potential obstacles to transplant xenogeneic or stem-cell-derived allogeneic cells might be removed with the use of encapsulation strategies. However, hypoxia near the gel's center may develop, shortening the time these cells are active. Hydrogels that improve vascularization and *in vivo* tissue development via microporous architectures or biodegradation face difficulties from the immune response. Another strategy for controlling the behavior and function of transplanted cells is to include therapeutic chemicals and particles into these hydrogels. Hydrogel-based tissue engineering will keep progressing toward therapeutic restoration of tissue function as long as researchers in both fields work to combine the best features of both techniques [67].

4.7 Using Hydrogels for Gene Therapy in Tissue Engineering-Based Drug

Regardless of the fact that cell delivery offers the cells required for collagen synthesis directly, it is possible that other chemicals that promote pro-regenerative activities will need to be added in order to enhance the activity of such cells. Protein administration has been used to stimulate repair processes, however, the results of the many clinical studies examining its use in tissue regeneration have been mixed, with the best results being shown with decellularized matrices. Therefore, new methods are required to trigger growth- or development-related gene expression programs. For example, the direct delivery of genes to regulate expression is one of the methods [68]. In order to promote pro-regenerative behaviors in transplanted or infiltrating host cells, this technique comprises transfecting host cells with genomes expressing for significant transcriptional, nourishing, or vascular endothelial growth factor again from scaffolding. The length of the expression, the activity of the delivery protein, and the delivery of numerous signals from the same hydrogel may all be modulated using this approach. Therefore, these advantages may be used to circumvent problems with vascularization and immunological rejection that compromise the therapeutic efficiency of cell-based treatments. Moreover, in comparison to viral approaches, non-viral methods avoid the immunogenicity problems that arise from repeated injections and insertional mutagenesis [69]. Obtaining adequate transgene expression to initiate a substantial therapeutic response has been the fundamental obstacle in bringing non-viral vectors to the clinical stage. Both depots that store plasmid DNA for gradual and steady release into surrounding tissue and DNA-loaded biomimetic scaffolds that facilitate cellular penetration into the scaffold have been studied for their potential roles in local gene distribution utilizing implantable or injectable hydrogels. The second strategy includes transfecting cells that will express the therapeutic gene locally before they infiltrate and degrade the scaffold. Design principles for scaffolds used in cell transport may often be used to gene delivery as well. To maximize transgene expression and, in turn, therapeutic benefit, Designing Genetic material hydrogels that consider the vector elements, hydrogel features that effect DNA loading and release, relations among cells and the scaffolding, as well as the gene being given is crucial [70].

4.8 Human Gene Therapy that Uses Hydrogel as an Alternative Method of Delivering Genetic Material to Patients

To develop effective treatments for human disorders, a distinct and crucial area of active research has been established around tissue engineering technologies based on a cargo (biomaterial) to control the release of a therapeutic compound for its enhanced spatial and temporal presentation in the targets. Natural (biocompatible and biodegradable) and synthetic (extremely reproducible) polymers make up today's biomaterials, which are adaptable to provide a variety of recombinant therapies and factors [71].The translation of tissue engineering strategies to the carefully regulated delivery of gene vectors is a fascinating field of research that holds the promise of overcoming the numerous drawbacks of conventional gene transfer techniques and improving the efficacy and security of human gene therapy. The modification of controlled release methods for gene delivery may [72]:

- Lengthen the time that therapeutic sequences spend in the bodies.
- Hide viral capsid epitopes and prevent harmful host immune responses.
- Reduce the amount of vector and sequencing needed for genome transfer and expression.
- Allow bigger genes to be transferred.
- Protect against deterioration by providing stability.

It is possible to produce gradient polymeric release or substrate-mediated release by enclosing the gene vector inside the cargo or immobilizing it on the cargo. Table 4.1 details the most extensively tried and proven gene therapy delivery techniques so far, such as hydrogels (vector diffusion process) and solid scaffolds (substrate vector transfer) [73].

Table 4.1 Controlled delivery of gene vector: Hydrogel [73, 74].

Materials	Vectors	Targets
Fibrin glue	Nonviral	Skin
	Lentiviral	Wound
	Adenoviral	Bone
	rAAV	Cartilage
Gelatin	Nonviral	Muscles
	Adenoviral	Tumors
Alginate	Adenoviral	Tumors
	rAAV	Cartilage
Collagen	Nonviral	Muscles
	Lentiviral	Wound, brain
	Adenoviral	Wound, vessels

(Continued)

Table 4.1 Controlled delivery of gene vector: Hydrogel [73, 74]. (*Continued*)

Materials	Vectors	Targets
Silk elastin like protein	Nonviral	Tumors
	Adenoviral	Tumors
Self-assembling peptides	Lentiviral	Cartilage
	rAAV	Cartilage
Pleuronic	Nonviral	Muscle
	Lentiviral	Wound, brain
	Adenoviral	Tumors
	rAAV	Tumors, cartilage
Poly lactic glycolic acid	Nonviral	Vessels
	Adenoviral	Tumors

4.9 Recent Advancement in Biocompatible Hydrogel

a) The development of synthetic materials with the same structures and features as tendons is of paramount importance. Human tendons have a tensile strength of around 10.0 MPa and a water content of about 60%, both of which are exceptional mechanical qualities that allow them to perform vital duties. However, although having similar water content, most synthetic hydrogels do not have the same qualified strength, swelling resistance, or biocompatibility as tendons. Researchers have developed a straightforward method using poly (vinyl alcohol) and tannic acid that involves series of successive procedures to solve the problem [75]. The process of making hydrogels involves several steps, such as the freezing and thawing aqueous Polyvinyl alcohol solutions to form crystals, stretching and dried in small spaces to stimulate anisotropic frameworks, dipping in TA solutions to produce numerous chemical bonds between PVA and TA, and dialyzing against water to formulate hierarchically anisotropic structures. PVA's hydrogen-bonded interaction with TA offers a mechanism for dissipating energy, and the alignment of PVA bundles adds to the material's high modulus. Due to the interplay between the hydrogel's chemical make-up and its carefully engineered architecture, the resulting material is very mechanically robust (with a tensile stress of 19.3 MPa and a toughness of 32.1 MJ/m3). Contrary to expectations, this hydrogel does not expand very much when exposed to water. Swelling is negligible after 7 days in distilled water phosphate- buffered saline (PBS), and SBF aqueous solutions, and it retains about the same strength and volume as when first immersed. Because of these qualities and its high cytocompatibility, it is an excellent material for use in biomedical and technological applications [76].

b) Reducing immune responses and regulating physio-electrical signals need a physical and electronic mismatched between implanted nanoelectronics to be as small as possible. However, due to the lack of appropriate materials, the design of such implanted microelectronics remains difficult. Here, scientists describe developing a biocompatible, electrically conductive, and mechanically robust alginate hydrogel ionotronic fiber by reacting ionic chelation- assisted wet-spinning with mechanical conditioning. Because of the interplay between these two mechanisms, alginate can self-assemble into a molecular network of highly aligned nanofibrils with a hierarchical structure strikingly similar to that of natural fibers [77]. Due to its favorable structure, AHIF has mechanical properties that resemble those of tissues, such as self-stiffening and pliability. Biomacromolecules, Ca^{2+} ions, and water are found in tissue, and they provide AHIF a tissue-like quality such as biocompatibility and conductivity. These benefits provide insight into the use of AHIFs in implanted devices that carry electricity. One purpose of the prototype AHIF is to provide physioelectrical modulation through noncontact electromagnetic induction. Physioelectrical-like signals from the AHIF are used with experimental and machine-learning advancements to recognize the shape and tension condition of the implanted device. An intelligent AHIF system may be useful in the domains of bioelectronics, IntelliSense, and human-machine interfaces [78].

c) For their biocompatibility and ability to imitate the ECM, multifunctional hydrogels have drawn a lot of attention in now a day. Hydrogels with photothermal characteristics were created using cross-linked networks that were safe for use on living organisms. To do this, they use a photothermal conversion hydrogel composed of chitosan, polyaniline, and laponite (COL). To create a hydrogel that was cross-linked using a azomethines black aniline was first grafted onto chitosan. The solution was then added to oxidized dextran. Subsequently, laponite (LAP), an aluminosilicate clay substance, was put into the hydrogel. The COL hydrogel demonstrated excellent degrading ability and exhibited swelling ratios of higher than 580% in different solutions. Both the cells (cell viability was above 97%) and animal levels showed that this composite hydrogel was biocompatible and had strong photothermal conversion capabilities. Under the illumination of a near-infrared laser, the COL hydrogel demonstrated a photothermal conversion efficiency of 23.7%. The COL hydrogel, in combination with LAP's osteogenic differentiation-inducing capabilities, might be used as a scaffold for bone tissue regeneration or to thermally inactivate malignancies [79].

d) Elegant drug delivery systems based on DNA hydrogels have been developed. Although DNA hydrogel production and application are still in their infancy, it exhibits many novel properties and hold the potential of addressing the shortcomings of present delivery strategies and materials. DNA hydrogels are stable as immobilization matrices for an extensive use of pharmaceuticals due to their porosity structure, numerous conjugation sites, and highly programmable sequence. The natural concept of base pairs makes DNA hydrogels an unmatched foundation for the ordered load of these drugs as construction blocks or tons of cargo, making them perfect for the delivery of therapeutic nucleic acids. In addition, strongly cross-linked networks limit enzyme penetration, safeguarding the biological activity of functioning proteins. Exciting progress has been made in the study of the transport of inorganic minerals and tiny molecules. DNA hydrogels have made significant strides as a drug delivery technology over the last two decades, but there are still many challenges that must be addressed before they can be used in clinical settings [80]. Obtaining DNA hydrogels with consistent purity at a cheap cost, high yield, and on a wide scale is the primary issue for biological applications. Therefore, better DNA hydrogel synthesis methods are needed urgently. Also, it's important to find ways to mass-produce DNA at a reasonable price. For this purpose, we used a highly parallel gene synthesis strategy predicated on DNA microchips and targeted oligonucleotide pool amplification. The result was a dramatic drop in the price of mass-producing DNA. Second, there has only been very little research done on DNA hydrogels' pharmacokinetics, which includes their *in vivo* distribution, circulation, metabolism, and excretion. Different methods of forming DNA hydrogels result in DNA hydrogels with varying resistance to nuclease hydrolysis. Improving hydrogel stability *in vivo* requires a thorough knowledge of the mechanism behind this behavior [81]. To slow the clearance of DNA hydrogels from the bloodstream, it is important to learn more about how their physicochemical features (such as surface charge, shape, and base compositions) affect their aggregation in the blood. The kinematics of DNA hydrogel release in reaction to external stimuli provide a fresh area for further research. Although many different triggers have been described, the release properties of DNA hydrogels still need to be elucidated by theoretical research. The regulated release behavior and the selection of responsive DNA hydrogels for certain conditions will benefit from more accurate kinetic research. Drugs should be made available on demand depending on the patient's location and the severity of their sickness. Finally, DNA hydrogels' immunostimulatory qualities in terms of human safety and therapeutic effectiveness need to be investigated. It is important to lessen the severity of the body's immunological response to exogenous nucleic acids and get rid of any potentially hazardous reaction remnants. DNA hydrogels, because of their many benefits, are inevitably going to play a vital role in the future of innovative therapeutics. We may also anticipate that the integration of DNA hydrogels with cutting-edge treatment modalities like CRISPR/Cas will pave the way for the development of further hydrogel-based platforms that are both effective and safe for the treatment of a broad variety of illnesses [82].

4.10 Applications of Hydrogel

Creating an Immunomodulatory Niche by Engineering and Characterization
There is emerging evidence that hydrogel vaccinations may provide healthier and more beneficial effects than bolus injection of the indentical medicines. Although the exact processes remain a mystery, it appears that important immune cells are infiltrating the hydrogel, which may be a contributing factor. These cells may be stimulated by the hydrogel itself, and outside signals, such as immunosuppression from a surrounding tumor, can be reduced [83]. This results in the establishment of a novel immunological microenvironment, also known as an immunomodulatory niche. The capacity of a cytokine (granulocyte-macrophage colony-stimulating factor to attract a crucial type of dendritic cell (DC) into hydrogel vaccines was investigated by the researchers. They discovered that if the GMCSF dosage was too high, the DCs wouldn't return to the lymph nodes in the area to convey the newly acquired antigen. Recent studies demonstrated the role played by other immunostimulatory substances in shaping the immunomodulatory niche. Hydrogel scaffolds may be created to stimulate cell infiltration in several ways, including the release of chemoattractants. In addition, the infiltrating cells may be reprogrammed depending on the hydrogel's contents, directing the immune response toward a specific biological goal. To increase the production of inflammatory cytokines that attract DCs, an *in situ* gelling mesoporous silica rod was created. Macrophages, another essential kind of cell that may live in a variety of phenotypic states, can be reprogrammed by hydrogels. Certain phenotypes are more ideal than others for mediating healing or treating illness in a specific biological context [84].

Hydrogels for Adoptive Cell Therapy

Injectable cell scaffolds have been used in immunology and immunotherapy in addition to tissue regeneration. Despite the paucity of studies involving injectable materials, It is clear that several ideas created for efficient cellular scaffolding in regenerative medicine may be applied identically to immune system. Hydrogels allow for the formation of an immune niche that has early similarities to that seen in lymphatic tissue. To further enhance specific biological functions, hydrogel materials may have stimulating medicines and signal transduction inserted or coupled to them. Delivering CAR-T cells is the only kind of T cell that has been studied concerning its growth and administration in adoptive cell therapy. The time it takes for cells to multiply before therapy has led researchers to concentrate on creating 3D hydrogel culture methods that can accelerate the process of T cell growth [85]. This has sparked research into developing self-assembled scaffolds or hydrogels that can simulate lymphatic tissue. For biomimicry to work, surface chemistry must be carefully tailored to facilitate intricate biological signals. If a silica microrod scaffold is coated with PEG or integrins, for instance, the scaffold will either promote or suppress inflammatory reactions, respectively [86]. In addition, hydrogel carriers are elucidating the mechanobiology of T cells, which may provide key signals for *in vivo* function and *ex vivo* cell multiplication. Recent research, for instance, found that even when porosity was maintained, the stiffness of an alginate hydrogel significantly affected T cell motility and activation level. This research shows that T cell activation might be greatly enhanced by using stiff gels (44 kPa) as opposed to soft gels (4 kPa). Increased production of cytokines and the appearance of activation markers on the surface of T lymphocytes are indicators of their heightened proliferation and activation (CD25). Though similar results have been shown in 2D, these 3D experiments suggest that stiffer hydrogels may lead to better *ex vivo* growth, which is now responsible for a substantial delay between patient cell collection and subsequent therapy. The results of adoptive cell treatment for patients might be enhanced if researchers could learn more from trials how to build T cell delivery gels that keep T cells active *in vivo* [87].

Traumatic Wound Healing

Wound healing, which requires a complicated cascade of reactions from a wide variety of cells, is a promising area for the use of regenerative medicine. Tissue injury and scarring may occur if the healing process is interrupted. Accordingly, systems that can prompt appropriate answers at an appropriate time and location are very beneficial. Physical and chemical signals are spatial temporal controlled by hydrogels, making them important not just as release mechanisms for regenerative chemicals but also as scaffolding for the infiltration of restorative cells [88].

Bone and Cartilage Repair

Osteo and ligament bioengineering seeks to improve the well-being of individuals with a variety of conditions, such as those caused by birth abnormalities, accidents, or aging. Because bone is rigid, hydrogels employed in this field must also be fairly rigid. In successful injectable systems, hydrogels with induced *in vivo* cross-linking gel formation and, finally, large elastic modulus are typically utilized. These hydrogels are liquid during the injection. More exogenously supplied cells are retained in stiffer hydrogels, and MSCs are induced to differentiate along osteogenic pathways. hydrogels containing chemical markers for osteogenesis, such as calcium salts or proteins called bone morphogenic, have been found to improve osteogenic differentiation and biointegration, despite bone's extremely distinctive growth factor and mineral composition. Similar results were shown in a rat cranial defect model where MSC distribution was improved by conjugating phosphate groups to a PNIPAM-based hydrogel, leading to increased osteogenic differentiation, biomineralization, and host integration [89].

Hydrogels Used in Medical Procedures

Hydrogels has been the subject of research into their possible use in reducing and avoiding surgical adhesions. In the aftermath of surgery, scar tissue may form fibrous bands called surgical adhesions (or postoperative adhesions) between an organ and its surrounding tissues [90]. Adhesions are a source of concern for more than 19 million people annually in the United States. Problems like this contribute billions of dollars to the cost of treating people in the United States every year. hyaluronan and cellulose gum are used in commercially available products like a film and a woven cloth (AdhereTech, Coloplast) to prevent adherence (CMC). Both are approved for intragastric use exclusively (Interceed, Ethicon). These barriers may be challenging to spread over through the targeted organs to achieve adequate surface coverage, despite being effective at preventing adhesion in between cells and organs of concern. It has also been reported that these sheet-like barriers are easily displaced by normal tissue motion and fall short of completely enclosing target domain with uneven surfaces or ones that are strongly rolled up, like the intestinal tract. This leaves these surfaces susceptible to the development of adhesions [91].

Medical Device Coatings Made of Hydrogel

Hydrogels are used in medical equipment as they can be engineered to be antibacterial and/or antifouling. They are a promising material for use in coatings for medical devices and "skin"-like scaffolding for electronics and sensors. Adding a hydrogel coating to a device may increase its functionality and biocompatibility by creating a more natural contact between the device and the human body. As these coatings are antifouling, antibacterial, and nonimmunogenic [92], they are particularly helpful for long-term implants in terms of biocompatibility.

Hydrogels may be easily integrated into devices since they can be attached to a wide range of surfaces either chemically (by polymerizing directly onto the surface) or physically (e.g., spray coating). It's worth noting that many of these techniques may be used to functionalize a wide range of geometries and forms. Many applicable procedures, such as coating by dipping, coating by spinning, spraying, or doctor blading, are frequently amenable to industrial scalability and may be employed on a wide variety of devices and surfaces. One study showed that acrylate-based films may be painted on surfaces by simply brushing or spraying the film onto the surface, and the film thereafter hardens into a hydrogel. Some similar work [93] used inkjet printing and spray coating to apply hierarchical nanostructured hydrogels on paper and glass.

Hydrogel Coatings as Antimicrobial Agents

Hydrogel coatings provide new ways to battle infections, a major problem for implanted medical devices. Antibacterial hydrogels may be used as "passive" coverings with intrinsic antimicrobial capabilities or moieties for the prophylactic prevention of adhering materials or organisms. Water - loving and quaternary ammonium moieties, which contribute to strong surface hydration and are resistant to protein adsorption, have been successfully incorporated into numerous medical devices. Catheters functionalized with these coatings have demonstrated resistance against Staphylococci, Gram-negative E. coli, and fungus (*C. albicans* and *C. neoformans*) bacteria. These coatings are made of cationic polycarbonate and the anti-fouling polymer poly (ethylene glycol) [94].

Use of Hydrogels in Anti-Fouling

Coatings rejection by the body is a key aspect influencing the longevity and usefulness of implantation, even though sensor-based pathogen are a severe cause of death attributable to medical equipment implantation. The foreign body response is the primary mechanism by which the body rejected an implantation. The body is trying to get rid of macro foreign bodies, therefore a sophisticated chain of cellular mechanisms that starts with surface fouled and ends with the creation of a fibrotic capsule takes place. Researchers have been looking at hydrogels as antifouling coatings [95] since the efficacy of antimicrobial hydrogel coatings relies in part on the avoidance of biofilm formation. Antifouling materials work to stop the immediate, nonspecific adherence of molecules to device surfaces that

Table 4.2 Therapeutic applications of parenteral hydrogels categorized by material class and broad scope [99].

Title of product	Kind of injection	Authorized therapeutic application	Authorization (year)
SpaceOAR Hydrogel	Percutaneous	to safeguard delicate tissues during treatment for prostate cancer	EMA (2010) FDA (2015)
Vantas	Subcutaneous	treatment of prostate cancer	FDA (2015)
Radiesse	Dermis	for the purpose of erasing face creases and folds	EMA (2004) FDA (2006)
Revanesse	Dermis	face creases and wrinkles ranging from mild to severe	FDA (2018)
Belotero balance	Dermis	face creases and wrinkles ranging from mild to severe	FDA (2019)
EUFLEXXA	Intra-articular	Knee osteoarthritis	FDA (2004) EMA (2005)
TraceIT® Hydrogel Tissue Marker	Percutaneous	better alignment of the soft tissues for image-guided treatment	FDA (2013)

happens when one material comes into contact with another material or complicated fluids. Fouling, which may cause thrombosis, inflammation, and obstruction, affects all implanted materials. The non-specific attachment of peptides causes fouling to begin when they are adsorbed upon a membrane, undergo conformational changes that expose regions of the protein that are ordinarily shielded in the native shape and so contribute to the protein's unfolding [96, 97]. These newly discovered locations may then serve as mediators when interacting with various immune system components. Creating materials with extremely strong interactions with water molecules to generate a protective layer of water to sterically limit access to the underlying material has proven to be our most successful approach to generate antifouling materials. These anti-biofouling properties have been linked to elements like interface packaging, wettability, electrostatic polarity, and the malleability of polymer chains. Materials with a tight hydration layer produce a physical and energetic barrier that restricts inappropriate adherence and raises the Gibbs free barrier required for peptides to bind. Hydrogels, because of their hydrophilicity, seem very promising for constructing this barrier, both with water and perhaps other solvents. In the case of a neural electrode, for example, hydrogel coatings might further enhance biocompatibility by creating most amenable contact with delicate brain tissues. It is not apparent, however, whether there has to be a trade-off between the coating's longevity and attempts to mimic the mechanical characteristics of its surrounding tissues. Hence, if the anti-biofouling properties of biopolymers are separated from their mechanical properties, it may be useful to create safe and biodegradable covers for various kinds of tissues [98]. Therapeutic applications of parenteral hydrogels are summarized in Table 4.2.

4.11 Current Hydrogels in Clinical Trials

New injection based hydrogel systems are tested extensively in the clinical research for their potential as therapeutic, diagnostic, and cosmetic agents. The objective of this part is to give a short outline of (a) the present hydrogel-based systems used in clinical studies, as indicated in Table 4.3, and (b) the major techniques aiming to incorporate enhanced functionality and increased bio polymer nature and bio-adsorbable hydrogel architectures.

Table 4.3 List of ongoing injection based hydrogel in clinical trials [100–103].

Name (sponsor company/ University)	Injection type	Indications	Clinical trials
Argiform (Research Centre BIOFORM)	Intra-articular	Knee osteoarthritis	NCT03897686(NA)
Aquamid (Henning Bliddal)	Intra-articular	Knee osteoarthritis	NCT03060421 (NA)
Algisyl-LVR® device (LoneStar Heart, Inc.)	Intra-myocardial	Heart failure and dilated cardiomyopathy	NCT01311791 (PhII/III)
The Second Affiliated Hospital of Chongqing Medical University	Uterine cavity	Asherman's syndrome	NCT03223454 (Ph I)
Neo-kidney augment (inRegen)	Kidney	Type 2 diabetes and chronic kidney disease	NCT02525263 (Ph II)
Absorbable Radiopaque Tissue Marker (Sidney Kimmel Comprehensive Cancer Center at Johns Hopkins)	Between pancreas and duodenum	Imaging of pancreatic adenocarcinoma	NCT03307564
Absorbable Radiopaque Tissue Marker (Washington University School of Medicine)	Resection bed	Imaging of oropharyngeal cancer	NCT03713021 (Ph I)

Applying Hydrogel Scaffolds to Cancer Treatment

Scientists are looking at hydrogels as a potential means of improving cancer treatment. There has been a lot of research into using synthetic hydrogels like TraceIT® and SpaceOAR® to image cancer cells and shield good and normal cells from harmful effect of radioactive material. With the use of MRI, CT, and/or ultrasound, malignant tissue may be seen for up to 3 months with the help of the TraceIT® hydrogel tissue marker, which is composed of PEG hydrogel microparticles comprising covalently bound iodine. During radiation therapy for malignant tissues, the SpaceOAR® hydrogel is used to shield nearby healthy cells from harm. This disparity in structural integrity between natural and synthetic hydrogels may explain why just one natural hydrogel has been tested clinically so far for use in cancer treatment. This suggests that there is potential for natural hydrogels to be used in cancer treatment if their stiffness and longevity can be enhanced [101].

Hydrogels for Urinary Incontinence

Bulkamid® hydrogel, which is based on synthetic polyacrylamide, has dominated clinical testing. It has received approval in 25 countries, including Europe. Multicenter clinical studies including 33 locations in the United States and Canada are actively investigating the possibility of its use. Natural injectable hydrogel treatments are not being studied in clinical trials at the moment, perhaps because they cannot yet attain sufficient stiffness to be used as a spacer for incontinence purposes [102].

Different Uses of Hydrogels

Preclinical studies on injectable hydrogels for ocular purposes have been substantial; nevertheless, no such products have yet entered clinical usage. The only therapy for eye, tyrosine phosphorylation inhibitor microcrystals in PEG hydrogel administered through intravitreal injection (AMD). A drug called FX-322, which consists of two patented small-molecule medications in a nonionic triblock co-polymer, has been given hasten naming by the FDA to treat sensorineural hearing loss. various inborn hydrogels are already on the market, but synthetic polyacrylamide-based hydrogels are now being researched for the treatment of face lipoatrophy. The variety of hydrogel-based beauty products evaluated in clinical trials suggests a most targeted approach to assessment is required [103].

4.12 Conclusions

Hydrogel features like porous nature, cell-based degradation, and Associated bioactive signals influence and cellular connections, multiplication, motility, and differentiation in addition to helping penetrate and transport cells. Today's as a delivery agent hydrogel technologies provide cells with a highly dynamic microenvironment. Hydrogels may be loaded with therapeutics like genes to induce sustained transgenic expression in infiltrated and grafted cells, and this expression might be enhanced by adjusting hydrogel properties. The intricacy of the tissue environment in both normal and pathological conditions makes managing hydrogel dynamics difficult. To enhance cell activity and viability during skin regeneration and transplanting, no one substance currently exists that satisfies all mechanical criteria. Long-term memory and differentiation may benefit from stiffness, but the disadvantage is that it may impede the healing process in the outset. Regulating cellular processes and guiding new tissue formation by alterations to the tissue environment are essential for successful regenerative therapy results. Peptide amphiphiles and self-assembling peptides, which regulate gel formation, decay process, and the appearance of cellular adhesion pattern, are also being investigated as potential hydrogel designs. Biomaterials with layered structures that resemble the radial arrangement of local tissue are an exciting area of study because they have the potential to address the mechanical requirements of transplantation. Directing cell fate and controlling wound healing and angiogenesis may be improved by using a layered hydrogel with a variety of ECM proteins and mechanical features. Multipotent stem cells (MSCs) and chondrocytes may exhibit zone-specific differentiation in multilayered scaffolding with physical strength with varying matrix components and mechanical stimulation. Hydrogel stiffness may be altered by cross-linking and implantation, which is a useful property. This has been done by the use of secondary chemistry between functional groups initiated by the first cross-linking process or through a second reaction method, happens on a longer interval than the initial inter-connected event. In order to optimize the regeneration process, the hydrogel's stiffness may vary over time. Recent studies of the host reaction to implanted biomaterials may be used to build substances which can actively influence the immune reaction to favor the tissue repairing and the survivability and functioning of any adjacent new repairing cells. Tuning material characteristics may promote implant site macrophage pro-remodeling vs. pro-inflammatory phenotypes and affect transplanted stem or parent cell activity and differentiation. Scaffolds may provide anti-inflammatory cytokines to decrease pro-inflammatory gene expression and increase pro-regenerative gene expression to promote tissue healing. Finally, CRISPR can modulate the immunological response of transplanted cells to prevent rejection. In conclusion, hydrogels may construct environments that support cell and tissue development and manage gene delivery to promote differentiation and enhance therapeutic results. There are numerous clinically and financially viable injectable hydrogels. Recent injectable hydrogel systems are under investigation in clinical settings and show continuous interest in the development of the existing applications. Due to their polymer, mechanical, solute transport, degradability, compatibility, and scale-up requirements, injectable hydrogels provide unique problems. Hydrogels are being used to improve healthcare, and more technologies are being tested on people. This technology might transfer genes and cells for a numerous of tissue engineering implication as hydrogel design issues are studied.

References

1. Lugin, M.L., Lee, R.T., Kwon, Y.J., Synthetically engineered adeno-associated virus for efficient, safe, and versatile gene therapy applications. *ACS Nano*, 14, 11, 14262–83, 2020.
2. Sharma, S. and Tiwari, S., A review on biomacromolecular hydrogel classification and its applications. *Int. J. Biol. Macromol.*, 162, 737–47, 2020.
3. Carballo-Pedrares, N., Fuentes-Boquete, I., Díaz-Prado, S., Rey-Rico, A., Hydrogel-based localized nonviral gene delivery in regenerative medicine approaches—An overview. *Pharmaceutics*, 12, 8, 752, 2020.
4. Wu, X., Wu, T., Liu, J., Ding, B., Gene therapy based on nucleic acid nanostructure. *Adv. Healthcare Mater.*, 9, 19, 2001046, 2020.
5. Li, F., Lyu, D., Liu, S., Guo, W., DNA hydrogels and microgels for biosensing and biomedical applications. *Adv. Mater.*, 32, 3, 1806538, 2020.
6. Varaprasad, K., Raghavendra, G.M., Jayaramudu, T., Yallapu, M.M., Sadiku, R., A mini review on hydrogels classification and recent developments in miscellaneous applications. *Mater. Sci. Eng.: C*, 79, 958–71, 2017.
7. Singhal, R. and Gupta, K., A review: Tailor-made hydrogel structures (classifications and synthesis parameters). *Polym.-Plast. Technol. Eng.*, 55, 1, 54–70, 2016.
8. Khansari, M.M., Sorokina, L.V., Mukherjee, P., Mukhtar, F., Shirdar, M.R., Shahidi, M., Shokuhfar, T., Classification of hydrogels based on their source: A review and application in stem cell regulation. *JOM*, 69, 8, 1340–7, 2017.
9. Kaith, B.S., Singh, A., Sharma, A.K., Sud, D., Hydrogels: Synthesis, classification, properties and potential applications—A brief review. *J. Polym. Environ.*, 29, 12, 3827–41, 2021.
10. Fu, J., Hydrogel properties and applications. *J. Mater. Chem. B*, 7, 10, 1523–5, 2019.
11. Zhu, J. and Marchant, R.E., Design properties of hydrogel tissue-engineering scaffolds. *Expert Rev. Med. Devices*, 8, 5, 607–26, 2011.

12. Dunbar, C.E., High, K.A., Joung, J.K., Kohn, D.B., Ozawa, K., Sadelain, M., Gene therapy comes of age. *Science*, 359, 6372, eaan4672, 2018.
13. Ma, C.C., Wang, Z.L., Xu, T., He, Z.Y., Wei, Y.Q., The approved gene therapy drugs worldwide: From 1998 to 2019. *Biotechnol. Adv.*, 40, 107502, 2020.
14. Ghasemiyeh, P. and Mohammadi-Samani, S., Hydrogels as drug delivery systems; pros and cons. *Trends Pharm. Sci.*, 5, 1, 7–24, 2019.
15. Mo, F., Jiang, K., Zhao, D., Wang, Y., Song, J., Tan, W., DNA hydrogel-based gene editing and drug delivery systems. *Adv. Drug Delivery Rev.*, 168, 79–98, 2021.
16. Chalanqui, M.J., Pentlavalli, S., McCrudden, C., Chambers, P., Ziminska, M., Dunne, N., McCarthy, H.O., Influence of alginate backbone on efficacy of thermo-responsive alginate-gP (NIPAAm) hydrogel as a vehicle for sustained and controlled gene delivery. *Mater. Sci. Eng.: C*, 95, 409–21, 2019.
17. Khan, F., Tare, R.S., Oreffo, R.O., Bradley, M., Versatile biocompatible polymer hydrogels: Scaffolds for cell growth. *Angew. Chem. Int. Ed.*, 48, 5, 978–82, 2009.
18. Shin, S. and Shea, L.D., Lentivirus immobilization to nanoparticles for enhanced and localized delivery from hydrogels. *Mol. Ther.*, 18, 4, 700–6, 2010.
19. Oliva, N., Conde, J., Wang, K., Artzi, N., Designing hydrogels for on-demand therapy. *Acc. Chem. Res.*, 50, 4, 669–79, 2017.
20. Bulcha, J.T., Wang, Y., Ma, H., Tai, P.W., Gao, G., Viral vector platforms within the gene therapy landscape. *Signal Transduction Targeted Ther.*, 6, 1, 1–24, 2021.
21. Mendell, J.R., Al-Zaidy, S.A., Rodino-Klapac, L.R., Goodspeed, K., Gray, S.J., Kay, C.N., Boye, S.L., Boye, S.E., George, L.A., Salabarria, S., Corti, M., Current clinical applications of *in vivo* gene therapy with AAVs. *Mol. Ther.*, 29, 2, 464–88, 2021.
22. Youngblood, R.L., Truong, N.F., Segura, T., Shea, L.D., It's all in the delivery: designing hydrogels for cell and non-viral gene therapies. *Mol. Ther.*, 26, 9, 2087–106, 2018.
23. Chen, Z., Liu, F., Chen, Y., Liu, J., Wang, X., Chen, A.T., Deng, G., Zhang, H., Liu, J., Hong, Z., Zhou, J., Targeted delivery of CRISPR/Cas9-mediated cancer gene therapy via liposome-templated hydrogel nanoparticles. *Adv. Funct. Mater.*, 27, 46, 1703036, 2017.
24. Madry, H., Gao, L., Rey-Rico, A., Venkatesan, J.K., Müller-Brandt, K., Cai, X., Goebel, L., Schmitt, G., Speicher-Mentges, S., Zurakowski, D., Menger, M.D., Thermosensitive hydrogel based on PEO–PPO– PEO poloxamers for a controlled *in situ* release of recombinant adeno-associated viral vectors for effective gene therapy of cartilage defects. *Adv. Mater.*, 32, 2, 1906508, 2020.
25. Chen, W., Chen, H., Zheng, D., Zhang, H., Deng, L., Cui, W., Zhang, Y., Santos, H.A., Shen, H., Gene-hydrogel microenvironment regulates

extracellular matrix metabolism balance in nucleus pulposus. *Adv. Sci.*, 7, 1, 1902099, 2020.
26. Ramamoorth, M. and Narvekar, A., Non viral vectors in gene therapy-an overview. *J. Clin. Diagn. Res.: JCDR*, 9, 1, GE01, 2015.
27. Ginn, S.L., Amaya, A.K., Alexander, I.E., Edelstein, M., Abedi, M.R., Gene therapy clinical trials worldwide to 2017: An update. *J. Gene Med.*, 20, 5, e3015, 2018.
27. Han, S.O., Mahato, R.I., Sung, Y.K., Kim, S.W., Development of biomaterials for gene therapy. *Mol. Ther.*, 2, 4, 302–17, 2000.
28. Sinclair, A., Islam, S., Jones, S., Gene therapy: An overview of approved and pipeline technologies. Lundstrom K. Viral vectors in gene therapy. *Diseases*, 6, 2, 42, 2018.
29. Waehler, R., Russell, S.J., Curiel, D.T., Engineering targeted viral vectors for gene therapy. *Nat. Rev. Genet.*, 8, 8, 573–87, 2007.
30. Arabi, F., Mansouri, V., Ahmadbeigi, N., Gene therapy clinical trials, where do we go? An overview. *Biomed. Pharmacother.*, 153, 113324, 2022.
31. Hecht, J.E., Jounaidi, Y., Waxman, D.J., Construction of P450-expressing tumor cell lines using retroviruses, Department of Biology, Boston University, Boston, MA, in: *Gene Therapy of Cancer*, pp. 85–94, Humana Press.
32. Cohen, S.A., Simaan-Yameen, H., Fuoco, C., Gargioli, C., Seliktar, D., Injectable hydrogel microspheres for sustained gene delivery of antisense oligonucleotides to restore the expression of dystrophin protein in duchenne muscular dystrophy. *Eur. Polym. J.*, 7, 111038, 2022.
33. Ansar, R., Saqib, S., Mukhtar, A., Niazi, M.B., Shahid, M., Jahan, Z., Kakar, S.J., Uzair, B., Mubashir, M., Ullah, S., Khoo, K.S., Challenges and recent trends with the development of hydrogel fiber for biomedical applications. *Chemosphere*, 287, 131956, 2022.
34. Zhang, Q., Liu, Y., Yang, G., Kong, H., Guo, L., Wei, G., Recent advances in protein hydrogels: From design, structural and functional regulations to healthcare applications. *Chem. Eng. J.*, 6, 138494, 2022.
35. Ta, H.T., Dass, C.R., Larson, I., Choong, P.F., Dunstan, D.E., A chitosan hydrogel delivery system for osteosarcoma gene therapy with pigment epithelium-derived factor combined with chemotherapy. *Biomaterials*, 30, 27, 4815–23, 2009.
36. Mahand, S.N., Aliakbarzadeh, S., Moghaddam, A., Moghaddam, A.S., Kruppke, B., Nasrollahzadeh, M., Khonakdar, H.A., Polyoxazoline: A review article from polymerization to smart behaviors and biomedical applications. *Eur. Polym. J.*, 9, 111484, 2022.
37. Almawash, S., Osman, S.K., Mustafa, G., El Hamd, M.A., Current and future prospective of injectable hydrogels—Design challenges and limitations. *Pharmaceuticals*, 15, 3, 371, 2022.
38. Fu, B., Wang, X., Chen, Z., Jiang, N., Guo, Z., Zhang, Y., Zhang, S., Liu, X., Liu, L., Improved myocardial performance in infarcted rat heart by injection

of disulfide-cross-linked chitosan hydrogels loaded with basic fibroblast growth factor. *J. Mater. Chem. B*, 10, 4, 656–65, 2022.
39. Rey-Rico, A. and Cucchiarini, M., Supramolecular cyclodextrin-based hydrogels for controlled gene delivery. *Polymers*, 11, 3, 514, 2019.
40. Choi, S.W., Guan, W., Chung, K., Basic principles of hydrogel-based tissue transformation technologies and their applications. *Cell*, 184, 16, 4115–36, 2021.
41. Kim, S.H., Seo, Y.B., Yeon, Y.K., Lee, Y.J., Park, H.S., Sultan, M.T., Lee, J.M., Lee, J.S., Lee, O.J., Hong, H., Lee, H., 4D-bioprinted silk hydrogels for tissue engineering. *Biomaterials*, 260, 120281, 2020.
42. Sander, E.A., Grassl, E.D., Tranquillo, R.T., Hydrogel scaffolds for regenerative medicine, University of Iowa, UI · Department of Biomedical Engineering, in: *Biomaterials and Regenerative Medicine*, pp. 295–316, Cambridge University Press, 2015.
43. Bankoti, K., Rameshbabu, A.P., Datta, S., Maity, P.P., Goswami, P., Datta, P., Ghosh, S.K., Mitra, A., Dhara, S., Accelerated healing of full thickness dermal wounds by macroporous waterborne polyurethane-chitosan hydrogel scaffolds. *Mater. Sci. Eng.: C.*, 81, 133–43, 2017.
44. Mantha, S., Pillai, S., Khayambashi, P., Upadhyay, A., Zhang, Y., Tao, O., Pham, H.M., Tran, S.D., Smart hydrogels in tissue engineering and regenerative medicine. *Materials*, 12, 20, 3323, 2019.
45. Ding, X., Gao, J., Awada, H., Wang, Y., Dual physical dynamic bond-based injectable and biodegradable hydrogel for tissue regeneration. *J. Mater. Chem. B*, 4, 6, 1175–85, 2016.
46. Ding, X., Gao, J., Awada, H., Wang, Y., Dual physical dynamic bond-based injectable and biodegradable hydrogel for tissue regeneration. *J. Mater. Chem. B*, 4, 6, 1175–85, 2016.
47. Dong, L., Wang, S.J., Zhao, X.R., Zhu, Y.F., Yu, J.K., 3D-printed poly (ε-caprolactone) scaffold integrated with cell-laden chitosan hydrogels for bone tissue engineering. *Sci. Rep.*, 7, 1, 1–9, 2017.
48. Shi, D., Dhawan, V., Cui, X.T., Bio-integrative design of the neural tissue-device interface. *Curr. Opin. Biotechnol.*, 72, 54–61, 2021.
49. Jiang, Y., Chen, J., Deng, C., Suuronen, E.J., Zhong, Z., Click hydrogels, microgels and nanogels: Emerging platforms for drug delivery and tissue engineering. *Biomaterials*, 35, 18, 4969–85, 2014.
50. Zhang, Z., Injectable biomaterials for stem cell delivery and tissue regeneration. *Expert Opin. Biol. Ther.*, 17, 1, 49–62, 2017.
51. Zhang, S., Huang, D., Lin, H., Xiao, Y., Zhang, X., Cellulose nanocrystal reinforced collagen-based nanocomposite hydrogel with self-healing and stress-relaxation properties for cell delivery. *Biomacromolecules*, 21, 6, 2400–8, 2020.
52. Bjugstad, K.B., Lampe, K., Kern, D.S., Mahoney, M., Biocompatibility of poly (ethylene glycol)- based hydrogels in the brain: An analysis of the glial response across space and time. *J. Biomed. Mater. Res. Part A*, 95, 1, 79–91, 2010.

53. Neufeld, L. and Bianco-Peled, H., Designing a biocompatible hydrogel for the delivery of mesalamine. *Int. J. Pharm.*, 491, 1-2, 170–9, 2015.
54. Kouser, R., Vashist, A., Zafaryab, M., Rizvi, M.A., Ahmad, S., pH-responsive biocompatible nanocomposite hydrogels for therapeutic drug delivery. *ACS Appl. Bio Mater.*, 1, 6, 1810–22, 2018.
55. Bardajee, G.R. and Hooshyar, Z., One-pot synthesis of biocompatible superparamagnetic iron oxide nanoparticles/hydrogel based on salep: characterization and drug delivery. *Carbohydr. Polym.*, 101, 741–51, 2014.
56. Martens, T.P., Godier, A.F., Parks, J.J., Wan, L.Q., Koeckert, M.S., Eng, G.M., Hudson, B.I., Sherman, W., Vunjak-Novakovic, G., Percutaneous cell delivery into the heart using hydrogels polymerizing *in situ*. *Cell Transplant.*, 18, 3, 297–304, 2009.
57. Xu, Q., Sigen, A., Gao, Y., Guo, L., Creagh-Flynn, J., Zhou, D., Greiser, U., Dong, Y., Wang, F., Tai, H., Liu, W., A hybrid injectable hydrogel from hyperbranched PEG macromer as a stem cell delivery and retention platform for diabetic wound healing. *Acta Biomater.*, 75, 63–74, 2018.
58. Schmidt, J.J., Rowley, J., Kong, H.J., Hydrogels used for cell-based drug delivery. *J. Biomed. Mater. Res. Part A: An Official Journal of The Society for Biomaterials, The Japanese Society for Biomaterials, and The Australian Society for Biomaterials and the Korean Society Biomaterials*, 87, 4, 1113–22, 2008.
59. Ghasemiyeh, P. and Mohammadi-Samani, S., Hydrogels as drug delivery systems; pros and cons. *Trends Pharm. Sci.*, 5, 1, 7–24, 2019.
60. Xue, X., Hu, Y., Deng, Y., Su, J., Recent advances in design of functional biocompatible hydrogels for bone tissue engineering. *Adv. Funct. Mater.*, 31, 19, 2009432, 2021.
61. Kurdtabar, M., Nezam, H., Rezanejade Bardajee, G., Dezfulian, M., Salimi, H., Biocompatible magnetic hydrogel nanocomposite based on carboxymethylcellulose: synthesis, cell culture property and drug delivery. *Polym. Sci. Ser. B*, 60, 2, 231–42, 2018.
62. Ying, H., Yen, J., Wang, R., Lai, Y., Hu, Y., Cheng, J., Degradable and biocompatible hydrogels bearing a hindered urea bond. *Biomater. Sci.*, 5, 12, 2398–402.
63. Ferreira, H., Amorim, D., Lima, A.C., Pirraco, R.P., Costa-Pinto, A.R., Almeida, R., Almeida, A., Reis, R.L., Pinto-Ribeiro, F., Neves, N.M., A biocompatible and injectable hydrogel to boost the efficacy of stem cells in neurodegenerative diseases treatment. *Life Sci.*, 287, 120108, 2021.
64. Nguyen, Q.V., Park, J.H., Lee, D.S., Injectable polymeric hydrogels for the delivery of therapeutic agents: A review. *Eur. Polym. J.*, 72, 602–19, 2015.
65. Koch, F., Ekat, K., Kilian, D., Hettich, T., Germershaus, O., Lang, H., Peters, K., Kreikemeyer, B., A versatile biocompatible antibiotic delivery system based on self-assembling peptides with antimicrobial and regenerative potential. *Adv. Healthcare Mater.*, 8, 13, 1900167, 2019.

66. Seol, D., Magnetta, M.J., Ramakrishnan, P.S., Kurriger, G.L., Choe, H., Jang, K., Martin, J.A., Lim, T.H., Biocompatibility and preclinical feasibility tests of a temperature sensitive hydrogel for the purpose of surgical wound pain control and cartilage repair. *J. Biomed. Mater. Res. Part B: Appl. Biomater.*, 101, 8, 1508–15, 2013.
67. Li, G. and Zhao, Z., Advanced materials for high biocompatible hydrogel system. *Coatings*, 12, 9, 1357, 2022.
68. Mardpour, S., Ghanian, M.H., Sadeghi-Abandansari, H., Mardpour, S., Nazari, A., Shekari, F., Baharvand, H., Hydrogel-mediated sustained systemic delivery of mesenchymal stem cell-derived extracellular vesicles improves hepatic regeneration in chronic liver failure. *ACS Appl. Mater. Interfaces*, 11, 41, 37421–33, 2019.
69. Carballo-Pedrares, N., Fuentes-Boquete, I., Díaz-Prado, S., Rey-Rico, A., Hydrogel-based localized nonviral gene delivery in regenerative medicine approaches—An overview. *Pharmaceutics*, 12, 8, 752, 2020.
70. Jiang, X., Huang, X., Jiang, T., Zheng, L., Zhao, J., Zhang, X., The role of Sox9 in collagen hydrogel- mediated chondrogenic differentiation of adult mesenchymal stem cells (MSCs). *Biomater. Sci.*, 6, 6, 1556–68, 2018.
71. Ni, Y., Zhao, L., Xue, X., Guoying, X., Zhao, C., Research progress of hydrogel-mediated disease therapeutics. *J. Nanopart. Res.*, 23, 4, 1–1, 2021.
72. Hunt, J.A., Chen, R., van Veen, T., Bryan, N., Hydrogels for tissue engineering and regenerative medicine. *J. Mater. Chem. B*, 2, 33, 5319–38, 2014.
73. Yuan, Z., Tsou, Y.H., Zhang, X.Q., Huang, S., Yang, Y., Gao, M., Ho, W., Zhao, Q., Ye, X., Xu, X., Injectable citrate-based hydrogel as an angiogenic biomaterial improves cardiac repair after myocardial infarction. *ACS Appl. Mater. Interfaces*, 11, 42, 38429–39, 2019.
74. Yan, C., Wang, X., Xiang, C., Wang, Y., Pu, C., Chen, L., Jiang, K., Li, Y., Applications of functionalized hydrogels in the regeneration of the intervertebral disc. *BioMed. Res. Int.*, 2021, 1–19, 2021.
75. Mathiyalagan, R., Kariyarath Valappil, A., Yang, D.C., Kang, S.C., Thambi, T., Gene regulations upon hydrogel-mediated drug delivery systems in skin cancers—An overview. *Gels*, 8, 9, 560, 2022.
76. Li, Y., Wang, X., Han, Y., Sun, H.Y., Hilborn, J., Shi, L., Click chemistry-based biopolymeric hydrogels for regenerative medicine. *Biomed. Mater.*, 16, 2, 022003, 2021.
77. Truong, N.F., Kurt, E., Tahmizyan, N., Lesher-Pérez, S.C., Chen, M., Darling, N.J., Xi, W., Segura, T., Microporous annealed particle hydrogel stiffness, void space size, and adhesion properties impact cell proliferation, cell spreading, and gene transfer. *Acta Biomater.*, 94, 160–72, 2019.
78. Li, Y., Tian, H., Chen, X., Hyaluronic acid based injectable hydrogels for localized and sustained gene delivery. *J. Control. Release: Off. J. Control. Release*, 213, e140–1, 2015.
79. Mo, F., Jiang, K., Zhao, D., Wang, Y., Song, J., Tan, W., DNA hydrogel-based gene editing and drug delivery systems. *Adv. Drug Delivery Rev.*, 168, 79–98, 2021.

80. Mo, F., Jiang, K., Zhao, D., Wang, Y., Song, J., Tan, W., DNA hydrogel-based gene editing and drug delivery systems. *Adv. Drug Delivery Rev.*, 168, 79–98, 2021.
81. Li, Z., Yue, M., Liu, Y., Zhang, P., Qing, J., Liu, H., Zhou, Y., Advances of engineered hydrogel organoids within the stem cell field: A systematic review. *Gels*, 8, 6, 379, 2022.
82. Zhang, Y., Zhu, L., Tian, J., Zhu, L., Ma, X., He, X., Huang, K., Ren, F., Xu, W., Smart and functionalized development of nucleic acid-based hydrogels: Assembly strategies, recent advances, and challenges. *Adv. Sci.*, 8, 14, 2100216, 2021.
83. Vega, S.L., Kwon, M.Y., Burdick, J.A., Recent advances in hydrogels for cartilage tissue engineering. *Eur. Cells Mater.*, 33, 59, 2017.
84. Zhang, Y.S. and Khademhosseini, A., Advances in engineering hydrogels. *Science*, 356, 6337, eaaf3627, 2017.
85. Tavakoli, J. and Tang, Y., Hydrogel based sensors for biomedical applications: An updated review. *Polymers*, 9, 8, 364, 2017.
86. Cascone, S. and Lamberti, G., Hydrogel-based commercial products for biomedical applications: A review. *Int. J. Pharm.*, 573, 118803, 2020.
87. Kesharwani, P., Bisht, A., Alexander, A., Dave, V., Sharma, S., Biomedical applications of hydrogels in drug delivery system: An update. *J. Drug Delivery Sci. Technol.*, 66, 102914, 2021.
88. Daly, A.C., Riley, L., Segura, T., Burdick, J.A., Hydrogel microparticles for biomedical applications. *Nat. Rev. Mater.*, 5, 1, 20–43, 2020.
89. Chai, Q., Jiao, Y., Yu, X., Hydrogels for biomedical applications: Their characteristics and the mechanisms behind them. *Gels*, 3, 1, 6, 2017.
90. Cao, H., Duan, L., Zhang, Y., Cao, J., Zhang, K., Current hydrogel advances in physicochemical and biological response-driven biomedical application diversity. *Signal Transduction Targeted Ther.*, 6, 1, 1–31, 2021.
91. Liu, Y. and Hsu, S.H., Synthesis and biomedical applications of self-healing hydrogels. *Front. Chem.*, 6, 449, 2018.
92. Galante, R., Pinto, T.J., Colaco, R., Serro, A.P., Sterilization of hydrogels for biomedical applications: A review. *J. Biomed. Mater. Res. Part B: Appl. Biomater.*, 106, 6, 2472–92, 2018.
93. Liu, G., Ding, Z., Yuan, Q., Xie, H., Gu, Z., Multi-layered hydrogels for biomedical applications. *Front. Chem.*, 6, 439, 2018.
94. Larrañeta, E., Henry, M., Irwin, N.J., Trotter, J., Perminova, A.A., Donnelly, R.F., Synthesis and characterization of hyaluronic acid hydrogels crosslinked using a solvent-free process for potential biomedical applications. *Carbohydr. Polym.*, 181, 1194–205, 2018.
95. Palmese, L.L., Thapa, R.K., Sullivan, M.O., Kiick, K.L., Hybrid hydrogels for biomedical applications. *Curr. Opin. Chem. Eng.*, 24, 143–57, 2019 Jun 1.
96. Du, H., Liu, W., Zhang, M., Si, C., Zhang, X., Li, B., Cellulose nanocrystals and cellulose nanofibrils based hydrogels for biomedical applications. *Carbohydr. Polym.*, 209, 130–44, 2019.

97. Kang, D.H., Kim, D., Wang, S., Song, D., Yoon, M.H., Water-insoluble, nanocrystalline, and hydrogel fibrillar scaffolds for biomedical applications. *Polym. J.*, 50, 8, 637–47, 2018.
98. Cho, H., Jammalamadaka, U., Tappa, K., Nanogels for pharmaceutical and biomedical applications and their fabrication using 3D printing technologies. *Materials*, 11, 2, 302, 2018.
99. Parhi, R., Cross-linked hydrogel for pharmaceutical applications: A review. *Adv. Pharm. Bull.*, 7, 4, 515–30, 2017.
100. Mohapatra, S., Mirza, M.A., Hilles, A.R., Zakir, F., Gomes, A.C., Ansari, M.J., Iqbal, Z., Mahmood, S., Biomedical application, patent repository, clinical trial and regulatory updates on hydrogel: An extensive review. *Gels*, 7, 4, 207, 2021.
101. Mandal, A., Clegg, J.R., Anselmo, A.C., Mitragotri, S., Hydrogels in the clinic. *Bioeng. Transl. Med.*, 5, 2, e10158, 2020.
102. Itkonen Freitas, A.M., Mentula, M., Rahkola-Soisalo, P., Tulokas, S., Mikkola, T.S., Tension-free vaginal tape surgery versus polyacrylamide hydrogel injection for primary stress urinary incontinence: A randomized clinical trial. *J. Urol.*, 203, 2, 372–8, 2020.
103. Soleymani, S., Zargaran, A., Farzaei, M.H., Iranpanah, A., Heydarpour, F., Najafi, F., Rahimi, R., The effect of a hydrogel made by Nigella sativa L. @ on acne vulgaris: A randomized double-blind clinical trial. *Phytother. Res.*, 34, 11, 3052–62, 2020.

5

Progress and Prospects for Non-Viral Gene Therapy

Shashimala Tiwari

India Health Action Trust, Uttar Pradesh Technical Support Unit, Lucknow, India

Abstract
The creation of the gene delivery vector is fundamental to the achievement of gene therapy. Recently, straightforward and secure process of gene transfection into target cells utilizing bare DNA has been developed by integrating numerous somatic approaches, such as ultrasound, electroporation, hydrodynamic pressure, and gene gun. In order to rise the usefulness and cell specificity of gene transfer, chemical methods have been used. New gene carrier molecules have been created that make it easier for DNA to outflow from endosome and enter cytoplasm. There have also been reports of a number of functional polymers that allow for the regulated release of DNA in response to a transform in the environment. To lessen the immunological response brought on by plasmid DNA, strategies such as using PCR fragments, sequential injection, and plasmids with fewer CpG patterns have been created. Building a durable gene expression system is another crucial aspect of nonviral gene therapy. The reported plasmid systems include self-replicating, tissue-specific expression, and integrating ones. The effectiveness, specificity, and temporal control of nonviral vectors have significantly improved as a result of advancements in delivery techniques and the intelligent design of DNA itself.

Keywords: Gene transfection, electroporation, hydrodynamic pressure, tissue-specific expression, gene therapy, polymer, self-replicating, immunological response

Email: shashitiwari634@gmail.com

Rishabha Malviya, Sonali Sundram and Neelam Jain (eds.) Integration of Biomaterials for Gene Therapy, (117–138) © 2023 Scrivener Publishing LLC

5.1 Introduction

The issues that arise with viral paths, such as endogenous virus recombination, carcinogenic consequences, and unanticipated immune response, should be avoided by using nonviral vectors. Nonviral vectors also benefit from use, large-scale manufacture, and the absence of a particular immune reaction.

These methods divided into two broad groups:

1. delivery of bare DNA using physical techniques like electroporation and the gene gun, and
2. distribution using a chemical carrier like cationic polymer or lipid. We cover techniques in these two areas and concentrate on the development over the previous 2 years in this review.

Naked DNA transfer by physical means: to address the concern of safety and to achieve effective *in vivo* gene expression. The mechanical techniques covered in this section are numerous. Shortest injection of bare plasmid DNA into the tissue is simplest technique of administering DNA [1]. The safest way is to use naked DNA without any carrier molecules. The assessment of the safety of complicated formations and their related difficulties require little consideration. The heart's muscle, the urological organs, Skeletal muscle, the liver, thyroid, the skin, and tumors are among the sites of direct injection so far [1]. Another practical method for delivering genes is systemic injection.

However, the manifestation level and zone after inoculation of naked DNA are often constrained. To increase effectiveness, numerous physical interventions have been achieved. Many techniques, including electroporation, hydrodynamics (high pressure injection), bio-ballistic (gene gun), ultrasound, and others, have been developed (Figure 5.1) [2]. After injecting bare DNA into cells, electroporation, which uses regulated electric fields to induce cell permeabilization, is cast-off to improve gene absorption into cells [3]. Additionally, electroporation can produce expression that lasts a long time and can be applied to different tissues.

5.2 Definition

The genetic alteration of cells to have a favorable outcome [4] or the repair or rebuilding of genetic material damaged in order to treat disease is a crucial aspect of gene therapy [5]. Although Martin Cline made the 1st attempt to change human DNA in 1980, the 1st nuclear gene transfer in

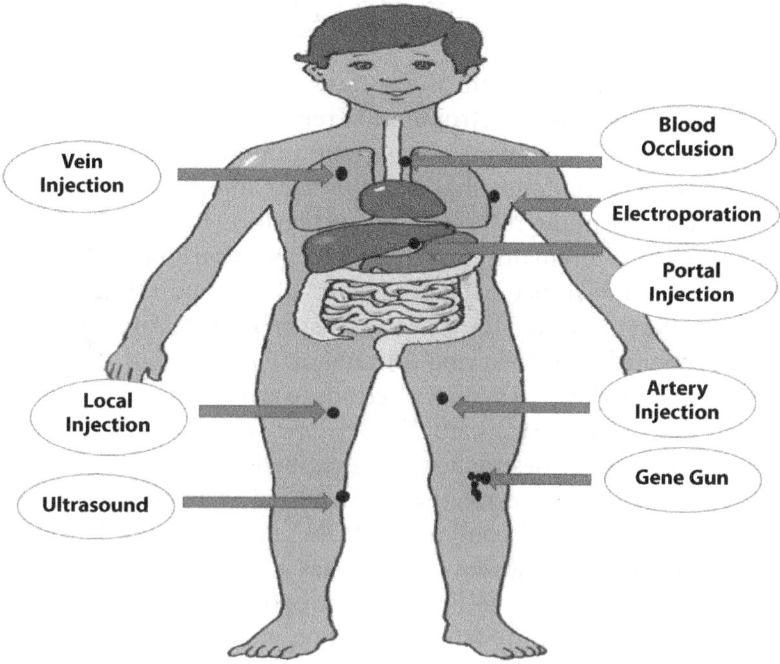

Figure 5.1 Different methods for injecting bare DNA and enhancing techniques.

humans—which was authorized by the National Institutes of Health—was accomplished in May 1989.

Goal of gene therapy is to treat a hereditary condition from the inside out. If a particular gene mutation results in construction of a protein that is malfunctioning and leads to the development of a congenital disease, for example, gene therapy used to issue a copy of the gene that does not include damaging mutation and creates a functional protein (typically recessively). It is mentioned as gene replacement therapy and is used to address inherited retinal issues [6, 7].

5.3 Technology Overview for Non-Viral Gene Delivery

The ease of administration makes the skin one of the best targets. For DNA vaccination, Drabick *et al.* [8] developed the cutaneous transfection technique. For the manifestation of hepatitis B surface antigen, erythropoietin [8] and IL-12, variables including DNA dose, electrode form, and number, duration, and electrical field strength, have been optimized on behalf of electroporation [9].

Gene expression in the skin is also favored by great ionic strength in the injection media [10]. Electroporation remains a viable option for muscle. The majority of recent papers focus on immunological applications. By electro-poring muscle following intramuscular injection of bare plasmid DNA, powerful immune responses to the hepatitis B surface antigen and the HIV gag protein remained acquired for DNA vaccination [11].

Studies have shown that cytokines like IL-12 [12] and IFN-a [13] have a therapeutic effect in slowing the growth of distant tumors. Additionally, IL-12 was electroporated with following intratumor injection [14]. Recently, our group reported employing a syringe electrode to achieve the same transfection efficiency while using a significantly less electric field than a traditional electrode [15]. Thus, tissue destruction triggered by the electric field is reduced [16]. Heller *et al.* described electrically facilitated DNA transfer to hepatocellular cancer in liver. [17] Local plasmid DNA injection is used in each electroporation technique. However, our team has just in recent times confirmed effective Gene transmission to liver by electroporation after injecting naked DNA into the tail vein. [18] Systemic injection provides benefit of delivering genes to the liver extra similarly when compared to local DNA injection.

Straight gene delivery keen on tissues or cells is possible with a gene cannon. Bypassing the endosomal compartment, gold particles coated with DNA can be shot directly past cell membrane and into cytoplasm and even nucleus. For vaccination and immunotherapy, majority of published efforts over the past two years have focused on introducing antigen or cytokine genes into the skin [19, 20] or liver [21–23]. The modest penetration of DNA into the tissue by this approach is a drawback, though.

Cell membrane's permeability to huge molecules like plasmid DNA can be increased by ultrasound. In fact, after injecting DNA into the tissue, an increase in gene expression was seen after applying an ultrasonic wave to the area [24, 25].

The flexibility and safety of ultrasonic application make it a great tool for therapeutic gene transfer. It was recently revealed that using microbubbles in addition to ultrasound could boost gene expression even more. The threshold for cavitation by ultrasonic energy is lowered by microbubbles, also known as ultrasound contrast agents. The most common kind of microbubbles employed were Optison (Mallinckrodt, San Diego, USA) or albumin microbubbles filled with perfluoropropane. Before being injected, plasmid DNA was added, and then ultrasound was used to irradiate it. Currently, this method is employed to transport genes to mice vascular cells [26–30] muscle [30, 31], and foetal tissue [32].

A powerful gene transfer can be induced in internal organs, particularly the liver, by hydrodynamic injection, which is fast injection of a huge amount of naked DNA solution. Hepatocytes take up bare plasmid DNA via a receptor-mediated mechanism, according to a theory put forth by Budker *et al.* [33] A few DNA receptors have been discovered in a variety of soft tissue, but their purpose is still unknown. Given that the liver has an elastic structure and can hold a considerable volume of solution, it has been suggested that the inoculated DNA solution collects primarily [34].

Additionally, it has been suggested that the main mechanism for the extremely effective expression in the liver is the pressure-induced breaching of the endothelium barrier. Our team recently reported that external abdominal massage, followed by small-volume DNA injection through the tail vein, can improve liver gene expression. [35] The finding implies that the endothelium barrier may be affected by mechanical stretching with regard to DNA uptake by hepatocytes. Other tissues can be transfected using this pressure-mediated technique. Wolff's team demonstrated that fast-speed, large-volume injections into the arteries of the leg muscles or the portal vein of the liver resulted in elevated gene expression in corresponding organ [33, 36].

Considerable Gene expression in liver can be attained through momentarily limiting blood flow across the organ just after exterior intravenous injection of bare DNA. [37] The manifestation level in the liver was elevated when blood supply was cut off at vena cava, portal vein, or hepatic artery. According to Budker *et al.*'s [33] postulated receptor-mediated mechanism or a nonreceptor-mediated pathway, it is conceivable that give a shot DNA is internalized into hepatic cells. However, DNA binding to the cover of hepatic cells maybe therefore poor that, under normal physiological conditions, DNA may be easily disconnected and washed away by blood flow. The DNA can only steadily interact with receptor and enter cells when the blood flow is momentarily interrupted [37].

Shortly after a peripheral intravenous DNA injection, blood flow through the diaphragm was briefly blocked to induce a comparable acceptance of DNA through the diaphragm muscle cells [38].

5.4 Chemical Carriers for Gene Transfer: Establishing Effective *In Vivo* Gene Delivery

Novel carriers have been established to deliver functionally and obtain high stages of Gene expression. Gene carriers divided into five groups: In order to attain constant or controlled expression, DNA can be released into the tissue in four different ways:

1. those intended to issue DNA into tissue,
2. those intended to promote transport of DNA to cytosol,
3. those intended to separate from DNA in cytosol, and
4. those intended to target delivery to particular cell types. Gene delivery mostly employs lipids and polymers.

5.5 Types of Gene Delivery

5.5.1 Lipid-Mediated Gene Delivery

Most used procedures for delivering genes into cells is liposome-based gene transport, which was first defined by Felgner in 1987. Many cationic lipids, including quaternary ammonium detergents, lipid products of polyamines, and diacylglycerol and derivatives of cholesterol, were described in the 1990s. To improve efficacy through the aforementioned modifications and for particular *in vivo* applications, as the progress of innovative forms of lipid molecules look as if to be drenched [39]. We'll discuss a few fresh ideas that have emerged in the previous two years.

Thiol groups' reduction-oxidation (redox) sensitivity has been used to manipulate the synthesis of DNA-lipid complexes. A dimerizable cationic cleanser with free thiol, alkyl groups, and amine was reported by Dauty *et al.* [39]. This alkylated ornithinyl cysteine derivative combines by plasmid DNA to generate a complex. The complex is changed into stable nanometric particles through further thiol group oxidation to disulfides. The particle contains a single condensed plasmid DNA molecule that consistently has a diameter of <40 nm and displayed satisfactory transfection activity *in vitro.*

5.5.2 Peptide-Mediated Gene Delivery

Additionally, peptide gene carriers have been constructed using redox-sensitive thiols. McKenzie *et al.* [40] created peptides like Cys-Trp-Lys18 that include a cysteine residue and an uninterrupted series of lysine residues. [40] In addition to condensing plasmid DNA, this peptide also spontaneously oxidizes the thiol group, creating a very stable complex with strong *in vitro* transfection activity. Without boosting DNA uptake by the cells, cross-linking peptide increased gene expression, which suggests that intracellular DNA release brought on through disulfide bond reduction was important [41].

5.5.3 Polymer-Mediated Gene Transfer

Branching and linear PEI/DNA complexes' capacity to transfer cells *in vitro* and *in vivo* at many amine/phosphates ratios and salt concentrations was comprehensively examined by Wightman *et al.* [42]. They demonstrated that linear PEI (22 kDa) salt-free DNA developments, which exhibited great transfection effectiveness in lung, were initially small but soon matured larger when salt was added. In compare, even when salt was introduced, the DNA compound of branching PEI (25 kDa), which had poor transfection efficacy in the majority of organs, stayed tiny. The dynamic structural transformation of complex at high salt concentrations like those observed in blood may account for higher competency of linear PEI *in vivo*. Future vector design may benefit from knowledge of relationship between linear PEI and DNA [42].

Biodegradable polymers have high biocompatibility and low toxicity as two of its characteristics [43]. Poly[a-(4-aminobutyl)-L-glycolic acid] (PAGA), a poly-L-lysine derivative in which the ester link is changed to an amide, is a biodegradable polymer created by Kim's team [44]. DNA is condensed before being released following hydrolysis by this biodegradable and water-soluble polymer. Combination presented improved *in vitro* gene transfection productivity and reduced cytotoxicity when compared to poly-L-lysine. Protein expression was evident to a great extent, and tumor growth was constrained [45].

Kim's and Leong's groups, who made cationic polyphosphoester [46] and cationic polyethylene glycol (PEG) from PEI and PEG, respectively, have just published new types of biodegradable polymers. The release of encapsulated DNA can be controlled using thermosensitive polymers in reaction to temperature changes that cause the hydrated polymer to swell or de-swell [47].

Temperature also affected how well *in vitro* transmission worked. The thermosensitive and biodegradable PEG-poly (D,L-lactic acid-co-glycolic acid)-PEG triblock co-polymer was created by Kim's team [48]. Plasmid DNA can be added to this non-ionic, hydrophilic polymer in the aqueous phase at 4-201C, and it displays temperature-dependent solution-to-gel transitions [49, 50]. The transition from a clarification to a gel takes place at temperatures higher than 30-31C, such as body temperature.

These nuclease-resistant, water-soluble polyion complex micelles work well as *in vivo* gene delivery systems. Therefore, despite fact that gene expression following injection into a mouse tail vein remained only detectable in the liver, DNA in the complex stayed together in blood stream for 30 minutes [51].

5.6 Reduction of Immunological Responses Through Alteration of Delivery Method or DNA Structure

Even though it is generally established that virus-mediated gene delivery results in more severe immune reactions than nonviral gene delivery, issues still exist. Immune cells such as macrophages and dendritic cells are able to identify the DNA/gene complex. When cationic liposomes are present, proinflammatory cytokines are rapidly induced. [52] This reaction results from the immune cells being stirred by plasmid DNA's unmethylated CpG motifs. Usage of PCR fragments by fewer CpG patterns [53] and active aiming of DNA to endothelium, which reduces collaboration through immune cells, are a few strategies that have been used to limit this inflammatory [54].

Additionally, Liu's group was the first to discover that consecutive injection of cationic liposomes monitored through naked plasmid DNA lowers inflammatory response [55]. Thus, 50–80% less proinflammatory cytokines (relative to lipoplexes) were detected when mouse tail veins were injected with Plasmid DNA 2–5 min next injection of cationic liposome, without changing lung gene expression levels.

5.7 To Enable Long-Lasting Gene Expression, Self-Replicating, Tissue-Specific, and Integrating Plasmid Expression Systems are Designed

In place of nonviral gene therapy, achieving persistent gene expression is also a crucial goal. Stable expression can be achieved using tissue-specific expression methods because they reduce the likelihood of the transgene provoking an immune response. When plasmid DNA hydrodynamically injected into mouse liver, it boosted factor IX gene expression (within the therapeutic range) and sustained these stages for at least 10 months [56].

In cells, the plasmid vector based on Epstein-Barr virus (EBV) is recognized to self-replicate. EBV nuclear antigen 1 (EBNA1) gene and oriP element are 2 genetic components from EBV that are present in it. By attaching to oriP, EBNA1 protein encourages plasmid replication in time with chromosomal DNA [57]. The EBNA1 similarly helps the plasmid DNA get localized in the nucleus. In cardiomyocytes [58], tumor suicide therapy [59] (linked to a polyamidoamine dendrimer), the β2-adrenergic receptor has been found to be long-term expressed [60].

5.8 Hybrid Vector Systems to Improve Transfection and Lessen Cytotoxicity

At the cutting edge of gene delivery, non-viral transmitters with effective gene transfection and low cytotoxicity proven to be a two-edged sword. A few of current initiatives to increase transfection efficiency while reducing cytotoxicity are peptide-lipid transmitters, inorganic-lipid transmitters, modified PEI transmitters, and Inorganic-hybrid transmitters [61].

5.8.1 Inorganic-Organic Hybrid Transmitters

Inorganic-organic hybrid transmitters, with moreover a platform or tailored gene delivery, to different cell types, have been discovered as a growing development in the new session of non-viral transmitters. Mannitol-group was one effort to be incorporated into the vector. The motivation for Ma *et al.*'s research was the use of the multi-hydroxyl molecule mannitol as an osmolyte in the clinic in 2020 [62]. For the 1st time, they developed mannitol-based CaP-MA organic-inorganic non-viral transmitters that organized cellular uptake and subsequent intracellular transferring for gene delivery.

By transferring genetic consignment through a non-destructive method and simulating caveolae-mediated cellular updating, the novel vectors with mannitol groups could prevent Gene degradation in lysosome. These vectors thereby demonstrated to have improved transfection, biocompatibility, and toxicity over unmodified CaP nanoparticles [63].

Utilizing amine-free vectors to combat toxicity caused by amines was another method of reducing cytotoxicity [63].

5.8.2 Improved PEI Vectors

Since PEI vectors feature non-degradable amide bonds that cause buildup and impairment of metabolic function in healthy cells, cytotoxicity is the biggest problem they face. Furthermore, it is well known that molecular weight of the polymer has a positive correlation with the degree of cytotoxicity [64]. Low molecular weight PEI (CD-PEI800) self-assembled nanoparticles (SNPs) grafted with cyclodextrin displayed good transfection effectiveness to Jurkat cells and low cytotoxicity, according to Yu *et al.* [65]. They hypothesized that the increased gene encapsulation effectiveness of SNPs was due in part to cationic hydrogel produced from CD-PEI800 [66].

Vector elements with DNA discovered to be lesser than those made with PEI that had not been changed, which altered cellular absorption favorably. The TAC cross-linking raised the polymer's cationic charge, enabling mandatory to the genetic material and encouraging endosomal egress, both of which improved total transfection effectiveness. The toxicity effects were further mitigated by the reduction in the amount of polymer added to the cell [67].

Combining arginine-modified PEI polymers (AnPn) with poly-lactic-co-glycolic acid (PLGA) nanoparticles is alternative method for maintaining PEI's biocompatibility and reducing its toxicity [68]. The inclusion of AnPn greatly increased the successful gene expression and nuclear localization of pDNA in primary human astrocytes.

5.8.3 Peptide-Lipid Vectors or Inorganic-Lipid

The low transfection efficiency of a lipoplex presents one difficulty. To boost polycation and improve transfection, Lebrón *et al.* created Cationic metallo-liposomes containing Ruthenium (Ru) and phospholipid complexes [69]. They also found that the hydrophobicity and length of the Ru-lipid had an effect on transfection. The vectors' successful transfection rate suggested that they might be used as nano-vectors in gene treatment. Their dimension and surface charge also affected compaction utilizing various DNA to lipid mass ratios, further avoiding the destruction of the nucleic acid. It's interesting to note that the Ru-lipid vectors showed fewer cytotoxicity toward cancer cells than toward normal cells, indicating potential for targeted gene therapy [69].

Another type of encouraging non-viral vectors for delivering nucleic acids into cells is peptide-based cationic lipids [70]. In order to introduce green fluorescent protein (GFP) into Hela and Hep-2 cells, both lipoplexes were employed as vectors. Although the chemical structures of the two peptide-lipid vectors are comparable, LOrn3 has more amino groups, leading to cellular internalization and a better proton buffering capability. As a result, LOrn3 was more effective in transfecting cells than LOrn1 and boosted endosome escape and DNA release. The study provides theoretical guidance for future peptide-lipid vector advancement research paths [70].

5.8.4 Effects of *In Vitro* and *In Vivo* Performance on Characteristics of Gene Vectors and Formulation Parameters

Once the therapeutic gene has connected to cell membrane, entered cell, and fled from endosomes into the cytoplasm, it must then be released from the vehicle substance. If delivered, the pDNA must also enter the nucleus through cytoplasm. The creation of a successful gene delivery system needs an understanding of how the properties of gene carriers and delivery systems affect the efficiency of gene transfection. Method for delivering non-viral gene is shown in Figure 5.2.

5.9 Vehicle Material

Important factors for effective gene delivery include organic structure and characteristics of the vehicle materials. In order to comprehend the impacts of molecular structures and to make strategy of gene vectors easier, interest has been aroused in establishing the structure activity relationship (SAR) [71]. Investigations into SAR for lipid and polymer materials have been conducted. For instance, when short-chain aminolipids were added to

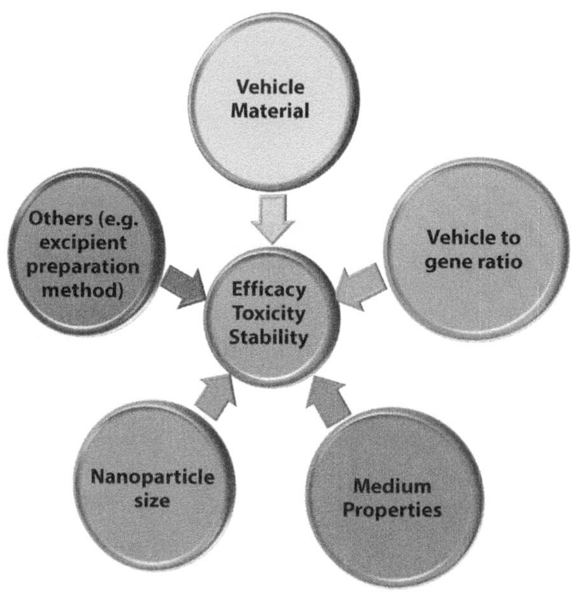

Figure 5.2 Method of delivering non-viral gene vectors.

DOTAP:chol-based lipoplexes, Buck *et al.* discovered that the longer lipid tails (C12) enhanced minicircle DNA transfection efficiency as compared to the shorter ones (C10) [71]. Different vehicle material is summarized in Figure 5.3.

In order to create a chain of cyclodextrin-threaded polyrotaxanes, different molecular weights of polyethylene glycols (PEG) were used [72]. Because research is ongoing, there isn't yet a set global standard for polymer-based vectors [73].

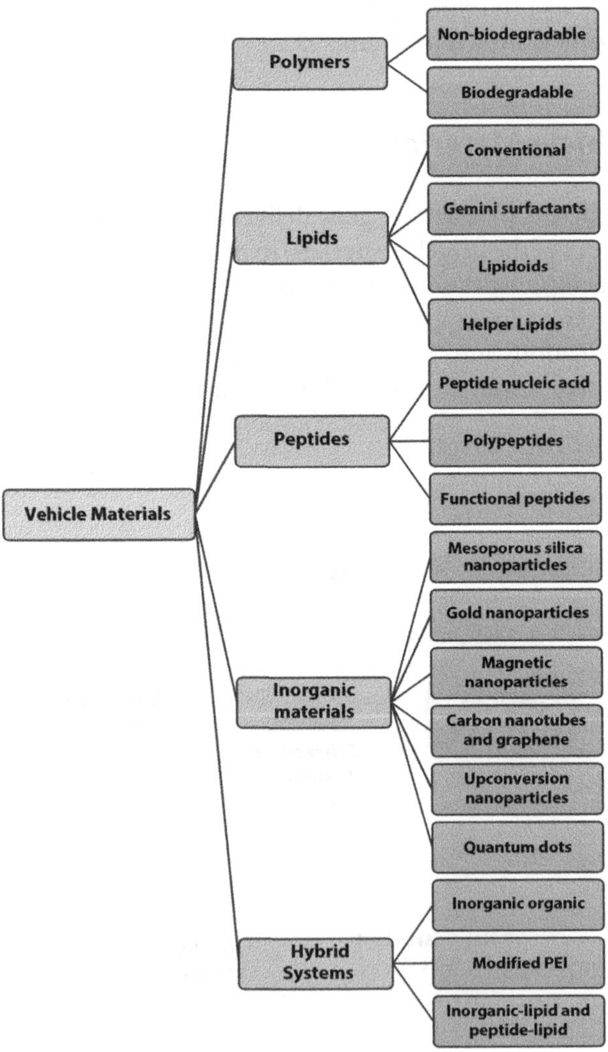

Figure 5.3 Vehicle materials.

5.9.1 The Proportion of Vehicle Components to Genes

Since it has potential to change the complexation potential and physicochemical characteristics of vehicle-gene complexes, the proportion of vehicle material to gene is critical to transfection effectiveness. Systems through N/P between 5 and 10 ensured lesser particle sizes, added positively charged planes, and greater pDNA encapsulation capacities [74], when tested using HeLa cells, resulted in better cell internalization, higher levels of p53 protein expression, and a significant amount of apoptosis. To transport CFTR gene for treatment of cystic fibrosis, Sainz-Ramos *et al.* created nioplex formulations based on 2,3-di(tetradecyloxy)propan-1-amine (DTPA) with 3 distinct lipids to DNA ratios [75].

5.9.2 Nanoparticle Size

Dimensions of the nanoparticles is a crucial variable that impacts each phase of the delivery process. The preferred size for lengthy blood circulation times is typically greater, around 100 nm, however a relatively smaller vector can penetrate cells more effectively. Renal filtration can remove particles smaller than 5 nm, while spleen and liver filtration can remove particles larger than 200 nm [76]. The pace and method of cellular absorption can also be impacted by particle size. However, because the size effect is so strongly correlated with the type of vector and cellular absorption pathway, bio-distribution, cytotoxicity, and cellular uptake of commercially presented gold nanoparticles by means of a size range of 5–50 nm were compared by scientist.

The HepG2 cancer cells' cellular absorption was found to rise with particle size, but to decrease with L02 normal cells. The most dangerous gold nanoparticles were 5 nm in size, while 50 nm nanoparticles had the longest blood circulation [77]. Additionally, Haddick *et al.* created a variety of 60–160 nm mesoporous silica nanoparticles. They observed that the 160 nm nanoparticles had the quickest cellular internalization and highest knockdown efficacy [78]. It was proposed that bigger particles might have more surface area of interaction with cell membranes, increasing endocytosis.

5.10 Further Effects

It is predicted that the formulation performance will be impacted by excipients utilized in a drug delivery system. Keil *et al.* created gene therapy nano-embedded microparticle (NEM) powder for the treatment of localized pulmonary disease [79]. Mannitol was found to increase the polyplex's particle size following reconstitution and create dried particles with noticeably larger aerodynamic diameters. Additionally, mannitol-containing dry powders showed limited solubility when compared to trehalose, making it impossible to assess its biological effects. According to Wickline *et al.*, hyaluronic acid unpredictably provided an additional consistent nanoparticle size dissemination deprived of aggregation to a cationic amphipathic peptide and RNA formulation [80].

The new hyaluronic acid nanoparticle formulations were 10-fold more potent than conventional nanoparticle formulations in inhibiting abdominal aortic aneurysms. In light of the evidence as a whole, choosing formulation compositions may be another factor to look into in order to enhance gene delivery systems' stability and transfection efficiency [79]. The method used to manufacture the nanoparticles can affect the qualities of the formulation in addition to the composition. The PEI-pDNA polyplexes were made using two different methods, according to Pezzoli *et al.* [81]. One involves adding pDNA to polymer solution one drop at a time in the "dropping" mode, after which complex solution is left standing. Another is the mixing mode, in which pDNA was vigorously pipetted up and down to combine with the polymer solution. The study showed that the larger nanoparticles produced by the DROPPING mode led to a noticeably better transfection efficiency [81].

5.11 Challenges and Prospects

Gene therapy has made strides in recent years and is presently being used to treat a variety of therapeutic illnesses. The systematic trials of providing Gene therapy using a non-viral vector were thoroughly examined by Yin and Anderson *et al.* in 2014 [82]. To permit transit to nucleus, nuclear uptake, and vector unpacking, a non-viral vector must have the capability to defend genetic material from endonuclease degradation. However, this typically requires a customized strategy. Additionally, delivery strategies utilizing the various therapeutic modalities' mechanisms of action should be developed. The substantially larger particle size adds a further barrier to efficacy since DNA must be distributed into nucleus in order to have a therapeutic outcome [82]. Additionally, different modalities have varied interactions between cells and vectors, which has a significant impact on transfection efficiency. It has not yet been possible to choose a non-viral vector delivery strategy using a 1 size fits all approach. The vector systems can also be fairly complex.

For instance, the Pfizer/BioNTech vaccine required the use of four lipids: distearoylphosphatidylcholine (DSPC), a phospholipid, cholesterol, and an ionizable cationic lipid to assist construct the structure of lipid nanoparticles and encapsulate negatively charged mRNA [83]. It takes a lot of work to screen a successful vector because vector formulations are so complex. A siRNA medication called patsiran is said to have required the screening of more than 300 ionizable lipids [84]. GuidTx recently created a high through put technique to test lipid nanoparticle structures, which might increase effectiveness of using non-viral vectors for increased precision in gene delivery [85].

Consider the multiple research opportunities provided by changes in cell membrane electrical charge between many types of targeted cells, which can be used to study and transform non-viral vectors for a given genetic payload to attain needed efficiency in a particular cell type. Developers should take toxicity into account. It's vital to keep in mind that although while non-viral gene vectors are less immunotoxic than viral gene vectors, the nanoparticles still have the potential to elicit immunological reactions [86].

For instance, it was discovered that chitosan causes the IL-1 reaction in human peripheral blood mononuclear cells, peritoneal macrophages, and mice dendritic cells [87]. Pro-inflammatory cytokine release increased as a result of the presence of silica nanoparticles. They have also been demonstrated to stimulate macrophages to release IL-1 [88]. The cells and viral materials affect immunotoxicity. It must therefore be assessed on a case-by-case basis.

It is not unexpected that the majority of latest study papers on novel non-viral vectors have concentrated on enhancement of vector presentation traits, including as transfection, magnitude, cytotoxicity, surface charge, and inter- and intracellular trafficking. Lesser articles, however, discussed product development-related studies, particularly when it came to a vector's physical and chemical steadiness during the creation of a formulation and product storage [88].

It would be worthwhile, for instance, to look into how the ionic strength and pH of formulation buffer affect steadiness of non-viral vectors because those variables may affect how effectively the vector is transfected or modify its potential therapeutic applications. Additionally, gene therapy formulations are frequently stored using refrigeration or freezing. As a result, freeze-thaw stress and the effects of elevated temperatures prior to delivery may reduce the efficacy of a gene therapy by, for example, causing alterations in nanoparticle size distributions. Tools available for research in this field are limited because to the lack of well-established stability-indicating methodologies for gene therapy product characterization [89]. Non-viral gene delivery technologies also add to difficulty of transporting medicine from the lab to the patient's bedside. In order to guarantee product quality and expedite product development, the proper critical quality attributes and essential process parameters must be determined [90, 91].

Many strategies are in the process of emerging. Recently, Chen *et al.* developed a hydrophobic prodrug-based lipid-based nanoparticle gene delivery technique that delivered simple means of avoiding immune-stimulatory effects of systemic administration of genetic therapeutic formulations [92]. Properties and implications of formulations for non-viral vector gene therapy on experimental presentations still have a great deal of room for investigation.

Sigilon therapeutics only used a non-viral vector in one phase I/II experiment to transfer genetically modified cell line [93]. Numerous pre-clinical studies and articles on other non-viral vectors have been published recently, which undeniably demonstrated how quickly the field of gene delivery is expanding [94, 95]. In spite of numerous obstacles, it's significant that both were created in such a short amount of time. These recent accomplishments have raised anticipation and hope for the future discovery of other secure non-viral vector-based gene therapy solutions [96].

5.12 Conclusion

The solution lies on greater research, as is the case in many scientific fields. We need to learn more about how non-viral vectors can improve upon the already well-established viral vectors that are currently used often if we want to see viral vectors replaced by them for widespread usage in delivering gene treatments.

Both viral and non-viral vectors have properties that make them effective for delivering vectors. If the problems with ineffective transfection in non-viral vectors are solved, then the advantages they offer in terms of scalability, lower production costs, and higher flexibility could provide a preferable substitute to well-liked viral vectors.

References

1. Nishikawa, M. and Huang, L., Nonviral vectors in the new millennium: Delivery barriers in gene transfer. *Hum. Gene Ther.*, 12, 861–870, 2001.
2. Li, S. and Ma, Z., Nonviral gene therapy. *Curr. Gene Ther.*, 1, 201–226, 2001.

3. Somiari, S. *et al.*, Theory and *in vivo* application of electroporative gene delivery. *Mol. Ther.*, 2, 178–187, 2000.
4. Kaji, E.H., Gene and stem cell therapies. *JAMA*, *285*, 5, 545–550, 2001.
5. Ermak, G., *Emerging medical technologies*, World Scientific, Singapore, 2015.
6. Maguire, A.M., Simonelli, F., Pierce, E.A., Pugh, E.N., Jr, Mingozzi, F. *et al.*, Safety and efficacy of gene transfer for Leber's congenital amaurosis. *N. Engl. J. Med.*, 358, 21, 2240–8, 2008.
7. MacLaren, R.E., Groppe, M., Barnard, A.R., Cottriall, C.L. *et al.*, Retinal gene therapy in patients with choroideremia: Initial findings from a phase 1/2 clinical trial. *Lancet*, 383, 9923, 1129–37, 2014.
8. Drabick, J.J. *et al.*, Cutaneous transfection and immune responses to intradermal nucleic acid vaccination are significantly enhanced by *in vivo* electropermeabilization. *Mol. Ther.*, 3, 249–255, 2001.
9. Maruyama, H. *et al.*, Skin-targeted gene transfer using *in vivo* electroporation. *Gene Ther.*, 8, 1808–1812, 2001.
10. Dujardin, N., Van Der Smissen, P., Preat, V., Topical gene transfers into rat skin using electroporation. *Pharm. Res.*, 18, 61–66, 2001.
11. Chesnoy, S. and Huang, L., Enhanced cutaneous gene delivery following intradermal injection of naked DNA in a high ionic strength solution. *Mol. Ther.*, 5, 57–62, 2002.
12. Widera, G. *et al.*, DNA vaccine delivery and immunogenicity by electroporation *in vivo*. *J. Immunol.*, 164, 4635–4640, 2000.
13. Hanna, E. *et al.*, Intramuscular electroporation delivery of IL-12 gene for treatment of squamous cell carcinoma located at distant site. *Cancer Gene Ther.*, 8, 151–157, 2001.
14. Li, S. *et al.*, Intramuscular electroporation delivery of IFN-a gene therapy for inhibition of tumor growth located at a distant site. *Gene Ther.*, 8, 400–407, 2001.
15. Tamura, T. *et al.*, Intratumoral delivery of interleukin 12 expression plasmids with *in vivo* electroporation is effective for colon and renal cancer. *Hum. Gene Ther.*, 12, 1265–1276, 2001.
16. Liu, F. and Huang, L., A syringe electrode device for simultaneous injection of DNA and electro transfer. *Mol. Ther.*, 5, 323–328, 2002.
17. Heller, L. *et al.*, Electrically mediated plasmid DNA delivery to hepatocellular carcinomas *in vivo*. *Gene Ther.*, 7, 826–829, 2000.
18. Liu, F. and Hunag, L., Electric gene transfer to the liver following systemic administration of plasmid DNA. *Gene Ther.*, *8*, 1531–1537, 1998.
19. Lin, M.T., Pulkkinen, L., Uitto, J., Yoon, K., The gene gun: Current application in cutaneous gene therapy. *Int. J. Dermatol.*, *39*, 161–170, 2000.
20. Davidson, J.M., Krieg, T., Eming, S.A., Particle-mediated gene therapy of wounds. *Wound Repair Regen.*, *8*, 452–459, 2000.
21. Muangmoonchai, R. *et al.*, Transfection of liver *in vivo* by biolistic particle delivery: Its use in the investigation of cytochrome P450 gene regulation. *Mol. Biotechnol.*, *20*, 145–151, 2002.

22. Kuriyama, S. *et al.*, Particle-mediated gene transfer into murine livers using a newly developed gene gun. *Gene Ther.*, *7*, 1132–1136, 2000.
23. Yoshida, S. *et al.*, Direct immunization of malaria DNA vaccine into the liver by gene gun protects against lethal challenge of Plasmodium berghei sporozoite. *Biochem. Biophys. Res. Commun.*, *271*, 107–115, 2000.
24. Newman, C.M., Lawrie, A., Brisken, A.F., Cumberland, D.C., Ultra-sound gene therapy: On the road from concept to reality. *Echocardiography*, *18*, 339–347, 2001.
25. Amabile, P.G. *et al.*, High-efficiency endovascular gene delivery via therapeutic ultrasound. *J. Am. Coll. Cardiol.*, *37*, 1975–1980, 2001.
26. Taniyama, Y. *et al.*, Local delivery of plasmid DNA into rat carotid artery using ultrasound. *Circulation*, *105*, 1233–1239, 2002.
27. Teupe, C. *et al.*, Vascular gene transfer of phosphomimetic endothelial nitric oxide synthase (S1177D) using ultrasound-enhanced destruction of plasmid-loaded microbubbles improves vasoreactivity. *Circulation*, *105*, 1104–1109, 2002.
28. Unger, E.C., Hersh, E., Vannan, M., McCreery, T., Gene delivery using ultrasound contrast agents. *Echocardiography*, *18*, 355–361, 2001.
29. Lawrie, A. *et al.*, Microbubble-enhanced ultrasound for vascular gene delivery. *Gene Ther.*, *7*, 2023–2027, 2000.
30. Song, J. *et al.*, Influence of injection site, microvascular pressure and ultrasound variables on microbubble-mediated delivery of microspheres to muscle. *J. Am. Coll. Cardiol.*, *39*, 726–731, 2002.
31. Shohet, R.V. *et al.*, Echocardiographic destruction of albumin microbubbles directs gene delivery to the myocardium. *Circulation*, *101*, 2554–2556, 2000.
32. Endoh, M. *et al.*, Fetal gene transfer by intrauterine injection with microbubble-enhanced ultrasound. *Mol. Ther.*, *5*, 501–508, 2002.
33. Budker, V. *et al.*, Hypothesis: Naked plasmid DNA is taken up by cells *in vivo* by a receptor-mediated process. *J. Gene Med.*, *2*, 76–88, 2000.
34. Siess, D.C. *et al.*, A human gene coding for a membrane-associated nucleic acid-binding protein. *J. Biol. Chem.*, *275*, 33655–33662, 2000.
35. Liu, F. and Huang, L., Noninvasive gene delivery to the liver by mechanical massage. *Hepatology*, *35*, 1314–1319, 2002.
36. Zhang, G. *et al.*, Efficient expression of naked DNA delivered intraarterially to limb muscles of nonhuman primates. *Hum. Gene Ther.*, *12*, 427–438, 2001.
37. Liu, F. and Huang, L., Improving plasmid DNA-mediated liver gene transfer by prolonging its retention in the hepatic vasculature. *J. Gene Med.*, *3*, 569–576, 2001.
38. Liu, F., Nishikawa, M., Clemens, P.R., Huang, L., Transfer of full- length Dmd to the diaphragm muscle of Dmd(mdx/mdx) mice through systemic administration of plasmid DNA. *Mol. Ther.*, *4*, 45–51, 2001.
39. Dauty, E., Remy, J.S., Blessing, T., Behr, J.P., Dimerizable cationic detergents with a low cmc condense plasmid DNA into nanometric particles and transfect cells in culture. *J. Am. Chem. Soc.*, *123*, 9227–9234, 2001.

40. McKenzie, D.L., Kwok, K.Y., Rice, K.G., A potent new class of reductively activated peptide gene delivery agents. *J. Biol. Chem.*, *275*, 9970–9977, 2000.
41. Park, Y., Kwok, K.Y., Boukarim, C., Rice, K.G., Synthesis of sulfhydryl cross-linking poly(ethylene glycol)-peptides and glyco- peptides as carriers for gene delivery. *Bioconjug. Chem.*, *13*, 232–239, 2002.
42. Wightman, L. *et al.*, Different behavior of branched and linear polyethylenimine for gene delivery *in vitro* and *in vivo*. *J. Gene Med.*, *3*, 362–372, 2001.
43. Lim, Y.B. *et al.*, Biodegradable polyester, poly[a-(4-aminobutyl)-L-glycolic acid], as a non-toxic gene carrier. *Pharm. Res.*, *17*, 811–816, 2000.
44. Koh, J.J. *et al.*, Degradable polymeric carrier for the delivery of IL-10 plasmid DNA to prevent autoimmune insulitis of NOD mice. *Gene Ther.*, *7*, 2099–2104, 2000.
45. Maheshwari, A. *et al.*, Soluble biodegradable polymer-based cytokine gene delivery for cancer treatment. *Mol. Ther.*, *2*, 121–130, 2000.
46. Ahn, C.H., Chae, S.Y., Bae, Y.H., Kim, S.W., Biodegradable poly (-ethylenimine) for plasmid DNA delivery. *J. Control. Release*, *80*, 273–282, 2002.
47. Wang, J., Mao, H.Q., Leong, K.W., A novel biodegradable gene carrier based on polyphosphoester. *J. Am. Chem. Soc.*, *123*, 9480–9581, 2001.
48. Kurisawa, M., Yokoyama, M., Okano, T., Gene expression control by temperature with thermo-responsive polymeric gene carriers. *J. Control. Release*, *69*, 127–137, 2000.
49. Jeong, B., Kim, S.W., Bae, Y.H., Thermosensitive sol-gel reversible hydrogels. *Adv. Drug Deliv. Rev.*, *54*, 37–51, 2002.
50. Jeong, B., Bae, Y.H., Kim, S.W., Drug release from biodegradable injectable thermosensitive hydrogel of PEG-PLGA-PEG triblock copolymers. *J. Control. Release*, *63*, 155–163, 2000.
51. Harada-Shiba, M. *et al.*, Polyion complex micelles as vectors in gene therapy – pharmacokinetics and *in vivo* gene transfer. *Gene Ther.*, *9*, 407–414, 2002.
52. Tousignant, J.D. *et al.*, Comprehensive analysis of the acute toxicities induced by systemic administration of cationic lipid:Plasmid DNA complexes in mice. *Hum. Gene Ther.*, *11*, 2493–2513, 2000.
53. Yew, N.S. *et al.*, Reduced inflammatory response to plasmid DNA vectors by elimination and inhibition of immunostimulatory CpG motifs. *Mol. Ther.*, *1*, 255–262, 2000.
54. Li, S. *et al.*, Targeted gene delivery to pulmonary endothelium by anti-PECAM antibody. *Am. J. Physiol. Lung Cell Mol. Physiol.*, *278L*, 504–511, 2000.
55. Tan, Y. *et al.*, Sequential injection of cationic liposome and plasmid DNA effectively transfects the lung with minimal inflammatory toxicity. *Mol. Ther.*, *3*, 673–682, 2001.
56. Miao, C.H. *et al.*, Inclusion of the hepatic locus control region, an intron, and untranslated region increases and stabilizes hepatic factor IX gene expression *in vivo* but not *in vitro*. *Mol. Ther.*, *1*, 522–532, 2000.
57. Chen, Z.Y. *et al.*, Linear DNAs concatemerize *in vivo* and result in sustained transgene expression in mouse liver. *Mol. Ther.*, *3*, 403–410, 2001.

58. Maruyama-Tabata, H. *et al.*, Effective suicide gene therapy *in vivo* by EBV-based plasmid vector coupled with polyamidoamine dendrimer. *Gene Ther.*, *7*, 53–60, 2000.
59. Tomiyasu, K. *et al.*, Direct intra-cardiomuscular transfer of beta2-adrenergic receptor gene augments cardiac output in cardio- myopathic hamsters. *Gene Ther.*, *7*, 2087–2093, 2000.
60. Cui, F.D. *et al.*, Highly efficient gene transfer into murine liver achieved by intravenous administration of naked Epstein–Barr virus (EBV)-based plasmid vectors. *Gene Ther.*, *8*, 1508–1513, 2001.
61. Stoll, S.M. *et al.*, Epstein–Barr virus/human vector provides high-level, long-term expression of a1-antitrypsin in mice. *Mol. Ther.*, *4*, 122–129, 2001.
62. Ma, X.X., Xu, J.L., Jia, Y.Y., Zhang, Y.X., Wang, W., Li, C., He, W., Zhou, S.Y., Zhang, B.L., Enhance transgene responses through improving cellular uptake and intracellular trafficking by bio-inspired non-viral vectors. *J. Nanobiotechnol.*, *18*, 1, 26, 2020.
63. Choi, E., Lee, J., Kwon, I.C., Lim, D.K., Kim, S., Cumulative directional calcium gluing between phosphate and silicate: A facile, robust and biocompatible strategy for siRNA delivery by amine-free non-positive vector. *Biomaterials*, *209*, 126–137, 2019.
64. Chen, Z., Lv, Z., Sun, Y., Chi, Z., Qing, G., Recent advancements in polyethyleneimine-based materials and their biomedical, biotechnology, and biomaterial applications. *J. Mater. Chem. B*, *8*, 15, 2951–2973, 2020.
65. Lu, S., Bao, X., Hai, W., Shi, S., Chen, Y., Yu, Q., Zhang, M., Xu, Y., Peng, J., Multi-functional self-assembled nanoparticles for pVEGF-shRNA loading and anti-tumor targeted therapy. *Int. J. Pharm.*, *575*, 118898, 2020.
66. Yu, Q., Zhang, M., Chen, Y., Chen, X., Shi, S., Sun, K., Ye, R., Zheng, Y., Chen, Y., Xu, Y., Peng, J., Self-assembled nanoparticles prepared from low-molecular-weight PEI and low-generation PAMAM for EGFRvIII-chimeric antigen receptor gene loading and T-cell transient modification. *Int. J. Nanomed.*, *15*, 483–495, 2020.
67. Steinman, N.Y., Campos, L.M., Feng, Y., Domb, A.J., Hosseinkhani, H., Cyclopropenium nanoparticles and gene transfection in cells. *Pharmaceutics*, *12*, 8, 768, 2020.
68. Proulx, J., Joshi, C., Vijayaraghavalu, S., Saraswathy, M., Labhasetwar, V., Ghorpade, A., Borgmann, K., Arginine-modified polymers facilitate poly (lactide-co-glycolide)-based nanoparticle gene delivery to primary human astrocytes. *Int. J. Nanomed.*, *15*, 3639–3647, 2020.
69. Lebrón, J.A., Ostos, F.J., López-López, M., Moyá, M.L., Sales, C., García, E., García-Calderón, C.B., García-Calderón, M., Peña-Gómez, M.J., Rosado, I.V., Balestra, F.R., Metallo-liposomes of ruthenium used as promising vectors of genetic material. *Pharmaceutics*, *12*, 5, 482, 2020.
70. Zhao, Y., Zhao, T., Du, Y., Cao, Y., Xuan, Y., Chen, H., Zhi, D., Guo, S., Zhong, F., Zhang, S., Interaction kinetics of peptide lipids-mediated gene delivery. *J. Nanobiotechnol.*, *18*, 1, 1–4, 2020.

71. Buck, J., Mueller, D., Mettal, U., Ackermann, M., Grisch-Chan, H.M., Thöny, B., Zumbuehl, A., Huwyler, J., Witzigmann, D., Improvement of DNA vector delivery of DOTAP lipoplexes by short-chain aminolipids. *ACS Omega*, *5*, 38, 24724–24732, 2020.
72. Ghodke, S.B., Parkar, J.N., Deshpande, A.R., Dandekar, P.P., Jain, R.D., Structure–activity relationship of polyester-based cationic polyrotaxane vector-mediated *in vitro* siRNA delivery: Effect on gene silencing efficiency. *ACS Appl. Bio Mater.*, *3*, 11, 7500–7514, 2020.
73. Routkevitch, D., Sudhakar, D., Conge, M.J., Varanasi, M., Tseng, S., Wilson, D., Green, J.J., Efficiency of cytosolic delivery with poly (beta-amino ester) nanoparticles is dependent on the effective pKa of the polymer. *ACS Biomater. Sci. Eng.*, *6*, 3411–3421, 2020.
74. Neves, A.R., Sousa, A., Faria, R., Albuquerque, T., Queiroz, J.A., Costa, D., Cancer gene therapy mediated by RALA/plasmid DNA vectors: Nitrogen to phosphate groups ratio (N/P) as a tool for tunable transfection efficiency and apoptosis. *Colloids Surf. B: Biointerfaces*, *185*, 110610, 2020.
75. Sainz-Ramos, M., Villate-Beitia, I., Gallego, I., Qtaish, N.A., Lopez-Mendez, T.B., Eritja, R., Grijalvo, S., Puras, G., Pedraz, J.L., Non-viral mediated gene therapy in human cystic fibrosis airway epithelial cells recovers chloride channel functionality. *Int. J. Pharm.*, *588*, 119757, 2020.
76. Zhou, Z., Liu, X., Zhu, D., Wang, Y., Zhang, Z., Zhou, X., Qiu, N., Chen, X., Shen, Y., Nonviral cancer gene therapy: Delivery cascade and vector nanoproperty integration. *Adv. Drug Deliv. Rev.*, *115*, 115–154, 2017.
77. Xia, Q., Huang, J., Feng, Q., Chen, X., Liu, X., Li, X., Zhang, T., Xiao, S., Li, H., Zhong, Z., Xiao, K., Size-and cell type-dependent cellular uptake, cytotoxicity and *in vivo* distribution of gold nanoparticles. *Int. J. Nanomed.*, *14*, 6957–6970, 2019.
78. Haddick, L., Zhang, W., Reinhard, S., Möller, K., Engelke, H., Wagner, E., Bein, T., Particle-size-dependent delivery of antitumoral miRNA using targeted mesoporous silica nanoparticles. *Pharmaceutics*, *12*, 6, 505, 2020.
79. Keil, T.W., Feldmann, D.P., Costabile, G., Zhong, Q., da Rocha, S., Merkel, O.M., Characterization of spray dried powders with nucleic acid-containing PEI nanoparticles. *Eur. J. Pharm. Biopharm.*, *143*, 61–69, 2019.
80. S.A. Wickline, H. Pan, C.T. Pham, H. Yan, Peptide-polynucleotide-hyaluronic acid nanoparticles and methods for polynucleotide transfection. United States Patent Application; 16/870,035, 2020.
81. Pezzoli, D., Giupponi, E., Mantovani, D., Candiani, G., Size matters for *in vitro* gene delivery: Investigating the relationships among complexation protocol, transfection medium, size and sedimentation. *Sci. Rep.*, *7*, 44134, 2017.
82. Yin, H., Kanasty, R.L., Eltoukhy, A.A., Vegas, A.J., Dorkin, J.R., Anderson, D.G., Non-viral vectors for gene-based therapy. *Nat. Rev. Genet.*, *15*, 8, 541–55, 2014.
83. McCoy, M., Lipids, the unsung COVID-19 vaccine component, get investment, in: *Chemical & Engineering News. American Chemical Society*, 2021,

https://cen.acs.org/business/outsourcing/Lipids-unsung-COVID-19-vaccine/99/web/2021/02. (Accessed 19 Mar 2022).
84. Let's talk about lipid nanoparticles. *Nat. Rev. Mater.*, *6*, 99, 2021.
85. Beam therapeutics announces acquisition of guide therapeutics, 2021, https://investors.beamtx.com/news-releases/news-release-details/beam-therapeutics-announces-acquisition-guide-therapeutics. (Accessed 19 Mar 2022).
86. Jones, C.H., Chen, C.K., Ravikrishnan, A., Rane, S., Pfeifer, B.A., Overcoming nonviral gene delivery barriers: Perspective and future. *Mol. Pharm.*, *10*, 11, 4082–98, 2013.
87. Dinarello, C.A., Overview of the IL-1 family in innate inflammation and acquired immunity. *Immunol. Rev.*, *281*, 1, 8–27, 2018.
88. Bueter, C.L., Lee, C.K., Wang, J.P., Ostroff, G.R., Specht, C.A., Levitz, S.M., Spectrum and mechanisms of inflammasome activation by chitosan. *J. Immunol.*, *192*, 12, 5943, 2014.
89. Pandey, R.K. and Prajapati, V.K., Molecular and immunological toxic effects of nanoparticles. *Int. J. Biol. Macromol.*, *107*, 1278–93, 2018.
90. Crommelin, D.J., Mastrobattista, E., Hawe, A., Hoogendoorn, K.H., Jiskoot, W., Shifting paradigms revisited: Biotechnology and the pharmaceutical sciences. *J. Pharm. Sci.*, *109*, 1, 30–43, 2020.
91. Quality-by-design approach to manufacturing cell and gene therapies, CCRM, 2020, https://www.contractpharma.com/csd/profile/centre-for-commercialization-of-regenerative-medic/view_quality-by-design-approach-to-manufacturing-cell-and-gene-therapies/ (Accessed 12 Mar 2022).
92. Gavin, D.K., Advanced topics: Successful development of quality cell and gene therapy products, https://www.fda.gov/media/80404/download (Accessed 13 Mar 2022).
93. Chen, S., Zaifman, J., Kulkarni, J.A., Zhigaltsev, I.V., Tam, Y.K., Ciufolini, M.A. *et al.*, Dexamethasone prodrugs as potent suppressors of the immunostimulatory effects of lipid nanoparticle formulations of nucleic acids. *J. Control. Release*, *286*, 46–54, 2018.
94. Ramamoorth, M. and Narvekar, A., Nonviral vectors in gene therapy- an overview. *J. Clin. Diagn. Res.*, 9, 1, GE01–6, 2015.
95. Polack, F.P., Thomas, S.J., Kitchin, N., Absalon, J., Gurtman, A., Lockhart, S. *et al.*, Safety and efficacy of the BNT162b2 mRNA COVID-19 vaccine. *N. Engl. J. Med.*, *383*, 27, 2603–15, 2020.
96. Moderna announces primary efficacy analysis in phase 3 cove study for its COVID-19 vaccine candidate and filing today with U.S. FDA for emergency use authorization, 2020, https://investors.modernatx.com/news-releases/news-release-details/moderna-announces-primary-efficacy-analysis-phase-3-cove-study (Accessed 11 Dec 2022).

6

Nanoparticles for Tumor Gene Therapy: Recent Advances and Perspective

R. Shivhare[1]*, V. Sabale[1], A. Ingole[1] and Neelam Jain[2]

[1]Dadasaheb Balpande College of Pharmacy, Besa, Nagpur (MH), India
[2]Faculty of Pharmacy, Oriental University, Indore (MP), India

Abstract

In contrast to usual therapy, gene therapy proposes an array of benefits for cancer medication which encompasses higher efficacy and sensitivity, very little off-target toxicity, and delivery of genetic factors that all together target cancer tumorigenesis, relapse, and resistance to the drug. Despite the discovery of novel approaches such as gene suppression, antisense therapy, RNA interference, gene, and genome editing to gene modulation efficient targeting of cells and tissues it remains a challenge. Nanoparticulate gene carriers made from a variety of materials are gaining a lot of attention due to their potential benefits of very low toxicity and immunogenicity. Nanoparticles are loaded with nucleic acids and swiftly pass via the cell membrane due to their small size (approximately 100 nm). Numerous studies on nanoparticles have been published which include cationic chemicals and inorganic nanoparticles that are inappropriate for systemic distribution due to their rapid enzymatic breakdown by serum nucleases. Extensive and innovative changes to these nanoparticles have been researched to improve their systemic stability and use them as gene-delivery vehicles. Nanoparticles demonstrated excellent properties for gene delivery among non-viral vectors like the potential to target a definite tissue or cell, protection against nuclease degradation of the target gene, improved deoxyribonucleic acid stability, and improved transfection affectivity or safety. This chapter highlights the role of nanoparticles in tumor gene therapy representing the recent studies on physicochemical properties and mechanisms of action of nanoparticles to use them for tumor genetic treatment.

Keywords: Gene therapy, nanoparticles, non-viral vectors, nucleic acids

**Corresponding author*: shivharer4@gmail.com

Rishabha Malviya, Sonali Sundram and Neelam Jain (eds.) Integration of Biomaterials for Gene Therapy, (139–182) © 2023 Scrivener Publishing LLC

6.1 Introduction

Tumor suppressor gene play important role in controlling cell proliferation, differentiation, and apoptosis as well as maintaining genomic integrity. The development of tumors are linked with dropping of functions of tumor suppressor gene. Therefore, the utility of tumor suppressor genes for the treatment of cancer intensely studied through experimental and clinical studies. Several tumor suppressor gene have been shown to be directly delivered to tumorous cells, causing cell cycle arrest, apoptosis, and in some cases inhibiting tumor growth through bystander effects. Numerous research have further demonstrated the potential for synergistic therapeutic benefits when tumor suppressor gene therapy is combined with traditional anticancer therapy [1].

Gene therapy is the transmission of unique genetic material in order to change the coding of a gene product or to flip the organic and elemental properties of tissues in the treatment of various disorders.

Recombinant therapeutic peptide boundaries like poor biological availability, uncertainty, extreme poisoning, low clearance rates, and expensive fabrication fees are all conquered by means of gene therapy. Gene treatment options can deal with illnesses by using introducing new genes, deactivating or knocking out disorder-inflicting genes, or repairing broken genes with healing ones [2]. Gene therapy is "the management of genetic material to alter or manage the expression of a gene product or to alter the biological characteristics of living cells for therapeutic use," according to the FDA [3]. There are many uses for gene therapy, including gene substitution, vaccination, and takedown for genetic diseases such as cancer hemophilia, hypercholesterolemia, and neurodegenerative disorders. All of these uses have specific requirements for how the gene should be administered [4]. Gene delivery systems control the use of a gene encoding system based on plasmids, delivering the encoding gene to the host tissue that regulates the expression of a gene in the intended organ, as well as a gene that expresses vital therapeutic peptides [5]. The principle goal of gene therapy for genetic issues is augmentation gene therapy, also referred to as persevering with gene expression of the inherited gene at levels sufficient for therapeutic use.

The most frequent type of transfer involves a balanced replica of a mutated gene. The use of RNA interference or gene editing tools is additionally used to prevent the gene expression that is harmful. Furthermore, it is theoretically possible to apply gene editing strategies to restore a mutated gene to its specific genomic site via base modification or homologous recombination with a donor template, although this has yet no longer been investigated in clinical trials [6]. Figure 6.1 describes the strategies used for gene therapy:

Gene therapies are governed by the Food and Drug Regulations in Canada and are incorporated into the definition of "drug" in accordance with the Food and Drug Act. Gene therapies are governed as biological

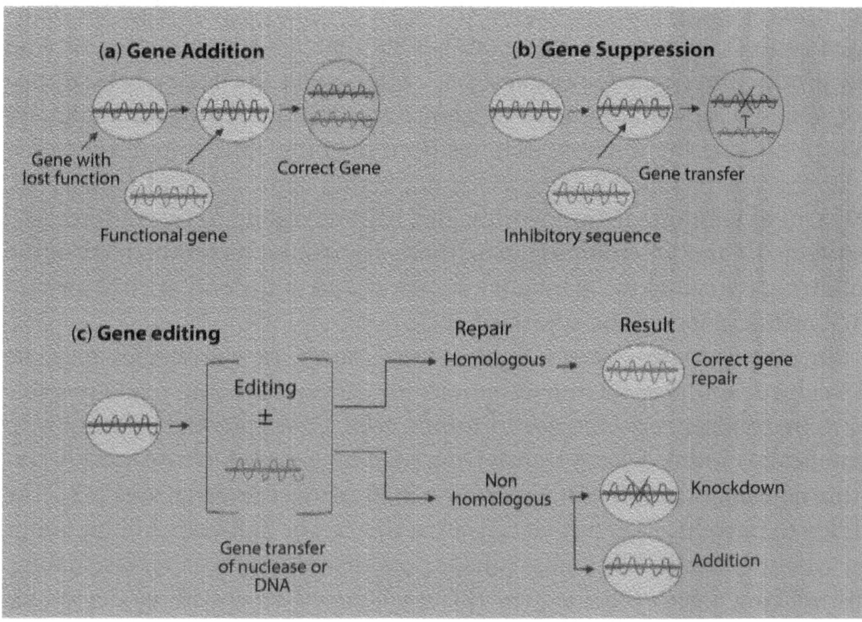

Figure 6.1 The strategies used for gene therapy: (a) By imparting a functional copy of a diseased gene in trans, gene addition (augmentation) targets to restore normal cellular characteristics (this is, without affecting the diseased gene itself, which remains inside the cell) this approach has been used to treat primary immunodeficiencies like X-linked intense blended immunodeficiency as well as a disease like Leber congenital amaurosis, Hemophilia and spinal muscular atrophy (model examples of ocular, liver and central nervous system indications). (b) In a few other cases, including Huntington's disorder, the cellular characteristic is lost because of poisonous protein accumulation. Gene suppression uses RNA interference to decrease the expression of the mutated gene so as to restore cell normal functioning. (c) Even though nucleases are not always necessary for gene editing, causing DNA double-stranded damage at the target site typically increases the effectiveness of gene-specific editing in mammalian cells. The result of genome modification will depend on which DNA repair mechanism is chosen over another. Once cleavage of DNA is completed, the damage is reunited by non-homologous end joining, which can also bring about gene knock-down if a repair may additionally bring about homologous recombination is used to correct the mutated gene *in situ*. A third possibility is the insertion of the DNA template through non-homologous end joining, which would add genes as opposed to correct them.

drug therapies in the two countries like Canada and the United states. Gene therapies are governed differently in some jurisdictions, including Europe [7]. In order to the foreign genetic material to attain the target cells, it needs to be administered as a specific genetic material (such as Ribonucleic acid and Deoxyribonucleic acid) using the "vector" or "carrier".

Since modified versions of natural viruses are an effective way to introduce genetic material inside the cell, they are typically used as vectors in gene therapies. Replacement gene therapy's objective is to replace the damaged gene(s), increase the gene availability that can treat diseases, or stop the production of the faulty gene(s) [7, 8]. In order to effectively treat cancer, gene therapy mainly aims to suppress or kill the growth of malignant cells [9]. Gene therapy research has been ongoing since a long time. Ten gene therapy products had received marketing approval as of January 28, 2018, in at least one nation worldwide [10]. According to recent forecasts, between 12 and 14 emerging gene therapies will be recommended in the following two years for regulatory approval, and at the end of 2022 around 40 new therapies will be approved [11].

To make progress in the fight against cancer, new strategies must be developed, and new treatment options for cancer are being made possible by gene transfer technologies to tumor cells. One potential strategy is to genetically modify cells by transferring suicide genes, which are quickly creating new opportunities for application in the cancer research area [12]. The ability to introduce a gene that can either directly or indirectly inhibit tumor growth into tumor cells makes suicide gene therapy particularly fascinating. Introducing a gene Suicide gene therapy is based on encoding a cytotoxic protein into tumor cells. There are two ways to accomplish this [13]:

I. Indirect gene therapy uses enzyme-activated prodrugs. That permits a prodrug to be converted into a toxic drug in cells.
II. Direct gene therapy with a toxin gene, whose expression can alter cell membrane stability and reduce tumor cell viability, or repair mutated proapoptotic genes, which are usually tumor suppressor genes that include cell suicide under normal circumstances. Table 6.1 includes a list of recent suicide gene therapy patents from 2002 to 2022.

6.2 Technologies for Gene Delivery

The major challenge afterward was the invention of genetic engineering. Which aids genetic treatment, has been determine the safety and affectivity

Table 6.1 Patents on suicide gene therapy for cancer.

Title and patent number	Year	Enzyme	Prodrug	Drug	References
"Regulation of Human carboxypeptidase A." WO2002052020	2002	Carboxypeptidase A	Methotrexate-a-peptides	Methotrexate (MTX)	[14]
"NitroreductaseEnzyme" US 2004/0014191A1	2004	Nitroreductase (NTR)	CB1954	4-Hydroxylamine (4HX)	[15]
"Recombinant B-lactamase-RGD-fusion protein and application thereof in medicine." N101824406	2009	β - lactamase enzyme	Cephalosporin Doxorubicin and 7-(4-carboxy butanamido)–Cephalosporin mustard	Cephalosporin Doxorubicin & Carboxybuta namido)-Cephalospori n mustard	[16]
"A cyanogenic system's ability to work in concert with another oxidative inducing system to treat tumors. US20100028366	2010	Linamarase	Linamarin	Cyanide	[17]

(Continued)

Table 6.1 Patents on suicide gene therapy for cancer. (*Continued*)

Title and patent number	Year	Enzyme	Prodrug	Drug	References
Cytochrome p450 reductase for the enhancement of p450-based anticancer gene therapy." EP1017835	2011	Cytochrome p450	Cyclophosphami de and Ifosfamide	Phosphorami de mustard and Acrolein	[18]
"Vaccinia virus for gene-directed enzyme prodrug therapy." Wo2014202977	2014	Carboxy-peptidase G2	N,N-[(2- chloroethyl) (mesyloxy-ethyl) amino] benzoyl-L-glutamic acid	N,N-[(2-chloroethyl) (2-mesyloxyethyl) amino] benzoic acid	[19]
"Gene therapy vectors and cytosine deaminases." US20140242033	2014	Cytosinedeaminase	5-Flurocytosine	5-Flurorouracil	[20]
"Purine nucleoside phosphorylase as an enzymatic activator of nucleoside prodrugs." US8628767	2014	Purine nucleoside phosphorylase	6-Methylpurine-2-deoxyriboside	6- Methylpurine	[21]

(*Continued*)

Table 6.1 Patents on suicide gene therapy for cancer. (*Continued*)

Title and patent number	Year	Enzyme	Prodrug	Drug	References
"Improved thymidine kinase gene" WO20141532258 "Vector simultaneously expressing dodecameric trail and HSV-TK suicide gene, and anticancer stem cell therapeutic agent using same."	2015 2014	Herpes simplex virus thymidine kinase	Ganciclovir	Ganciclovirtriphosphate	[22, 23]
Suicide gene therapeutic agent for publication classification brain tumors using pluripotent stem cells. US 2021/0369785 A1	2021	Cytosinedeaminase	5-Flurocytosine	5-Flurouracil	[24]

of genetic products. A perfect carrier is non-immunogenic, secure, capable of delivering genetic material to a particular tissue, able to provide space for a sufficiently large foreign gene, and able to achieve adequate transgenic expression to repair the mutated gene. Genetic products are delivered using viral vectors, Bactofection and non-viral vectors, as shown in Figure 6.2 (Chemical and physical) [25].

The selection of the vectors is a very crucial step in achieving gene therapy.

- Gene Delivery Using Viral Vectors
  The original and best-known vectors for inserting genes into target tissues were viruses. Vectors like viruses enable it to infect almost all cells without endangering their viability. The unique properties of viruses make them acceptable for clinical gene transfer. Endocytosis is triggered by interactions between virus surface proteins and host receptors. Once inside the nucleus, viruses release their genome to start expressing their genes [26, 27].
- Gene Delivery with Chemical-Based Nonviral Vectors
  Better and longer-lasting gene encoding is shown by viral-vector-based gene transfer, but it has drawbacks like immunogenicity, less cell-specificity, carcinogenicity, high cost, and inability to deliver large genome sizes. Because they are easier to produce, deliver a large genome, and are generally safe, non-viral methods have better advantages [28–30].

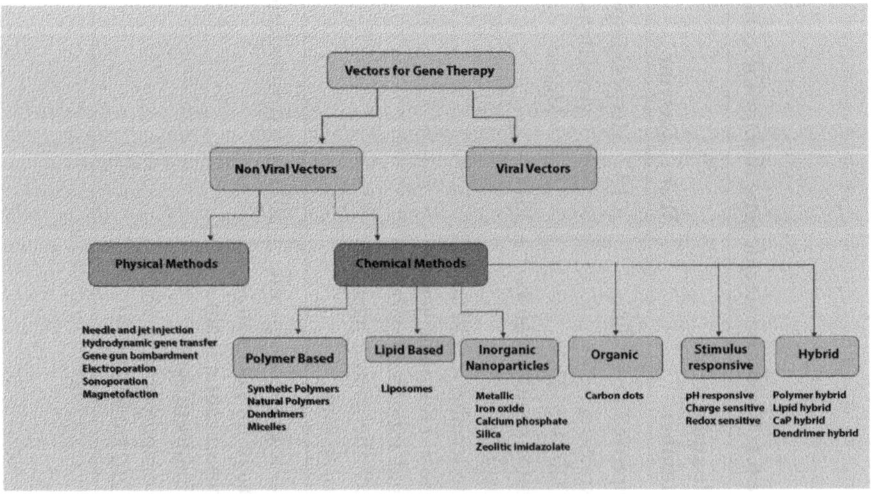

Figure 6.2 A summary of the gene therapy delivery methods [2].

- Gene Transfer Mediated by Bacteria
  Gene Transfer Mediated by Bacteria is referred to as "Bactofection."
  Certain bacteria target tumor cells specifically, causing interference with ribonucleic acid (RNAi) and gene silencing by obstructing ribonucleic acid (RNA) activity, such as synthesis of proteins. Intracellular bacteria like listeria monocytogenes, *Salmonella* spp., *Bifid bacterium longum*, *E. coli*, *Yersinia enterocolitica*, and *Shigella flexneri* can deliver plasmids, cytotoxic agents and prodrug converting enzymes into the target cell, as shown by numerous *in vivo* and *in vitro* studies [31].

6.3 Cancer Treatment with Gene Therapy

Cancer develops as a consequence of a disruption in the normal apoptosis and proliferation processes. Multiple mechanisms of cell death, a novel therapeutic agent with an unique mode of action, and synergy with conventional management are all necessary for advancements in cancer therapy. These are all characteristics apply to gene therapies. Immunotherapy, pro-apoptotic gene therapy, oncolytic virotherapy, siRNA therapy, suicide gene therapy, gene directed-enzyme prodrug therapy, anti-angiogenic gene therapy, pro-apoptotic gene therapy, and nanotechnology-based gene therapy are a few of the gene therapy techniques that have been developed to treat cancer [32].

6.4 Gene Therapy Using Nanotechnology

Nanomedicine, a rapidly developing science that aims to circumvent cancer through tailored drug/gene delivery, promises to address the drawbacks of traditional treatments. In order to effectively deliver a specific medicine or gene, nanoparticles must improve the circulation time, increase bioavailability, decrease the likelihood of immune system-based identification, and deliver the gene regulator precisely. Nanoparticles are minuscule crystals that are smaller than 100nm in size, or limited to the three dimensions, with an integrated interfacial layer that is usually made up of ions, inorganic molecules, or organic compounds. In nanomedicine, where nanoparticles are used to improve drug transport and activity, they are of great scientific interest (Figure 6.3). Hydrophobic anticancer medications can be delivered

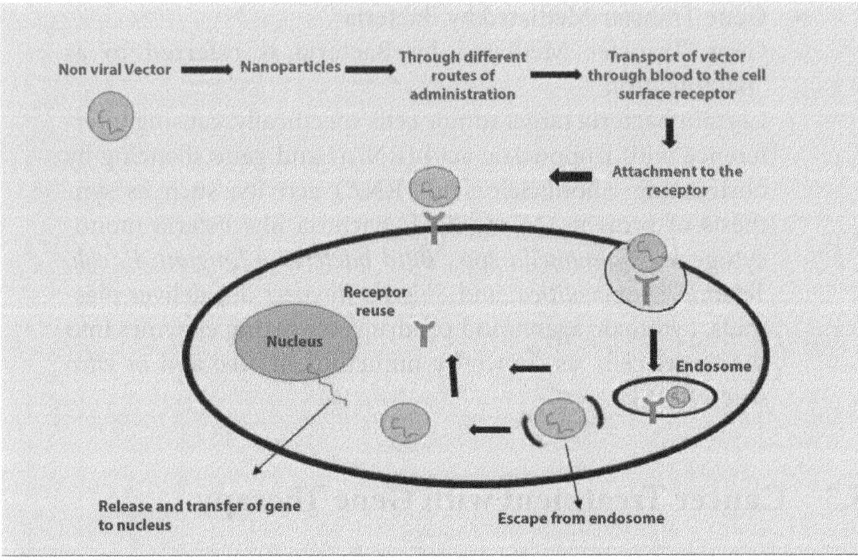

Figure 6.3 Mechanism of internalization of the nonviral vector carrying into cell and nucleus.

to precise locations in the body with fewer difficulties thanks to nanoparticles' characteristically small size and unique coating, which also minimizes immune system opsonization. There are several different types of nanoparticles including polymeric, lipid-based metallic, magnetic, dendrimeric, micelles, stimulus-responsive and hybrid nanoparticles [33].

6.4.1 Polymeric Nanoparticles

Polymeric nanoparticles are fine in size lying within the range of 1 and 1000nm. They may entrap or adsorb the substance on the surface of the polymeric core [34]. In order to address selectivity toward the target cells and the capability of gene transduction, polymeric gene delivery technologies have garnered considerable interest. The most often employed method among the various polymer materials is the incorporation of DNA into condensed particles based on cationic lipids or cationic polymers. When negatively charged DNA interacts with cationic polymers that include many positively charged amine groups in their backbones, the result is the self-assembly and condensing of DNA or RNA into compact NPs. The contained nucleic acid medications are extremely well protected from nucleases by the charge-neutralized core's compact shape. The cores should remain stable once the genes have been encased in polycations, enabling simple cellular absorption and localization inside the cell. Thus, from the point of administration to the final localization in the cell nucleus, these vectors should unavoidably overcome the many challenges. Here are some noteworthy developments in polymeric nanoparticles for gene delivery [35].

Poly (β-amino-ester) nanoparticles were used to transport a completely CpG-free plasmid containing mutant herpes simplex virus type 1 sr39 thymidine kinase DNA into human HCC cells. A build-up of 9-(4-18F-fluro-3-hydroxymethylbutyl) Guanine (18F-FHBG) for *in vivo* study and the prodrug Ganciclovir to kill cancer cells are made possible by Sr39 transfection. Targeting was carried out utilizing a CpG-free human α–fetoprotein (AFP) promoter (CpGf-AFP-sr39). When expression was limited to hepatocellular carcinoma cells that produced AFP, it was possible to transfect orthotopic hepatocellular carcinoma xenografts only in those cells. Treatment with CpGf-AFP-sr39 NP reduced tumor size by 62%, and positron emission tomography could identify therapeutic gene expression (PET) [36]. An effective, adaptable method for precisely controlling the spatiotemporal transfer of therapeutic micro RNAs (miR-21 and miR-100) and the TK-p53-NTR triple therapeutic gene, which were together loaded in PLGA-PEG-PEI polymeric nanoparticles to improve the delivery to two types of tumor models i.e. Triple negative breast cancer (TNBC) and hepatocellular carcinoma (HCC), with ultrasonic mediation of destruction of BR38 MBs. The analysis of BALB/c mice's 4T1 (TNBC) subcutaneous tumors by *in vivo* imaging and *ex vivo* tissue analysis of BALB/c mice's 4T1 (TNBC) subcutaneous tumors show a $19 \pm 0.5\%$ (p 0.01) growth in miRNA delivery and a $48 \pm 1.79\%$ (p 0.001) growth in TK-p53-NTR transport after ultrasonic treatment, that leads to a $48 \pm 6.98\%$ (p 0.01) decrease in tumor growth when compared to similar tumors in absence of ultrasonic treatment [37].

Researchers investigated polymeric nanoparticles for Gene transfer of herpes simplex virus type I thymidine kinase (HSVtk) blended with the prodrug Ganciclovir (GCV) in a malignant glioma model to make nanoparticle formulations more efficient than the most well-known marketed reagent Lipofectamine 2000. Nanoparticles harboring HSVtk DNA caused 100% cancer death *in vitro* in the two glioma cell lines when combined with GCV exposure. The mice treated with nanoparticles were found to show an improved survival rate ($p = 0.0012$ vs. control) [38]. In a mouse xenografts model using polymer-based nanoparticles (NPs), human granulocyte colony-stimulating factor (GCSF)-carrying CD34 cells were found to exhibit increased engraftment capability of human HSCs that could also be employed for *in vivo* administration. Polymer-based NP found to have the potential of delivering mRNA and nucleases to control the current restrictions of HSC gene editing [39].

6.4.2 Lipid Nanoparticles

Lipid-based nanoparticles are lipid moieties formed for encapsulation/entrapment of therapeutic drug delivery. They are nowadays being formulated as non-viral mRNA vectors by improving their stability and increased transfection with guaranteed delivery to the site of action. Lipid nanoparticle (LNP) used for Cas9 mRNA and sgRNAs delivery with the application of a new amino ionizable lipid. CRISPR-LNPs intracerebral route of administration contrary to PLK1 (sgPLK1-cLNPs) into orthotopic glioblastoma allowed gene editing resulting in tumor growth inhibition and increased survival. Whereas EGFR-targeted sgPLK1-cLNPs by intraperitoneal showed improved uptake by spread ovarian tumors, with a positive effect on survival rate [40]. The main biological elements of triple negative breast cancer are adipocytes (TNBC). Targeted lipid-protamine-DNA (LPD) nanoparticles are used to deliver the plasmid DNA encoding the C-C Motif Chemokine Ligand 2 trap (pCCl2) selectively to the tumor microenvironment (TME) in order to improve the immunosuppressive TME and locally express the (CCL2) trap. Significantly, this approach exhibits improved therapeutic efficacy and substantial tumor growth inhibition when compared to the commercially available CCL2 antibody [41]. Lipid base NP for transporting Cas9 messenger RNA (mRNA) and guide RNA help to edit the Angptl3 gene using CRISPR-Cas9. This process was more effective as compared to MC-3 LNP which has already been approved by FDA and acts as a standard. The Lipid base NP showed better ANGPTL3 gene knockdown in C57BL/6 mice liver, leading to decreased serum concentrations of ANGPTL3 protein [42].

6.4.3 Inorganic Nanoparticle

Metal cores formed of inorganic metal or metal oxide make up the metallic nanoparticles (MNPs), which are typically encased in an organic or inorganic substance or metal oxide shell. The limitations of traditional chemotherapy may be overcome by metal nanoparticles (NPs). According to reports, metal NPs have a potent and advantageous role in cancer therapy by improving drug transport, gene silencing, and targeting. More effective control over the accumulation of energy in tumors is provided by functionalized metal NPs with targeting ligands [43].

6.4.4 Gold Nanoparticles

Gold nanoparticles (AuNPs) are metallic NPs with additional therapeutic effects of gold and are being used in biotechnological for various therapeutic uses. This section contains reviews of the advances in gold nanoparticles and their applications in gene therapy. Rigidized Y-shaped triangular DNA bricks were developed with sticky ends for siRNA delivery using gold NP. A siRNA encapsulated NP, siRNA/Ap-CS, is produced after aptamers are organized on the outer surface, and 100% tumor growth inhibition is obtained, revealing remarkable prospects for future cancer target therapy [44]. Gold nanoparticles (AuNPs) containing doxorubicin and siRNA were developed to target the anti-apoptotic gene Bcl-2, to increase the effectiveness of treatment for advanced breast cancer while ensuring effective drug internalization. Moreover, the co-delivery AuNPs technology enabled the synchronized transport of Bcl-2 siRNA and Dox, enhancing apoptosis and inhibiting cancer cell proliferation. Additionally, chemotherapy plus Bcl-2 gene silencing resulted in less breast cancer cell colony formation and migration [45]. Ultra-small gold nanoparticles were assembled by using a complimentary strand and the triplex-forming oligonucleotide sequence. The generated nanostructures, which resembled sunflowers, had a high capacity for photothermal conversion and near-infrared (NIR) absorption. The large particles can be reduced in particle size due to NIR radiation, resulting in a modified c-myc oncogene silencing sequence carrying ultra-small particles. These nanoparticles could then be directed at the cell nucleus while simultaneously regulating the time before incubation required for their self-assembly and also for NIR irradiation [46].

6.4.5 Calcium Phosphate Nanoparticles

Calcium phosphate an inorganic mineral present in bones and teeth can also be used to formulate nanoparticles for efficient gene transfer for desired antitumor effect. These CaP nanoparticles show effective delivery of drugs, DNA, RNA, proteins and enzymes into tumor cells. Furthermore, functionalized CaP also shows pH–dependent solubility that can be advantageous for the effective delivery and release of genetic substances [47]. Calcium phosphate lipid coated NP (LCPNP) and layered double hydroxide NP (LDHNP) formulated to deliver siRNA for suspending T lymphocytes to obtain functional planned cell death. These LDHNP and LCPNP efficiently carried a DNA segment to mouse T cell lines (EL4) with a short time span. The LCP was proved to be superior to LDH [48]. LCPNP were found to be more efficient than commercial transfection agent Oligofectamine, in maintaining the integrity and stability of encapsulated siRNA in a culture medium. This LCP was found to increase the cellular uptake of siRNA and decrease *in vitro* cell growth of breast cancer cell lines [49]. Calcium phosphate NP encapsulated with liposomes containing siRNA to silence VEGF expressions in MCF-7 were obtained and evaluated by real time quantitative polymerase chain reaction and western blot test. The treatment showed a prominent reduction in tumor development and angiogenesis in MCF-7 xenograft mice. PLCP mediating VEGF siRNA may reduce 60% to 80% VEGF expression *in vitro* which was found to be much higher by Lipofectamine 2000 [50].

6.4.6 Iron Oxide Nanoparticles

Magnetofection using green fluorescent protein DNA-bearing polyethyleneimine coated SPIONs using four rare earth magnets was developed. This magnetofection was found to be effectively reducing transfection time in MCF-7 cells [51]. Supramagnetic iron oxide NP having iron oxide core modified with galactose and polyethyleneimine (Gal-PEI-SPIO) nanoparticles to deliver siRNA in the liver cancer cells. These nanoparticles (Gal-PEI-SPIO) were found effective in protecting the siRNA from degradation due to enzymes and increasing the half-life. Simultaneously, increase uptake by Hepa1–6 cells was observed [52]. Low molecular weight PEI and PEG nanogels loaded with transforming growth factors-1 (TGF-1) siRNA and ultra-small ferric oxide NP showed less tumor proliferation in mice sarcoma model and improved T1 mediated MR imaging. The ferric oxide PEI-PEG nanogels NGs helps to combine TGF-1 and siRNA by electrostatic combination and effectively deliver siRNA to the tumor cells to silence TGF-1, which plays a major role in tumor inhibition and invasion *in vitro* and *vivo* for subcutaneous carcinoma and lung cancer [53]. Due in part to their excellent biological stability and compatibility, magnetic effect, and functional surface changes, these IONPs have shown their brilliant potential for very effective gene delivery targets the malignant tissues [54].

6.4.7 Magnetic Nanoparticle

Transfection of the gene by lipid-based techniques on pDNA is utilized in conjunction with magnetic nanoparticles to induce breast cancer-active CD95 cell expression. Its interaction with ligand CD95L results in apoptosis (FasL). In Cancer cells, CD95's death domain (procaspase-8 and c-Flip) is altered making them resistant to apoptosis caused by CD95.

pDNA transfection to activate the apoptotic pathway thereby dramatically enhancing apoptosis [55]. A magnetic field applied externally increases transfection efficiency due to the paramagnetic properties of the polyethyleneimine (PEI) coated iron oxide superparamagnetic nanoparticles thus introducing a high positive surface charge, encouraging genetic material binding. Meanwhile, LIPUS stimulation increases cell permeability. The action of LIPUS along with the externally applied magnetic field allows the gene (in our case, Cherry Red or GFP plasmid) to enter inside the cell. Higher cell viability led to greater HEK 293 cell transfection efficiency [56].

To increase the biocompatibility of nanoparticles, magnetic nanoparticles were coupled with polyethyleneimine, and then polyethylene glycol. siRNA that targets a disintegrin and metalloproteinase 10 (ADAM 10) was loaded in magnetic NP. The findings unmistakably showed a notable decline in cell growth, which increase specific siRNA levels. Confocal imaging verified that nanoparticles loaded with siRNA were delivered intracellularly to the tumor cell cytoplasm [57]. The magnetic nanoparticles ferric oxide-polyethyleneimine that can be employed as a vector to treat tumors with gene and also for magnetic fluid hypothermia were created by altering Fe3O4 nanoparticles with Polyethyleneimine (PEI). Fe3O4@PEI nanoparticles with recombinant vectors [HRE], p[HRE], and AFP-HSVTK. The tumor mass in the treatment group with the combination was considerably less compared to the single treatment group, according to AFP-Luc analysis (all P-0.05) and the tumor cell intracellular structure displayed cell-killing morphology [58]. C-myc antisense oligodeoxynucleotides (ODNs) were targeted in breast cancer cell lines using silk polyethyleneimine NPs (SPP) and magnetic silk polyethyleneimine core-shell NP(MSPPs). Both SPPs and MSPPs were effective at introducing ODNs into target cells and greatly reducing cell proliferation. High ODN uptake efficacy was attained by employing MSPPs as carriers by the process of magnetofection, displaying a markedly improved uptake impact against those with non-magnetofection. When compared to human skin fibroblast cells, the inhibitory impact of both SPPs and MSPPs was much stronger against targeted breast cancer cells [59].

6.4.8 Zeolitic Imidazolate (ZI)

The PER base DNA machine is activated surviving mRNA present intracellularly and continuously manufactures Bcl-2 antisense DNA, allowing it to carry out both gene therapy and the imaging of surviving mRNA. ZIF-8 NP is used by the DNA machine to co-deliver the parts to the living, such as a primer, hairpin, and strand displacement polymerase. The findings show that ZIF-8 NPs are capable of preventing protease attack and nuclease degradation on polymerases and nucleic acid probes while still killing cancer cells [60]. A hybrid Zeolitic-Imidazolate nanoparticle (HP) decorated with a cancer cell membrane was effectively created to deliver Cisplatin drug (DDP) and Oleanolic acid ingredient, with promising findings for the platform co-treatment of drugs in bladder cancer (SW780). These co-loaded nanoparticles may increase apoptosis while reducing multidrug resistance more effectively than other therapies, suggesting that it would be a promising DDS for the co-delivery of several treatments [61]. ZIF-C is a polymorph of ZIF-8 and MOF polymorph of Zeolite Imidazolite Framework-8 (ZIF-8) MOF nanoparticles known as is used to deliver gene to produce killing effect when a protein gene ribosomal protein is overexpressed. The ability of ZIF-C induced cell death at cellular and cytoplasmic levels by using RNA interference to deliver RPSA-specific siRNA while using ZIF-C. The use of ZIF-C to deliver RPSA-specific CRISPR/Cas9 plasmids by CRISPR/Cas9 shows the genomic viability of ZIF-C-induced knockdown. ZIF-C is coated with epigallocatechin-gallate to further achieve specific targeting (EGCG). Cellular transfection experiments show that RPSA-targeting nucleic acids supplied by ZIF-C are gradually expressed for up to 96 hours. Genomic cleavage detection and quantitative polymerase chain assays show a progressive KD with a 20% drop in RPSA expression [62].

6.4.9 Silica Nanoparticles

Mesoporous silica nanoparticles have been a popular agent for the intracellular delivery of siRNA. Mesoporous silica nanoparticle co-delivery will enhance the efficacy of cancer treatments. A combinational method, with improved effect due to synergistic activity, while restricting the effect of survivin was revealed by co-delivery of antitumor medications together with survivin siRNA in the silica NP and survivin siRNA in the MSNs for the treatment of lung cancer [63]. The effectiveness of magnetically targeted, magnetic silica NP controlled in shape along with suicide gene delivery based on hyperthermia for the treatment of hepatocellular cancer was explored. MSNs with rod shape had increased drug content, entrapment capacity, release mechanism, and magnetically accelerated gene transport in comparison to sphere-like MSNs [64].

6.4.10 Dendrimer

A wide range of positive characteristics of dendrimers, including strong and effective gene insertion with reduced toxic effects, have encouraged further research into carriers for genetic materials. Because adjustable functional groups are easily attached to their clearly defined branching molecular architecture, a variety of chemical alterations are possible. These aim to provide stealth or targeting capabilities after delivery, for instance, or to increase the gene condensation capacity while lowering toxicity [65]. The following section covers different dendrimers and their applications in gene therapy.

In order to create nanoclusters, cross-linking of generation 3 (G3) poly (amidoamine) (PAMAM) dendrimers were done using dithiobutyric acid. This process was triggered by 1-(3-dimethyaminopropyl)-3-ethyl carbodiimide hydrochloride. Impairing cell cycle at G1 phase *in vitro* at the G1 phase by downregulating Cyclin-D1 and CDK-4 protein with the level of mRNA expression and upregulating p53 and p21 protein and increased mRNA level. These NCs can condense pDNA, encoding enhanced green fluorescent protein and tumor suppressor genes with increased gene transfection efficiency [66]. Third-generation PAMAM dendrimers attached to GX1 peptide gold nanorods were developed to combine gene and photothermal therapy (PTT) of colon cancer (HCT-8 cells). These dendrimers work as a gene delivery vehicle for the FAM172A gene, which controls the growth and demise of colon cancer cells with exceptional transfection efficiency [67]. Using different PEG grade and dendrimer generations, it has been tried to PEGylate the diaminobutyric polypropylenimine (DAB) dendrimer in an effort to reduce cytotoxicity while increasing DNA entrapment, better uptake, and transfection effectiveness. The greatest level of galactosidase gene expression was attained at a dendrimer: DNA ratio of 20:1. When compared to unmodified dendrimer, these delivery systems considerably reduced cytotoxicity with high transfection effectiveness in B16F10-Luc cells by a factor of more than 3.4 [68]. lncRNA MEG3 (pMEG3) gene which acts as a tumor suppressor was delivered to CRPC cells via (PAMAM) dendrimer linked with an EpDT3 aptamer targeting EpCAM. PAMAM and PAMAM-PEG-EpDT3/pMEG3 NP endocytosis in CRPC cells were stimulated by the EpDT3 aptamer. Both *in vivo* and *in vitro*, the later demonstrated a substantial anti-CRPC impact [69]. Fluorinated PAMAM dendrimer (G4- F735) significantly increased the amount of cell death compared to a range of transfection agents such as PEI, SuperFect, and Lipofectamine 2000. The 3D spherical multicellular particles produced by MDA-MB-231 cells were more easily destroyed and their proliferation *in vivo* was inhibited by the G4-F735/pTRAIL combination as compared to PEI/pTRAIL [70].

A novel ultrasound-targeted microbubble destroying based on gold nanoparticles for entrapment of dendrimers is used for gemcitabine and miR-21. Cancer cells are capable of absorbing the Gem-gold dendrimers/miR-21i and ultrasound-targeted NP helps to improve the cellular uptake using ultrasound power of 0.4W/cm2 for the increase in cell penetration, significantly reducing tumor volume, and improving blood perfusion in xenografted pancreatic tumors [71]. To create CD-PAMAM conjugates, carbon dots (CD) from sweet lemon peel were combined with several generations of PAMAM dendrimers (CDPs). To specifically target v 3 integrin, which is overexpressed in TNBC, the RGDS peptide was further conjugated to CDP. When Cu (II) was present, CDP3 selectively quenched fluorescence with a 93% quenching efficiency [72].

6.4.11 DOTAP-mPEG-PCL (DMP) Micelles

A different approach to treating cancer has been made possible by gene therapy for delivering siRNA. Method of delivery affects the gene delivery. Mouse cancer models were used to analyze the anticancer effects of DMP/siMcl1 and DMP/siBcl-xl complexes. DMP micelles were successfully used in colon cancer therapy by DNA plasmid gene transfer research after being modified with cationic DOTAP lipid. They successfully stop the *in vitro* development of C26 cell lines for colon cancer [73].

In order to create the DMP/phBimS combination, which has an anti-tumor impact, the interaction between DOTAP and the mPEG-PCL polymer was coupled with the apoptosis-inducing BimS gene. By triggering endocytosis mediated by caveolin and lipid raft, DMP effectively delivered plasmid into SCC15 cells with the best possible internal stability. When plasmids containing the BimS gene are loaded, which induces apoptosis, the DMP/phBimS complex demonstrated a definite apoptosis-related anticancer effect of SCC15 *in vitro* [74]. A novel interleukin22 (IL-22) -based cancer gene therapy strategy with a novel gene carrier DMP made with mPEG polycaprolactone with DOTAP lipid by combination. Several anti-proliferative strategies, including IL-22 inhibiting, apoptosis triggering, lymphocyte infiltration, and angiogenesis inhibition, were identified in healthy tissues despite the absence of pathological alterations [75]. DOTAP-mPEG-PCL, a compatible and degradable micelle made up of DMP created by the self-assembly of the copolymers of mPEG and polycaprolactone and cationic lipid DOTAP, is an effective method for delivering mRNA. These micelles have increased transport efficiency on mouse colon cancer cells and are very effective at condensing and safeguarding mRNA molecules. An increased mRNA level was seen in C26 cells after binding with the IL-22BP-encoding mRNA, which had a significant inhibitory effect on C26 colon cancer models [76].

6.4.12 Stimulus Responsive Nanoparticles

To create miniature Nano drug delivery systems (NDDSs), siRNA was loaded onto polyamidoamine (PAMAM) dendrimers. Stimuli-responsive drug delivery systems (DDSs) could transport and release drugs more precisely in response to specific stimuli, such as external triggers, dual or multiple stimuli, or distinct tumor microenvironments, effective treatment and decreased side effects. Cancer therapy stimulus-responsive DDSs may react to a variety of external and endogenous stimuli, like pH, oxygen, light, and magnetic field [77]. This section included advances in stimulus-responsive nanoparticles for gene therapy. For potential cancer treatment, a co-loaded supramolecular NP was made with redox-responsive poly-ethyleneimine–cyclodextrin (roFPC) containing doxorubicin and human telomerase reverse transcriptase siRNA. The nanotherapeutic system provides simultaneous and efficient gene transfection. Next, the adamantine-doxorubicin combination (Ad-Dox) was loaded into roFPC [78]. In order to create small-sized PAM/siRNA through electric charge exchange, siRNA was loaded onto polyamidoamine (PAMAM) dendrimers. This was done to create Nano drug delivery systems (NDDSs), which were then sealed with a pH -responsive copolymer of PEG/PLL/DMA. The acidic reaction of the PLM copolymers transforms from a negative charge to a positive charge when in the tumor microenvironment because of the cleavable amide bond between mPEG-PLL and DMA. Thus larger-sized NP having a negative charge converted to positive and nanosized NP penetrates the tumor efficiently in response to pH change [79]. The surface charge of Niosomal Curcumin was reduced to +15mV and the diameter was increased to nearly 68nm by adding microRNA-34a (miR-34a). The effects of free Curcumin, free miR-34a, Niosomal Curcumin (NCur), Niosomal miR-34a (NmiR), Niosomal Curcumin + miR34a(NCur-miR), also known as co delivery, and Niosomal Curcumin + Niosomal miR-34a, also known as co-administration, on the cytotoxicity was evaluated. When Curcumin and miR-34a were given together to cancer model organisms, the rate of tumor inhibition was higher than in other groups [80].

6.4.13 Carbon Dots

DNA is shielded from enzymatic degradation via electrostatic conjugation of Arginine-CQDs with carboplex, a pseudo homogeneous carrier. It's interesting to note that, arginine-CQDs have shown a superior capability than chitosan to cross biological barriers and distribute into cells. While having a somewhat higher transfection efficiency than PEI polymer, the arginine- CQDs/pEGFP (W/W) nano-complex likewise shows no discernible toxicity [81]. Carbon dots are investigated as a non-viral carrier for drug or DNA delivery through desired chemical changes.

Tryptophan (Trp) served as a source of nitrogen and carbon. The carbon dots with polyethylimine NP, are used for siRNA delivery into the gastric cancer cell line. PEI (polyethyleneimine) was used as the adsorbent. The outcomes showed that the nanocarrier had excellent biocompatibility, significantly increased cellular siRNA delivery, and effectively knocked down Survivin protein, impairing the G1 phase and resulting in apoptosis [82].

6.4.14 Hybrid Nanoparticles

Monotherapy nanomedicines have not led to higher response rates than traditional chemotherapy in human clinical studies, despite encouraging preclinical efficacy data. To solve these problems, researchers are creating nanocarriers with improved tumor deposition and controlled release features that can carry out various therapies. In ideal situations, various mechanisms of action for each medicine or therapy method would be used to achieve synergistic results [82]. Some of the hybrid nanoparticles are explained in the following sections.

An enhanced DMP-039, a system for delivery of mRNA, was created by combining a peptide capable to enter cell, cRGD-R9, with a cationic DMP structure. It was created by combining DOTAP lipid with polyethylene glycol the self-assembly of DOTAP lipid and methoxy poly(ethylene glycol)-poly(ε-caprolactone) polymer. High mRNA distribution efficiency was demonstrated by the produced DMP-039 hybrid nanoparticles during transmembrane transit through a variety of routes. A mBim/DMP-039 complex was injected in two models of colon cancer for suppression using a suicide gene [83]. Chitosan hybrid NP was made by self-assembling two amphiphilic copolymers, PEG and PEI-PDLLA. Curcumin (Cur) and small interfering RNA targeting lnc were two of these (DSPE-mPEG). Multiplication and movement of HT-29 cells were effectively stopped, while apoptosis was most frequently induced by CSNP [84].

In order to combine the advantages like biocompatible and stable nature with increased drug loading and release control, improved half-life and effect, hybrid nanoparticles containing combinations of lipids and polymers have been produced [85]. A chitosan-containing siRNA in the core surrounded by lipid-containing Cannabidiol was combined in the form of a nanocarrier. It has the capability to prevent or lessen inflammation and decrease the expression of the HTT gene, which causes Huntington's disease. The hybrid nanoparticle was found to successfully treat Huntington's disease by decreasing mutant HTT expression with decreased contents of cytokine IL-6 [86]. When HOXB7 siRNA is administered via calcium phosphate nanoparticles coated with copolymers, MCF7 cells' sensitivity to Tamoxifen is restored. The siRNA molecule provided by the nanoparticles aids in this decrease in HOXB7 gene expression. A decrease in breast cancer cell viability after 15 hours, which did not happen in the same amount of Tamoxifen treatment alone, shows that gene silencing and the cell viability assay have a significant link [87].

Hybrid NP containing lipid and polymers for effectual and focused CRISPR/Cas9 plasmids delivery to target a gene MGMT resistant to the drug Temozolomide (TMZ), to shield the gene against enzyme attack. These NP may have the ability to target GBM cells and increase their sensitivity to TMZ by transfecting them with pCas9/MGMT to decrease the genetic expression of MGMT. Using *in vivo* FUS irradiation, MBs-LPHNs-cRGD complexes were able to locally and safely increase BBB permeability. They also made it easier for mice with orthotopic tumors to gather nanoparticles to the tumor site, which enhanced the therapeutic benefits of TMZ on glioblastoma [88]. The DOTAP, cholesterol, and PLGA-made nanoparticle F-31 was treated on the surface with the aptamer A6. This aptamer acts on breast cancer cells' overexpressed Her-2 receptors on their surface. The absorption of doxorubicin by resistant 4T1-R is slowly than 4T1- S which is sensitive to doxorubicin, in part due to increased drug efflux expression on the cell surface of resistant cells. Thus the aptamer-based hybrid nanoparticles containing lipids and polymers help in down-regulating the gene by targeting P-gp siRNA, thus increasing the localization of doxorubicin in mice breast cancer cells [89]. Using two frequently used treatments, cell death siRNA and -tocopheryl succinate, a novel calcium carbonate/phosphate nanoparticles coated was created that targets the folic acid (FA) receptor and has increased drug entrapment, with folic acid improving drug uptake by cells. The anti-proliferative effects of the therapeutic nanoparticles on B16F0 melanoma cell proliferation were only marginally synergistic. As part of the inhibition's mechanism, phase and cell cycles are stopped resulting in cell death [90].

Alendronate and plasmid deoxyribonucleic acid is delivered simultaneously with Calcium phosphate nanoparticles in an effort toward promoting cell uptake of two drugs and may provide a synergistic therapeutic impact on similar or distinct cell culture. These DNA-loaded CaP nanoparticles were created using a pH-responsive PEG-Alendronate conjugation. These particles have a potent cytotoxic potential on macrophages and a good transfection effect in cancer cells [91].

To increase the tumor-suppressive effect of co-delivery, the first-line chemotherapy medication paclitaxel (PTX) and a small interfering RNA targeting disintegrin and metalloproteinase 10 (ADAM10) were simultaneously packed in shell-core unique nanoparticles. In particular, siADAM10 is coated in CaP to create the hydrophilic shell and copolymer DSPE-PEG2K is packed with PTX to create the hydrophobic core. When siADAM10 is released from the outer shell of the CaP-PTX/siADAM10 NPs to exert its gene-inhibitory impact and improve the response of tumor cells to anticancer therapy, the inner PTX is then released from the NPs to exercise its tumor-killing action, allowing them to target the tumor site A novel cross-linking technique is introduced that can successfully stabilize CaP nanoparticles and the release of small interfering ribonucleic acid in response to stimuli. We created a polysaccharide derivative of hyaluronic acid called hyaluronic acid with a thiol group having redox-responsive delivery. Calcium phosphate (CaP) effectively condensed siRNA through electrostatic contact to create a positively charged inner "core." As a result, both luciferase and Bcl2 genes were effectively silenced *in vitro* by the hybrid delivery system at a rate of 80%. B16F10 cells had severe apoptosis with a high transfection efficiency when Bcl2 was silenced [92]. Table 6.2 summarizes about nanoparticles used in cancer therapy for gene delivery.

6.5 Challenges and Future Aspects

The delivery mechanisms show a noteworthy influence on the effectiveness of genetic therapy. Genetic and NP engineering will likely be combined in gene therapy in the future. As transfection reagents, cationic lipids, and polymers are frequently utilized. Each of them has benefits and drawbacks. Despite the excitement surrounding recent advances in gene therapy, this area is relatively young. A comparatively small number of human trials of gene treatment are either complete or are being carried out, despite the fact that *in vitro* studies and animal, tests have demonstrated the effectiveness of NPs as gene delivery vehicles.

Additionally, the complexes must be stable before to cellular internalization thanks to the interaction between the gene and cationic carriers. The complexes then frequently settle in endosomes. A crucial stage is an endosomal egress, which can be accelerated by the polyplexes' proton sponge effect or the lipoplexes' ability to fuse lipids. When this happens, the laden nucleic acids diffuse throughout the cytoplasm before being freed from the complexes and translocate into the nucleus. The entire delivery procedure could fail if there are issues in any one of the steps.

Table 6.2 Nanoparticles used for gene delivery in cancer therapy.

Type	Delivery vehicle	Therapeutic modality	Target	Reference
Natural/Synthetic polymer	Chitosan NP	Messenger RNA	Increased transfection	[93]
	Chitosan NP	mRNA	Delivery of mRNA to macrophages	[94]
	Chitosan heparin NP	Small interfering RNA	Effective pH-responsive delivery	[95]
	Chitosan hyaluronic acid dialdehyde		Targeting CD44 for bladder tumor	[96]
	Alkylated 6AC-100derived copolymers	Small interfering RNA	B16 melanoma cell line knockdown	[97]
	Milk exosome		Lung tumor reduction	[98]
	Exosome	MessengerRNA	Glioma model tumor reduction	[99]
				[100]
				[101]
	Chitosan NP		Treatment for cystic fibrosis	[102]

(Continued)

Table 6.2 Nanoparticles used for gene delivery in cancer therapy. (*Continued*)

Type	Delivery vehicle	Therapeutic modality	Target	Reference
	Poly(beta-amino-ester) (PBAE) nanoparticles	Thymidine kinase (sr39) DNA	Reduced tumor size	[36]
	PLGA-PEG-PEI polymer nanoparticles	microRNAs and tumor killer p53 NTR	Better tumor inhibition	[37]
	Poly(β-amino ester) (PBAE)	Thymidine kinase (HSVtk)	Survival of the treated mice improved	[38]
	polymer-based NP delivery	mRNA andnucleases	Improved engraftment capability of human hscs	[39]
	Polyethylene glycol PBGA NP	Messenger ribonucleic acid	Increased gene expression	[103]
	hPBAE		Used for inhalation to lung	[104]
	Cholesteryl polyethylene glycol polyaspartamide		mRNA successfully delivered into the mouse model	[105]
	Polyethylene glycol polyplex		Effective protein delivery in mice brain	[106]
	Polyethylene glycol polyaspartamide		mRNA delivery to brain mice model	[107]

(*Continued*)

Table 6.2 Nanoparticles used for gene delivery in cancer therapy. (*Continued*)

Type	Delivery vehicle	Therapeutic modality	Target	Reference
	polyaspartamide with cyclohexylethyl		Effective Brain Targeting	[108]
	Polyaspartamide		Editing of a gene sequence in brain	[109]
	Fluorinated polyethyleneimine	Small interfering ribonucleic acid (Si RNA)	siRNA delivery in the lungs increased	[110]
	Conjugated PEG polyplexed	Messenger RNA	Inhalation route of mRNA delivery	[111]
	Polyethylene glycol polyaspartamide		mRNA delivery to the kidney	[112]
	Polyethyleneimine Poly-γ-glutamic acid		Increase liver and spleen protein expression	[113]
	Combination antiretroviral NP		Reduction of lymphoma	[114]
	Poly (Beta-AminoEster) NP		Effective immune therapy	[115]

(*Continued*)

Table 6.2 Nanoparticles used for gene delivery in cancer therapy. (*Continued*)

Type	Delivery vehicle	Therapeutic modality	Target	Reference
	Polyethyleneimine modified with tyrosine	Small interfering ribonucleic acid (Si RNA)	Effective antitumor activity	[116]
	Extracellular vesicle-modified		Tumor growth inhibition in prostate carcinoma xenograft	[117]
	Poly (Beta-Amino Ester)		Reduction in glioblastoma	[118]
	Poly (Beta-Amino Ester)		Treatment for glioblastoma	[119]
	Polyglucin spermidine	Messenger RNA	COVID Vaccine	[120]
	Polyethyleneimine vitamin E copolymer		COVID mRNA vaccine	[121]
	Polyethlenimine polyaspartamide/GlcAm		mRNA release with the help ofATP	[122]
Lipid nanoparticles	CRISPR-LNPs	Cas9 mRNA and sgRNAs	Inhibited tumor growth and increased survival	[40]

(*Continued*)

Table 6.2 Nanoparticles used for gene delivery in cancer therapy. (*Continued*)

Type	Delivery vehicle	Therapeutic modality	Target	Reference
	Lipid-protamine- DNA (LPD) nanoparticles	Plasmid DNA encoding the Chemokine Ligand 2	Improved therapeutic efficacy with substantial tumor growth inhibition	[41]
	CRISPR-Cas9 LNP	Cas9 messenger RNA (mRNA) and guide RNA	Profound reductions in serum ANGPTL3 protein	[42]
Inorganic materials	PAMAM-G5D-grafted AuNP	Messenger ribonucleic acid	Transfection effectiveness increase	[123]
	Selenium Nanoparticle		mRNA targeted to hepatic cells	[124]
	Gold Nanoparticles	DNA	HeLa cell growth inhibited	[125]
	Chitosan- functionalized selenium nanoparticles		Provide immunotherapy in cancer	[126]
	gold nanoparticle	siRNA	Tumor growth inhibition	[44]
	gold nanoparticles	siRNAs	Effective drug internalization	[45]
	gold nanoparticles	c-myc	Effective drug internalization	[46]

(*Continued*)

Table 6.2 Nanoparticles used for gene delivery in cancer therapy. (*Continued*)

Type	Delivery vehicle	Therapeutic modality	Target	Reference
	Calcium phosphatelipid NP coated withlipid and double layered hydroxide	Small interfering RNA	Greater cellular uptake	[47]
	Calcium phosphate nanoparticles coated with lipids	siRNA	Improve cellular accumulation	[49]
	liposome-encapsulated calcium phosphate nanoparticles	Small interfering RNA	Reduce VEGF expression *in vitro*	[50]
	Superparamagnetic iron oxide (SPIO) nanoparticles	DNA	Reduced the transfection time	[51]
	Galactose-Polyethelenimine-superparamagnetic iron oxide	Small interfering RNA	Tumor tissue residence increases	[52]
	Fe3O4/PEI-PEG NGs	siRNA	Inhibits cancer cell growth	[53]
	magnetic nanoparticles	Plasmid DNA	Enhanced transfection efficiency	[55]

(*Continued*)

Table 6.2 Nanoparticles used for gene delivery in cancer therapy. (*Continued*)

Type	Delivery vehicle	Therapeutic modality	Target	Reference
	iron oxide superparamagnetic nanoparticles, coated with polyethyleneimine (PEI)	GFP plasmid	Greater HEK 293 cell transfection efficiency	[56]
	Magnetic nanoparticles witha cationic polymer, polyethyleneimine, and polyethylene glycol	siRNA	The decline in cell viability	[57]
	magnetic nanoparticles Fe3O4@PEI	siRNA	Tumor mass reduction	[58]
	Silk-Polyethelenimine NP/magnetic-silk Polyethyleneimine NP	c-myc antisense oligodeoxynucleotides	Improved cellular uptake	[59]
	Zeolite imidazolate framework	RNA primer	Prevent protease attacks and nuclease degradation on polymerases	[60]
	zeolitic-imidazolate nanoparticle	RNA	Drop in RPSA expression	[62]

(*Continued*)

Table 6.2 Nanoparticles used for gene delivery in cancer therapy. (*Continued*)

Type	Delivery vehicle	Therapeutic modality	Target	Reference
	Mesoporous silica nanoparticles	siRNA	Enhanced therapeutic efficacy	[63]
	magnetic mesoporous silica nanoparticles	siRNA	Accelerated gene transport	[64]
Dendrimers	(PAMAM) dendrimers	pDNA	Upregulating p53 and p21 protein	[66]
	dendrimers (PAMAM, G3) attached gold nanorods	FAM172A	Outstanding transfection efficiency	[67]
	diaminobutyric polypropylenimine (DAB) dendrimer	DNA	High transfection effectiveness	[68]
	(PAMAM) dendrimer	lncRNA MEG3	Substantial anti-CRPC impact	[69]
	fluorinated poly (amidoamine)	mRNA	Inhibiting tumor growth	[70]
	dendrimer (G4-F735)			
	Gemcitabine gold dendrimer NP/Micro RNA-21li	Micro RNA-21	Significantly reduce tumor volume	[71]
	CD-PAMAM dendrimers	RGDS peptide	High transfection effectiveness	[72]

(*Continued*)

Table 6.2 Nanoparticles used for gene delivery in cancer therapy. (*Continued*)

Type	Delivery vehicle	Therapeutic modality	Target	Reference
Micelles	DOTAP-mPEG-PCL (DMP)	siMcl1 and siBcl-xl	Improved anticancer effect	[73]
	DMP/phBimS	Plasmid	High transfection effectiveness	[74]
	DOTAP-mPEG-PCL (DMP)	interleukin2 2 (IL-22)	Lymphocyte infiltration and angiogenesis inhibition	[75]
	DOTAP-mPEG-PCL (DMP)	mRNA	Significant inhibitory effect on C26 colon cancer models	[76]
Stimulus responsive nanoparticles	Redox-responsivefolate Polyethyleneimine β CD	hTERT siRNA	Effective gene transfection	[78]
	pH-triggered dendrimers	siRNA	High tumor-penetrating ability	[79]
	Charge-sensitive niosome	microRNA-34a	Higher tumor inhibition	[80]
Carbon dots	Arginine- CQDs/pEGFP (W/W) nanocomplex	pEGFP	Higher transfection efficiency	[81]
	PEI (polyethyleneimine)-adsorbed Cdots nanoparticles (Cdots@PEI)	SurvivinsiRNA	increase in cellular delivery	[82]

(*Continued*)

Table 6.2 Nanoparticles used for gene delivery in cancer therapy. (*Continued*)

Type	Delivery vehicle	Therapeutic modality	Target	Reference
Hybrid system	Zr-based metal-organic framework	Messenger Ribonucleic acid	mRNA transfection increased	[127]
	Supramolecular metal-organic framework	Messenger Ribonucleic acid	mRNA level analysis and imaging	[128]
	DOTAP/protamine		Improved cellular transfection	[129]
	Zr-based metal-organic framework	Messenger Ribonucleic acid	Better effect on ovarian cancer	[130]
	Zeolitic imidazolate frameworks-8	Messenger Ribonucleic acid	Suppress breast tumor growth	[131]
	Cell-penetrating peptide Functionalized DMP/PEG	Messenger Ribonucleic acid	Colon cancer growth reduction	[132]
	Polylactic glycolic acid		Better immune effect after vaccination	[133]
	Polylactic/cell penetrating peptide		Better effect of mRNA vaccine	[134]
	DMP-039 hybrid nanoparticles	Messenger Ribonucleic acid	Higher tumor inhibition	[83]

(*Continued*)

Table 6.2 Nanoparticles used for gene delivery in cancer therapy. (*Continued*)

Type	Delivery vehicle	Therapeutic modality	Target	Reference
	Polymeric hybrid nanoparticles (CSNP)	Small interfering RNA	The highest rate of apoptosis	[84]
	chitosan-based matrix hybrid nanoparticles	siRNA	Reduce the expression of the Huntington's Disease gene	[86]
	Calcium phosphate nanoparticles	siRNA	Reduction in breast cancer viability	[87]
	Lipid-polymer hybrid nanoparticles (LPHNs-cRGD)	CRISPR/Cas9	Improved the therapeutic benefits of TMZ	[88]
	Lipid-based hybrid nanoparticles	P-gp siRNA	Boost the intracellular trafficking of doxorubicin	[89]
	Lipid-coated calcium carbonate/phosphate nanoparticle	siRNA	The G1 phase of the cell cycle isimpaired	[90]
	CaP nanoparticles	Plasmid DNA	Good transfection efficiency	[91]
	CaP nanoparticles	siRNA	High transfection efficiency	[92]

There has been a lot of work done in this field that has helped with the transport of nucleic acids utilizing nanoparticles. During the last few years, biodegradable materials have been extensively used in the formation of nanoparticles, as they help in making them safer, nontoxic, non-immunogenic, and very stable with high transfection effectiveness. With better flexibility involved in using NP-mediated gene therapy, the most effective set of variables can be chosen. Therefore, these nanoparticles for gene transfer will prove useful and improve the effectiveness of the genes' ability to fight cancer.

References

1. Fang, B. and Roth, J.A., Tumor suppressing gene therapy. *Cancer Biol. Ther.*, 2, supl, 114–20, 2003.
2. Belete, T.M., The current status of gene therapy for the treatment of cancer. *Biol.: Targets Ther.*, 15, 67, 2021.
3. US Food and Drug Administration, What is gene therapy, 2018. https://www.fda.gov/BiologicsBloodVaccines/CellularGeneTherapyProducts/ucm573960.htm.
4. Ginn, S.L., Amaya, A.K., Alexander, I.E., Edelstein, M., Abedi, M.R., Gene therapy clinical trials worldwide to 2017: An update. *J. Gene Med.*, 20, 5, e3015, 2018.
5. Sung, Y.K. and Kim, S.W., The practical application of gene vectors in cancer therapy. *Integr. Cancer Sci. Ther.*, 5, 1–5, 2018.
6. Anguela, X.M. and High, K.A., Entering the modern era of gene therapy. *Annu. Rev. Med.*, 70, 1, 273–88, 2019.
7. Sinclair A, Islam S, Jones S. Gene therapy: An overview of approved and pipeline technologies. In: *CADTH Issues in Emerging Health Technologies. Ottawa (ON): Canadian Agency for Drugs and Technologies in Health,* 1–23, 2018.
8. Dunbar, C.E., High, K.A., Joung, J.K., Kohn, D.B., Ozawa, K., Sadelain, M., Gene therapy comes of age. *Science*, 359, 6372, eaan4672, 2018.
9. Duffy, M.R., Fisher, K.D., Seymour, L.W., Making oncolytic virotherapy a clinical reality: The European contribution. *Hum. Gene Ther.*, 28, 11, 1033–46, 2017.
10. Marsden, G., Towse, A., Pearson, S.D., Dreitlein, B., Henshall, C., Gene therapy: Understanding the science, assessing the evidence, and paying for value, 2017.
11. MIT NEWDIGS Initiative, Existing gene therapy pipeline likely to yield dozens of approved products within five years 2017F211, v011, 2017.
12. Karjoo, Z., Chen, X., Hatefi, A., Progress and problems with the use of suicide genes for targeted cancer therapy. *Adv. Drug Deliv. Rev.*, 99, Pt A, 113–128, 2016.

13. Navarro, S.A., Carrillo, E., Griñán-Lisón, C., Martín, A., Perán, M., Marchal, J.A., Boulaiz, H., Cancer suicide gene therapy: A patent review. *Expert Opin. Ther. Pat.*, 26, 9, 1095–104, 2016.
14. Z. Zhu, Regulation of human carboxypeptidase A. WO2002052020, 2003.
15. N. Minton, G. Anlezark, T. Vaughan, Nitroreductase enzymes. US20040014191, 2004.
16. Recombinant beta-lactamase-RGD-fusion protein and application thereof in medicine. N101824406, 2010.
17. M.I. Rojo, V.G.E. Barreras, R. Gargini, Synergistic effect between a cyanogenic system and another oxidative inducing system for treating tumors. US20100028366, 2010.
18. D.J. Waxman and L. Chen, Cytochrome p450 reductase for the enhancement of p450-based anti-cancer gene therapy. EP1017.
19. R. Marais and C. Springer, Tommasini S Vaccinia virus for gene-directed enzyme prodrug therapy. WO2014202977, 2014.
20. H.E. Gruber, D.J. Jolly, O.D. Perez *et al.*, Gene therapy vectors and cytosine deaminases. US 20140242033, 2014.
21. W.B. Parker and E.J. Sorscher, Purine nucleoside phosphorylase as enzymatic activator of nucleoside prodrugs. US8628767, 2014.
22. J.P. Levy, R.A. Reed, J. Mcnulty *et al.*, Improved thymidine kinase gene. WO 2014153258, 2014.
23. Y.C. Sung, S.W. Kim, S.J. Kim *et al.*, Vector simultaneously expressing dodecameric trail and hsv-tk suicide genes, and anticancer stem cell therapeutic agent using same. US 20140369979, 2014.
24. M. Toda, H. Okano, H. Miyoshi, R. Tamura, inventors; Keio University, assignee, Suicide gene therapeutic agent for brain tumors using pluripotent stem cell. United States patent application US 16/765,257, 2021.
25. Mali, S., Delivery systems for gene therapy. *Indian J. Hum. Genet.*, 19, 1, 3, 2013.
26. Sung, Y.K. and Kim, S.W., Recent advances in the development of gene delivery systems. *Biomater. Res.*, 23, 1, 8, 2019.
27. Humphreys, I.R. and Sebastian, S., Novel viral vectors in infectious diseases. *Immunology*, 153, 1, 1–9, 2018.
28. Ramamoorth, M. and Narvekar, A., Non viral vectors in gene therapy-an overview. *J. Clin. Diagn. Res.*, 9, 1, GE01, 2015.
29. Buck, J., Grossen, P., Cullis, P.R., Huwyler, J., Witzigmann, D., Lipid-based DNA therapeutics: Hallmarks of non-viral gene delivery. *ACS Nano*, 13, 4, 3754–3782, 2019.
30. Yang, R., Chen, F., Guo, J., Zhou, D., Luan, S., Recent advances in polymeric biomaterials-based gene delivery for cartilage repair. *Bioact. Mater.*, 5, 4, 990–1003, 2020.
31. Pilgrim, S., Stritzker, J., Schoen, C. *et al.*, Bactofection of mammalian cells by Listeria monocytogenes: Improvement and mechanism of DNA delivery. *Gene Ther.*, 10, 24, 2036–2045, 2003.

32. Li, T., Kang, G., Wang, T., Huang, H.E., Tumor angiogenesis and anti-angiogenic gene therapy for cancer. *Oncol. Lett.*, 16, 1, 687–702, 2018.
33. Mirza, Z. and Karim, S., Nanoparticles-based drug delivery and gene therapy for breast cancer: Recent advancements and future challenges [Internet]. *Semin. Cancer Biol.* Elsevier Ltd, 69, 226–237, 2021.
34. Zielinska, A., Carreiró, F., Oliveira, A.M., Neves, A., Pires, B., Nagasamy Venkatesh D., *et al.* Polymeric nanoparticles: production, characterization, toxicology and ecotoxicology. *Molecules*, 25, 16, 3731, 2020.
35. Lin, G., Zhang, H., Huang, L., Smart polymeric nanoparticles for cancer gene delivery. *Mol. Pharm.*, 12, 2, 314–21, 2015.
36. Vaughan, H.J., Zamboni, C.G., Hassan, L.F., Radant, N.P., Jacob, D., Mease, R.C. *et al.*, Polymeric nanoparticles for dual-targeted theranostic gene delivery to hepatocellular carcinoma. *Sci. Adv.*, 8, 29, 1–13, 2022.
37. Kumar, S.U., Wang, H., Telichko, A.V., Natarajan, A., Bettinger, T., Cherkaoui, S. *et al.*, Ultrasound triggered co-delivery of therapeutic microRNAs and a triple suicide gene therapy vector by using biocompatible polymer nanoparticles for improved cancer therapy in mouse models. *Adv. Ther.*, 4, 5, 1–21, 2021.
38. Mangraviti, A., Tzeng, S.Y., Kozielski, K.L., Wang, Y., Jin, Y., Gullotti, D. *et al.*, Polymeric nanoparticles for nonviral gene therapy extend brain tumor survival *in vivo*. *ACS Nano*, 9, 2, 1236–49, 2015.
39. El-Kharrag, R., Berckmueller, K.E., Madhu, R., Cui, M., Campoy, G., Mack, H.M., Wolf, C.B., Perez, A.M., Humbert, O., Kiem HP, R.S., Efficient polymer nanoparticle-mediated delivery of gene editing reagents into human hematopoietic stem and progenitor cells. *Mol. Ther.*, 30, 6, 2186–98, 2022.
40. Rosenblum, D., Gutkin, A., Kedmi, R., Ramishetti, S., Veiga, N., Jacobi A.M., *et al.* CRISPR- Cas9 genome editing using targeted lipid nanoparticles for cancer therapy. *Sci Adv*, 6, 47, p. eabc9450, 2020.
41. Liu, Y., Tiruthani, K., Wang, M., Zhou, X., Qiu, N., Xiong, Y. *et al.*, Tumor-targeted gene therapy with lipid nanoparticles inhibits tumor-associated adipocytes and remodels the immunosuppressive tumor microenvironment in triple-negative breast cancer. *Nanoscale Horiz.*, 6, 4, 319–29, 2021.
42. Qiu, M., Glass, Z., Chen, J., Haas, M., Jin, X., Zhao, X., *et al.* Lipid nanoparticle-mediated codelivery of Cas9 mRNA and single-guide RNA achieves liver-specific *in vivo* genome editing of Angptl3. *Proc. Natl. Acad. Sci.* U.S.A., 118, 10, p. e2020401118, 20.21.
43. Sharma, A., Goyal, A.K., Rath, G., Recent advances in metal nanoparticles in cancer therapy. *J. Drug Targeting* [Internet], 26, 8, 617–32, 2018.
44. Xue, C., Hu, S., Gao, Z.H., Wang, L., Luo, M.X., Yu, X., *et al.* Programmably tiling rigidified DNA brick on gold nanoparticle as multi-functional shell for cancer-targeted delivery of siRNAs. *Nat. Commun.*, 12, 1, 2928, 2021.
45. Tunç, C.Ü. and Aydin, O., Co-delivery of Bcl-2 siRNA and doxorubicin through gold nanoparticle-based delivery system for a combined cancer therapy approach. *J. Drug Deliv. Sci. Technol.*, 74, 103603, 2022.

46. Huo, S., Gong, N., Jiang, Y., Chen, F., Guo, H., Gan, Y., *et al.* Gold-DNA nanosunflowers for efficient gene silencing with controllable transformation. *Sci Adv.*, 5, 10, p. eaaw6264, 2019.
47. Khalifehzadeh, R. and Arami, H., Biodegradable calcium phosphate nanoparticles for cancer therapy. *Adv. Colloid Interface Sci.* [Internet], 279, 102157, 2020.
48. Wu, Y., Gu, W., Li, L., Chen, C., Xu, Z.P., Enhancing PD-1 gene silence in T lymphocytes by comparing the delivery performance of two inorganic nanoparticle platforms. *Nanomaterials*, 9, 2, 1–12, 2019.
49. Tang, J., Li, L., Howard, C.B., Mahler, S.M., Huang, L., Xu, Z.P., Preparation of optimized lipid- coated calcium phosphate nanoparticles for enhanced *in vitro* gene delivery to breast cancer cells. *J. Mater. Chem. B*, 3, 33, 6805–12, 2015.
50. Chen, J., Sun, X., Shao, R., Xu, Y., Gao, J., Liang, W., VEGF siRNA delivered by polycation liposome-encapsulated calcium phosphate nanoparticles for tumor angiogenesis inhibition in breast cancer. *Int. J. Nanomed.*, 12, 6075–88, 2017.
51. Zuvin, M., Kuruoglu, E., Kaya, V.O., Unal, O., Kutlu, O., Yagci Acar, H. *et al.*, Magnetofection of green fluorescent protein encoding DNA-bearing polyethyleneimine-coated superparamagnetic iron oxide nanoparticles to human breast cancer cells. *ACS Omega*, 4, 7, 12366–74, 2019.
52. Yang, Z., Duan, J., Wang, J., Liu, Q., Shang, R., Yang, X. *et al.*, Superparamagnetic iron oxide nanoparticles modified with polyethyleneimine and galactose for siRNA targeted delivery in hepatocellular carcinoma therapy. *Int. J. Nanomed.*, 13, 1851–65, 2018.
53. Peng, Y., Gao, Y., Yang, C., Guo, R., Shi, X., Cao, X., Low-molecular-weight poly(ethylenimine) nanogels loaded with ultrasmall iron oxide nanoparticles for T1-weighted MR imaging-guided gene therapy of sarcoma. *ACS Appl. Mater. Interfaces*, 13, 24, 27806–13, 2021.
54. Zhang, J., Zhang, T., Gao, J., Biocompatible iron oxide nanoparticles for targeted cancer gene therapy: A review. *Nanomaterials*, 12, 19, 3323, 2022.
55. Basoglu, H., Goncu, B., Akbas, F., Magnetic nanoparticle-mediated gene therapy to induce Fas apoptosis pathway in breast cancer. *Cancer Gene Ther.*, 25, 5–6, 141–7, 2018.
56. Zhang, W., Abdelrasoul, G.N., Savchenko, O., Abdrabou, A., Wang, Z., Chen, J., Ultrasound- assisted magnetic nanoparticlebased gene delivery. *PLoS One* [Internet], 15, 9 September, 1–16, 2020.
57. Panday, R., Abdalla, A.M.E., Yu, M., Li, X., Ouyang, C., Yang, G., Functionally modified magnetic nanoparticles for effective siRNA delivery to prostate cancer cells *in vitro*. *J. Biomater. Appl.*, 34, 7, 952–64, 2020.
58. Yuan, C., An, Y., Wang, L., Magnetic nanoparticles Fe3O4@PEI induced targeted suicide gene therapy combined with magnetic fluid hyperthermia on hepatoma xenograft. *Chin. J. Cancer Biother.*, 25, 9, 913–9, 2018.

59. Song, W., Gregory, D.A., Al-janabi, H., Muthana, M., Cai, Z., Zhao, X., Magnetic-silk/polyethyleneimine core-shell nanoparticles for targeted gene delivery into human breast cancer cells. *Int. J. Pharm.* [Internet], 555, 322–36, 2019.
60. Zhang, J., He, M., Nie, C., He, M., Pan, Q., Liu, C. *et al.*, Biomineralized metal-organic framework nanoparticles enable a primer exchange reaction-based DNA machine to work in living cells for imaging and gene therapy. *Chem. Sci.*, 11, 27, 7092–101, 2020.
61. Chen, D., Cai, L., Guo, Y., Chen, J., Gao, Q., Yang, J. *et al.*, Cancer cell membrane-decorated zeolitic-imidazolate frameworks codelivering cisplatin and oleanolic acid induce apoptosis and reversed multidrug resistance on bladder carcinoma cells. *ACS Omega*, 5, 2, 995–1002, 2020.
62. Poddar, A., Pyreddy, S., Carraro, F., Dhakal, S., Rassell, A., Field, M.R., Reddy, T.S, Doherty, C.M., Shukla, R., ZIF polymorphs for nucleic acid delivery and targeted knockdown of gene expression in prostate cancer. *Cancer Chem. Commun.*, 56, 98, 1–29, 2020.
63. Dilnawaz, F. and Sahoo, S.K., Augmented anticancer efficacy by si-RNA complexed drug-loaded mesoporous silica nanoparticles in lung cancer therapy. *ACS Appl. Nano Mater.*, 1, 2, 730–40, 2018.
64. Wang, Z., Chang, Z., Lu, M., Shao, D., Yue, J., Yang, D. *et al.*, Shape-controlled magnetic mesoporous silica nanoparticles for magnetically-mediated suicide gene therapy of hepatocellular carcinoma [Internet]. *Biomaterials. Elsevier Ltd*, 154, 147–157, 2018.
65. Santander-Ortega, M.J., de la Fuente, M., Lozano, M.V., Tsui, M.L., Bolton, K., IFU and AGS, Optimisation of synthetic vector systems for cancer gene therapy – The role of the excess of cationic dendrimer under physiological conditions. *Curr. Top. Med. Chem.*, 14, 9, 1172–81, 2014.
66. Mekuria, S.L., Li, J., Song, C., Gao, Y., Ouyang, Z., Shen, M. *et al.*, Facile formation of PAMAM dendrimer nanoclusters for enhanced gene delivery and cancer gene therapy. *ACS Appl. Bio Mater.*, 4, 9, 7168–75, 2021.
67. Ye, L., Chen, Y., Mao, J., Lei, X., Yang, Q., Cui, C., Dendrimer-modified gold nanorods as a platform for combinational gene therapy and photothermal therapy of tumors. *J. Exp. Clin. Cancer Res.* [Internet], 40, 1, 1–16, 2021.
68. Somani, S., Laskar, P., Altwaijry, N., Kewcharoenvong, P., Irving, C., Robb, G. *et al.*, PEGylation of polypropylenimine dendrimers: Effects on cytotoxicity, DNA condensation, gene delivery and expression in cancer cells. *Sci. Rep.*, 8, 1, 1–13, 2018.
69. Tai, Z., Ma, J., Ding, J., Pan, H., Chai, R., Zhu, C. *et al.*, Aptamer-functionalized dendrimer delivery of plasmid-encoding lncrna meg3 enhances gene therapy in castration-resistant prostate cancer. *Int. J. Nanomed.*, 15, 10305–20, 2020.
70. Wang, Y., Wang, M., Chen, H., Liu, H., Zhang, Q., Cheng, Y., Fluorinated dendrimer for TRAIL gene therapy in cancer treatment. *J. Mater. Chem. B* [Internet], 4, 7, 1354, 2016.

71. Lin, L., Fan, Y., Gao, F., Jin, L., Li, D., Sun, W. *et al.*, UTMD-promoted co-delivery of gemcitabine and miR-21 inhibitor by dendrimer-entrapped gold nanoparticles for pancreatic cancer therapy. *Theranostics*, 8, 7, 1923–39, 2018.
72. Ghosh, S., Ghosal, K., Mohammad, S.A., Sarkar, K., Dendrimer functionalized carbon quantum dot for selective detection of breast cancer and gene therapy. *Chem. Eng. J.* [Internet], 373, 468–84, 2019.
73. Lu, Y., Zhong, L., Jiang, Z., Pan, H., Zhang, Y., Zhu, G., *et al.* Cationic micelle-based siRNA delivery for efficient colon cancer gene therapy. *Nanoscale Res. Lett.*, 14, 1–9, 2019.
74. Ma, P., Li, J., Gao, Y., Wu, J., Men, K., Li, C. *et al.*, Local and systemic delivery of the BimS gene nano-complex for efficient oral squamous cell carcinoma therapy. *Int. J. Nanomed.*, 17, 2925–41, 2022.
75. Men, K., Huang, R., Zhang, X., Zhang, R., Zhang, Y., Peng, Y. *et al.*, Delivery of interleukin-22 binding protein (IL-22BP) gene by cationic micelle for colon cancer gene therapy. *RSC Adv.*, 8, 30, 16537–48, 2018.
76. Li, J., Men, K., Gao, Y., Wu, J., Lei, S., Yang, Y. *et al.*, Single micelle vectors based on lipid/block copolymer compositions as mRNA formulations for efficient cancer immunogene therapy. *Mol. Pharm.*, 18, 11, 4029–45, 2021.
77. Liang, J., Yang, B., Zhou, X., Han, Q., Zou, J., Cheng, L., Stimuli-responsive drug delivery systems for head and neck cancer therapy. *Drug Deliv.* [Internet], 28, 1, 272–84, 2021.
78. Mousazadeh, H., Bonabi, E., Zarghami, N., Stimulus-responsive drug/gene delivery system based on polyethyleneimine cyclodextrin nanoparticles for potential cancer therapy. *Carbohydr. Polym.*, 276, 34823779, 2022.
79. Shi, M., Wang, Y., Zhao, X., Zhang, J., Hu, H., Qiao, M. *et al.*, Stimuli-responsive and highly penetrable nanoparticles as a multifunctional nanoplatform for boosting nonsmall cell lung cancer siRNA therapy. *ACS Biomater. Sci. Eng.*, 7, 7, 3141–55, 2021.
80. Abtahi, N.A., Naghib, S.M., Ghalekohneh, S.J., Mohammadpour, Z., Nazari, H., Mosavi, S.M. *et al.*, Multifunctional stimuli-responsive niosomal nanoparticles for co-delivery and co-administration of gene and bioactive compound: *In vitro* and *in vivo* studies. *Chem. Eng. J.* [Internet], 429, June 2021, 132090, 2022.
81. Rezaei, A. and Hashemi, E., A pseudohomogeneous nanocarrier based on carbon quantum dots decorated with arginine as an efficient gene delivery vehicle. *Sci. Rep.* [Internet], 11, 1, 1–10, 2021.
82. Wang, Q., Zhang, C., Shen, G., Liu, H., Fu, H., Cui, D., Fluorescent carbon dots as an efficient siRNA nanocarrier for its interference therapy in gastric cancer cells. *J. Nanobiotechnol.*, 12, 1, 1–12, 2014.
83. Gao, Y., Men, K., Pan, C., Li, J., Wu, J., Chen, X. *et al.*, Functionalized dmp-039 hybrid nanoparticle as a novel mrna vector for efficient cancer suicide gene therapy. *Int. J. Nanomed.*, 16, July, 5211–32, 2021.

84. Jia, F., Li, Y., Deng, X., Wang, X., Cui, X., Lu, J. *et al.*, Self-assembled fluorescent hybrid nanoparticles-mediated collaborative lncRNA CCAT1 silencing and curcumin delivery for synchronous colorectal cancer theranostics. *J. Nanobiotechnol.* [Internet], 19, 1, 1–15, 2021.
85. Mohanty, A., Uthaman, S., Park, I.K.. Utilization of polymer-lipid hybrid nanoparticles for targeted anti-cancer therapy. *Molecules*, 25, 19, 4377, 2020.
86. Fihurka, O., Sava, V., Sanchez-Ramos, J., Dual-function hybrid nanoparticles with gene silencing and anti-inflammatory effects. *Nanomedicine*, 17, 9, 577–90, 2022.
87. Valle ABC dos, S., Gualberto, A.C., Ferreira, K.C.B., Creczynski-Pasa, T.B., Gameiro, J., Tavares, G.D. *et al.*, HOXB7 siRNA delivered by hybrid nanoparticles and the co-therapy with tamoxifen: Promising strategy against hormone receptor-positive breast cancer, vol. 69, 2021.
88. Yang, Q., Zhou, Y., Chen, J., Huang, N., Wang, Z., Cheng, Y., Gene therapy for drug-resistant glioblastoma via lipid-polymer hybrid nanoparticles combined with focused ultrasound. *Int. J. Nanomed.*, 16, 185–99, 2021.
89. Chandra, S., Nguyen, H., Wiltz, K., Hall, N., Chaudhry, S., Olverson, G. *et al.*, Aptamer-functionalized hybrid nanoparticles to enhance the delivery of doxorubicin into breast cancer cells by silencing P-glycoprotein. *J. Cancer Treatmet Diagn.*, 4, 1, 1–13, 2020.
90. Wu, Y., Gu, W., Xu, Z.P., Enhanced combination cancer therapy using lipid-calcium carbonate/phosphate nanoparticles as a targeted delivery platform. *Nanomedicine*, 14, 1, 77–92, 2019.
91. Bisso, S., Mura, S., Castagner, B., Couvreur, P., Leroux, J.C., Dual delivery of nucleic acids and PEGylated-bisphosphonates via calcium phosphate nanoparticles. *Eur. J. Pharm. Biopharm.*, 142, 142–52, 2019.
92. Zhou, Z., Li, H., Wang, K., Guo, Q., Li, C., Jiang, H. *et al.*, Bioreducible cross-linked hyaluronic acid/calcium phosphate hybrid nanoparticles for specific delivery of siRNA in melanoma tumor therapy. *ACS Appl. Mater. Interfaces*, 9, 17, 14576–89, 2017.
93. Lee, W.J., Kim, K.J., Hossain, M., Cho, H.Y., Choi, J.W., DNA–gold nanoparticle conjugates for intracellular MiRNA detection using surface-enhanced Raman spectroscopy. *Biochip J.*, 16, 1, 33–40, 2022.
94. Nguyen, M.A., Wyatt, H., Susser, L., Geofrion, M., Rasheed, A. *et al.*, Delivery of microRNAs by chitosan nanoparticles to functionally alter macrophage cholesterol efux *in vitro* and *in vivo*. *ACS Nano*, 13, 6491–6505, 2019.
95. Pilipenko, I., Korzhikov-Vlakh, V., Sharoyko, V., Zhang, N., Schäfer-Korting, M., Rühl, E., Zoschke, C., Tennikova, T., pH-sensitive chitosan–heparin nanoparticles for effective delivery of genetic drugs into epithelial cells. *Pharmaceutics*, 11, 7, 317, 2019.
96. Liang, Y., Wang, Y., Wang, L., Liang, Z., Li, D., Xu, X., Chen, Y., Yang, X., Zhang, H., Niu, H., Self-cross-linkable chitosan-hyaluronic acid dialdehyde nanoparticles for CD44-targeted siRNA delivery to treat bladder cancer. *Bioact. Mater.*, 6, 2, 433–46, 2021.

97. Erdene-Ochir, T., Ganbold, T., Zandan, J., Han, S., Borjihan, G., Baigude, H., Alkylation enhances biocompatibility and siRNA delivery efficiency of cationic curdlan nanoparticles. *Int. J. Biol. Macromol.*, 143, 118–25, 2020.
98. Aqil, F., Munagala, R., Jeyabalan, J., Agrawal, A.K., Kyakulaga, A.H., Wilcher, S.A., Gupta, R.C., Milk exosomes-natural nanoparticles for siRNA delivery. *Cancer Lett.*, 449, 186–95, 2019.
99. Yang, Z., Shi, J., Xie, J., Wang, Y., Sun, J. *et al.*, Large-scale generation of functional mRNA-encapsulating exosomes via cellular nanoporation. *Nat. Biomed. Eng.*, 4, 69–83, 2020, https://doi.org/ 10.1038/s41551-019-0485-1.
100. Kojima, R., Bojar, D., Rizzi, G., Hamri, G.C., El-Baba, M.D., Saxena, P., Ausländer, S., Tan, K.R., Fussenegger, M., Designer exosomes produced by implanted cells intracerebrally deliver therapeutic cargo for Parkinson's disease treatment. *Nat. Commun.*, 9, 1, 1–0, 2018.
101. Li, Z., Zhao, P., Zhang, Y., Wang, J., Wang, C., Liu, Y., Yang, G., Yuan, L., Exosome- based Ldlr gene therapy for familial hypercholesterolemia in a mouse model. *Theranostics*, 11, 6, 2953, 2021.
102. Kolonko, A.K., Efing, J., González-Espinosa, Y., Bangel-Ruland, N., van Driessche, W., Goycoolea, F.M., Weber, W.M., Capsaicin-loaded chitosan nanocapsules for wtCFTR-mRNA delivery to a cystic fibrosis cell line. *Biomedicines*, 8, 9, 364, 2020.
103. Miyazaki, T., Uchida, S., Nagatoishi, S., Koji, K., Hong, T., Fukushima, S., Tsumoto, K., Ishihara, K., Kataoka, K., Cabral, H., Polymeric nanocarriers with controlled chain flexibility boost mRNA delivery *in vivo* through enhanced structural fastening. *Adv. Healthcare Mater.*, 9, 16, 2000538, 2020.
104. Patel, A.K., Kaczmarek, J.C., Bose, S., Kauffman, K.J., Mir, F., Heartlein, M.W., DeRosa, F., Langer, R., Anderson, D.G., Inhaled nanoformulated mRNA polyplexes for protein production in lung epithelium. *Adv. Mater.*, 31, 8, 1805116, 2019.
105. Yoshinaga, N., Uchida, S., Naito, M., Osada, K., Cabral, H., Kataoka, K., Induced packaging of mRNA into polyplex micelles by regulated hybridization with a small number of cholesteryl RNA oligonucleotides directed enhanced *in vivo* transfection. *Biomaterials*, 197, 255–67, 2019.
106. Koji, K., Yoshinaga, N., Mochida, Y., Hong, T., Miyazaki, T., Kataoka, K., Osada, K., Cabral, H., Uchida, S., Bundling of mRNA strands inside polyion complexes improves mRNA delivery efficiency *in vitro* and *in vivo*. *Biomaterials*, 261, 120332, 2020.
107. Chan, L.Y., Khung, Y.L., Lin, C.Y., Preparation of messenger RNA nanomicelles via non-cytotoxic PEG-polyamine nanocomplex for intracerebroventicular delivery: A proof-of-concept study in mouse models. *Nanomaterials*, 9, 1, 67, 2019.
108. Kim, H.J., Ogura, S., Otabe, T., Kamegawa, R., Sato, M., Kataoka, K., Miyata, K., Fine- tuning of hydrophobicity in amphiphilic polyaspartamide derivatives for rapid and transient expression of messenger RNA directed toward genome engineering in brain. *ACS Cent. Sci.*, 5, 11, 1866–75, 2019.

109. Abbasi, S., Uchida, S., Toh, K., Tockary, T.A., Dirisala, A., Hayashi, K., Fukushima, S., Kataoka, K., Co-encapsulation of Cas9 mRNA and guide RNA in polyplex micelles enables genome editing in mouse brain. *J. Controlled Release*, 332, 260–8, 2021.
110. Xue, L., Yan, Y., Kos, P., Chen, X., Siegwart, D.J., PEI fluorination reduces toxicity and promotes liver-targeted siRNA delivery. *Drug Deliv. Transl. Res.*, 11, 1, 255–60, 2021.
111. Grun, M.K., Suberi, A., Shin, K., Lee, T., Gomerdinger, V. *et al.*, PEGylation of poly(amine-co-ester) polyplexes for tunable gene delivery. *Biomaterials*, 272, 120780, 2021.
112. Oyama, N., Kawaguchi, M., Itaka, K., Kawakami, S., Efficient messenger RNA delivery to the kidney using renal pelvis injection in mice. *Pharmaceutics*, 13, 11, 1810, 2021.
113. Hamada, E., Kurosaki, T., Hashizume, J., Harasawa, H., Nakagawa, H., Nakamura, T., Kodama, Y., Sasaki, H., Anionic complex with efficient expression and good safety profile for mRNA delivery. *Pharmaceutics*, 13, 1, 126, 2021.
114. Haabeth, O.A., Blake, T.R., McKinlay, C.J., Tveita, A.A., Sallets, A., Waymouth, R.M., Wender, P.A., Levy, R., Local delivery of Ox40l, Cd80, and Cd86 mRNA kindles global anticancer immunity. *Cancer Res.*, 79, 7, 1624–34, 2019.
115. Zhang, F., Parayath, N.N., Ene, C.I., Stephan, S.B., Koehne, A.L., Coon, M.E., Holland, E.C., Stephan, M.T., Genetic programming of macrophages to perform anti-tumor functions using targeted mRNA nanocarriers. *Nat. Commun.*, 10, 1, 1–6, 2019.
116. Ewe, A., Noske, S., Karimov, M., Aigner, A., Polymeric nanoparticles based on tyrosine-modified, low molecular weight polyethyleneimines for siRNA delivery. *Pharmaceutics*, 11, 11, 600, 2019.
117. Zhupanyn, P., Ewe, A., Büch, T., Malek, A., Rademacher, P., Müller, C., Reinert, A., Jaimes, Y., Aigner, A., Extracellular vesicle (ECV)-modified polyethyleneimine (PEI) complexes for enhanced siRNA delivery *in vitro* and *in vivo*. *J. Controlled Release*, 319, 63–76, 2020.
118. Karlsson, J., Rui, Y., Kozielski, K.L., Placone, A.L., Choi, O., Tzeng, S.Y., Kim, J., Keyes, J.J., Bogorad, M.I., Gabrielson, K., Guerrero-Cazares, H., Engineered nanoparticles for systemic siRNA delivery to malignant brain tumours. *Nanoscale*, 11, 42, 20045–57, 2019.
119. Kozielski, K.L., Ruiz-Valls, A., Tzeng, S.Y., Guerrero-Cázares, H., Rui, Y., Li, Y., Vaughan, H.J., Gionet-Gonzales, M., Vantucci, C., Kim, J., Schiapparelli, P., Cancer-selective nanoparticles for combinatorial siRNA delivery to primary human GBM *in vitro* and *in vivo*. *Biomaterials*, 209, 79–87, 2019.

120. Karpenko, L.I., Rudometov, A.P., Sharabrin, S.V., Shcherbakov, D.N., Borgoyakova, M.B., Bazhan, S.I., Volosnikova, E.A., Rudometova, N.B., Orlova, L.A., Pyshnaya, I.A., Zaitsev, B.N., Delivery of mRNA vaccine against SARS-CoV-2 using a polyglucin: Spermidine conjugate. *Vaccines*, 9, 2, 76, 2021.
121. Ren, J., Cao, Y., Li, L., Wang, X., Lu, H., Yang, J., Wang, S., Self-assembled polymeric micelle as a novel mRNA delivery carrier. *J. Controlled Release*, 338, 537–47, 2021.
122. Yoshinaga, N., Uchida, S., Dirisala, A., Naito, M., Osada, K., Cabral, H., Kataoka, K., mRNA loading into ATP-responsive polyplex micelles with optimal density of phenylboronate ester cross-linking to balance robustness in the biological milieu and intracellular translational efficiency. *J. Controlled Release*, 330, 317–28, 2021.
123. Mbatha, L.S., Maiyo, F., Daniels, A., Singh, M., Dendrimer-coated gold nanoparticles for efficient folate-targeted mRNA delivery *in vitro*. *Pharmaceutics*, 13, 6, 900, 2021.
124. Singh, D. and Singh, M., Hepatocellular-targeted mRNA delivery using functionalized selenium nanoparticles *in vitro*. *Pharmaceutics*, 13, 3, 298, 2021.
125. Yeom, J.H., Ryou, S.M., Won, M., Park, M., Bae, J., Lee, K., Inhibition of xenograft tumor growth by gold nanoparticle-DNA oligonucleotide conjugates-assisted delivery of BAX mRNA. *PLoS One*, 8, 9, e75369, 2013 Sep 20.
126. Maiyo, F. and Singh, M., Folate-targeted mRNA delivery using chitosan-functionalized selenium nanoparticles: Potential in cancer immunotherapy. *Pharmaceuticals*, 12, 4, 164, 2019.
127. Sun, P., Li, Z., Wang, J., Gao, H., Yang, X., Wu, S., Liu, D., Chen, Q., Transcellular delivery of messenger RNA payloads by a cationic supramolecular MOF platform. *Chem. Commun.*, 54, 80, 11304–7, 2018.
128. Liu, N., Zou, Z., Liu, J., Zhu, C., Zheng, J., Yang, R., A fluorescent nanoprobe based on azoreductase-responsive metal–organic frameworks for imaging VEGF mRNA under hypoxic conditions. *Analyst*, 144, 21, 6254–61, 2019.
129. Siewert, C.D., Haas, H., Cornet, V., Nogueira, S.S., Nawroth, T., Uebbing, L., Ziller, A., Al-Gousous, J., Radulescu, A., Schroer, M.A., Blanchet, C.E., Hybrid biopolymer and lipid nanoparticles with improved transfection efficacy for mRNA. *Cells*, 9, 9, 2034, 2020.
130. He, C., Lu, K., Liu, D., Lin, W., Nanoscale metal–organic frameworks for the co-delivery of cisplatin and pooled siRNAs to enhance therapeutic efficacy in drug-resistant ovarian cancer cells. *J. Am. Chem. Soc.*, 136, 14, 5181–4, 2014.
131. Zhao, H., Li, T., Yao, C., Gu, Z., Liu, C., Li, J., Yang, D., Dual roles of metal–organic frameworks as nanocarriers for miRNA delivery and adjuvants for chemodynamic therapy. *ACS Appl. Mater. Interfaces*, 13, 5, 6034–42, 2021.

132. Gao, Y., Men, K., Pan, C., Li, J., Wu, J., Chen, X., Lei, S., Gao, X., Duan, X., Functionalized DMP-039 hybrid nanoparticle as a novel mRNA vector for efficient cancer suicide gene therapy. *Int. J. Nanomed.*, 16, 5211, 2021.
133. Yang, J., Arya, S., Lung, P., Lin, Q., Huang, J., Li, Q., Hybrid nanovaccine for the co-delivery of the mRNA antigen and adjuvant. *Nanoscale*, 11, 45, 21782–9, 2019.
134. Coolen, A.L., Lacroix, C., Mercier-Gouy, P., Delaune, E., Monge, C., Exposito, J.Y., Verrier, B., Poly (lactic acid) nanoparticles and cell-penetrating peptide potentiate mRNA-based vaccine expression in dendritic cells triggering their activation. *Biomaterials*, 195, 23–37, 2019.

7

Effective Gene Transfer with Non-Viral Vectors

Anil Kumar Mavi[1*], Sonal Gaur[2], Neelesh Kumar[3], Avanish Kumar Shrivastav[3], Sankha Bhattacharya[4], Sateesh Belemkar[5], Saurabh Maru[6†] and Dhruv Kumar[7‡]

[1]Department of Pulmonary Medicine, Vallabhbhai Patel Chest Institute, University of Delhi, Delhi, India
[2]Department of Bioscience and Biotechnology, Banasthali Vidyapith, Jaipur, India
[3]Department of Biotechnology, Delhi Technological University, Delhi, India
[4]Department of Pharmaceutics, School of Pharmacy and Technology Management, SVKM's NMIMS, Babulde, Banks of Tapi River, Shirpur, Maharashtra, India
[5]Department of Pharmacology, Shobhaben Pratapbhai Patel School of Pharmacy & Technology Management, SVKM's NMIMS, V.L. Mehta Road, Vile Parle(W), Mumbai, India
[6]Department of Pharmacology, School of Pharmacy and Technology Management, SVKM's NMIMS, Babulde, Banks of Tapi River, Shirpur, Maharashtra, India
[7]School of Health Sciences and Technology (SoHST), UPES University, Dehradun, Uttarakhand, India

Abstract

Gene therapy has long been considered a viable treatment for serious disorders, whether they are acquired by genetic predisposition or inherited. A nucleic acid-based medication must, however, reach into the target cell while avoiding a range of biological defenses that have evolved through time. Therefore, the development of a trustworthy delivery vector is crucial for the accomplishment of gene therapy. Viral and non-viral vectors have been the main targets of research into gene delivery vehicles. Due to the recent success of the non-viral delivery of nucleic acids, there has been a significant increase in the amount of research and medicines connected to nucleic acids that have been produced during the past ten years. This system composed of physical and chemical components. Non-viral

**Corresponding author*: anilmavi100@gmail.com
†*Corresponding author*: skmaru@gmail.com
‡*Corresponding author*: dhruvbhu@gmail.com

Rishabha Malviya, Sonali Sundram and Neelam Jain (eds.) Integration of Biomaterials for Gene Therapy, (183–222) © 2023 Scrivener Publishing LLC

vectors provide many benefits, including safety, cost-efficiency, accessibility, simple production, efficient cell/tissue targeting, a minimal immune response, and the ability to employ plasmids at chosen molecular weights. In this chapter, we try to investigate *in vivo* challenges obstructing non-viral nucleic acid delivery various diseases and the strategies can be used to overcome from them. Further improvements to increase the efficiency of non-viral vectors are needed before to see any remarkable clinical results.

Keywords: Viral vector, non-viral vector, nucleic acids, gene therapy

7.1 Introduction

Gene therapy can be novel method that utilizes genes to treat or prevent various diseases. Instead of utilizing medications or surgery to treat a problem, doctors may be able to use the gene therapy procedure to implant a gene into the patient's cell. Some scientists and medical experts are looking at various gene therapy methods, including: i) a disease-causing mutant gene is replaced with a healthy gene; ii) a mutated gene is "knocked out" or rendered inactive; and iii) new genes are inserted into the cells to prevent any disorders from developing.

Gene therapy is a capable alternative for treating a variety of illnesses, including cancer, hereditary abnormalities, and some viral infections; nevertheless, the procedure is dangerous and is still being studied to make it safe before it can be effective. Only conditions for which there are currently no other treatment options have been studied using gene therapy. Genetic substances enter host cells' nucleus to activate gene expression [1–3].

In order for a patient's somatic cells to create precise curative proteins to modify genetic disorders, gene therapy has evolved to give genetic information. For the system to be designed successfully, it is important to have a comprehensive knowledge of the interactions between the targeting cell and the gene delivery system. A gene delivery process which directs the delivery of the gene expression plasmid to a particular site inside the body is one of the three components that make up gene delivery systems. The other two are a gene that encodes a specific therapeutic protein and a plasmid-based gene expression system that regulates a gene's function within the target cell [4, 5]. The foreign genetic material must endure in the host cells for the gene delivery method to be effective [6–8].

Viral-based vectors first appeared as a transgene expression technique in the 1980s [9]. In order to protect chimpanzees against Hepatitis B, the

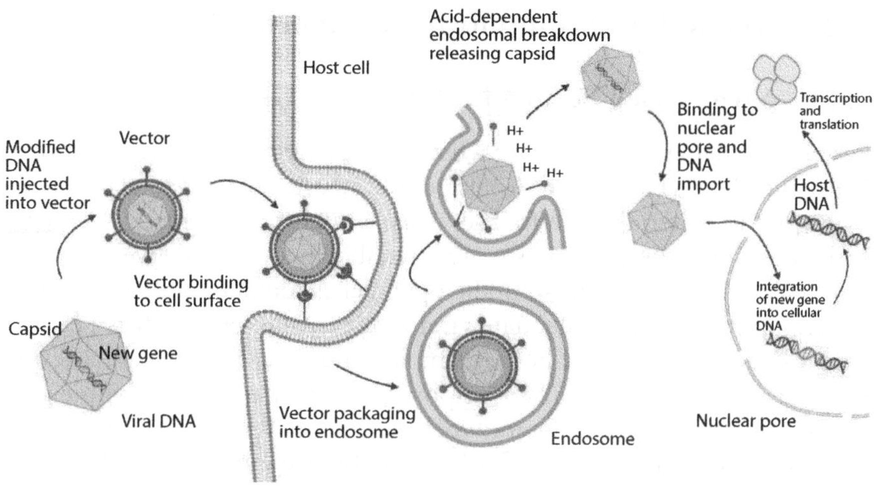

Figure 7.1 Systems for delivering genes. As part of the viral gene delivery method, the trans-gene is first incorporated inside the viral DNA. This altered DNA is then inserted into the viral vector. When the vector reaches its target, receptors on cell membranes of target cells bind to it. The vector is then squished into endosomes after being taken up by the cell, and these are later destroyed by acid to release the capsid holding the altered DNA. The changed gene is eventually incorporated into the DNA of the target cell when the capsid moves towards the nucleus which further connects with the nuclear pores and finally moves inside the nucleus. The desired protein is then produced through transcription and translation, which triggers gene expression.

vaccinia virus was utilized as a vaccine vector in 1984 [9]. Contrarily, gene delivery systems were the first to report on changing cellular phenotypes in response to DNA exposures in the body, as demonstrated in Figure 7.1 [10, 11].

The lack of efficient gene carrying vectors which satisfy all of the following requirements is the major obstacle to medical research in gene therapy:

- Protection of genetic cargo or trans-gene from endonucleases' destructive effects,
- Release of genetic material to the desired location, such as the cytoplasm or nucleus of a cell,
- Low likelihood of inducing genotoxicity or unfavorable immunological reactions,
- Affordable and practical accessibility for patients [11].

7.2 System Development for Delivering Genes

7.2.1 Virus-Based Vectors for Delivering Genes

Systems for delivering genes by viruses rely on their capacity to inject their DNA into host cells. These profit from a virus's capacity to replicate its own genetic material. Because the virus's structure prevents DNA's lyposome from degrading it, viruses are a reliable method of delivering genes [12]. Gene delivery systems bundle open genes intended for delivery into replication. From the inadequate viral particles and viral vectors [13, 14]. Adeno-associated viruses, herpes simplex and Adenoviruses are a few of the viruses that are utilized to transport genes. Somatic and germline gene delivery methods are 2 different types of gene delivery systems. Germline gene therapy cannot be used ethically, notwithstanding the potential of germline gene delivery technologies [15, 16]. In actuality, somatic cell modification is the sole gene delivery method available for humans.

7.2.2 Non-Viral Vector for Delivery Genes

For non-viral driven delivery systems, physico-chemical techniques have recently been developed [17]. Naked DNA, cationic polymers, Lipid, chromosomes, plasmids, and conjugate complexes are examples of biocompatible materials that are less likely to cause an immunological reaction in the body than non-viral vectors [18]. Non-viral vectors offer long-lasting gene expression but are challenging to use due to their labor-intensive manufacture. The dangers allied with using vectors, such as the potential for insertional mutagenesis and the ability to provoke an immunological response, have also hampered their development.

7.3 Methods for Non-Viral Vector for Delivery of Genes

Researchers have been searching for perfect vectors based on non-viral technique in order to bring forth sustained gene expression without inducing unfavorable inflammatory, immunological responses which are also absolutely benign for this goal. Due to their great efficiency, adaptability, non-immunogenicity, stability, and substantial loading capacity, nonviral vectors have become more important in gene delivery research [19]. There are two basic categories for nonviral vectors used in gene transfer, namely physical and chemical systems. Chemical methods include the use of

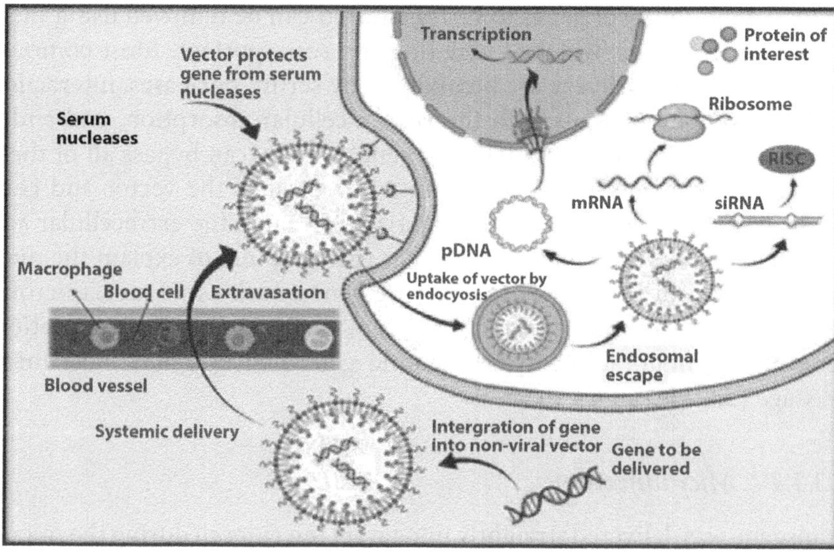

Figure 7.2 A nonviral method of gene delivery. It demonstrates the polymer, lipid-based systems, number of in organic materials used to deliver nucleic acid into the cells [20, 21].

polymer- and lipid-based systems, as well as a number of inorganic materials. Physical methods include sonoporation, microinjection, electroporation, magnetoporation, gene gun and needle injection. These chemical reactions frequently exhibit cationic characteristics. They have the ability to unite with negatively charged DNA through electrostatic interactions. This traditional interaction results in a general positive charge being added to the vector-gene pair. As a result of this positively charged substance being pulled to and crossing the negatively charged molecules of the cell membrane, the cell is internalized. The transgene must then be moved to the proper location, such as the cell cytoplasm (for DNA) or nucleus, while preventing lysosome and endosome destruction (for mRNA and siRNA, respectively). This nonviral gene delivery method is described in Figure 7.2 [20, 21].

7.3.1 Physical Techniques for Delivering Non-Viral Genes

The difficulties in delivering genes using viral vectors have drawn the attention of researchers. Numerous clinical studies have been conducted to find novel and creative physical methods of gene delivery which are superior to other chemical techniques or viral vectors. The intracellular and extra cellular obstacles to gene delivery that impede techniques for delivering genes and limit the ability of genes to be transfected can be removed using these physical approaches, which is why they are so important. Most common problems of gene delivery are breakdown by serum nucleases, interaction of the blood components with the vector, cellular absorption, and endosomal escape. Physical techniques of gene delivery can bypass all of these difficulties because they use mechanisms that allow the vector and gene to reach the target location without passing through the extracellular and intracellular environments. This chapter uses graphics to explain the most modern physical processes of delivering genes, such as jet gun, microinjection, needle injection, magnetoporation, nucleofection, electroporation, gene gun, sonoporation, hydrodynamic gene delivery, and mechanical massage [22, 23].

7.3.1.1 Microinjection

Transgene was delivered straightly into target interior cell during the microinjection procedure. The medical researcher Barber initially showed and documented this technique. In this method, a very tiny needle, between 0.5 and 5 microns in diameter, is infused with a genetic material-containing fluid. After that, this fluid is continuously monitored under a microscope while it is injected into a single cell [21]. Large-sized DNA can be delivered using the microinjection method, which is also simple and reasonably priced [24]. It is a successful, non-toxic technique of delivering genes. The need to handle each cell separately, the stringent handling guidelines, and the low levels of gene persistence and expression are some drawbacks of this method. The shortcoming of this method is that it only needs a very minimal amount of genetic material to be inserted to prevent harm to the intracellular environment and cellular membrane. Furthermore, this method only works with somewhat large cells, like oncocytes [25].

7.3.1.2 Needle Injection

Straightly insert bare genetic material into the mark location is involved in this. In 1990, Wolf and colleagues published the first report on the needle injection technique. They injected mouse muscle fibers with naked DNA intra-muscularly [26]. When the gene is directed to a specific location, such as blood, brain, or eye cells, this method's extraordinary simplicity makes it useful for use in gene therapy. This technology can be applied to DNA-based gene therapy and immunization processes thanks to its favorable properties. It can transfer several forms of RNA, including siRNA, in addition to naked DNA [27]. The process is also significantly less harmful and kind to the biological environment. However, there are several downsides to the needle injection method, such as relatively little gene expression when injecting naked genetic material in to an IV.

The invention of the high-pressure delivery mechanism known as hydrodynamic gene transfer, covered above in this section, has now overcome this issue [28].

7.3.1.3 Jet Gun

Since 1947 drug administration techniques like jet injection have been applied in medical settings. It is a technique that does not require the use of needles and spreads the therapeutic substance or gene across the target cells [29]. With the help of this technique, a very fine jet of genetic material may be transported at a very high pressure to the desired location. The homogeneity of the genetic composition dispersion depends on the amount of pressure applied. The tissue damage generated by the jet injection conceivably tolerated if the desired gene expression and gene delivery is accomplished. The example of a method requirement for illness like cardiac treated using gene therapy is: 110 m/s nozzle jet velocity, a separation of 20–25 cm, a volume of injection of 100–500l, and a pressure of 150–250 kPa. The depth of tissue penetration and the effectiveness of this approach are significantly influenced by the jet parameter selection. In order to change these settings, this process is flexible, which significantly reduces tissue damage caused by this technique [30]. Gene therapy uses are genetic vaccinations or targeting of suicide genes which encode DNA for anticancer treatment have both made use of jet gun or injectable technologies. The jet gun method has been used to genetically correct a number of skin issues. It has been found that employing a jet cannon to target a variety of tissues, such as the epidermis, fat, mammary tissues, and muscles, is beneficial. Only a tiny number of possible side effects have been linked to this therapy thus far, including minor bleeding at the injection site, localized heat or inflammation, and edema [31].

7.3.1.4 Gene Gun

This method of gene therapy, commonly referred to as ballistic gene delivery or injection, involves repeatedly pelting the intended tissues or cells with DNA-coated particles. The varied target cells that can be genetically altered using this method include the skin, mucosa, or tumor cells that have been surgically exposed. In order to efficiently pass cell membranes and achieve cellular internalization, where the genes or nucleic acids may be carried to the cell cytoplasm or nucleus, heavy metals like W or Au are coated with targeted gene. Depending on the physical qualities of these payload particles, the efficiency with which genes are carried to target regions may vary. Particle size, shape, dosage, surface morphology, and release speed are a few examples [32]. This technique is practical and safe, but it has a number of drawbacks that make it difficult to apply for gene transfer. This technique's biggest flaw is that it only offers a few milliliters of depth for gene transfer, making it more suitable for therapies requiring localized injection. Additionally, due to the transient nature of the gene expression it generates, this approach can only be used on cells that exhibit active cell division and rapid proliferation. In order to increase the efficacy of this therapy, researchers have developed a number of approaches, such as rapid carrier preparation techniques, tissue injury minimization techniques, and process acceleration techniques [33].

7.3.1.5 Electroporation

Electro-permeabilization, also known as 'electrofection' or 'electroporation', which widely employed to transferring to a marked area in case of both *in vitro* and *in vivo* applications since its inception in 1982 [34] and 1991 [35]. This approach is usually utilized to deliver chemicals which are not permeable to cell membranes. This approach works by passing an electric current through the cells of the target spot, causing the temporary formation of nanometric-sized holes. These holes permit the passage of negatively charged genetic components, such as DNA or other genetic materials, to the cytoplasm or nucleus of the cell. Depending on the kind of target cells, different durations and intensities of electric pulses are delivered. The skeletal muscles provide as an illustration of this, where membrane pores are first formed after a relatively large electric current is delivered. The genetic material is then guided across these pores by some low-intensity pulses, creating an electrophoretic effect [36]. Depending on the target organ, a wide range of electrodes are available for use and selection. These include nonpenetrating charged needles, multi electrode arrays, caliper electrodes, plate electrodes, and customized defibrillator pads [37]. Pores are created by electric pulses, because they form quickly, genetic makeup must be near at that precise moment in order for it to successfully pass through the pores and reach the cytosol. This technique is primarily employed *in vitro* to transport nucleic acids, but it is also applied *in vivo* to treat cancers of the skin, liver, etc. Moreover, it is primarily in adequate to local targeting and poses a serious threat to cell toxicity and apoptosis [38].

7.3.1.6 Nucleofection

Nucleofection, which is a technique comparable to electroporation, is a somewhat more sophisticated and effective approach than electrofection. This technique delivers genes by exposing target cells to electric pulses that create membrane breaks that let genetic material and cell cytoplasm pass through. The key distinction is that nucleofection directly injects outside material within the cell, increasing the progression efficiency and allowing transferring of gene to cells that are often difficult to transfect. Nucleofection's rapid DNA distribution and low cell density are two of its key benefits. However, numerous studies have found an elevated rate of mortality among participants in these types of gene therapy clinical trials. Thus, more studies are needed for more applications of this technique while focusing on reducing its risk [39].

7.3.1.7 Sonoporation

The same basic idea that under lies electroporation also governs sonoporation, commonly referred to as sonofection. In order to allow targeted gene to enter the cell that cause gene expression, pores are made in the cell membrane using ultrasonic waves. In this procedure, high-intensity ultrasonic pulses are frequently used. Moreover, efficiency of this procedure and volume of genes which flow through the pores are significantly influenced by the intensity of the ultrasonic waves utilized. Cavitation is the primary mechanism by which ultrasonic waves work. Acoustic field mediates the development as well as dissolution of spaces with extremely low pressure in this process. These bubbles or voids oscillate and expand, which causes them to burst and release a significant quantity of energy. Sonoporation results from this, and the temperature and pressure in the area also increase. Involvement of gene cargo into the cytosol is then caused by hole formation as a result of genetic material passing across cell membranes [40]. Researchers have been attempting to improve the cavitation process by using nucleation agents such as acoustic contrast, agents, and microbubbles that increase membrane permeability and, as a result, achieve successful gene transfer. The fundamental benefit of this method is that it is noninvasive, making it appropriate in situations where gene therapy is not localized or is focused at deeper tissues [41]. This method conceivably used in local gene therapy. A contrast agent as well as the targeted gene is delivered into the bloodstream under these conditions, and then ultrasonic waves are employed. Local gene delivery is made possible by combining the contrast agent with specialized binding agents or ligands that target the specific tissue for which they have affinity [42].

7.3.1.8 Hydrodynamic Gene Transfer

By using high level hydrostatic pressure as a driving element, this approach promotes the insertion of targeted gene across the membranes. Rapid injection of a DNA solution into the mouse tail vein at a volume greater than 8% of body weight resulted in sufficient transgenic expression in organs such as the heart, liver, kidneys and lungs. The highest level of gene expression was discovered in the liver. The DNA material can enter the cell membranes through these pores and cause intra cellular gene activation when the DNA solution is quickly injected. This rapid injection causes a buildup of pressure at the injection site [43]. Although this technique has showed promise for delivering genes to rodents, it has not yet been widely used for transferring genes to humans. Furthermore, when exceptionally large volumes of genetic solution are injected, high-pressure accumulation in hepatocytes may cause cardiac dysfunction and liver congestion indirectly by increasing the activity of hepatic enzymes. To overcome the limitations of this approach, researchers have created catheters for gene transfer in larger animals such as pigs. Even though they have seen better outcomes, more research is still needed to make this approach practical for using in human gene delivery [44].

7.3.1.9 Magnetoporation

The magnetic field leads the flow of targeted gene through membranes and inside the cytoplasm, also known as magnetic transfection or magnetofection. Vectors made of nucleic acids and magnetic nanoparticles are affected by this magnetic field. These magnetic vectors are created by mixing gene vectors based on lipid or polymers with magnetic nanoparticles. Iron oxide is typically employed to give nanoparticles their magnetic characteristic. An electromagnet is employed to direct the gene-carrying magnetic vector toward the exact target cells that require gene expression [45]. The ability of genetic material to penetrate cell membranes is improved by the application of a magnetic field, according to studies [46], and it can as well increase degree of gene contact with the interior environment of the cell, which improves transfection and gene expression [47]. Compared to electroporation, magnetoporation provides a number of benefits. The main benefit of this procedure is that electrodes do not need to come into contact with the target site's cells directly, which makes it more convenient and noninvasive. Additionally, this technique has a greater chance of penetrating anatomical regions and deeper tissues that are generally unreachable with electroporation. It takes much less time and money as well. According to studies [48] melanoma and tumor cells [49], and pulmonary metastasis [50], magnetoporation has been employed to deliver genes to cardiac tissues.

7.3.1.10 Mechanical Massage

Liu and colleagues found that after injecting naked DNA into mouse liver cells, physically massaging the injection site improved gene expression. They provided this explanation for the phenomenon: light pressure applied during massage temporarily disturbs the membranes of liver cells, allowing genetic matter to pass past the membrane and go through the cell cytoplasm to give gene expression [51]. A significantly enhanced therapeutic outcome was obtained when hepatic growth factor was administered using this method to treat fulminant hepatic failure brought on by endotoxins. However, rather than being tested on people, this procedure has only been done on small animals like mice. For this procedure to be practical for people, more study is necessary [52].

7.3.2 Chemical Methods for Delivering Non-Viral Genes

Recently, the field of gene delivery has seen an increase in use of chemical, non-viral vectors. In comparison to viral systems, these methods have a number of benefits, including immunity, safety, the capacity to transfect vast amount of gene as well as decreased toxicity. Cationic lipids, polymers, and inorganic components including carbon nanotubes, quantum dots, metal nanoparticles, and silica-based systems make up the bulk of these cationic systems. By using electrostatic forces, these cationic chemicals can create stable complexes with genetic material, such as negatively charged nucleic acids. These complexes can prevent issues with gene transport such cellular absorption, transportation of genetic material inside the cytoplasm or nucleus. They stand a fair chance of establishing unrestricted gene expression which is long-lasting, sustainable, as well as successfully protecting the genetic material from environmental degradation [53]. The numerous polymers, lipids, and inorganic parts that makeup chemical systems for nonviral gene delivery are described in detail below.

7.3.2.1 Polymer-Based Nano-Vectors

Large, complex chemicals known as polymers are created by joining together several monomers, or repeating components. These polymers can be categorized into one of two groups: natural or synthetic polymers. Chitosan, peptides, and certain proteins are examples of natural polymers. Polyethyleneimine and cyclodextrins are two examples of synthetic polymers. These polymeric gene delivery techniques frequently have cationic characteristics. Electrostatic interactions allow them to attach to negatively charged genetic material. These provide the vector gene complex, commonly referred to as a "polyplex," a general positive charge that enables it to swiftly cross cell membranes and deliver genetic material to the proper cytoplasm or nucleus Figure 7.3 [54, 55].

Gene delivery polymers each have special qualities of their own. They can also be modified to develop desirable traits and increase the effectiveness of the gene delivery. These polymers can also be given targeting ligands to make them unique to a certain receptor located at the target region. Additionally, they contain shielding agents like polyethylene glycol (PEG) added, which improve the time the vector circulates in the blood and permit the polyplex to safely access the target cells. Various polymers are marketed in forms that are prepared for use in gene delivery [84]. Cationic polymers like poly-L-lysine (PLL), polyethyleneimines (PEI), and poly2-N-dimethylaminoethyl methacrylate are used in some to

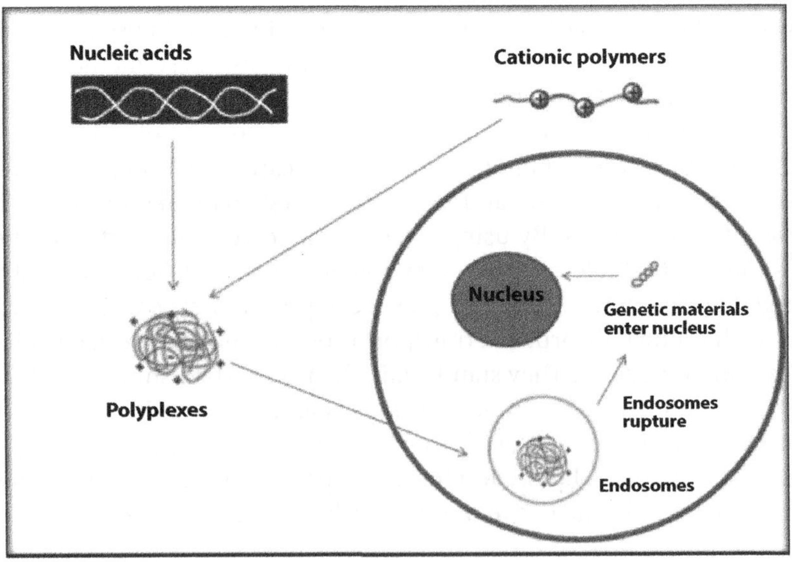

Figure 7.3 Polymer based nanovector. When these cationic polymers combine with nucleic acid likes DNA by using an electrostatic interaction they make a polyplexes. These polyplexes can carry both positive and negatives charges on it. It can cross the cell membrane by carrying specific target ligands to a particular site [54, 55].

create polyplexes for the delivery of genes (PDMAEMA). Among other structural variants, these polymers can be found in linear, hyperbranched, star-shaped, branched, and dendritic architectures [85]. These structures each exhibit special characteristics that are exploited to make cationic polymers for the transport of genes [86]. The cytotoxic properties of these polymers, however, restrict their use in gene therapy. According to studies, the size and surface charge of these polymers contribute to their cytotoxicity. These polymers accumulate in healthy body cells due to their high molecular weight as well as inefficiency to breakdown extra charged groups and bonds, which disrupts normal cell physiology and results in cell cytotoxicity [87]. These polymers' intricate structure also makes it difficult for them to clear, which can add to the complexity. These materials are no longer employed as cationic polymers because of the aforementioned problems. Instead, researchers are concentrating on the usage of biodegradable materials that can form polyplexes as well as carry targeted gene for gene therapy [84]. Earlier, a variety of therapeutic modalities, such as gene therapy, regenerative medicine, biotechnology, genetic engineering, and tissue engineering, have made use of biodegradable polymers. Additionally, they have been employed as spreaders, vectors, and medicine delivery systems. Biodegradable polymers are more suited for gene transport among all of these uses because they enable repeated administration and usage, which is crucial in gene therapy [88]. Depending on where they come from, biodegradable polymers can be either natural or artificial. Before choosing a specific polymer for gene delivery, its unique properties, benefits, and drawbacks are researched and thoroughly studied for both natural and manmade biodegradable polymers. The compatibility of natural biodegradable polymers with biological surroundings, safe bioactivity levels, and capability for proteolytic disintegration by cell activation are among their primary advantages [89]. However, extracting, purifying, identifying, and characterizing natural polymers obtained from nature can be strenous, resulting in variations in each batch of items created from them [90]. Because of incorrect identification and assessment of polymer composition, such issues may eventually pose trouble in gene therapy. This may have a seriously detrimental impact on biomedicine.

Additionally, these polymers' high levels of bioactivity may cause a potent inflammatory or immunological response that, in some circumstances, could be fatal [91]. Synthetic biodegradable polymers' are suitable for biological environment because of the primary benefits like their inherent inertness and chemical structure that enables structural alterations in accordance with the needs of a particular therapy. In contrast to natural polymers, this feature also shows significant regularity in every batch of synthetic polymers manufactured [92]. However, one drawback of these polymers is that they are biologically inactive, which makes it difficult to prepare them for a particular biological therapy. Various functional groups and chemical moieties are being incorporated into synthetic polymers by scientists and medical researchers in order to employ them for therapeutic techniques and targeted therapies [93].

Due to the presence of chemical linkages that are susceptible to hydrolysis in their backbone, polyesters are a form of biodegradable synthetic polymer. The nomenclature of the related polymers, such as polyesters, polyamides, polyurethanes, polyanhydrides, and polycarbonates, results from these chemical linkages, which are esters, carbonates, anhydrides, urethanes, and amides. Polyesters, which are primarily aliphatic in nature, have been extensively used in medicine for sutures, gene therapy, drug delivery systems, and genetic engineering. For possible uses in gene delivery, a broad variety of polyester polymers are being studied. The most often utilized polymers are PBAE PVL, PPE, PHP, and PAGA [91]. PPE polymers have a structure that is similar to nucleic acids and compatible with biological enzymes. They are ideal for delivering genes because of this characteristic. Although these polymers have excellent biodegradable qualities, the preparation of gene transfer is hampered by their intricate molecular architectures. To do this, scientists have discovered a number of methods to alter these polymers, resulting in increased transfection effectiveness and decreased toxicity. The development of a copolymer micellar system, which exhibits far better transgene delivery than straightforward polyester polymers, is an illustration of this. In numerous investigations, siRNA was delivered to target cells by forming micelles out of three distinct polyester polymers for the treatment of hypoxic tumors and other cancers. They demonstrated that using micelles made of these polymers as vectors may successfully deliver the transgene to the intended region and induce successful gene expression in addition to effectively encapsulating and protecting the gene cargo [91].

Polyurethanes are well suited for use in tissue manufacturing and gene delivery because of their flexibility, biocompatibility, and elasticity. When combined with other cationic polymers, polyurethanes can be used for gene delivery applications just like other synthetic biodegradable polymers. Yang and colleagues gave a demonstration of this. To work as a vector for the transport of miRNA for the treatment of tumors like brain, they created a combination of polyurethane and PEI. In addition to being exceedingly efficient at delivering the genetic material, this combination was also capable of glioblastoma growth inhibition and was highly effective at lung cancer gene therapy [92].

In addition to being biodegradable, polycarbonates have improved biocompatibility, reduced cytotoxicity, and tunable mechanical characteristics, making them effective gene delivery vectors. Frere and colleagues gave a demonstration of chemical modification of polycarbonates for gene transfer. They created a number of fresh, biocompatible, and biodegradable polycarbonate-based vectors with guanidinium and morpholino functionalizations. These vectors were created with intention of giving siRNA to HeLa cells to cure cervical cancer. However, they demonstrated that the chemical change increased the transfection and gene transport capabilities of the polycarbonate vectors [91]. They became effective vehicles for nonviral gene delivery thanks to chemical alterations and blend with copolymers of polycarbonates, which also improved the capacity for gene transfection and biodegradability while lowering their cytotoxicity and immune response side effects [93].

7.3.2.2 Protein-Based Vectors

Proteins and polysaccharides are the two source of naturally biodegradable polymers. Polypeptide molecules makeup proteins. These have a folded 3-dimensional structure. These are the building blocks of several biological tissues, including the vascular, skin, hair, and musculoskeletal systems of all living being as well as those of cattle, equines, and porcine animals. Additional sources of these proteins include blood plasma and microbial fermentation [56]. These protein-based polymers' superior biocompatibility and biodegradability make them exceptional vectors for nonviral gene transfer. Gelatin and albumin are the two most often employed protein-based vectors for the transportation of genes. Gelatin polymers are perfect for the transport of genes because of their outstanding antigenicity. Gelatin-based nanoparticle vectors for the delivery of polymerized siRNA were created by Kim and colleagues. These nanoparticles successfully transported siRNA genetic material to melanoma cells, resulting in proper gene expression in tumor-bearing mouse models [57]. They also preserved the siRNA genetic information from destruction in the environment. Chougule and colleagues developed a gelatin-based nano-carrier formulation to deliver STAT6 siRNA to A459 cancer cells and inhibit the growth and production of the STAT6 gene. The A459 cancer cells were effectively silenced and eliminated, proving the utility of gelatin-based nanovectors for gene delivery [58].

For nonviral gene transfer, albumin has also been combined with a few other cationic polymers. This is due to the fact that albumin lacks charges like positive on the surface and requires cationic to impart its charged character. Plasmid DNA and PEI were combined before being inserted into albumin by many research team to create a polyplex. Albumin was the main component discovered to give this effective transfection ability, and these nanovectors are very effective in transportation of targeted gene to HeLa cells in the form of plasmid DNA [58].

7.3.2.3 Polysaccharide-Based Vectors

Complex molecules comprised of glucose units linked together by glycosidic bonds are referred to as polysaccharides. These polymeric molecules are capable of immunological detection as well as very effective cell signaling. These characteristics have led researchers to examine these substances for application in nonviral gene transfer. Because these molecules include reactive functional groups on their glucose units, they are open to chemical modification, resulting in the generation of polysaccharide derivatives that can be employed for a number of gene delivery applications. Chitosan, cyclodextrin, dextran, and hyaluronic acid are the most widely used biodegradable polysaccharide-based vectors [59]. The most popular polysaccharides for gene delivery are chitosan and its derivatives because they are appropriate for cationic polymer based gene therapy and include arbitrarily distributed N-acetyl-D-glucosamine and-1-4-linked D-glucosamine groups [60]. Since chitosan has a low pKa value and is less soluble than other materials, it cannot interact with genes. Chitosan is typically chemically altered to increase its effectiveness in moving and transfecting genetic material in order to address this problem. By altering the chitosan backbone, Researcher and his research team created trimethyl chitosan based nano-vectors. They found that this technique successfully delivered plasmid DNA to cancer cells [61].

7.3.2.4 Liposomes

Lipid based nano-vectors are now three primary lipid based systems for delivery of gene that have emerged from research on the subject. These include other forms of liposomes, micro vesicular systems, and systems that resemble high-density lipoprotein. Each of these systems possesses unique characteristics, and each system's capacity to carry genetic matter for the delivery of genes is governed by various elements including bio-distribution, drug release characteristics, pharmacokinetic features, and release kinetics. Recently, the potential for using lipid-based nanotechnologies to build up vaccines for SARS-CoV-2wasexplored. The use of lipid-based nanoparticles was crucial in this case. Even the initial COVID-19 vaccine was an mRNA-based vaccination given using lipid nanoparticle

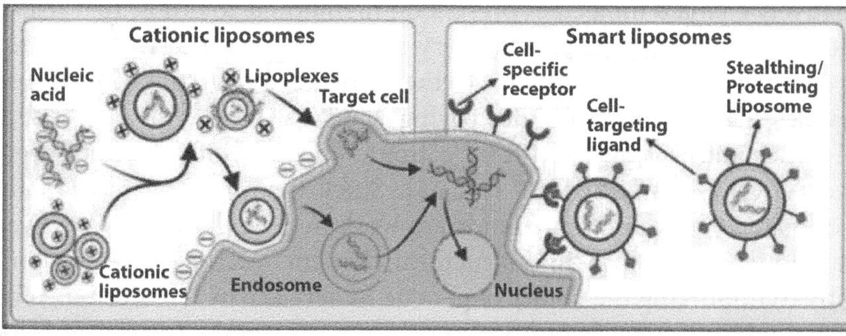

Figure 7.4 Lipid based nano-vectors. Liposomes are specialized types of nanoparticles that contain more than one lipid bilayers and it is of 2 types 1) cationic and 2) smart liposomes. Both these liposomes having common a bilipid layer that surrounding an aqueous core but also contains positive charges all around it in case of cationic liposomes. In case of smart liposomes, it is having a stealthing/protective which binds to cell specific receptors [64].

vectors when it entered the clinical trial phase [62]. The COVID-19 vaccine from Moderna is based on lipid nanoparticles. Other lipid-based nano particulate systems, such as cationic nano emulsions and liposomes, are also being researched for the purpose of developing a SARS-CoV-2 vaccine, can be seen in Figure 7.4 [63].

Gene delivery using non-viral vectors based on lipids. It is specialized macro- or nanoscale particles that have an aqueous center and one or more lipid bilayers encircling them. By incorporation an ethanolic solution containing the lipids present in aqueous solution which have the genetic material, the process of liposome production is based on self-assembly. Since the 1970s, it has been used as transport systems for genes and drugs. Liposomes' effectiveness as a gene-carrying vector depends on a variety of physical and chemical individuality, like composition, size, net charge, loading efficiency, and stability [64]. Although there are various diverse kinds of liposomes, the two main forms significant for genetic material transfer: cationic and smart/Trojan horse liposomes.

7.3.2.4.1 Cationic Liposomes

This specific kind of liposome has a net positive charge (Figure 7.4), which enables it to combine with negatively charged components and nucleic acids through electrostatic interactions. This contact causes the bigger nucleic acid particles to become trapped inside the lipoplex structure. In order to come in *in vitro* surrounding with positively charged surface, cationic liposomes can be easily changed. This allows them to connect with negatively charged molecules present on membrane, foremost to transportation of genes. The nucleic acid-carrying lipoplex can transmit the gene to the cytoplasm or nucleus thanks to the fusogenic properties of liposomes, preventing endosomes from destroying it [65].

Felgner and colleagues utilized cationic liposomes for the first time as gene delivery vectors in1987 [64, 65]. Since then, other scientists have investigated various cationic lipids are work in gene therapy and delivery. The numerous cationic lipid structural elements each contribute significantly to improving the ability of genetic material to be transfected. When the cationic headgroups of liposomes contain several amines, imidazole, or guanidine groups, the total cationic character of the lipoplex entity can be increased. Such multivalent cationic liposomes are better at preventing lysosomal destruction of the genetic material. However, an excessively positive charge might result in a very tight binding linking the nucleic acids and the liposomes, making it difficult to detach the genetic material from the liposomes and release it into the cytoplasm. Multivalent liposomes may jeopardize the stability of the liposome-nucleic acid complex, resulting in cytotoxicity, as well as being prone to micelle formation. Because lipid-containing lipoplexes are more hard to digest, removing them after gene delivery is time-consuming [64]. Using cationic liposomes instead of the incredibly complex Trojan horse liposomes has many advantages, including the astounding efficiency with which cationic liposomes deliver nucleic acids in an *in vitro* environment, the fact that there is no restriction on the size of genetic material that can be delivered using cationic liposomes, the ease and simplicity of manufacturing, and the cost effectiveness. Liposomes are widely available for purchase in the pharmaceutical business that is prepared to be used for gene delivery [64]. Cationic lipoplexes are undesirable for intravenous gene delivery due to their potential to aggregate and form micelles, instability, propensity to interact with blood components, and poor dispersion within the body. These flaws not only decrease their transfection competence other than also alter the charges of lipoplexes, causing them to build up in different organs and potentially having chronic hazardous effects [65].

To make these silica-based devices competent of conjugating with the negatively charged (DNA) and delivering it to specified locations, cationic compounds can be added to the surface. They can be customized to acquire stimulus receptive characteristics, which qualify them for use in non-invasive transportation method of genes. The potential of these systems to connect with blood cell membranes and cause hemolysis [65] and metabolic alterations that can lead to melanoma [64] are their two key advantages as seen by research.

7.3.2.4.2 Smart Liposomes

Any liposomal method is good for delivering genes since it can flow for longer duration in the blood and main particular cells. Because of these properties, the method can deliver the therapeutic dose to the target region while minimizing the negative effects caused by nucleic acid interaction with healthy cells. Trojan horse or cunning liposomes, also known as stealth liposomes, often exhibit these qualities. They can evade the immune system and produce undesired immunological reactions because they stay in the bloodstream for a long time. Blood circulation can be increased by adding hydrophilic polymers to the surface of liposomes, like polyethylene glycol. As a result, the reticuloendothelial system is unable to opsonize or remove the liposomal complex from blood components or other circulating blood entities [66].

Smart liposomes differ from previous delivery technologies that rely on targeting ligands on nano vector surfaces because they can transfer of genetic matter to a precise cell population or group. The transferrin protein is an illustration of one of these ligands; it is especially utilized to target cancer cells that have an overexpressed transferrin receptor [66]. Additionally, intelligent liposomes are able to adapt the effectiveness of their gene delivery in response to various stimuli. Liposomal instability frequently results from stimuli like pH, light, magnetic field, temperature [64] redox processes, and ultrasound, which cause the discharge of gene cargo from these carriers. Systems for controlled release that will only transport their payload upon reply to a specific stimulus at the aim site can be created using the capacity to respond to stimuli [66].

7.3.2.5 Inorganic Materials

Different inorganic materials are used to create a huge number of inorganic nanoparticles. These nanoparticles' physical and chemical characteristics are solely determined by their composition.

Most widely used inorganic nanoparticles for drug delivery are iron oxides, quantum dots, metallic compounds, silica-based compounds and carbon nanotubes found in graphene-based systems. These systems offer several benefits which make them viable for the delivery of vectors. These consist of outstanding biocompatibility, minimal cytotoxicity, simple handling, and simple functionalization [67]. The additional benefits and drawbacks of using inorganic materials for gene transfer are mentioned below.

Metal nanoparticles: Recently, metal nanoparticles have proven to be good vectors for delivering genes. Gold and silver are the two metals that are most frequently used in the production of these nano particles. These nanoparticles possess unique characteristics that set them apart from other nonviral gene delivery technologies and improve them [64]. They have a particular plasmon resonance on their surface that makes them more receptive to their surroundings. They can be utilized for particular targeting and other uses due to their easily adjustable surface functionality and composition [67]. These compounds can be made in an easy way. They have excellent biocompatibility and low cytotoxicity. These particles can be easily followed after injection using a variety of techniques, like FRET (fluorescence resonance energy transfer).

- Quantum Dots: Crystals formed of semiconductor resources are known as quantum dots. They range in size from one to twenty nanometers. They are made up of several chemicals that number in the hundreds to thousands and are grouped in groups or clusters. These substances might be ternary, like InGaAs and InGaN, or binary, like SiC and GaAs. Quantum dots are mostly used as fluorescent probes, and they have a number of benefits over other types of conventional probes. They have a greater brightness, a longer fluorescence period, and an excessive amount of photostability [67]. As the main component of the gene-carrying vector, quantum dots are utilized for the delivery of genes. On the surface of this core, nucleic acids and other targeted ligands are affixed. This results in a combination of the genes' therapeutic potential and the quantum dots' fluorescent property, making this system theranostic. These quantum dots can be used in this technique to track the supplied nucleic acids after they enter the body.
- Graphene sheets are stacked to form cylindrical, tube-like structures to form carbon nano tubes. As the base of their construction, they can use multiwalled tubes with a diameter of 4–30 nm or single-walled cylinder with a 0.4–3 nm diameter. These systems have many advantages, such as a large drug carrying capacity, chemical inertness, and a propensity to give regulated drug release. Additionally, they enable the bond formation and affection of fluorescent probes, main ligands, and other biomolecules. They can distribute drugs and genes more effectively this way, and probes can be used to track the delivery process. Additionally, when they combine with nucleic acids, they create very stable complexes that guarantee their safe delivery to the intended cells. These molecules have the problem of being very unsuitable and

soluble in aqueous conditions, making biological use more difficult.

- Silica-Based Systems: For the transport of genetic materials, silicon-based chemicals like silicon dioxide, have been widely used. This technique can produce nano particles that can have their surface charges, sizes, shapes, and other characteristics changed so that they can connect to various ligands and other biomolecules for aim to drug delivery or gene therapy. Therefore, these nanoparticles can be modified to have qualities like prolonged blood circulation, effective transfection, reduced toxicity, increased drug loading capacity, and improved cellular internalization. Additionally, these compounds don't need complicated manufacturing processes and are readily available in the commercial market. These systems have the main compensation over other carrier systems in that they can store a lot of DNA or genetic matter which are well-matched with both hydrophilic and hydrophobic substances [65].

7.4 Delivery System

7.4.1 Distribution Methods Using Non-Viral Nucleic Acids are Used for Localized Delivery

Additionally, negatively charged Nucleic acid (NA) is electrostatically repulsed by negatively charged moieties that are linked to the cell membrane. Non-viral delivery mechanisms either complex and reverse the charge of NA or physically enable NA to enter the cell directly in order to increase cellular absorption efficiency. The difficulty with these delivery strategies, therefore, is in both preventing NA breakdown within the micro- or nanoparticles and achieving effective intracellular release [68]. NA's high molecular weight and negative charge are significant barriers to successful cellular absorption. Particle sizes of up to 200 nm are required for the majority of cell types, which are a lot smaller than the hydrodynamic diameter of DNA, which are a few thousand base pairs [69].

7.4.2 Polymeric Systems (Polyplex)

Dendrimers, chitosan, and poly-ethylenimine (PEI) are a few of the cationic polymers that have been explored the most as non-viral vectors for

NA [70]. They have shown excellent *in vitro* efficacy, and some of them are offered commercially as transfection agents (JetPEI, Polyplus transfection SA; ExGen500, Biomol GmbH). When nucleic acid and appropriate cationic polymers are blend under the right parameters, positively charged polymeric nanocomplexes (polyplexes) with sizes ranging from 10 to 100 nm spontaneously form. Through electrostatic contact, endocytosis, and endosomal escape, polyplexes enhance cell adhesion [71–76]. The ability of polymeric systems to be easily modified, which enables the direct design and synthesis of multifunctional transportation methods, is one of their main benefits for non-viral dissemination. One of these changes is the alteration of hydrophilic polymers to conceal surface charge either before or after complex formation with NA [77]. Additionally, targeted moieties can be attached directly [78, 79] or via grafted hydrophilic polymers [80]. Another potential alteration that could allow for altering the stability of polyplexes in different conditions to specific needs is biodegradable cross-linking, such as using bio reducible linkers [80, 81]. However, one of the major difficulties that was frequently shown to connect with transfection effectiveness was [82–84], is cytotoxicity of polycations. This problem was addressed by several groups, for example, by adding bio reducible cross-links to the polymer's backbone, which may have enhanced excretion and improved breakdown [85, 86] and [87] provides a full assessment of the existing state of polymeric non-viral vectors.

7.4.2.1 Lipid-Based Systems (Lipoplex)

Anionic nucleic acid interacting with the surface of cationic liposomes results in the formation of lipoplexes, which are aggregated and multilamellar complexes. Cationic liposomes' high zeta potential, low transfection efficiency, and frequently insufficient resistance to lysosomal NA degradation are disadvantages of using them as a delivery system for NA. One of the first non-viral vectors to be utilized to deliver NA into cells was a lipid-based one. *In vitro* condensation and transfection of several cell types were both shown by Felgner *et al.* in 1987 using the synthetic cationic lipid N-[1-(2, 3-dioleyloxy)propyl]-N,N,N-trimethylammonium chloride (DOTMA) [88]. Almost all of the pDNA complexed with it, and it was revealed that these complexes fuse with the cell membrane, allowing for successful intracellular delivery of the cargo [88, 89]. To cause the interaction with the cell surface, positively charged lipoplexes are thought to electrostatically interact with negatively charged residues at the cell surface, such as sialic acid. Furthermore, hydrophobic interactions contribute in cargo delivery and cell membrane fusion. In this scenario, neutral helper lipids such as cholesterol or dioleoyl phosphatidylethanolamine (DOPE) stabilize liposomes and facilitate membrane fusion, boosting transfection effectiveness [90]. Later research discovered that clathrin-mediated endocytosis is necessary for effective transfection and that NA's direct translocation across the cell membrane is only marginally significant [75, 91]. However, cytotoxicity is a significant issue, just like with polymeric systems. The dose given, the kind of reagent or cell type, and the charge ratio between NA and cationic lipid were usually found to be related to toxicity [92]. By reducing the effects on non-target cells, the local injection of lipoplexes can assist in addressing these toxicity issues. Immunoliposomes, stabilized liposomes, and stimuli-responsive liposomes are only a few examples of how the fundamental idea of lipofection has been improved and modified. For a thorough overview of modern lipoplex systems and references [92, 93], the reader is directed [93, 94].

7.4.2.2 Lipopolyplex

To merge the beneficial qualities of lipoplexes and polyplexes, lipopolyplexes, or ternary complexes made of cationic polymers, cationic liposomes, and NA, were produced. Lipopolyplexes were found to considerably boost transfection efficiency when compared to polyplexes, in addition to increases in particle size, repeatability, and stability in the presence of serum [94, 95]. Lipopolyplexes with much lower immunostimulatory potential compared to conventional lipopolyplexes were created by further supplementing with yaluronic acid or heparin [96]. Recently, structurally stabilized lipo-polyplexes were created [97] by coating lipoplexes alternately with polyethylenimine and poly (acrylicacid). Comparing these lipo-polyplexes to other polyplexes, lipoplexes, and commercial lipofectamine, they demonstrated enhanced transfection efficiency and decreased cytotoxicity. Therefore, with less cytotoxicity than conventional lipo-and polyplexes, lipo-polyplexes are an effective substitute [97].

7.4.3 Delivery Methods for Nucleic Acids with Localized Release Naked Nucleic Acids Delivered Locally

7.4.3.1 Direct Method

Straight governance of NA as a liquid is possibly the simplest technique of local administration. It is particularly appealing because to the ease of formulation manufacturing and administration, and it offers excellent NA delivery under the correct conditions. Shenetal [98] looked into the possibility of treating choroidal neovascularization with direct injection of stabilized siRNAs. A mouse model of choroidal or retinal neovascularization was treated with local intravitreal or periocular injections of Sirna-027, a chemically stabilized siRNA targeting vascular endothelial growth factor receptor 1 (VEGFR1). Sirna027 injection dramatically reduced VEGFR1 levels of mRNA by 40e57%. Labeled siRNA was also able to penetrate the sclera and choroidea after intraocular injection, and it was detectable in the retina for at least five days. After intravitreal injection, labeled siRNA was detected in retinal cells for 4–5 days. Finally, Sirna-027 therapy effectively suppressed the neovascularization of the eye in mouse models of choroidal and retinal neovascularization. Later, twenty six participants with choroidal neovascularization caused by neovascular age-related macular degeneration took part in a phase 1 clinical study to assess the safety and feasibility of Sirna-027 intravitreal inoculation [99].

7.4.3.2 Composite Method

The vector, which condenses NA and aids in transfecting target cells, and the matrix, which regulates the discharge of the vector and provides structural hold, are often linked in the creation of drug delivery systems for the localized release of NA. This is in addition to the physical techniques discussed below as well as the application of bare NA outlined above. Local administration of NA or NA/vector combinations provides a solution in the simplest situation without the need for a supporting matrix. If in any case we used the matrix, might be NA attached to the surface of the matrix which included within the matrix (immobilization) [99]. It is the main and the majority popular systems in the field of localized therapy, including tissue manufacturing, is hydrogel. Hydrogels are the perfect scaffolds for targeted NA delivery due to their well-hydrated, tissue-like environment, versatility in chemical composition and modification, and often gentle, aqueous manufacturing conditions. Hydrophilic polymers can be semi-synthetic, natural, or synthetic origin are commonly used to create hydrogels through cross-linking or self-assembly. The hydrogel's pore size, rate of degradation, or physical or chemically connected to vector system with hydrogel components can all regulate how quickly encapsulated NA is released. We will emphasize the benefits and difficulties of several hydrogel-based non-viral vector delivery methods in the sections that follow. The reader is directed to for a thorough examination of the properties and applications of semi-synthetic and natural polymers for drug administration. Natural polymers such as collagen, fibrin, or alginate were frequently used to create NA-releasing hydrogels. In this aspect, natural hydrogels may act as a reservoir for NA vectors, with release governed by the pore size of the hydrogel, interactions with the matrix, and/or degradation. Fibrin [60, 61], Collagen [57–59], alginate [62, 63], hyaluronic acid and chitosan, have all been used as the foundation for the development of such systems enhanced transfection efficacy, together *in vitro* and *in vivo*, was typically the outcome of the continuous release of NA from these biocompatible scaffolds [65, 66]. Furthermore, it has frequently been revealed the cytotoxicity of non-viral vectors can be greatly decreased by prolonged release of polyplexes or lipoplexes [59, 67]. Variations in the chemistry of the vector and hydrogel might alter the vector's retention within the hydrogel and, consequently, the rate at which it is released. For instance, Cohen-Sacks *et al.* added poly-L-lysine to a collagen hydrogel matrix to enhance non-specific electrostatic interactions that improved plasmid DNA retention [57]. The hydrogel matrix can also be chemically cross-linked with NA or vector, i.e. polyplexes. By directly contrasting the two methods of cross-linking plasmid DNA with the Scaffold and simple drenched of already cross-linked hydrogel, Capito *et al.* demonstrated that the latter method allows for the reduction of burst release and establishes a link between DNA release and hydrogel degradation [58]. Similar to this, photocross linked dextranhydro gel was attached covalently to PEI via ester linkages to control the persistent release of siRNA/PEI complexes [68]. However, it was shown that aggregation of non-viral vectors upon incorporation into hydrogels constituted a significant constraint, leading to loss of activity. Caged nanoparticle encapsulation, a technique created recently by Lei *et al.*, enables concentrated non-viral vector nanoparticles to be placed into different hydrogels without aggregating [61, 69]. Polyplexes were lyophilized with agarose and sucrose as part of the encapsulating process, which also involved reconstituting the lyophilized powder with a hydrogel precursor and cross-linking to create hydrogel.

7.4.3.3 Physical Methods

Along with polymer- and liposome-based non-viral methods, other physical techniques have been tried for the localized delivery of NA in an effort to enhance safety, effectiveness, and targeting. Hydrodynamic techniques, iontophoresis, microneedles, ultrasound, and electroporation are examples of physical means for delivering NA [95–99]. The latter two techniques, which are used the most commonly, are covered in further feature are in following sections.

7.4.3.3.1 Ultrasound

The physical technique that is most frequently used for localized delivery of nucleic acids is probably ultrasound-facilitated delivery (also known as sonoporation). To efficiently absorb various chemicals into cells, sonoporation relies on the momentary creation of holes in the cell membrane in the presence of micro-bubbles [100]. Among the commercially available microbubble formulations that have received FDA approval for use in people are SonoVue from Bracco International, Optison from Mallinckrodt, and Definity from Lantheus Medical Imaging. As ultrasonic contrast agents, microbubbles have long been used. These microbubbles are known to be drug delivery systems for non-permeable compounds such macromolecular medicines and NA in addition to their usage as contrast agents [101, 102].

Various researchers demonstrated *in vitro* which utilizing PEI/DNA complexes in addition to sonoporation with Optison microbubbles considerably increased transfection efficiency compared to using either treatment alone [100]. Similar to this, increased lipoplex cytoplasmatic absorption through cell membrane holes was seen. Additionally, siRNA-liposome complexes, bare siRNA, and siRNA-polycation complexes were successfully transfected using ultrasound-assisted delivery. These findings imply that employing ultrasound to deliver NA *in vivo* may improve its efficacy. After injecting microbubbles containing naked plasmid DNA encoding for luciferase into rat muscles, Pislaru *et al.* found that ultrasound exposure increased the efficiency of gene transport of the skeletal muscle as compared to the same formulation without ultrasound treatment [101]. Sonoporation tests using adenoviral gene transfer revealed a dose-related increase in the efficacy of ultrasonic aided transfection when compared to the control group. Significant variability in transfection efficacy were observed in different muscles, which was unanticipated and related to the inadequacies of IM injection as well as the variable dosages per volume employed in the individual muscles [102].

7.4.3.3.2 Electroporation

Applying an external electrical field can also make the cell membrane more permeable to non-permeable, macromolecular medicines. In comparison to polymeric and liposomal transfection vectors, electroporation has a number of benefits, as well as the capability to achieve elevated transfection effectiveness in primary cells, fewer safety issues, ease of use, and a minimal impact of cell line on efficiency [103]. Electroporation research has been studied for more than 30 years [104], and as a result, various systems are now old commercially. Numerous clinical investigations have examined electroporation, demonstrating that the technique can be fruitfully used in clinical treatment [105].

Paganin-Gioanni *et al.* [104] recently studied siRNA electroporation on a single-cell level to better understand its fundamental mechanism. In this study, siRNA was directly transferred across the cell membrane by electrotransfer of NA, which was immediately detected by confocal microscopy. The 2 opposing sides of the cell in front of the electrodes had electro permeabilization, but the side in front of the cathode only experienced siRNA entry, indicating that electrophoresis facilitate selector transfer. Interestingly, it was discovered that the mechanism of siRNA electro transfer differs somewhat from plasmid DNA transfer. While DNA of plasmid 1st created long-lasting patches on siRNA appeared to penetrate the cell membrane promptly and freely during electroporation, but DNA appeared to enter the plasma membrane and then translocate into the cytoplasm in a second stage many minutes after electroporation. For the transfer of DNA, electroporation is well known, and significant research has produced both optimal electroporation devices and e parameters. However, fewer studies have looked into the best RNA delivery techniques. Given that DNA and RNA have different basic methods of electroporation-enhanced NA transfer on a cellular level, it is possible to speculate that the ideal transdermal transfer conditions may vary greatly. DNA and siRNA were electroporated *in vivo* through the skin investigated by various researchers [105]. This work shown that a) low voltage electroporation (10V range) produced the most robust transfection for plasmid DNA and b) the electrotransfer of siRNA was successful when the same conditions were used for plasmid DNA. Overall, no spot inflammation or tissue damage was seen during this approach, suggesting the greater tolerability of low voltage electroporation.

7.5 Current Methods for Nonviral Gene Delivery: Benefits and Drawbacks

Because of their high transfection efficiency and ability to reach the cytosol without going through endocytosis, viral vectors have emerged as the gold standard for changing gene expression *in vivo*, especially when compared to synthetic transfection approaches such as lipo/polyplex-based carriers [106]. Viral vectors have significant drawbacks because of the ongoing risk of triggering immune reactions, which makes repeating the procedure challenging, the constrained size of the molecular cargo due to capsid size restrictions, and potential biosafety issues for clinical applications. This is true even though viral vectors can achieve stable (when using adeno-associated viruses) or brief (when using adenoviruses) cell transfection [107, 108]. Numerous non-viral physical transfection techniques, including those based on electroporation, artificial nano carriers, and electroexosomes/EVs, have been developed to get over these restrictions [109]. The poor transfection effectiveness and electrotoxicity of various electroporation-based methods, such as bulk electroporation, which exposes the whole cell surface to a high-intensity electric field, as well as the toxicity of nanocarriers (e.g., gold/tungsten) for biolistic transfection methods are some of these methods' remaining drawbacks [105–110]. Electroporation methods based on nano channel shave become an effective way to get around these restrictions. Because the electric field is only delivered to a very short part of the cell membrane similar to the area of the nanochannel, only the cells in contact with the nanochannels are permeable using this sort of technology. Nanochannel membranes concentrate the high electric field applied to the cell membrane. Unlike the stochastic profile found with bulk electroporation, this characteristic improves cell survival and leads in a higher transmembrane potential with increased transfection efficiencies and better control over molecular cargo transfer [110]. Using the same underlying physical principles, by researcher in 2017 have extra lately made it possible to directly reprogram cells in tissues *in vivo* using Tissue Nano-Transfection for regenerative purposes [108–113].

7.6 Current Barriers for Non-Viral Vectors

In order to achieve good gene delivery, a quantity of procedures must be followed during the process of delivering genetic matter to neurons using non-viral vectors. Early delivery process inhibition will negatively impact subsequent downstream events and reduce the capacity for gene delivery. This means that while certain design techniques can be used to get around specific biological obstacles. The process involved in the gene delivery process must be considered and given special attention to ensure that they do not impair the non-viral vector's ability to overcome previous difficulties. To appropriately create and implement Prior to use reasoning design techniques, the major biological issues confronting non-viral vectors must be recognized and acknowledged. The major biological barriers found thus far include significant transit to neurons in the central and peripheral nervous systems, cellular uptake of the vector by the target cell, endosomal escape, and nuclear escape entrance following DNA release [114, 115]. The vectors must be able to effectively pass the blood-brain barrier during delivery to the central nervous system after systemic injections, which serves to keep undesirable molecules and materials out of the brain and spinal-cord area. While other invasive injection techniques can be used to go beyond this barrier, a perfect nonviral vector would be systemically injected and have the capacity to get past this biological barrier on its own [116]. When it enters the brain, the vector must be able to go around the area without being picked up by neighboring glial cells or held by the ECM. These cationic surface charges on traditional cationic polyplexes and lipoplexes can interact with proteoglycans, glycoproteins, hyaluronan, glycosaminoglycans, and chondroitin.

Sulphate in the surrounding ECM, sequestering these particles before they reach the plane of the target cells [117, 118]. In addition to acting as an electrostatic barrier, these functional structures in the ECM also act as a steric barrier that must be broken by using tiny, extremely strong nanoparticles to prevent sequestration [118–120]. Once within the target cell, the vector must be able to connect with the cell surface or associated proteins, either by fundamental electrostatic contacts or receptor-mediated interactions. The non-viral vector's many characteristics will ultimately affect how it interacts with the cellular surface and is internalized [121]. Nanoparticles are typically kept in an endosomal compartment once they have been taken up by the cell. To achieve high transgenic expression, these compartments must be successfully escaped into the cellular cytoplasm. Lysosomes are formed as the endosome expands and eventually combines with vesicles containing hydrolytic enzymes. Here, genetically modified vectors that couldn't make it into the cytoplasm are enzymatically broken down into tiny parts and expelled as waste from the cell [121, 122].

The genetic material or the vector carrying it must be able to diffuse or be actively carried through the cytoplasm of the cell to the nucleus and through the nuclear envelope following the endosomal escape in order to produce a functional response. The nucleus needs some kind of active transport since larger vector and genetic material complexes have slower diffusion properties [123, 124]. The kind and amount of genetic material that may reach the nucleus in non-diverging cells is further restricted by the nuclear membrane's pores, which are fewer than 40 nm in diameter, without the aid of nuclear transport proteins [125, 126].

7.7 Possibilities for Enhancing the Non-Viral Vector Delivery System

We might speculate that the vectors, whether viral or non-viral, should be designed based on its effectiveness and ability to deliver significant clinical results, based on the documented evidence sand published results.

1. Interaction ability with serum components effectively without sacrificing the therapeutic component
2. Proper biodistribution and circulation in the body
3. Evasion of macrophages and the immune system
4. The cell's capacity for targeting
5. Contact with the surface of the cell
6. Breaking through the cell membrane's defenses
7. The ability of intracellular trafficking (release from endosomes and escape from degradation by nucleases)
8. Nuclear persistence
9. Keeping gene expression constant (time dependent)
10. Transition to succedenic cells (progeny cells)
11. The capacity to transcript

7.8 Conclusion

Given the significant advancements made in the earlier period of time in a variety of fields related to the expansion of novel gene delivery systems and the expansion of our considerate of genomic sand structural biology, there is reason to believe that non-viral vectors will soon be routinely used in medical practice as gene delivery systems to combat a number of deadly diseases. Non-viral vectors exhibit relevant advantages over their viral vector counterparts. The ability to diagnose human diseases has improved, thanks to advances in genome sequencing and genetic analysis, but treatment benefit is still generally ineffectual. The greatest obstacle to using gene therapy to treat human diseases is the inability to discover the perfect vector. The use of non-viral vectors has full-fledged dramatically over the past few years. Before observing any noteworthy clinical benefits, additional transfection efficiency increases are required. For instance, because they lack the restrictions imposed by biological agents, their synthesis is simpler and less expensive. They are additionally less confined by genetic packing capacity and are better handled by the host immune response. Non-viral vectors are classified as pharmaceuticals rather than biological agents by a regulatory agency, which improves clinical application. Non-viral vectors, in fact, have had a difficult time making it into clinical practice. Low transfection efficiency, brief gene expression effects, difficulties reaching target cells after *in vivo* administration, a lack of standardized protocols for an unbiased evaluation of non-viral vector performance, and issues with the commercialization process are just a few of the concerns that restrict the use of non-viral vectors in clinical care. These accomplishments may be based on our comprehension of limiting factors and solutions to them. We might be able to better comprehend intra cellular trafficking and make advancements if we develop specialized intracellular probes. To create are liable, effective, and safe gene delivery method, it may be helpful to combine non-viral and viral free biological vectors.

7.9 Future Relevance

Viral vectors continue to be the most common method of gene delivery in the 110 future gene therapy trials described as recruiting or not yet recruiting on Clinical Trials.gov. Sigilon Therapeutics' only phase I/II trial used a non-viral vector to transport a genetically altered cell line [127]. In the meantime, there have been a lot of preclinical studies and publications in recent years on various non-viral vectors, which definitely showed how swiftly the area of gene delivery is developing.

Despite numerous obstacles, the Pfizer-BioNTechor Moderna in urgent situation utilized the vaccines of COVID-19 both utilized mRNA-based lipid nanoparticles to great success and achieved respective COVID-19 prevention rates of 94.5% and 95% [128, 129]. It's significant that both were created in such a short amount of time. These recent accomplishments have raised anticipation and hope for the future discovery of other secure non-viral vector-based gene therapy solutions.

References

1. Guo, X. and Huang, L., Recent advances in non-viral vectors for gene delivery. *Acc. Chem. Res.*, 45, 7, 971–9, 2012.
2. Jones, C.H., Chen, C.K., Ravikrishnan, A., Rane, S., Pfeifer, B.A., Overcoming nonviral gene delivery barriers: Perspective and future. *Mol. Pharm.*, 10, 4082–98, 2013.

3. Friedmann, T. and Roblin, R., Gene therapy for human genetic disease. *Science*, 175, 949–55, 1972.
4. Han, S.-O., Mahato, R.I., Sung, Y.K., Kim, S.W., Development of biomaterials for gene therapy. *Mol. Ther.*, 2, 302–17, 2000.
5. Mahato, R.I., Smith, L.C., Rolland, A., Pharmaceutical perspectives of nonviral gene therapy. *Adv. Genet.*, 41, 95–156, 1999.
6. Park, D.W., Hoffman, A.S., Pu, S., Stayton, P.S., Design and development of polymers for gene delivery. *Nat. Rev. Drug Discovery*, 4, 581–593, 2005.
7. Mali, S., Delivery systems for gene therapy. *Indian J. Hum. Genet.*, 19, 3–8, 2013.
8. Moss, B., Smith, G.L., Gerin, J.L., Purcell, R.H., Live recombinant vaccinia virus protect schimpanzees against hepatitis B. *Nature*, 311, 5981, 67–9, 1984.
9. Yellepeddi, V.K., Vectors for non-viral gene delivery-clinical and biomedical applications. *Austin Ther.*, 2, 1, 1014, 2015.
10. Kim, S.W., Biomaterials to gene delivery. *J. Control. Release*, 155, 2, 116–118, 2011, https://doi.org/10.1016/j.jconrel.2011.03.023.PMID:21457735.
11. Hill, A.B., Chen, M., Chen, C.-K., Pfeifer, B.A., Jones, C.H., Over coming gene-delivery hurdles: Physiological considerations for nonviral vectors. *Trends Biotechnol.*, 34, 91–105, 2016.
12. Singh, B.N., Prateeksha, A., Gupta, V.K., Chen, J., Atanasov, A.G., Organic nanoparticle-based combinatory approaches for gene therapy. *Trends Biotechnol.*, 35, 12, 1121–4, 2017.
13. Nimesh, S., Halappanavar, S., Kaushik, N.K., Kumar, P., Advances in gene delivery systems. *Biomed. Res. Int.*, 2015, 610342, 2015, https://doi.org/10.1155/2015/610342PMID:25918717.
14. Lodish, H., Berk, A., Zipursky, S.L., Matsudaira, P., Baltimore, D., Darnell, J., *Molecular cell biology*, 4th ed., W.H. Freeman, NewYork, 2000, www.ncbi.nlm.nih.gov/books/NBK21599.
15. Resnik, B. and Langer, P.J., Human germline gene therapy reconsidered. *Hum. Gene Ther.*, 12, 1449–58, 2001.
16. McDonough, P.G., The ethics of somatic and germline gene therapy. *Ann. N. Y. Acad. Sci.*, 816, 378–82, 1997.
17. Uddin, S.N. and Islam, K.K., Cationic polymers and its uses in non-viral gene delivery systems: A conceptual research. *Trends Med. Res.*, 1, 2, 86–99, 2006, https://doi.org/10.3923/tmr.2006.86.99.
18. Remes, A. and Williams, D.F., Immune response in biocompatibility. *Biomaterials*, 13, 11, 731–43, 1992.
19. Liu, L., Yang, J., Men, K., He, Z., Luo, M., Qian, Z., Wei, X., Wei, Y., Current status of non viral vectors for gene therapy in China. *Hum. Gene Ther.*, 29, 110–120, 2018.
20. Hidai, C. and Kitano, H., Nonviral gene therapy for cancer: A review. *Diseases*, 6, 57, 2018.

21. Jones, C.H., Hill, A., Chen, M., Pfeifer, B.A., Contemporary approaches for non viral gene therapy. *Discovery Med.*, 19, 447–454, 2015.
22. Mehier-Humbert, S. and Guy, R.H., Physical methods for gene transfer: Improving the kinetics of gene delivery into cells. *Adv. Drug Deliv. Rev.*, 57, 733–753, 2005.
23. Jinturkar, K.A., Rathi, M.N., Misra, A., Gene delivery using physical methods, in: *Challenges in Delivery of Therapeutic Genomics and Proteomics*, pp. 83–126, Elsevier, Amsterdam, The Netherlands, 2011.
24. Hosokawa, Y., Iguchi, S., Yasukuni, R., Hiraki, Y., Shukunami, C., Masuhara, H., Gene delivery process in a single animal cell after femto second laser microinjection. *Appl. Surf. Sci.*, 255, 9880–9884, 2009.
25. King, R., Gene delivery to mammalian cells by microinjection, in: *Gene Delivery to Mammalian Cells*, pp. 167–173, Springer, Berlin/Heidelberg, Germany, 2004.
26. Wolff, J.A., Malone, R.W., Williams, P., Chong, W., Acsadi, G., Jani, A., Felgner, P.L., Direct gene transfer into mouse muscle *in vivo*. *Science*, 247, 1465–1468, 1990.
27. Lui, V.W.Y., Falo, L.D., Jr., Huang, L., Systemic production of IL-12 by naked DNA mediated gene transfer: Toxicity and attenuation of transgene expression *in vivo*. *J. Gene Med.*, 3, 384, 2001.
28. Springer, M.L., Rando, T.A., Blau, H.M., Gene delivery to muscle. *Curr. Protoc. Hum. Genet.*, 31, 13.14. 11–13.14. 19, 2001.
29. Wendell, D.M., Hemond, B.D., Hogan, N.C., Taberner, A.J., Hunter, I.W., The effect of jet parameters on jet injection, in: *Proceedings of the 2006 International Conference of the IEEE Engineering in Medicine and Biology Society*, New York, NY, USA, 30 August–3 September 2006, pp. 5005–5008.
30. Fargnoli, A., Katz, M., Williams, R., Margulies, K., Bridges, C.R., A needleless liquid jet injection delivery method for cardiac gene therapy: A comparative evaluation versus standard routes of delivery reveals enhanced therapeutic retention and cardiac specific gene expression. *J. Cardiovasc. Transl. Res.*, 7, 756–767, 2014.
31. Lysakowski, C., Dumont, L., Tramèr, M.R., Tassonyi, E., A needle-free jet-injection system with lidocaine for peripheral intravenous cannula insertion: A randomized controlled trial with cost-effectiveness analysis. *Anesth. Analg.*, 96, 215–219, 2003.
32. Chuang, Y.-C., Chou, A.-K., Wu, P.-C., Chiang, P.-H., Yu, T.-J., Yang, L.-C., Yoshimura, N., Chancellor, M.B., Gene therapy for bladder pain with gene gun particle encoding pro-opio melano cortinc DNA. *J. Urol.*, 170, 2044–2048, 2003.
33. Dutta, S., Ray, B., Raut, S., Sahoo, C., Non-viral gene therapy: Technology and application. *Res. J. Sci. Technol.*, 13, 13–22, 2021.
34. Neumann, E., Schaefer-Ridder, M., Wang, Y., Hofschneider, P., Gene transfer into mouse lyoma cells by electroporation in high electric fields. *EMBO J.*, 1, 841–845, 1982.

35. Titomirov, A.V., Sukharev, S., Kistanova, E., *In vivo* electroporation and stable transformation of skin cells of new born mice by plasmid DNA. *Biochim. Biophys. Acta (BBA)-Gene Struct. Expression*, 1088, 131–134, 1991.
36. Heller, L.C. and Heller, R., *In vivo* electroporation for gene therapy. *Hum. Gene Ther.*, 17, 890–897, 2006.
37. Young, J.L. and Dean, D.A., Electroporation-mediated gene delivery. *Adv. Genet.*, 89, 49–88, 2015.
38. Lambricht, L., Lopes, A., Kos, S., Sersa, G., Préat, V., Vandermeulen, G., Clinical potential of electroporation for gene therapy and DNA vaccine delivery. *Expert Opin. Drug Delivery*, 13, 295–310, 2016.
39. Aslan, H., Zilberman, Y., Arbeli, V., Sheyn, D., Matan, Y., Liebergall, M., Li, J.Z., Helm, G.A., Gazit, D., Gazit, Z., Nucleofection based *ex vivo* nonviral gene delivery to human stem cells as a plat form for tissue regeneration. *Tissue Eng.*, 12, 877–889, 2006.
40. Delalande, A., Postema, M., Mignet, N., Midoux, P., Pichon, C., Ultrasound and micro bubble-assisted gene delivery: Recent advances and on going challenges. *Ther. Deliv.*, 3, 1199–1215, 2012.
41. Liang, H., Tang, J., Halliwell, M., Sonoporation, drug delivery, and gene therapy. *Proc. Inst. Mech. Eng. Part H J. Eng. Med.*, 224, 343–361, 2010.
42. Unger, E.C., Hersh, E., Vannan, M., Matsunaga, T.O., McCreery, T., Local drug and gene delivery through microbubbles. *Prog. Cardiovasc. Dis.*, 44, 45–54, 2001.
43. Suda, T. and Liu, D., Hydrodynamic gene delivery: Its principles and applications. *Mol. Ther.*, 15, 2063–2069, 2007.
44. Fabre, J., Grehan, A., Whitehorne, M., Sawyer, G., Dong, X., Salehi, S., Eckley, L., Zhang, X., Seddon, M., Shah, A., Hydrodynamic gene delivery to the pig liver via an isolated segment of the inferiorvenacava. *Gene Ther.*, 15, 452–462, 2008.
45. Du, X., Wang, J., Zhou, Q., Zhang, L., Wang, S., Zhang, Z., Yao, C., Advanced physical techniques for gene delivery based on membrane perforation. *Drug Deliv.*, 25, 1516–1525, 2018.
46. Zhang, H., Lee, M.-Y., Hogg, M.G., Dordick, J.S., Sharfstein, S.T., Gene delivery in three-dimensional cell cultures by superparamagnetic nanoparticles. *ACS Nano*, 4, 4733–4743, 2010.
47. Meacham, J.M., Durvasula, K., Degertekin, F.L., Fedorov, A.G., Physical methods for intracellular delivery: Practical aspects from laboratory use to industrial-scale processing. *J. Lab. Autom.*, 19, 1–18, 2014.
48. Li, W., Ma, N., Ong, L.L., Kaminski, A., Skrabal, C., Ugurlucan, M., Lorenz, P., Gatzen, H.H., Lützow, K., Lendlein, A., Enhanced thoracic gene delivery by magnetic nanobead-mediated vector. *J. Gene Med. A Cross-Discip. J. Res. Sci. Gene Transf. Its Clin. Appl.*, 10, 897–909, 2008.
49. Prosen, L., Markelc, B., Dolinsek, T., Music, B., Cemazar, M., Sersa, G., Mcam silencing with RNA interference using magnetofection has antitumor effect in murine melanoma. *Mol. Ther.-Nucleic Acids*, 3, e205, 2014.

50. Li, Z., Xiang, J., Zhang, W., Fan, S., Wu, M., Li, X., Li, G., Nanoparticle delivery of anti-metastatic NM23-H1 gene improves chemotherapy in a mouse tumor model. *Cancer Gene Ther.*, 16, 423–429, 2009.
51. Liu, F. and Huang, L., Non invasive gene delivery to the liver by mechanical massage. *Hepatology*, 35, 1314–1319, 2002.
52. Liu, F., Lei, J., Vollmer, R., Huang, L., Mechanism of liver gene transfer by mechanical massage. *Mol. Ther.*, 9, 452–457, 2004.
53. Wang, W., Li, W., Ma, N., Steinhoff, G., Non-viral gene delivery methods. *Curr. Pharm. Biotechnol.*, 14, 46–60, 2013.
54. Khalid, M. and El-Sawy, H.S., Polymeric nano particles: Promising platform for drug delivery. *Int. J. Pharm.*, 528, 675–691, 2017.
55. Rose, V.L., Mastrotto, F., Mantovani, G., Phosphonium polymers for gene delivery. *Polym. Chem.*, 8, 353–360, 2017.
56. Cavalu, S., Antoniac, I.V., Mohan, A., Bodog, F., Doicin, C., Mates, I., Ulmeanu, M., Murzac, R., Semenescu, A., Nanoparticles and nanostructured surface fabrication for innovative cranial and maxillo facial surgery. *Materials*, 13, 5391, 2020.
57. Lee, S.J., Yhee, J.Y., Kim, S.H., Kwon, I.C., Kim, K., Biocompatible gelatin nano particles for tumor-targeted delivery of polymerized siRNA in tumor-bearing mice. *J. Control. Release*, 172, 358–366, 2013.
58. Youngren, S.R., Tekade, R.K., Gustilo, B., Hoffmann, P.R., Chougule, M.B., STAT6siRNA matrix-loaded gelatin nano carriers: Formulation, characterization, and *ex vivo* proof of concept using adeno carcinoma cells. *BioMed. Res. Int.*, 2013, 858946, 2013.
59. Hu, Y., Li, Y., Xu, F.-J., Versatile functionalization of polysaccharides via polymer grafts: From design to biomedical applications. *Acc. Chem. Res.*, 50, 281–292, 2017.
60. Cavalu, S., Bisboaca, S., Mates, I.M., Pasca, P.M., Laslo, V., Costea, T., Fritea, L., Vicas, S., Novel formulation based on chitosan arabic gum nanoparticles entrapping propolis extract production, physico-chemical and structural characterization. *Rev. Chim.*, 69, 3756–3760, 2018.
61. Kean, T., Roth, S., Thanou, M., Trimethylated chitosansas non-viral gene delivery vectors: Cytotoxicity and transfection efficiency. *J. Control. Release*, 103, 643–653, 2005.
62. Shin, M.D., Shukla, S., Chung, Y.H., Beiss, V., Chan, S.K., Ortega-Rivera, O.A., Wirth, D.M., Chen, A., Sack, M., Pokorski, J.K., COVID-19 vaccine development and a potential nanomaterial path forward. *Nat. Nanotechnol.*, 15, 646–655, 2020.
63. Malabadi, R.B., Meti, N.T., Chalannavar, R.K., Applications of nano technology in vaccine development for coronavirus (SARSCoV-2) disease (COVID-19). *Int. J. Res. Sci. Innov.*, 8, 191–198, 2021.
64. Drummond, D.C., Noble, C.O., Hayes, M.E., Park, J.W., Kirpotin, D.B., Pharmacokinetics and *in vivo* drug release rates in liposomal nanocarrier development. *J. Pharm. Sci.*, 97, 4696–4740, 2008.

65. Ewert, K.K., Zidovska, A., Ahmad, A., Bouxsein, N.F., Evans, H.M., Mc Allister, C.S., Samuel, C.E., Safinya, C.R., Cationic liposome–nucleic acid complexes for gene delivery and silencing: pathways and mechanisms for plasmid DNA and siRNA, in: *Nucleic Acid Transfection*, Bielke, W., Erbacher, C. (eds). vol. 296,, pp. 191–226, Springer, Berlin, Heidelberg, 2010. https://doi.org/10.1007/128_2010_70
66. Siafaka, P.I., ÜstündagOkur, N., Karavas, E., Bikiaris, D.N., Surface modified multifunctional and stimuli responsive nano particles for drug targeting: Current status and uses. *Int. J. Mol. Sci.*, 17, 1440, 2016.
67. Riley, M.K. and Vermerris, W., Recent advances in nano materials for gene delivery—Are view. *Nanomaterials*, 7, 94, 2017.
68. Xu, L. and Anchordoquy, T., Drug delivery trends in clinical trials and translational medicine: Challenges and opportunities in the delivery of nucleic acid-based therapeutics. *J. Pharm. Sci.*, 100, 38e52, 2011.
69. Bertin, A., Polyelectrolyte complexes of DNA and polycations as gene delivery vectors, in: *Polyelectrolyte Complexes in the Dispersed and Solid State II: Application Aspects*, vol. 256, p. 103e195, 2014.
70. Germershaus, O., Mao, S.R., Sitterberg, J. *et al.*, Gene delivery using chitosan, trimethyl chitosan or polyethylenglycol- graft-trimethyl chitos an block copolymers:establishment of structure-activity relationships *in vitro*. *J. Control. Release*, 125, 145e154, 2008.
71. Boussif, O., Lezoualch, F., Zanta, M.A. *et al.*, A versatile vector for gene and oligonucleotide transfer into cells in culture and *in-vivo* e polyethylenimine. *Proc. Natl. Acad. Sci. U. S. A.*, 92, 7297e7301, 1995.
72. Mao, S.R., Sun, W., Kissel, T., Chitosan-based formulations for delivery of DNA and siRNA. *Adv. Drug Deliv. Rev.*, 62, 12e27, 2010.
73. Dufes, C., Uchegbu, I.F., Schatzlein, A.G., Dendrimers in gene delivery. *Adv. Drug Deliv. Rev.*, 57, 2177e2202, 2005.
74. Benjaminsen, R.V., Mattebjerg, M.A., Henriksen, J.R. *et al.*, The possible "proton sponge" effect of polyethylenimine (PEI) does not include change in lysosomal pH. *Mol. Ther.*, 21, 149e157, 2013.
75. Rehman, Z.U., Hoekstra, D., Zuhorn, I.S., Mechanism of polyplex- and lipoplex-mediated delivery of nucleic acids: Real-time visualization of transient membrane destabilization without endosomal lysis. *ACS Nano*, 7, 3767e3777, 2013.
76. Petersen, H., Fechner, P.M., Martin, A.L. *et al.*, Polyethylenimine-graft-poly(ethylene glycol) copolymers: Influence of copolymer block structure on DNA complex ation and biological activities as gene delivery system. *Bioconjug. Chem.*, 13, 845e854, 2002.
77. Ogris, M., Brunner, S., Schuller, S. *et al.*, PEGylated DNA/transferrin-PEI complexes:Reduced interaction with blood components, extended circulation in blood and potential for systemic gene delivery. *Gene Ther.*, 6, 595e605, 1999.

78. Germershaus, O., Merdan, T., Bakowsky, U. *et al.*, Trastuzumab-polyethylenimine-polyethylene glycol 8conjugates for targeting Her2-expressing tumors. *Bioconjug. Chem.*, 17, 1190e1199, 2006.
79. Germershaus, O., Neu, M., Behe, M. *et al.*, HER2 targeted polyplexes: The effect of polyplex composition and conjugation chemistry on *in vitro* and *in vivo* characteristics. *Bioconjug. Chem.*, 19, 244e253, 2008.
80. Neu, M., Germershaus, O., Mao, S. *et al.*, Cross-linked nanocarriers based upon poly(ethyleneimine)for systemic plasmid delivery: *In vitro* characterization and *in vivo* studies in mice. *J. Control. Release*, 118, 370e380, 2007.
81. Neu, M., Germershaus, O., Behe, M. *et al.*, Bioreversibly cross-linked polyplexes of PEI and high molecular weight PEG show extended circulation times *in vivo*. *J. Control. Release*, 124, 69e80, 2007.
82. Florea, B.I., Meaney, C., Junginger, H.E. *et al.*, Transfectione fficiency and toxicity of polyethylenimine in differentiated Calu-3 and non differentiated COS-1 cell cultures. *AAPS PharmSci*, 4, 12, 2002. https://doi.org/10.1208/ps040312.
83. VandeWetering, P., Cherng, J.Y., Talsma, H. *et al.*, Relation between transfection efficiency and cytotoxicity of poly(2-(dimethylamino) ethylmethacrylate)/plasmid complexes. *J. Control. Release*, 49, 59e69, 1997.
84. Breunig, M., Lungwitz, U., Liebl, R. *et al.*, Breaking up the correlation between efficacy and toxicity for non viral gene delivery. *Proc. Natl. Acad. Sci. U. S. A.*, 104, 14454e14459, 2007.
85. Peng, Q., Zhong, Z.L., Zhuo, R.X., Disulfide cross-linked polyethylenimines (PEI) prepared viathiolation of low molecular weight PEI as highly efficient gene vectors. *Bioconjug. Chem.*, 19, 499e506, 2008.
86. Yin, H., Kanasty, R.L., Eltoukhy, A.A. *et al.*, Non-viral vectors for gene-based therapy. *Nat. Rev. Genet.*, 15, 541e555, 2014.
87. Felgner, P.L., Gadek, T.R., Holm, M. *et al.*, Lipofection e a highly efficient, lipid-mediated DNA-transfection procedure. *Proc. Natl. Acad. Sci. U. S. A.*, 84, 7413e7417, 1987.
88. Konopka, K., Stamatatos, L., Larsen, C.E. *et al.*, Enhancement of human-immunodeficiency-virus type-1 infection by cationic liposomes e the role of cd4, serum and liposome cell-interactions. *J. Gen. Virol.*, 72, 2685e2696, 1991.
89. Felgner, J.H., Kumar, R., Sridhar, C.N. *et al.*, Enhanced gene delivery and mechanism studies with a novel series of cationic lipid formulations. *J. Biol. Chem.*, 269, 2550e2561, 1994.
90. Zuhorn, I.S., Kalicharan, R., Hoekstra, D., Lipoplex-mediated transfection of mammalian cell soccurs through the cholesterol - dependent clathrin-mediated pathway of endocytosis. *J. Biol. Chem.*, 277, 18021e18028, 2002.
91. Dass, C.R., Cytotoxicity issues pertinent to lipoplex-mediated gene therapy *in-vivo*. *J. Pharm. Pharmacol.*, 54, 593e601, 2002.
92. DelIarduya, C.T., Sun, Y., Duezguenes, N., Gene delivery by lipoplexes and polyplexes. *Eur. J. Pharm. Sci.*, 40, 159e170, 2010.

93. Ozpolat, B., Sood, A.K., Lopez-Berestein, G., Liposomal siRNA nanocarriers for cancer therapy. *Adv. Drug Deliv. Rev.*, 66, 110e116, 2014.
94. Li, S.D. and Huang, L., Targeted delivery of antisense oligodeoxy nucleotide and small interference RNA into lung cancer cells. *Mol. Pharm.*, 3, 579e588, 2006.
95. Garcia, L., Urbiola, K., Duzgunes, N. *et al.*, Lipopolyplexes as nano medicines for therapeutic gene delivery. Nano medicine: Infectious diseases, immunotherapy, diagnostics, antifibrotics, toxicology and gene medicine. *Meth. Enzymol.*, 509, 327e338, 2012.
96. Chono, S., Li, S.D., Conwell, C.C. *et al.*, An efficient and low immune stimulatory nanoparticle formulation for systemic siRNA delivery to the tumor. *J. Control. Release*, 131, 64e69, 2008.
97. Jain, S., Kumar, S., Agrawal, A.K. *et al.*, Enhanced transfection efficiency and reduced cytotoxicity of novel lipid-polymerhybrid nano-plexes. *Mol. Pharm.*, 10, 2416e2425, 2013.
98. Shen, J., Samul, R., Silva, R.L. *et al.*, Suppression of ocular neovascularization with siRNA targeting VEGF receptor 1. *Gene Ther.*, 13, 225e234, 2006.
99. Kaiser, P.K., Symons, R.C.A., Shah, S.M. *et al.*, RNAi-based treatment for neovascular age-related macular degeneration by Sirna-027. *Am. J. Ophthalmol.*, 150, 33e39, 2010.
100. Miller, D.L., Pislaru, S.V., Greenleaf, J.E., Sonoporation: Mechanical DNA delivery by ultra-soniccavitation. *Somatic Cell Mol. Genet.*, 27, 115e134, 2002.
101. Bao, S.P., Thrall, B.D., Miller, D.L., Transfection of a reporter plasmid into cultured cells by sonoporation *in vitro*. *Ultrasound Med. Biol.*, 23, 953e959, 1997.
102. Duvshani-Eshet, M. and Machluf, M., Therapeutic ultrasound optimization for gene delivery:A key factor achieving nuclear DNA localization. *J. Control. Release*, 108, 513e528, 2005.
103. Geng, T., Zhan, Y.H., Wang, H.Y. *et al.*, Flow-through electroporation based on constant voltage for large-volume transfection of cells. *J. Control. Release*, 144, 91e100, 2010.
104. Bodles-Brakhop, A.M., Heller, R., Draghia-Akli, R., Electroporation for the delivery of dna-based vaccines and immunotherapeutics: Current clinical developments. *Mol. Ther.*, 17, 585e592, 2009.
105. Daya, S. and Berns, K.I., Gene therapy using adeno-associated virus vectors. *Clin. Microbiol. Rev.*, 21, 583–593, 2008.
106. Joshi, C.R., Labhasetwar, V., Ghorpade, A., Destination brain: The past, present, and future of therapeutic gene delivery. *J. Neuroimmune Pharmacol.*, 12, 51–83, 2017.
107. Ziello, J.E., Huang, Y., Jovin, I.S., Cellular endocytosis and gene delivery. *Mol. Med.*, 16, 222–229, 2010.
108. Wu, Y., Terp, M.C., Kwak, K.J., Gallego-Perez, D., Nana-Sinkam, S.P., Lee, L.J., Surface-mediated nucleic acid delivery by lipoplexes prepared in micro well arrays. *Small*, 9, 2358–2367, 2013.

109. Wang, W., Li, W., Ma, N., Steinhoff, G., Non-viral gene delivery methods. *Curr. Pharm. Biotechnol.*, 14, 46–60, 2013.
110. Gallego-Perez, D., Otero, J.J., Czeisler, C., Ma, J., Ortiz, C., Gygli, P. *et al.*, Deterministic transfection drives efficient non viral reprogramming and uncovers reprogramming barriers. *Nanomedicine*, 12, 399–409, 2016.
111. Gallego-Perez, D., Pal, D., Ghatak, S., Malkoc, V., Higuita-Castro, N., Gnyawali, S. *et al.*, Topical tissuenano-transfection mediates non-viral stroma reprogramming and rescue. *Nat. Nanotechnol.*, 12, 974–979, 2017.
112. Al-Dosari, M.S. and Gao, X., Non viral gene delivery: Principle, limitations, and recent progress. *AAPS J.*, 11, 671, 2009.
113. Boukany, P.E., Morss, A., Liao, W.C., Henslee, B., Jung, H., Zhang, X. *et al.*, Nano channel electroporation delivers precise amounts of biomolecules into living cells. *Nat. Nanotechnol.*, 6, 747–754, 2011.
114. Maguire, C.A., Ramirez, S.H., Merkel, S.F., Sena-Esteves, M., Breakefield, X.O., Gene therapy for the nervous system: Challenges and new strategies. *Neurotherapeutics*, **11**, 4, 817–839, 2014.
115. Bulcha, J.T., Wang, Y., Ma, H., Tai, P.W.L., Gao, G., Viral vector platforms with in the gene therapy landscape. *Signal Transduct. Target. Ther.*, **6**, 1–24, 2021.
116. Pérez-Martínez, F.C., Guerra, J., Posadas, I., Ceña, V., Barriers to non-viral vector-mediated genedelivery in the nervous system. *Pharm. Res.*, **28**, 1843–1858, 2011.
117. Shi, N. and Pardridge, W.M., Noninvasive gene targeting to the brain. *Proc. Natl. Acad. Sci. U. S. A.*, **97**, 7567–7572, 2000.
118. Mead, B.P. *et al.*, Targeted gene transfer to the brain via the delivery of brain-penetrating DNA nano particles with focused ultrasound. *J. Control. Release*, **223**, 109–117, 2016.
119. Sykováand, E. and Nicholson, C., Diffusion in brain extra cellular space. *Physiol. Rev.*, **88**, 4, 1277–1340, 2008.
120. Zimmermann, D.R. and Dours-Zimmermann, M.T., Extracellular matrix of the central nervous system:From neglect to challenge. *Histochem. Cell Biol.*, **130**, 4, 635–653, 2008.
121. Barros, C.S., Franco, S.J., Müller, U., Extracellular matrix: Functions in the nervous system. *Cold Spring Harb. Perspect. Biol.*, **3**, 1, a005108, 2011.
122. Nance, E.A. *et al.*, A dense poly (ethylene glycol) coating improves penetration of large polymer icnano particles within braint issue. *Sci. Transl. Med.*, **4**, 149ra119, 2012.
123. Kenny, G.D. *et al.*, Multifunctional receptor-targeted nanocomplexes for the delivery of therapeutic nucleic acids to the brain. *Biomaterials*, **34**, 9190–9200, 2013.
124. Mastorakos, P. *et al.*, Biodegradable brain-penetrating DNA nano complexes and their use to treat malignant brain tumors. *J. Control. Release*, **262**, 37, 2017.

125. Xu, E., Saltzman, W.M., Piotrowski-Daspit, A.S., Escaping the endosome: Assessing cellular trafficking mechanisms of non-viral vehicles. *J. Control. Release*, **335**, 465–480, 2021.
126. Dowty, M.E., Williams, P., Zhang, G., Hagstromand, J.E., Wolff, J.A., Plasmid DNAentry into postmitotic nuclei of primary rat myotubes. *Proc. Natl. Acad. Sci. U. S. A.*, **92**, 4572–4576, 1995.
127. https://clinicaltrials.gov/ct2/show/NCT04541628.Accessed20Oct2022. Returntoref93inarticle.
128. Polack, F.P., Thomas, S.J., Kitchin, N., Absalon, J., Gurtman, A., Lockhart, S. *et al.*, Safety and efficacy of the BNT162b2 mRNA COVID-19v vaccine. *N. Engl. J. Med.*, 383, 27, 2603–15, 2020.
129. https://investors.modernatx.com/news-releases/news-release-details/moderna-announces-primary-efficacy-analysis-phase-3-cove-study. Accessed20Oct2022.

8

Utilization of Chitosan for Gene Delivery

Johnson Olaleye Oladele[1,2]

[1]*Biochemistry Unit, Department of Chemical Sciences, Faculty of Sciences, Kings University, Ode-Omu, Nigeria*
[2]*Phytochemistry and Phytomedicine Research Unit, Royal Scientific Research Institute, Osun State, Nigeria*

Abstract
Gene therapy is a novel and promising therapeutic alternative that uses genes for treating a variety of illnesses, including hereditary abnormalities, cancer, some viral infections and so on. The three components that constitute gene delivery systems include a plasmid-based gene expression system that regulates a gene's function within the target cell, a gene that encodes a specific therapeutic protein, and a gene delivery system that regulates the delivery of the gene expression plasmid to a specific site within the body. The main methods for delivering genes include the use of non-viral or viral vectors. Even though viral gene carriers make good transfection vehicles, their significance as gene carriers in gene therapy is constrained by their carcinogenic and mutagenic features. Recently, non-viral gene carriers have developed significantly as substitute for viral gene carriers in genetic biotechnology. Owning to their *in vivo* stability, high gene encapsulation, and enhanced transfection effectiveness, cationic polymers (e.g. chitosan) were widely exploited as alternative gene carriers for gene delivery. Chitosan, when used as a vehicle for gene transfer, appear not hazardous. By overcoming the systemic hurdles to gene delivery, chitosan appears to have outstanding physicochemical characteristics that are advantageous for nucleic acid transport. When chitosan interacts electrostatically with nucleic acids, it easily forms complexes, microspheres, or nanoparticles. This chapter presents recent advances on role of chitosan in gene delivery and therapy.

Keywords: Gene therapy, cationic polymers, chitosan, gene delivery system

**Corresponding author*: oladelejohn2007@gmail.com; jo.oladele@kingsuniversity.edu.ng

Rishabha Malviya, Sonali Sundram and Neelam Jain (eds.) Integration of Biomaterials for Gene Therapy, (223–246)

8.1 Introduction

Gene therapy is a novel method that uses genes to treat or prevent various diseases. Instead of utilizing medications or surgery to treat a medical problem, doctors may be able to use the gene therapy procedure to implant a gene into the patient's cell. A number of approaches to gene therapy are being investigated by some scientists and medical professionals, including i) introducing new genes into the cells to guard against any diseases; ii) "knocking out" or inactivating a mutated gene; and iii) replacing a disease-causing mutated gene with a healthy gene.

Gene therapy is a promising alternative for treating a variety of illnesses, including hereditary abnormalities, cancer, and some viral infections; nevertheless, the procedure is dangerous and is still being studied to make it safe before it can be effective. Only conditions for which there are currently no other treatment options have been studied using gene therapy. To activate gene expression, genetic molecules enter the nucleus of host cells [1–3]. In order for a patient's somatic cells to produce precise therapeutic proteins to modify genetic disorders, gene therapy has evolved to give genetic information. A thorough understanding of the interactions between the targeting cell and the gene delivery system is necessary for a successful design of the system. A plasmid-based gene expression system that regulates a gene's function within the target cell, a gene that encodes a specific therapeutic protein, and a gene delivery system that regulates the delivery of the gene expression plasmid to a specific site within the body make up the three components that constitute gene delivery systems [4, 5].

The foreign genetic material must remain stable inside the host cells for the gene delivery method to work successfully [6–8]. The main methods for delivering genes include the use of non-viral or viral vectors. Viral-based vectors were first developed as a transgene expression technique in the 1980s [9]. In order to protect chimpanzees against Hepatitis B, the vaccinia virus was utilized as a vaccine vector in 1984 [9]. Contrarily, non-viral gene delivery systems were the first to report on changing cellular phenotypes by exposure to DNA in the body [10, 11]. Even though viral gene carriers make good transfection vehicles, their significance as gene carriers in cancer gene therapy is constrained by the carcinogenic and mutagenic features [12].

Recently, non-viral gene carriers have developed significantly as substitute for viral gene carriers in genetic biotechnology [13]. Polymer-based carriers, liposomes, and various kinds of nanoparticles are types of non-viral gene carriers used in delivering gene. Because of its great transfection efficacy and simplicity of manufacture, liposome is among them one of the most researched gene carriers. The main drawbacks of liposomes, however, include their limited *in vivo* stability, non-specific toxicity, short shelf life, and poor encapsulation efficiency [14–16]. The instantaneous blood clearance occurrence brought on by continuous liposome treatment boosted its elimination from the body [17]. PEGylation (PEG: poly-ethylene glycol) seems to enhance the bioavailability of liposomes. Owning to their *in vivo* stability, high gene encapsulation, and enhanced transfection effectiveness, cationic polymers were widely exploited as alternative gene carriers for gene delivery [18].

Cationic polymers like chitosan, which have a lot of free amine groups, can effectively compress DNA or oligonucleotides. These polymers' high cationic charge densities enable improved intracellular nanoparticle trafficking via endosomal disruption, but they can cause unintended cellular toxicity [19]. Chitosan is an interesting exception since, when used as a vehicle for gene transfer, it doesn't appear to be hazardous. By overcoming the systemic hurdles to gene delivery, chitosan appears to have outstanding physicochemical characteristics that are advantageous for nucleic acid transport. When chitosan interacts electrostatically with nucleic acids, it easily forms complexes, microspheres, or nanoparticles [20]. Chitosan has been investigated more and more as a gene delivery technology in chemotherapy due to its promising properties.

This chapter presents recent developments in the creation of gene delivery systems, especially focused on role of chitosan in gene delivery and other non-viral gene delivery methods. The physicochemical approaches, such as the physical method and chemical method, have also been treated for the non-viral gene delivery systems. The gene delivery methods of cationic biomedical polymers, including poly(L-lysine) (PLL), polyethyleneimines (PEI), and polysaccharides derivatives, have been studied in terms of several types of gene delivery systems.

8.2 Cationic Polymers-Based Gene Delivery Systems

Researchers studying gene delivery systems have examined the possible advantages of polymeric carriers for genes in cationic biopolymers including chitosan and liposomes derivatives [21, 22]. The polymeric systems can conceal the negative DNA charges and compress the big genes into small molecular structures when used in conjunction with cationic polymers for the delivery of genes. A compound known as polyplex makes up the cationic polymer-based nucleic acid. Clinical testing of "lipoplexes," cationic non-viral lipid-based gene carriers, is now underway [23–27]. One important targeting delivery system for gene therapy is the gene complex.

The majority of studies concentrate on the impact of ligands that are covalently coupled to the DNA complex. The targeted ligands can combine with the cationic polymers. One potential possibility is poly(L-lactide), a polymer that is frequently used to connect the targeted ligands [28–31].

Under a wide range of circumstances, the cationic polymer condensation of plasmid DNA experiences compaction. The condensing agents have typically been multivalent cationic polymers [32, 33]. Researchers have looked at the tertiary structures that resemble non-condensing plasmid DNA in complexes with low hydrophobized stearyl-poly(L-lactide) [34]. Systems for attaching hydrophilic segments, such as hyaluronic acid, dextran, polyethylene glycol, or hydrophobic stearyl chains, are still possibly beneficial to DNA condensation [35–38].

8.2.1 Poly(L-lysine)-Based Gene Delivery Systems

One of the cationic polymers for gene transfers is poly(L-lysine) (PLL). At pH 7, PLL is a homo-polypeptide that belongs to the class of cationic polymers. A positively charged hydrophilic amino group can be found in PLL. In order to facilitate the co-adsorption of plasmid DNA, Replica Particles (RP) based on PLL cross-linked via a homo-bi-functional linker have been created for gene delivery systems [39]. For the *in vivo* transport of genes to the liver, poly-L-lysine/DNA polyplexes have been stabilized [40]. Poly-L-lysine-based gene delivery systems have been synthesized and characterized [41]. In mouse lung endothelial cells, N-terminally modified poly(L-lysine)-antibody conjugate had been employed as a carrier for targeted gene delivery [42]. Non-viral gene delivery methods have been made using biodegradable poly(ethylene glycol)-co-poly(L-lysine)-g-histidine multi-block copolymers [43]. In polymeric gene delivery, poly(-L-lysine) and copolymers for gene transfer were discussed and compiled [44]. Additionally, polycation copolymers of the poly-L-lysine-graft-PEG-comb type had been produced for gene delivery systems [45]. First created and characterized were nano-particle DNA carriers with poly(L-lysine) grafted polysaccharide copolymers and terplex DNA delivery systems as a gene carrier [46, 47]. For gene delivery systems, new non-viral DNA delivery vectors of the terplex systems and poly(-L-lysine)-grafted polyethylene glycol were also created [48–51]. The field of biomedical applications has recently undergone clinical examinations [52–63].

8.2.2 Polysaccharides-Based Gene Delivery Systems

On the basis of grafted oligo-amine residues for natural polysaccharides, a variety of biodegradable cationic polymers have been produced [64]. The grafting concept enables three-dimensional interaction with anionic surfaces of double strand DNA chains by attaching side chain oligomers to either linear or branched hydrophilic polysaccharides. The polysaccharide-based gene delivery technology suggests that the cationic polysaccharide's structure has a vital impact in the transfection activity for gene delivery. The mucosal transport of complexes such plasmids, oligonucleotides, proteins, and peptides had been explored using colloidal polysaccharide particles [65]. Nanoparticles made of natural polysaccharides have also been investigated as methods of delivering drugs and genes [66]. The mechanism for creating polysaccharide-based nanoparticles included hydrophobic polysaccharide self-assembly, covalent and ionic cross-linking, and electrolyte complexing. Natural polysaccharides like pectin and chitosan have been investigated as carriers for drugs or genes [67]. Polycationic derivatives of poly-N-acetyl-D-glucosamine, the positively charged molecule with significant muco-adhesion, have been strongly advised by chitosan families. The homo-galacturonan (HG) and rhamnogalacturonan (RG-I) structural segments of pectin are composed of both smooth and hairy regions.

8.2.3 Poly(ethylenimine)s-Based Gene Delivery Systems

In most cases, polyamines that turn cationic under physiologic conditions are the cationic polymers used for gene complexation. Numerous nitrogen atoms, including primary, secondary, and tertiary amine groups, are found in poly(ethylenimine) (PEI). These amine groups have the ability to increase their ability to act as buffers. These result in an endosomal or lysosomal rupture and, via the proton sponge effect, escape into the cytoplasm [68, 69]. The other functional molecules can receive some buffering capacity enhancement from PEI. PEI conjugated poly(crystamine bis(acrylamide)-diaminohexane) [poly(CBA-DAH)] was created to improve transfection efficiency and reduce weight ratio. Multiple disulfides make up Poly(CBA-DAH), and an intracellular reducing agent like glutathione can break them in the cytoplasm (GSH) [70–73]. The triple-peptide that makes up GSH was created in the cytosol from precursor amino acids [74]. To confirm the successful production of vectors and polyplexes with pDNA, the design of PEI-conjugated bio-reducible polymer for efficient gene transport had been examined. The improved buffering ability of other compounds including imidazole, histidine, and PEI was also demonstrated in other investigations [75–79]. There have been studies on biodegradable poly(ethylenimine)s for plasmid DNA delivery [80]. Additionally, water-soluble, branched poly(Ethylenimine)-cholesterol lipo-polymers for gene transport were created [81]. Gene delivery was made possible by modified linear polyethylenimine/cholesterol conjugates for DNA complexation [82]. A gene delivery approach using linear polyethylenimine-cholesterol/DNA complexes had been investigated for tumor effectiveness and biodistribution [83]. Biodegradable polyethylenimine with acid-labile links had been created and tested as a gene delivery method [84]. The effectiveness of reducible poly(amidoethylenimine)s for gene delivery has been studied [85–88].

8.3 Chitosan and Its Derivatives in Gene Delivery Systems

Chitosan is one of the chemotherapeutic agents obtained from chitin, a marine biopolymer isolated from crustaceans. Chitosan is derived from chitin via a partial deacetylation of chitin under strong alkaline conditions. Chitosan is readily soluble in acidic solutions but poorly soluble in water. It is cationic in nature and has (1 → 4)-2-amino-2-deoxy-β-D-glucan (D-glucosamine) and (1 → 4)-2-acetamido-2-deoxy-β-D-glucan (N-acetyl D-glucosamine) as its main repeat units. Due to its notable qualities for the development of a safe and effective drug delivery tool, such as high biodegradability and biocompatibility, chitosan has been widely used as a potential non-viral delivery vector for low molecular weight pharmaceuticals and for biomacromolecules [89, 90]. In cancer treatment, chitosan has been used as a potential immune-adjuvant for cancer vaccines [91], chemotherapeutic drugs, and carrier for gene therapeutics.

In cancer treatment, gene therapy has been widely employed and it promises to be an effective technique to combat cancer cells. Gene that has been reported to involve in the pathogenesis of cancer is manipulated or altered via introduction of exogenous genetic material into the tumor cells or the milieu [92]. Genetic materials such as siRNAs, plasmids, microRNAs, and so on are generally unstable; therefore, the chosen carrier/vector must be able to protect the gene materials from nucleases. Chitosan-based NPs are one of the most studied vectors used to carry genetic material to tumor sites due to their surface positive charges, biodegradability, and biocompatibility [93].

A group of scientists formulated microfibrillar-associated protein 5 (MFAP5)-targeting-siRNA integrated into chitosan NPs and administered to mice induced with ovarian cancer. The results from the study established that the treatment effectively suppressed invasion of the tumor, metastasis, and growth [94]. Cancer-associated fibroblasts (CAFs) have been shown to play a pivotal role in tumorigenesis but has the ability to suppress tumor progression. CAFs also expressed high level of MFAP5 in cancer cells which is secreted into the extracellular matrix of cancers [94]. Furthermore, ovarian cancer cells overexpress Gab2 gene which enhance cell proliferation via activation of PI3K pathway to regulate E-cadherin expression. Wang *et al.* also developed Gab2 siRNA incorporated into chitosan–polyethyleneimine NPs for the treatment of ovarian cancer. They efficiently suppressed Gab2 gene by means of chitosan–PEI–Gab2 siRNA NPs delivery into the SKOV3 cells, which showed good antitumor and inhibitor effect and early and late apoptosis [95].

Steg *et al.* [96] synthesized chitosan-incorporated siRNA and docetaxel NPs and elucidated their efficacies on animal model and different cancer cell lines. Chitosan-incorporated siRNA and docetaxel NPs were documented to reduce viability and proliferation of ovarian cancer cell lines, promote apoptosis down-regulated Jagged1. It was also reported that the treatment also prevented chemo-resistance to taxane and decreased the weight of the tumor [96]. In another experimental study, Yang *et al.* incorporated pshRNA plasmid into chitosan NPs aiming to reverse paclitaxel resistance ovarian cancer cells via MDR1 gene. Chitosan- incorporated pshRNA plasmid was found to effectively reduced MDR1 mRNA level and decreased the IC50 of paclitaxel. The results revealed that the synthesized NP discharged the genetic material, enhanced reactivity of the cancer cells to paclitaxel and downregulated expression of MDR1 gene [97].

Another group of researchers developed diethylaminoethyl chitosan to facilitate the reactions between siRNA and chitosan [98]. The synthesized chitosan called PEG-folic acid enhanced the physical properties of chitosan-siRNA NPs and knockdown the mRNA expression level of TNF-α in macrophages and improved the silencing efficiency in SKOV-3 cell lines [98]. Recently, Kim *et al.* [99] designed PLXDC1 siRNA-incorporated chitosan NPs coated with hyaluronic acid and targeted to the CD44 receptor on A2780 endothelial cells. This complex significantly decreased and inhibited mRNA expression of PLXDC1 as an angiogenic gene, enhanced apoptosis and suppressed cell proliferation [99].

8.3.1 Procedures of Preparation

Chitosan-based gene delivery systems are typically made using one of two formulation techniques: either simple complexation or ionic gelation. Through a straightforward complexation of the chitosan polymer with siRNA or DNA, cationic chitosan and anionic nucleic acids interact electrostatically [100–102]. Chitosan produces complexes in micro- or nanosized particles at the correct N/P ratios. For optimal condensation and complex formation, it is necessary to optimize the molecular weight,

DDA, pH, and occasionally an external force like gentle stirring. A popular technique for creating cross-linked nanoparticles is ionic gelation. Instead of completely relying on electrostatic interactions, siRNA or DNA is entrapped during ionic gelation [103]. The ionic gelation procedure makes use of cross-linkers that are significantly more negatively charged than chitosans, like hyaluronic acid, thiamine pyrophosphate, and tripolyphosphate. Cross-linking slows down the release of trapped nucleic acids even while it increases the stability of the nanoparticles. However, this linking technique may be helpful in gene delivery when nucleic acids need to be released slowly and continuously over time.

8.3.2 Chitosan's PEG Modification in Gene Delivery System

Poly-ethylene glycol (PEG) polymer conjugation with chitosan is a typical strategy to increase chitosan's blood-circulation duration and make it more water soluble [104]. PEG-coated nanoparticles are frequently referred to as "stealth" nanoparticles. Chitosan's hydrophilicity is increased by the neutral polymer PEG, which also causes the reticulo-endothelial system clearance to take longer to complete while in circulation. This raises the possibility that the chitosan-gene delivery system may passively accumulate in tumor locations by raising permeability and retention effect (EPR) in a time-dependent manner. However, the EPR effect usually only has an impact on nanoparticles with particle sizes below 200 nm. Additionally, PEGylation may change how gene treatments are distributed within cells and lessen the cationic chitosan's charge-based attraction to net negatively charged cell membranes. This issue has been resolved by attaching targeting ligands or stimuli-responsive polymers to nanoparticles, which transport nucleic acids to tumor cells at their receptor sites. While conjugating targeting ligands boosts the delivery of gene treatments to particular cells, PEG lengthens the half-life of nanoparticles in circulation. PEG also serves as a linker molecule for the addition of specific ligands to nanoparticles. Chan *et al.* [105] developed a chitosan gene delivery system with PEG-folate modification for targeted delivery to folic acid receptor-overexpressing tumor cells [105]. This DNA-carrying chitosan nanoparticle technology demonstrated reduced cytotoxicity toward typical HEK 293 (Human embryonic kidney cells 293) cells in addition to improving the water solubility upon PEG addition. In a recent study, transferrin (Tf)-functionalized chitosan nanoparticles were shown to be useful [106]. Tf was conjugated to chitosan using PEG. In order to create target-specific, circulating, and water-soluble nanoparticles, PEG modification is a crucial step.

Chitosan is a potential gene delivery technology overall for both *in vivo* and *in vitro* applications, but it has to have a number of formulation factors adjusted. Chitosan's potential can be increased by altering its structure or incorporating other polymers, which increases its target selectivity, desired intracellular release, and *in vivo* stability of gene therapies.

8.3.3 Derivatives of Chitosan in Gene Delivery

For applications involving gene delivery, chitosan's limited solubility in water at physiological pH is a disadvantage. Transfection at physiological pH may result in early siRNA dissociation into the medium without achieving effective cellular transfection because siRNA or DNA complexation with chitosan requires the protonation of its free amines, which is only achievable at acidic pH [107]. Another problem is the chitosan's sluggish release of nucleic acid components, which could reduce the effectiveness of transfection. To address water insolubility and ineffective gene transport, chitosan derivatives were thus created by chemically altering the structure of chitosan or by grafting polymers with different characteristics. Chitosan is frequently modified through quaternization, which involves alkylating tertiary amines using a variety of techniques to increase gene transfection [108, 109]. N-2-hydroxypropyl trimethyl ammonium chloride chitosan (HACC), a quaternary ammonium salt crystal, was recently employed as an example to transfer genes to human cells [110]. The HACC particles' lack of cytotoxicity and comparable transfection efficacy to liposome/pDNA complexes suggest that they have promise as an innovative method of gene delivery. Another chitosan derivative known as chitosan-HOBT is well-known for its capacity to transport siRNA in a secure and effective manner [111]. Chitosan-HOBT had strong gene silencing effectiveness, could condense siRNA, and generated stable complexes. However, its complete ability to treat cancer cells using gene therapy has not yet been accomplished. A dendronized chitosan derivative was employed in a different investigation [112]. It was made by treating 6-azido-6-deoxychitosan with a propargyl focal point poly(amidoamine) dendron. These unique dendronized chitosan/DNA complexes demonstrated improved gene transfection efficacy in human kidney and nasopharyngeal cancer cells as compared to PEI non-viral vector. For the purpose of delivering genes to cancer cells, kind of hybrid chitosan (MixNCH) was made using N, N-dimethyl-2-chloroethylamine hydrochloride, and 2-chloroethylamine hydrochloride using a different derivatization method [113]. MixNCH nanoparticles efficiently reduced cell growth, transfected HepG2 cancer cells, and demonstrated favorable physicochemical properties for gene delivery.

One of the extensively researched chitosan quarternized derivatives used in gene delivery applications is trimethyl chitosan (TMC) [114–118].

Trimethylation has numerous benefits, one of which is the ability to boost chitosan's solubility in physiological pH. Trimethyl chitosan nanoparticles significantly lessen the potential for aggregation and pH dependence of nucleic acid complexation when compared to chitosan polyplexes [114]. TMC nanoparticles significantly increased the transfection efficacy of pDNA when compared to chitosan polyplexes, according to investigations on mouse embryonic fibroblasts cells [114]. The same study also showed that adding PEG to TMC reduced particle size in physiological pH, improved particle stability, and lessened the toxicity seen with TMC that has not been changed. Finally, compared to TMC that had not been changed, PEG-TMC nanoparticles increased pDNA transfection efficacy by a factor of ten.

In a study by Zheng *et al.* [115], the efficacy of gene delivery was increased by conjugating targeting moieties to TMC. The folate on the surface of the TMC nanoparticles allowed for the delivery of pDNA to specific locations in the folate receptor-overexpressed HeLa contaminant carcinoma (KB) and human ovarian cancer (SKOV3) cells. The TMC nanoparticles successfully condensed pDNA. The ability of TMC nanoparticles to carry medications has also been employed to deliver medications and genes to cancer cells. In a recent study, high mobility group protein 2 (HMGA-2) siRNA and doxorubicin were simultaneously transfected into the triple-negative MDA-MB-231 (human breast cancer) cell line using a TMC nanoparticle method with success [118]. The simultaneous administration of siRNA that suppressed HMGA-2 gene expression enhanced the anti-cancer effects of doxorubicin. All of these investigations emphasize the significance of a number of variables, including targeting capability, toxicity, stability, particle size, and the adjustments needed for TMC-based gene delivery systems to successfully transfect genes.

8.3.4 Factors that Affect Chitosan-Based Gene Delivery

DNA or siRNA electrostatic binding to chitosan depends critically on the difference between the two molecules' charges [119]. The ratio of positively charged nitrogen in free amines in chitosan to negatively charged phosphates in nucleic acids is known as the nitrogen-to-phosphate (N/P) ratio, may have an impact on how well the chitosan polymer condenses and shields DNA or siRNA. According to reports, chitosan-DNA complexes are more stable when the N/P ratio is modest, whereas exceptionally increased N/P ratios result in poor transfection effectiveness [120, 121].

Other crucial elements that affect the steadiness of nucleic acid-chitosan complexes include molecular weight and deacetylation levels. The molecular weight of chitosan ranges widely, from very small to very large, and has a significant influence on its capacity to deliver into cells functionally active siRNA [122]. In comparison to a lower molecular weight (LMW) chitosan formulation, a 213-KDa chitosan preparation demonstrated improved absorption and transfection efficacy with pDNA (plasmid DNA) in cancerous cells Huang *et al.* [123]. According to this research, the LMW chitosan was less effective at preserving and condensing DNA and was unable to hold onto DNA after dilution [123]. However, there are reports of LMW chitosan having improved gene delivery and cell uptake in cancerous cells when compared to formulations that use higher molecular weight chitosan [124]. The improved cell absorption and transfection effectiveness can be due to LMW chitosan's comparatively weak DNA binding, which makes it easier to dissociate than greater molecular weight chitosan. According to reports, when chitosan nanoparticles were transformed with targeting ligands [125, 126] or with other polymers [127], LMW chitosan generated tiny nanoparticles and shown a considerable transfection efficacy for siRNA or DNA polyplexes. According to reports, both *in vivo* and *in vitro*, this compound was not harmful. These studies show that chitosan's biological and physicochemical properties are strongly influenced by its molecular weight. However, there is no consensus in the literature about the ideal chitosan molecular weight to select for the highest transfection effectiveness. The chitosan utilized for biomedical applications should have an advanced level of stability and a suitable molecular weight, which seems to be easier to accomplish with LMW chitosan and offers a balance between intracellular release and DNA protective capability [129], according to the aforementioned studies.

The level of deacetylation in chitosan has significant roles in influencing how well it forms stable complexes with DNA or siRNA, together with its molecular weight. Since high degree of deacetylation (DDA) is associated with extra free amines and enhanced positive charge for effective DNA binding, the majority of chitosan employed in gene transfer nanotechnology has DDA. Additionally, studies indicate that nucleic acid binding effectiveness in chitosan reduced with decreasing DDA, leading to partial complex development [128]. This is as result of few primary amine groups in chitosan with low DDA are readily available to interact electrostatically with negatively charged nucleic acids. According to research, for effective complex formation with DNA, DDA% in chitosan should be over 65% [130]. When chitosan is complexed with DNA, it has similarly been demonstrated that the particle size decreases with an increase in DDA% [123].

The pH of the transfection medium must also be considered since the protonation of amine groups in chitosan requires an acidic pH range

(5.0–6.0). Chitosan's ability to bind DNA is enhanced by this low acidic pH, which also improves transfection effectiveness [125]. The steadiness of a cationic gene delivery method is significantly influenced by the presence of serum [125]. Surprisingly, chitosan-based DNA transfection is enhanced in the existence of serum compared to other cationic polymers [131]. Sato *et al.* (2001) investigated the transfection effectiveness of pDNA/chitosan complexes in the presence of serum (0%–50%). According to their research, 50% serum in the medium provided the least transfection effectiveness, whereas 20% serum conditions produced the best [124].

The size and charge density of the DNA molecule and the siRNA molecule, respectively, have an impact on the formation of a complex with identical-length chitosan. Due to its bigger size and higher charge density compared to short siRNA (19–25 bp), plasmid DNA can bind electrostatically more strongly [119]. When taken as a whole, all of the aforementioned factors need to be carefully taken into account in order to create an effective chitosan-based gene delivery system. In addition to chitosan-gene complex physicochemical properties, the cell type may affect transfection efficacy. Therefore, the chitosan-gene complexes should be customized when working with a cell line that is challenging to transfect. This includes utilizing chitosan derivatives and adding other polymers to promote the highest transfection effectiveness in the particular cell line.

8.4 Chitosan as Chemotherapeutic Drugs

Chitosan has been widely used in pharmaceutical and other related fields such as in the production of drug carrier in microparticle system, site specific drug delivery, tablets with slow release of drugs, carrier of vaccine delivery and gene therapy, formulation of binder in wet granulation, hydrogels, and disintegrant owning to its nontoxicity and biocompatibility properties [132]. Chitosan enhances bioavailability of polar drugs and aids their movement via epithelial surfaces due to the cationic property of chitosan and its ability to bind to negatively charged surfaces including basement and mucosal membranes [132, 133]. Chitosan permeation enhancing mechanism has facilitated the anticancer activities reported both *in vitro* and *in vivo* [132]. Nam and Shon [134] reported the antimetastatic activity of chitosan, its ability to decrease the activity and amount of MMP9 protein and suppressed the migration of these cells through a Matrigel coated membrane when administered at high doses to human breast carcinoma cells (MDA-MB-231).

Apoptosis has been implicated as one of the anticancer mechanisms of action of chitosan [135, 136]. Chitosan induces tumor cells death via pro-caspase activation which facilitated by the exterior of the cell to speed-up the cleavage of cascade and amplify the death signaling pathway [135]. Chitosan oligosaccharide has been documented to Inhibit MMP-9, which play a pivotal function in tumor growth in Lewis Lung Carcinoma cells. It also inhibited the rate of DNA synthesis due to increase in the level of p21 and decrease in cyclin A and CDK-2 as well as decreased in the number of cells in S phase and inhibited cell proliferation [137].

Permanent arrest of cell cycle leading to cellular senescence is one of the main antitumor mechanisms [138]. Low molecular weight chitosan has been reported to induce cell cycle arrest at G1 and S- phase which resulted to cell senescence associated with over secretion of reactive oxygen species. This process is triggered by increased production of TGF-β which in turns activate Smads 2/3, Smad 4, p15, and p21 in stepwise manner before the ultimate release of ROS [138]. G1 arrest by low molecular weight chitosan is suggestive of the mechanism that prevent the cells from entering S phase in a manner independent of p53 involves the changes in protein expression [139]. Chitosan nanoparticles have been revealed to suppress human hepatoma cells proliferation via cell necrosis by decrease in MMP, permeation through the cell membrane, neutralization of its surface charge, and induction of lipid peroxidation *in vitro* [140].

The tumor growth inhibitory and antitumor properties of chitosan have been strongly linked to its immunomodulatory ability [141]. Seo *et al.* [142] revealed that in the presence of Interferon cell signaling pathway-γ (IFN-γ), chitosan mediated activation of murine peritoneal macrophages to destroy tumor cells. Similarly, a group of scientists reported that the maturation of killer T cells and splenic T-lymphocytes by increased secretion of interleukin-1 and interleukin-2 is one of the tumor growth inhibitory and anticancer mechanisms of chitosan [143]. Chitosan oligosaccharide exhibited immunomodulatory activities by inducing immune response that directly destroy cancer cells and pathogenic microbes and facilitate cytotoxic environment by secretion of interleukin-1 and tumor necrosis factor-α cytokines to activate T-cells and Natural killer (NK)-cells which ultimately suppress the growth of cancer cells [144].

Furthermore, chitosan has showed to have the ability to enhance the biodistribution and accumulation of anticancer drug in tumor cells. It has been documented in scientific study that chitosan nanoparticles loaded with mifepristone facilitated controlled drug delivery of the drug, improved its oral bioavailability and ultimately enhanced the drug anticancer activity [145]. Chitosan displayed antiangiogenic activities by interfering with the systemic regulation of both antiangiogenic and proangiogenic factors [146]. It has been reported that growth of human hepatocellular carcinoma can be suppressed chitosan nanoparticles via a mechanism which leads to impairment of vascular endothelial growth factor receptor 2 levels [147].

8.5 Conclusion

The creation of the ideal gene delivery method is essential for the success of gene therapy. Non-immunogenic vectors with cell selectivity are used in the gene delivery system in therapeutic settings in order to provide sufficient transgene expression to produce the desired effect. With analytical software that examines the patterns of gene expression, DNA or RNA microarrays used in various types of gene sequencing can identify the hundreds of genes. In order to administer a potential vaccine, a hybrid biosynthetic vector was created using the gene delivery system. By interacting it with *E. coli*, the vector gets over the usual obstacles to gene delivery. To make some vectors, *E. coli* are combined with synthetic polymers. A major possibility for treating some diseases resulting from genetic disorders, cancers, and other illnesses is provided by the gene therapy system. The non-viral vector has advantages in that it is relatively inexpensive to reproduce and has a low immunogenicity. The non-viral vector has no restriction on the amount of DNA it can pack and modify with ligands to target a particular cell. Animal gene transfer has been explored and evaluated using gene delivery devices based on synthetic polymeric materials. One of the potential carriers of the gene into the desired cells' nuclei safely is plasmid DNA. The development of non-viral vector-based gene delivery systems has also emerged as a fresh, potential therapeutic approach. Finally, the development of clinically applicable vectors for use in fighting elusive diseases like Alzheimer's diseases, cancer, AIDS, etc. is a key objective for research on gene delivery systems. The host cells can efficiently receive genetic material via these vectors. The focus of the subsequent research steps will likely be to develop DNA and RNA molecular technologies which can be used as standard treatment options in biomedical applications.

References

1. Guo, X. and Huang, L., Recent advances in non-viral vectors for gene delivery. *Acc. Chem. Res.*, 45, 7, 971–9, 2012.

2. Jones, C.H., Chen, C.K., Ravikrishnan, A., Rane, S., Pfeifer, B.A., Overcoming nonviral gene delivery barriers: Perspective and future. *Mol. Pharm.*, 10, 4082–98, 2013.
3. Friedmann, T. and Roblin, R., Gene therapy for human genetic disease. *Science*, 175, 949–55, 1972.
4. Han, S.-O., Mahato, R.I., Sung, Y.K., Kim, S.W., Development of biomaterials for gene therapy. *Mol. Ther.*, 2, 302–17, 2000.
5. Mahato, R.I., Smith, L.C., Rolland, A., Pharmaceutical perspectives of nonviral gene therapy. *Adv. Genet.*, 41, 95–156, 1999.
6. Park, D.W., Hoffman, A.S., Pu, S., Stayton, P.S., Design and development of polymers for gene delivery. *Nat. Rev. Drug Discovery*, 4, 581–593, 2005.
7. Nayerossadat, N., Maedeh, T., Ali, P.A., Viral and nonviral delivery systems for gene delivery. *Adv. Biomed. Res.*, 1, 27–31, 2012.
8. Mali, S., Delivery systems for gene therapy. *Indian J. Hum. Genet.*, 19, 3–8, 2013.
9. Moss, B., Smith, G.L., Gerin, J.L., Purcell, R.H., Live recombinant vaccinia virus protects chimpanzees against hepatitis B. *Nature*, 311, 5981, 67–9, 1984.
10. Yellepeddi, V.K., Vectors for non-viral gene delivery- clinical and biomedical applications. *Austin Ther.*, 2, 1, 1014, 2015.
11. Kim, S.W., Biomaterials to gene delivery. *J. Control. Release*, 155, 2, 116–118, 2011, https://doi.org/10.1016/j.jconrel.2011.03.023.
12. Nayerossadat, N., Maedeh, T., Ali, P.A., Viral and nonviral delivery systems for gene delivery. *Adv. Biomed. Res.*, 1, 27, 2012.
13. Yin, H., Kanasty, R.L., Eltoukhy, A.A., Vegas, A.J., Dorkin, J.R., Anderson, D.G., Non-viral vectors for gene-based therapy. *Nat. Rev. Genet.*, 15, 541–555, 2014.
14. Kedmi, R., Ben-Arie, N., Peer, D., The systemic toxicity of positively charged lipid nanoparticles and the role of Toll-like receptor 4 in immune activation. *Biomaterials*, 31, 6867–6875, 2010.
15. Tao, W., Mao, X., Davide, J.P., Ng, B., Cai, M., Burke, P.A., Sachs, A.B., Sepp-Lorenzino, L., Mechanistically probing lipid-siRNA nanoparticle-associated toxicities identifies Jak inhibitors effective in mitigating multifaceted toxic responses. *Mol. Ther.*, 19, 567–575, 2011.
16. Whitehead, K.A., Langer, R., Anderson, D.G., Knocking down barriers: Advances in siRNA delivery. *Nat. Rev. Drug Discovery*, 8, 129–138, 2009.
17. Ishida, T., Harada, M., Wang, X.Y., Ichihara, M., Irimura, K., Kiwada, H., Accelerated blood clearance of PEGylated liposomes following preceding liposome injection: Effects of lipid dose and PEG surface-density and chain length of the first-dose liposomes. *J. Control. Release*, 105, 305–317, 2005.
18. Cun, D., Jensen, L.B., Nielsen, H.M., Moghimi, M., Foged, C., Polymeric nanocarriers for siRNA delivery: Challenges and future prospects. *J. Biomed. Nanotechnol.*, 4, 258–275, 2008.

19. Singha, K., Namgung, R., Kim, W.J., Polymers in small-interfering RNA delivery. *Nucleic Acid Ther.*, 21, 133–147, 2011.
20. Mao, S., Sun, W., Kissel, T., Chitosan-based formulations for delivery of DNA and siRNA. *Adv. Drug Deliv. Rev.*, 62, 12–27, 2010.
21. Ginn, S.L., Amaya, A.K., Alexander, I.E., Edelstein, M., Abedi, M.R., Gene therapy: Clinical trials worldwide to 2017: An update. *J. Gene Med.*, 20, e3015, 1–16, 2018.
22. Huli-Curtis, S.L., Uusi-Kerttula, H., Jones, R., Hanna, L., Chester, J.D., Parker, A.L., Evaluation of CD46 re-targeted adenoviral vectors for clinical ovarian cancer intraperitoneal therapy. *Cancer Gene Ther.*, 23, 229–34, 2016.
23. Blaese, R.M., Culver, K.W., Miller, A.D., Carter, C.S., Fleisher, T. *et al.*, T lymphocytedirected gene therapy for ADA-SCID: Initial trial results after 4 years. *Science*, 270, 475–80, 1995.
24. Brenner, M.K., Rill, D.R., Moen, R.C., Krance, R.A., Mirro, J. Jr., *et al.*, Gene-marking to trace origin of relapse after autologous bone-marrow transplantation. *Lancet*, 341, 85–6, 1993.
25. Hacein-Bey-Abina, S., Von Kalle, C., Schmidt, M., McCormack, M.P., Wulffraat, N. *et al.*, LMO2-associated clonal T cell proliferation in two patients after gene therapy for SCID-X1. *Science*, 302, 415–9, 2003.
26. Gomez, C.E., Najera, J.L., Krupa, M., Esteban, M., The poxvirus vectors MVA and NYVAC as gene delivery systems for vaccination against infectious diseases and cancer. *Curr. Gene Ther.*, 8, 97–120, 2008.
27. Berges, B.K., Wolfe, J.H., Fraser, N.W., Transduction of brain by herpes simplex virus vectors. *Mol. Ther.*, 15, 20–9, 2007.
28. Stone, D., Novel viral vector systems for gene therapy. *Viruses*, 2010, 2, 1002–7, 2010.
29. Manno, C.S., Pierce, G.F., Arruda, V.R., Glader, B., Ragni, M. *et al.*, Successful transduction of liver in hemophilia by AAV-factor IX and limitations imposed by the host immune response. *Nat. Med.*, 12, 342–7, 2006.
30. Kabanov, A.V. and Kabanov, V.A., DNA complexes with polycations for the delivery of genetic material into cells. *Bioconjug. Chem.*, 6, 7–20, 1995.
31. Buwalda, S.J., Dijkstra, P.J., Feijen, J., Poly(ethylene glycol)-poly(L -lactide) star block copolymer hydrogels cross-linked by metal-ligand coordination. *J. Polym. Sci. Part A*, 50, 9, 1783–93, 2012.
32. Bloomfield, V., A condensation of DNA by multivalent cations: Considerations on mechanism. *Biopolymers*, 31, 1471–81, 1991.
33. Bloomfield, V.A., DNA condensation. *Curr. Opin. Struct. Biol.*, 6, 334–41, 1996.
34. Kim, J.S., Maruyama, A., Akaike, T., Kim, S.W., Terplex DNA delivery system as a gene carrier. *Pharm. Res.*, 5, 116–21, 1998.
35. Toncheva, V., Wolfert, M.A., Dash, P.R., Oupicky, D., Ulbrich, K. *et al.*, Novel vectors for gene delivery formed by self-assembly of DNA with poly(L-lysine) grafted with hydrophilic polymers. *Biochim. Biophys. Acta Gen. Rev.*, 1380, 354–68, 1998.

36. Maruyama, A., Katoh, M., Ishihara, T., Akaike, T., Comb-type polycations effectively stabilize DNA triplex. *Bioconjug. Chem.*, 8, 3–6, 1997.
37. Katayose, S. and Kataoka, K., Water-soluble polyion complex associates of DNA and poly(ethylene glycol) poly(L-lysine) block copolymer. *Bioconjug. Chem.*, 8, 702–7, 1997.
38. Kim, J.S., Maruyama, A., Akaike, T., Kim, S.W., *In vitro* gene expression on smooth muscle cells using a terplex delivery system. *J. Control. Release*, 47, 51–9, 1997.
39. Zhang, X., Oulad-Abdelghani, M., Zelkin, A.N., Wang, Y., Haîkel, Y., Mainard, D., Voegel, J.C., Caruso, F., Benkirane-Jessel, N., Poly(L-lysine) nanostructured particles for gene delivery and hormone stimulation. *Biomaterials*, 31, 7, 1699–706, 2010.
40. Kwoh, D.Y., Coffin, C.C., Lollo, C.P., Jovenal, J., Banaszczyk, M.G., Mullen, P., Phillips, A., Amini, A., Fabrycki, J., Bartholomew, R.M., Brostoff, S.W., Carlo, D.J., Stabilization of poly-L-lysine/DNA polyplexes for *in vivo* gene delivery to the liver. *Biochim. Biophys. Acta*, 1444, 171–90, 1999.
41. Choi, J.S., Joo, D.K., Kim, C.H., Kim, K., Park, J.S., Synthesis of a barbell-like triblock copolymer, poly(L-lysine) dendrimer-block-poly(ethylene-glycol)-block-poly(llysine) dendrimer, and its self-assembly with plasmid DNA. *J. Am. Chem. Soc.*, 122, 474–80, 2000.
42. Trubetskoy, V.S., Torchilin, V.P., Kennel, S.J., Huang, L., Use of N-terminal modified poly(L-lysine)-antibody conjugate as a carrier for targeted gene delivery in mouse lung endothelial cells. *Bioconjug. Chem.*, 3, 4, 323–7, 1992.
43. Bikram, M., Cheol-Hee Ahn, C.-H., Chae, S.-Y., Lee, M., Yockman, J.W., Kim, S.W., Biodegradable poly(ethylene glycol)-co-poly(L-lysine)-g-histidine multi-block copolymers for non-viral gene delivery. *Macromolecules*, 37, 5, 1903–16, 2004.
44. Martinez-Fong, D., Mullersman, J.E., Purchio, A.F., Armendariz-Borunda, J., Martinez-Hernandez, A., Non-enzymatic glycosylation of poly-L-lysine: A new tool for targeted gene delivery. *Hepatology*, 20, 6, 1602–8, 1994.
45. Banaszczyk, M.G., Kwoh, D.Y., Carlo, D.J., Poly-L-lysine-graft-PEG-comb-type polycation copolymers for gene delivery. *J. Macromol. Sci. Part A*, 36, 7–8, 1061–84, 1999.
46. Maruyama, A., Ishihara, T., Kim, J.-S., Kim, S.W., Akaike, T., Nanoparticle DNA carrier with poly(L-lysine) grafted polysaccharide copolymer and poly(D,L-lactic acid). *Bioconjug. Chem.*, 8, 735–40, 1997.
47. Kim, J.-S., Maruyama, A., Akaike, T., Kim, S.W., Terplex DNA delivery system as a gene carrier, pharm. *Research*, 15, 116–20, 1998.
48. Kim, J.-S., Kim, B.I., Maruyama, A., Akaike, T., Kim, S.W., A new non-viral DNA delivery vector: The terplex system. *J. Control. Release*, 53, 175–80, 1998.
49. Choi, Y.H., Liu, F., Kim, J.S., Choi, Y.K., Park, J.S., Kim, S.W., Polyethylene glycol-grafted poly-L-lysine as polymeric gene carrier. *J. Control. Release*, 54, 39–43, 1998.

50. Choi, Y.H., Liu, F., Park, J.S., Kim, S.W., Lactose-poly(ethylene glycol)-grafted polyL-lysine as hepatoma cell-targeted gene carrier. *Bioconjug. Chem.*, 9, 708–12, 1998.
51. Choi, Y.H., Liu, F., Park, J.S., Kim, S.W., Characterization of a targeted gene carrier, lactose-polyethylene glycol-grafted poly-L-lysine, and its complex with plasmid DNA. *Hum. Gene Ther.*, 10, 2657–61, 1999.
52. Park, T.G., Jeong, J.H., Kim, S.W., Current status of polymeric gene delivery systems. *Adv. Drug Deliv. Rev.*, 58, 4, 467–86, 2006.
53. van der Meel, R., Vehmeijer, L., Kok, R.J., Storm, G., van Gaal, E.V., Ligandtargeted particulate Nano-medicines undergoing clinical evaluation: Current status, in: *Intracellular Delivery III*, A. Prokop and V. Weissig (Eds.), Berlin/Heidelberg, Germany, pp. 163–200, Springer, 2015, Chap.7.
54. Bodles-Brakhop, A.M., Heller, R., Draghia-Akli, R., Electroporation for the delivery of DNA-based vaccines and Immunotherapeutics: Current clinical developments. *Mol. Ther.*, 17, 4, 585–92, 2009.
55. Pulkkanen, K.J. and Herttuala, S.Y., Gene therapy for malignant glioma: Current clinical status. *Mol. Ther.*, 12, 4, 585–98, 2005, https://doi.org/10.1016/j. ymthe.2005.07.357.
56. Rainov, N.G., A phase III clinical evaluation of herpes simplex virus type 1 thymidine kinase and ganciclovir gene therapy as an adjuvant to surgical resection and radiation in adults with previously untreated glioblastoma multiforme. *Hum. Gene Ther.*, 11, 17, 2389–401, 2000, https://doi.org/10.1089/104303400750038499.
57. Young, L.S., Searle, P.F., David Onion, D., Mautner, V., Viral gene therapy strategies: From basic science to clinical application. *J. Pathol.*, 208, 299–318, 2006, https://doi.org/10.1002/path.1896.
58. Yoshida, J., Mizuno, I.M., Wakabayashi, T., Interferon-β gene therapy for cancer: Basic research to clinical application. *Cancer Sci.*, 95, 11, 858–65, 2004.
59. Weichselbaum, R.R., Kufe, D.W., Hellman, S., Rasmussen, H.S., King, C.R., Fischer, P.H., Mauceri, H.J., Radiation-induced tumour necrosis factor-α expression: Clinical application of transcriptional and physical targeting of gene therapy. *Lancet Oncol.*, 3, 11, 665–7, 2002.
60. Amer, M.H., Gene therapy for cancer: Present status and future perspective. *Mol. Cell. Ther.*, 2, 27–32, 2014, https://doi.org/10.1186/2052-8426-2-27.
61. Jayant, R.D., Sosa, D., Kaushik, A., Atluri, V., Vashist, A., Tomitaka, A., Nair, M., Current status of non-viral gene therapy for CNS disorders. *Expert Opin. Drug Deliv.*, 13, 10, 1433–45, 2016, https://doi.org/10.1080/17425247.2016.1188802.
62. Ginn, S.L., Amaya, A.K., Alexander, I.E., Edelstein, M., Abedi, M.R., Gene therapy clinical trials worldwide to 2017: An update. *J. Gene Med.*, 20, e3015 (1–16), 2018.
63. Nam, K., Jung, S., Nam, J.-P., Kim, S.W., Poly(ethylenimine) conjugated bioreducible dendrimer for efficient gene delivery. *J. Control. Release*, 220, 447–455, 2015.

64. Azzam, T., Eliyahu, H., Shapira, L., Linial, M., Barenholz, Y., Domb, A.J., Polysaccharide-oligoamine based conjugates for gene delivery. *J. Med. Chem.*, 45, 1817–24, 2002.
65. Janes, K.A., Calvo, P., Alonso, M.J., Polysaccharide colloidal particles as delivery systems for macromolecules. *Adv. Drug Deliv. Rev.*, 47, 83–97, 2001.
66. Liu, Z., Jiao, Y., Wang, Y., Zhou, C., Zhang, Z., Polysaccharides-based nanoparticles as drug delivery systems. *Adv. Drug Deliv. Rev.*, 60, 1650–62, 2008.
67. Morris, G.A., Samil, M., Ernest, K.S., Adams, H.G., Polysaccharide drug delivery systems based on pectin and chitosan. *Biotechnol. Genet. Eng. Rev.*, 27, 1, 257–84, 2010.
68. Benjaminsen, R.V., Mattebjerg, M.A., Henriksen, J.R., Moghimi, S.M., Andersen, T.L., The possible "proton sponge" effect of polyethylenimine (PEI) does not include change in lysosomal pH. *Mol. Ther.*, 21, 1, 149–57, 2013.
69. Luu, Q.P., Shin, J.Y., Kim, Y.K., Islam, M.A., Kang, S.K., Cho, M.H., Cho, C.S., High gene transfer by the osmotic polysorbitol-mediated transporter through the selective caveolae endocytic pathway. *Mol. Pharm.*, 9, 80, 2206–18, 2012.
70. Doss, C.G., Debottam, S., Debajyoti, C., Glutathione-responsive nanotransporter-mediated siRNA delivery: Silencing the mRNA expression of Ras. *Protoplasma*, 250, 3, 787–92, 2013.
71. Hong, R., Han, G., Fernandez, J.M., Kim, B.J., Forbes, N.S., Rotello, V.M., Glutathionemediated delivery and release using monolayer protected nanoparticle carriers. *J. Am. Chem. Soc.*, 128, 4, 1078–9, 2006.
72. Oupicky, D. and Li, J., Bio-reducible polycations in nucleic acid delivery: Past, present, future trends. *Macromol. Biosci.*, 14, 7, 908–22, 2014.
73. Wen, H.Y., Dong, H.Q., Xie, W.J., Li, Y.Y., Wang, K., Pauletti, G.M., Shi, D.L., Rapidly disassembling nano-micelles with disulfide-linked PEG shells for glutathionemediated intracellular drug delivery. *Chem. Commun. (Camb)*, 47, 12, 3550–2, 2011.
74. Chakravarthi, S., Jessop, C.E., Bulleid, N.J., The role of glutathione in disulphide bond formation and endoplasmic-reticulum-generated oxidative stress. *EMBO Rep.*, 7, 3, 271–5, 2006.
75. Bello Roufai, M. and Midoux, P., Histidylated poly-lysine as DNA vector: Elevation of the imidazole protonation and reduced cellular uptake without change in the polyfection efficiency of serum stabilized negative polyplexes. *Bioconjug. Chem.*, 12, 1, 92–9, 2001.
76. Pack, D.W., Putnam, D., Langer, R., Design of imidazole-containing endosomolytic biopolymers for gene delivery. *Biotechnol. Bioeng.*, 67, 2, 217–23, 2000.
77. Pires, L.R., Oliveira, H., Barrias, C.C., Sampaio, P., Pereira, A.J., Maiato, H., Pego, A.P., Imidazole-grafted chitosan-mediated gene delivery: *In vitro* study on transfection, intracellular trafficking and degradation. *Nanomedicine (Lond)*, 6, 9, 1499–512, 2011.

78. Yang, Y., Xu, Z., Chen, S., Gao, Y., Gu, W., Chen, L., Li, Y., Histidylated cationic polyorganophosphazene/DNA self-assembled nanoparticles for gene delivery. *Int. J. Pharm.*, 353, 1–2, 277–82, 2008.
79. Zhang, X., Duan, Y., Wang, D., Bian, F., Preparation of arginine modified PEIconjugated chitosan copolymer for DNA delivery. *Carbohydr. Polym.*, 122, 53–9, 2015.
80. Ahn, C.H., Chae, S.Y., Bae, Y.H., Kim, S.W., Biodegradable poly(ethylenimine) for plasmid DNA delivery. *J. Control. Release*, 80, 273–8, 2002.
81. Wang, D.A., Narang, A.S., Kotb, M., Gaber, A.O., Miller, D.D., Kim, S.W., Mahato, R.I., Novel branched poly(ethylenimine)-cholesterol water-soluble lipopolymers for gene delivery. *Biomacromolecules*, 3, 1197–202, 2002.
82. Furgeson, D.Y., Chan, W.S., Yockman, J.W., Kim, S.W., Modified linear polyethylenimine-cholesterol conjugates for DNA complexation. *Bioconjug. Chem.*, 14, 840–5, 2003.
83. Furgason, D.Y., Yockman, J.W., Janat, M.M., Kim, S.W., Tumor efficacy and biodistribution of linear polyethylenimine-cholesterol/DNA complexes. *Mol. Ther.*, 9, 837–42, 2004.
84. Kim, Y.H., Park, J.H., Lee, M., Park, T.G., Kim, S.W., Polyethylenimine with acid-labile linkages as a biodegradable gene carrier. *J. Control. Release*, 103, 209–13, 2005.
85. Christensen, L.V., Chang, C.-W., Kim, W.J., Kim, S.W., Reducible poly(amido ethylenimine)s designed for triggered intracellular gene delivery. *Bioconjug. Chem.*, 17, 1233–40, 2006.
86. Christensen, L.V., Chang, C.-W., Yockman, J., Conners, W.R., Jackson, R.H., Zhong, Z., Feijen, J., Bull, D.A., Kim, S.W., Reducible poly(amido ethylenediamine) for hypoxia-inducible VEGF delivery. *J. Control. Release*, 118, 254–61, 2007.
87. Jeong, J.H., Christensen, L.V., Yockman, J.W., Zhong, Z., Engbersen, J.F.J., Kim, W.J., Feijen, J., Kim, S.W., Reducible poly(amido ethylenimine) direct to enhance RNA interference. *Biomaterials*, 28, 1912–7, 2007.
88. Jeong, J.H., Kim, S.H., Christensen, L.V., Feijen, J., Kim, S.W., Reducible poly(amido ethylenimine)-based gene delivey system for improved nucleus trafficking of plasmid DNA. *Bioconjug. Chem.*, 1, 2, 296–301, 2010.
89. Saikia, C., Gogoi, P., Maji, T.K., Chitosan: A promising biopolymer in drug delivery applications. *J. Mol. Genet. Med.*, S4, 006, 2015.
90. Lee, M., Nah, J.W., Kwon, Y., Koh, J.J., Ko, K.S., Kim, S.W., Water-soluble and low molecular weight chitosan-based plasmid DNA delivery. *Pharm. Res.*, 18, 427–431, 2001.
91. Highton, A.J., Girardin, A., Bell, G.M., Hook, S.M., Kemp, R.A., Chitosan gel vaccine protects against tumour growth in an intracaecal mouse model of cancer by modulating systemic immune responses. *BMC Immunol.*, 17, 39, 2016.
92. Cross, D. and Burmester, J.K., Gene therapy for cancer treatment: Past, present and future. *Clin. Med. Res.*, 4, 218–227, 2006.

93. Jayakumar, R., Chennazhi, K.P., Muzzarelli, R.A.A. *et al.*, Chitosan conjugated DNA nanoparticles in gene therapy. *Carbohydr. Polym.*, 79, 1–8, 2010.
94. Yeung, T.-L., Leung, C.S., Mok, S.C., CAF reprogramming inhibits ovarian cancer progression. *Cell Cycle*, 13, 3783–3784, 2014.
95. Wang, L., Wu, W., Wang, J. *et al.*, Highly efficient Gab2 siRNA delivery to ovarian cancer cells mediated by chitosan–polyethyleneimine nanoparticles. *J. Mater. Chem. B*, 4, 273–281, 2016.
96. Steg, A.D., Katre, A.A., Goodman, B. *et al.*, Targeting the notch ligand JAGGED1 in both tumor cells and stroma in ovarian cancer. *Clin. Cancer Res.*, 17, 5674–5685, 2011.
97. Yang, Y., Wang, Z., Li, M. *et al.*, Chitosan/pshRNA plasmid nanoparticles targeting MDR1 gene reverse paclitaxel resistance in ovarian cancer cells. *J. Huazhong Univ. Sci. Technol. [Med. Sci.]*, 29, 239–242, 2009.
98. de Souza, R.H.F.V., Picola, I.P.D., Shi, Q. *et al.*, Diethylaminoethylchitosan as an efficient carrier for siRNA delivery: Improving the condensation process and the nanoparticles properties. *Int. J. Biol. Macromol.*, 119, 186–197, 2018.
99. Kim, G.H., Won, J.E., Byeon, Y. *et al.*, Selective delivery of PLXDC1 small interfering RNA to endothelial cells for antiangiogenesis tumor therapy using CD44-targeted chitosan nanoparticles for epithelial ovarian cancer. *Drug Deliv.*, 25, 1394–1402, 2018.
100. Amaduzzi, F., Bomboi, F., Bonincontro, A., Bordi, F., Casciardi, S., Chronopoulou, L., Diociaiuti, M., Mura, F., Palocci, C., Sennato, S., Chitosan-DNA complexes: Charge inversion and DNA condensation. *Colloids Surf. B Biointerfaces*, 114, 1–10, 2014.
101. Bravo-Anaya, L.M., Soltero, J.F., Rinaudo, M., DNA/chitosan electrostatic complex. *Int. J. Biol. Macromol.*, 88, 345–353, 2016.
102. Liu, X., Howard, K.A., Dong, M., Andersen, M.Ø., Rahbek, U.L., Johnsen, M.G., Hansen, O.C., Besenbacher, F., Kjems, J., The influence of polymeric properties on chitosan/siRNA nanoparticle formulation and gene silencing. *Biomaterials*, 28, 1280–1288, 2007.
103. Ragelle, H., Vanvarenberg, K., Vandermeulen, G., Préat, V., Chitosan nanoparticles for siRNA delivery *in vitro*. *Methods Mol. Biol.*, 1364, 143–150, 2016.
104. Zhang, Y., Chen, J., Zhang, Y., Pan, Y., Zhao, J., Ren, L., Liao, M., Hu, Z., Kong, L., Wang, J., A novel PEGylation of chitosan nanoparticles for gene delivery. *Biotechnol. Appl. Biochem.*, 46, 197–204, 2007.
105. Chan, P., Kurisawa, M., Chung, J.E., Yang, Y.Y., Synthesis and characterization of chitosan-g-poly(ethylene glycol)-folate as a non-viral carrier for tumor-targeted gene delivery. *Biomaterials*, 28, 540–549, 2007.
106. Nag, M., Gajbhiye, V., Kesharwani, P., Jain, N.K., Transferrin functionalized chitosan-PEG nanoparticles for targeted delivery of paclitaxel to cancer cells. *Colloids Surf. B Biointerfaces*, 148, 363–370, 2016.
107. Ishii, T., Okahata, Y., Sato, T., Mechanism of cell transfection with plasmid/chitosan complexes. *Biochim. Biophys. Acta*, 1514, 1, 51–64, 2001.

108. Ouchi, T., Murata, J., Ohya, Y., Gene delivery by quaternary chitosan with antennary galactose residues, in: *Polysaccharide Applications; ACS Symposium Series*, vol. 737, American Chemical Society, Washington, DC, USA, pp. 15–23, 1999, Chapter 2.
109. Wei, W., Lv, P.P., Chen, X.M., Yue, Z.G., Fu, Q., Liu, S.Y., Yue, H., Ma, G.H., Codelivery of mTERT siRNA and paclitaxel by chitosan-based nanoparticles promoted synergistic tumor suppression. *Biomaterials*, 34, 3912–3923, 2013.
110. Li, G.F., Wang, J.C., Feng, X.M., Liu, Z.D., Jiang, C.Y., Yang, J.D., Preparation and testing of quaternized chitosan nanoparticles as gene delivery vehicles. *Appl. Biochem. Biotechnol.*, 175, 3244–3257, 2015.
111. Opanasopit, P., Techaarpornkul, S., Rojanarata, T., Ngawhirunpat, T., Ruktanonchai, U., Nucleic acid delivery with chitosan hydroxybenzotriazole. *Oligonucleotides*, 20, 127–136, 2010.
112. Deng, J., Zhou, Y., Xu, B., Mai, K., Deng, Y., Zhang, L.M., Dendronized chitosan derivative as a biocompatible gene delivery carrier. *Biomacromolecules*, 12, 642–649, 2011.
113. Zhong, J., Huang, H.L., Li, J., Qian, F.C., Li, L.Q., Niu, P.P., Dai, L.C., Development of hybrid-type modified chitosan derivative nanoparticles for the intracellular delivery of midkine-siRNA in hepatocellular carcinoma cells. *Hepatobiliary Pancreat. Dis. Int.*, 14, 82–89, 2015.
114. Germershaus, O., Mao, S., Sitterberg, J., Bakowsky, U., Kissel, T., Gene delivery using chitosan, trimethyl chitosan or polyethylenglycol-graft-trimethyl chitosan block copolymers: Establishment of structure–activity relationships *in vitro*. *J. Control. Release*, 125, 145–154, 2008.
115. Zheng, Y., Cai, Z., Song, X., Yu, B., Bi, Y., Chen, Q., Zhao, D., Xu, J., Hou, S., Receptor mediated gene delivery by folate conjugated N-trimethyl chitosan *in vitro*. *Int. J. Pharm.*, 382, 262–269, 2009.
116. Eivazy, P., Atyabi, F., Jadidi-Niaragh, F., Aghebati Maleki, L., Miahipour, A., Abdolalizadeh, J., Yousefi, M., The impact of the codelivery of drug-siRNA by trimethyl chitosan nanoparticles on the efficacy of chemotherapy for metastatic breast cancer cell line (MDA-MB-231). *Artif. Cells Nanomed. Biotechnol.*, 45, 5, 1–8, 2016.
117. Gao, Y., Wang, Z.Y., Zhang, J., Zhang, Y., Huo, H., Wang, T., Jiang, T., Wang, S., RVG-peptide-linked trimethylated chitosan for delivery of siRNA to the brain. *Biomacromolecules*, 15, 1010–1018, 2014.
118. Zheng, H., Tang, C., Yin, C., Exploring advantages/disadvantages and improvements in overcoming gene delivery barriers of amino acid modified trimethylated chitosan. *Pharm. Res.*, 32, 2038–2050, 2015.
119. Mao, S., Sun, W., Kissel, T., Chitosan-based formulations for delivery of DNA and siRNA. *Adv. Drug Deliv. Rev.*, 62, 12–27, 2010.
120. Alameh, M., Dejesus, D., Jean, M., Darras, V., Thibault, M., Lavertu, M., Buschmann, M.D., Merzouki, A., Low molecular weight chitosan nanoparticulate system at low N:P ratio for nontoxic polynucleotide delivery. *Int. J. Nanomed.*, 7, 1399–414, 2012.

121. Lavertu, M., Méthot, S., Tran-Khanh, N., Buschmann, M.D., High efficiency gene transfer using chitosan/ DNA nanoparticles with specific combinations of molecular weight and degree of deacetylation. *Biomaterials*, 27, 4815–4824, 2006.
122. Techaarpornkul, S., Wongkupasert, S., Opanasopit, P., Apirakaramwong, A., Nunthanid, J., Ruktanonchai, U., Chitosan-mediated siRNA delivery *in vitro*: Effect of polymer molecular weight, concentration and salt forms. *AAPS PharmSciTech*, 11, 64–72, 2010.
123. Huang, M., Fong, C.W., Khor, E., Lim, L.Y., Transfection efficiency of chitosan vectors: Effect of polymer molecular weight and degree of deacetylation. *J. Control. Release*, 106, 391–406, 2005.
124. Sato, T., Ishii, T., Okahata, Y., *In vitro* gene delivery mediated by chitosan. Effect of pH, serum, and molecular mass of chitosan on the transfection efficiency. *Biomaterials*, 22, 2075–2080, 2001.
125. Nimesh, S., Thibault, M.M., Lavertu, M., Buschmann, M.D., Enhanced gene delivery mediated by low molecular weight chitosan/DNA complexes: Effect of pH and serum. *Mol. Biotechnol.*, 46, 182–196, 2010.
126. Agirre, M., Zarate, J., Ojeda, E., Puras, G., Desbrieres, J., Pedraz, J.L., Low molecular weight chitosan (LMWC)-based polyplexes for pDNA delivery: From bench to bedside. *Polymers*, 6, 1727–1755, 2014.
127. Patil, S., Bhatt, P., Lalani, R., Amrutiya, J., Vhora, I., Kolte, A., Misra, A., Low molecular weight chitosan–protamine conjugate for siRNA delivery with enhanced stability and transfection efficiency. *RSC Adv.*, 6, 110951–110963, 2016.
128. Fernandes, J.C., Qiu, X., Winnik, F.M., Benderdour, M., Zhang, X., Dai, K., Shi, Q., Low molecular weight chitosan conjugated with folate for siRNA delivery *in vitro*: Optimization studies. *Int. J. Nanomed.*, 7, 5833–5845, 2012.
129. Köping-Höggård, M., Varum, K.M., Issa, M., Danielsen, S., Christensen, B.E., Stokke, B.T., Artursson, P., Improved chitosan-mediated gene delivery based on easily dissociated chitosan polyplexes of highly defined chitosan oligomers. *Gene Ther.*, 11, 1441–1452, 2004.
130. Köping-Höggård, M., Tubulekas, I., Guan, H., Edwards, K., Nilsson, M., Vårum, K.M., Artursson, P., Chitosan as a nonviral gene delivery system. Structure-property relationships and characteristics compared with polyethylenimine *in vitro* and after lung administration *in vivo*. *Gene Ther.*, 8, 1108–1121, 2001.
131. Erbacher, P., Zou, S., Bettinger, T., Steffan, A.M., Remy, J.S., Chitosan-based vector/DNA complexes for gene delivery: Biophysical characteristics and transfection ability. *Pharm. Res.*, 15, 1332–1339, 1998.
132. Gavhane, Y.N., Gurav, A.S., Yadav, A.V., Chitosan and its applications: A review of literature. *Int. J. Res. Pharm. Biomed. Sci.*, 4, 1, 312–331, 2013.
133. Thanou, M., Verhoef, J.C., Junginger, H.E., Chitosan and its derivatives as intestinal absorption enhancers. *Adv. Drug Deliv. Rev.*, 50, 1, S91–S101, 2001.

134. Namand, K.-S. and Shon, Y.-H., Suppression ofmetastasis of human breast cancer cells by chitosan oligosaccharides. *J. Microbiol. Biotechnol.*, 19, 6, 629–633, 2009.
135. Wimardhani, Y.S., Suniarti, D.F., Freisleben, H.J., Wanandi, S.I., Siregar, N.C., Ikeda, M.-A., Chitosan exerts anticancer activity through induction of apoptosis and cell cycle arrest in oral cancer cells. *J. Oral. Sci.*, 56, 2, 119–126, 2014.
136. Xu, Y., Wen, Z., Xu, Z., Chitosan nanoparticles inhibit the growth of human hepatocellular carcinoma xenografts through an antiangiogenic mechanism. *Anticancer Res.*, 29, 12, 5103–5109, 2009.
137. Shen, K.-T., Chen, M.-H., Chan, H.-Y., Jeng, J.-H., Wang, Y.-J., Inhibitory effects of chitooligosaccharides on tumor growth and metastasis. *Food Chem. Toxicol.*, 47, 8, 1864–1871, 2009.
138. Senturk, S., Mumcuoglu, M., Gursoy-Yuzugullu, O., Cingoz, B., Akcali, K.C., Ozturk, M., Transforming growth factor beta induces senescence in hepatocellular carcinoma cells and inhibits tumor growth. *Hepatology*, 52, 3, 966–974, 2010.
139. Vermeulen, K., van Bockstaele, D.R., Berneman, Z.N., The cell cycle: A review of regulation, deregulation and therapeutic targets in cancer. *Cell Proliferation*, 36, 3, 131–149, 2003.
140. Qi, L., Xu, Z., Chen, M., *In vitro* and *in vivo* suppression of hepatocellular carcinoma growth by chitosan nanoparticles. *Eur. J. Cancer*, 43, 1, 184–193, 2007.
141. Kuppusamy, S. and Karuppaiah, J., Screening of antiproliferative effect of chitosan on tumor growth and metastasis in T24 urinary bladder cancer cell line. *Austral.-Asian J. Cancer*, 12, 3, 145–149, 2013.
142. Seo, W.-G., Pae, H.-O., Kim, N.-Y. *et al.*, Synergistic cooperation between water-soluble chitosan oligomers and interferon-γ for induction of nitric oxide synthesis and tumoricidal activity in murine peritoneal macrophages. *Cancer Lett.*, 159, 2, 189–195, 2000.
143. Tokoro, A., Suzuki, K., Matsumoto, T., Mikami, T., Suzuki, S., Suzuki, M., Chemotactic response of human neutrophils to N-acetyl chitohexaose *in vitro*. *Microbiol. Immunol.*, 32, 4, 387–395, 1988.
144. Tsukada, K., Matsumoto, T., Aizawa, K. *et al.*, Antimetastatic and growth-inhibitory effects of N-acetylchitohexaose in mice bearing Lewis lung carcinoma. *Jpn. J. Cancer Res.*, 81, 3, 259–265, 1990.
145. Zhang, H., Wu, F., Li, Y. *et al.*, Chitosan-based nanoparticles for improved anticancer efficacy and bioavailability of mifepristone. *Beilstein J. Nanotechnol.*, 7, 1861–1870, 2016.
146. Jiang, Z., Han, B., Li, H., Yang, Y., Liu, W., Carboxymethyl chitosan represses tumor angiogenesis *in vitro* and *in vivo*. *Carbohydr. Polym.*, 129, 1–8, 2015.
147. Xu, Y., Wen, Z., Xu, Z., Chitosan nanoparticles inhibit the growth of human hepatocellular carcinoma xenografts through an antiangiogenicmechanism. *Anticancer Res.*, 29, 12, 5103–5109, 2009.

9

Nanoparticles as Gene Vectors in Tumor Therapy

Efstathia Triantafyllopoulou[1†], Orestis Kontogiannis[2†], Nefeli Lagopati[2], Natassa Pippa[1*] and Maria Gazouli[2]

[1]Department of Pharmaceutical Technology, Faculty of Pharmacy, Panepistimioupolis Zografou, National and Kapodistrian University of Athens, Athens, Greece
[2]Department of Basic Medical Science, Laboratory of Biology, School of Medicine National and Kapodistrian University of Athens, Athens, Greece

Abstract

Cancer is a group of diseases characterized by the uncontrolled growth of abnormal cells. It is the second leading cause of death globally. However, better prevention, screening, and treatment methods have increased survival rates for many types of cancer. Gene delivery is a scientific approach with several benefits in the areas of biomedical sciences and especially personalized medicine. On the other hand, nanotechnology offers several materials for the design and development of nanoparticles that can load genes and deliver them to the specialized tissues and cells. Commonly carriers for this purpose mainly include viral and nonviral vectors. Non-viral vectors involve-among others, inorganic, lipid- and polymer-based nanoparticles, which will be presented in terms of their ability to deliver and release genes. The nanoparticles are ideal carriers to overcome barriers and control the release profile of the genes. The aim of this chapter is to present the main categories, the technology, and the applications of nanoparticles as gene vectors in the field of tumor therapy as well as the limitations and the challenges of their usage. Nanoparticles as gene vectors in tumor therapy could be a new window for reducing the side effects and increase the effectiveness of oncotherapy.

Keywords: Gene delivery, polymeric nanoparticles, lipid nanoparticles, inorganic nanoparticles, cancer therapy, polyplexes, lipoplexes

**Corresponding author*: natpippa@pharm.uoa.gr
†*Equal contribution*

Rishabha Malviya, Sonali Sundram and Neelam Jain (eds.) Integration of Biomaterials for Gene Therapy, (247–280) © 2023 Scrivener Publishing LLC

9.1 Introduction

According to World Health Organization, "*Cancer is a broad group of diseases where abnormal cells grow out of control, spread to other parts of the body, and can invade and harm normal tissue. This process is known as metastasis and is a major contributor to cancer-related deaths. Other terms for cancer include neoplasm and malignant tumor. It is the second leading cause of death worldwide, resulting in approximately 9.6 million deaths (1 in 6 deaths) in 2018. The most common types of cancer in men are lung, prostate, colorectal, stomach, and liver cancers. For women, the most prevalent types are breast, colorectal, lung, cervical, and thyroid cancers.*" In simpler terms, cancer is a disease where abnormal cells grow rapidly and can harm normal body tissue. It can spread throughout the body and is the second biggest cause of death globally. However, advancements in prevention, screening, and treatment are leading to higher survival rates for many types of cancer. There are over 100 different forms of cancer, each named after the organ or tissue where it originates. Gene delivery is a useful scientific method with benefits in the fields of biomedical sciences and personalized medicine [1–5]. The great achievements in the field of molecular biology open new avenues in the treatment of different kind of tumors and cancers. The targeted gene therapy is the key point for the cellular and sub-cellular organelles delivery of the nucleic acids.

The pharmaceutical nanotechnology offers several materials for the design and the development of nanoparticles that can load genes and deliver them to the specialized tissues and cells [1–10]. The design and the development of a nanocarrier for gene therapy is the key factor for the safety, efficacy, and effectiveness of the final formulation. The immunogenicity is the most important side effect of the usage of the viral vectors [2–8]. For this reason, the application of drug delivery system in gene therapy opens new avenues in their fast clinical translation. Different types of nanosystems have been already used for delivery and the controlled release of nucleic acids. In some cases, these nanosystems allow the co-delivery of Active Pharmaceutical Ingredients (APIs), such anticancer molecules (doxorubicin, methotrexate, etc.) with nucleic acids achieving the best therapeutic outcome. The co-delivery of APIs and genes using nanoparticulate systems enhances the cellular uptake avoiding in parallel the API efflux, which is a very common phenomenon on the tumoral cell membranes. Additionally, the enhanced permeation and retention (EPR) effect which lead to the increase accumulation in the tumor is also driven by the small size (below to 100nm) of these nanovectors [4–8]. Except from size, some other physicochemical characteristics are also crucial for the successful delivery of genes. The lamellarity of vesicular carriers, the surface charge, and the surface coatings are also very important for their behavior *in vivo*. Namely, the positive surface charge leads o high loading efficiency of nucleic acids due to electrostatic interactions. The coatings, especially the PEGylation can extend the circulation time in the human plasma, offering "stealth properties". Furthermore, in the recent literature, several nanocarriers have been developed with stimulus responsive properties [3–5]. For example, these nanocarriers can alter their properties due to the microenvironment of the tumor, i.e., pH, temperature, glutathione concentration, etc. and release their cargo. By this approach, selectivity and targeting can be achieved. Last but not least, there is a research progress of gene nanovectors targeting abnormal glucose and lipid metabolism in tumors, too [4]. The main categories of these systems (stimulus or not) are lipid- and polymer- based nanoparticles and the inorganic nanocarriers. All these categories will be analyzed in depth in this chapter for the physicochemical characteristics and their loading and release capacities, as well as their efficacy in the treatment of different tumor types. The "nano-oncology" is a new term that has been already appeared in the literature and includes the delivery of APIs and genes [5]. In other words, "nano-oncology" is the most important approach for the treatment of different types of cancers with added value to their targeting, especially in the field of the gene therapy [5].

This chapter focuses on the technology and applications of nanoparticles as gene vectors in cancer therapy. It covers the main categories of nanoparticles used in oncotherapy, providing examples, and discussing the mechanisms for delivering and releasing nucleic acids. The limitations and challenges of using nanoparticles as gene vectors will be thoroughly examined. Nanoparticles have the potential to reduce side effects and improve the efficiency and effectiveness of cancer therapy.

9.2 Polymer-Based Nanocarriers: Their Technology and Recent Advances

Cancer patients today are mostly subjected to a therapeutic protocol involving chemotherapy, radiotherapy, and tumor extraction via the surgical route, treatments that are not always efficacious and more often than not lead to severe adverse effects, compromising the patient's vitality while promoting low patient compliance [11–14]. Especially in most cases where the indicated drugs are to be administered through the systematic route, due to low to none tissue selectiveness, the majority of the high toxicity compounds are distributed to healthy cells, rather than the malignant ones. To that end, it is of vital importance to be able to develop novel approaches towards the treatment of various forms of genetic abnormalities - both inherited and acquired - that accumulated lead to the formation and/or spread of cancerous cells, resulting in more favorable clinical outcomes.

Functionalized non-viral vectors of polymeric origin hold great promise in terms of being able to encapsulate and condense different types of genetic material, (synthetic DNA, DNA, RNA, iRNA, miRNA, pDNA, siRNA, tumor suppressive genes, RNA induced silencing complexes etc.) characterized by great diversities in terms of their length-size, stability and intracellular (or subcellular) interactions, since they represent chemically versatile, synthetic vehicles with easily modified properties, structures and functional groups, based on each individual case (Figure 9.1) [15–17]. As of today, polymeric nanoparticles are the most utilized vehicles in preclinical non-viral vector gene transfection [18].

Amongst the mostly researched and used materials in terms of polymer based gene delivery in the nanoscale are polymeric nanoparticles of cationic nature (formulating electrostatic interactions with negatively charged nucleic acids-polyplexes) of various natures, polyethyleneimine (PEI), Poly(L-Lysine) (PLL), Poly(lactic-co-glycolic acid) (PLGA), Poly(β-amino esters) (PBAE), dendrimers (mostly derived from polyamidoamine – PAMAM dendriplexes), Poly[2-(dimethylamino)ethyl methacrylate] (PDMAEMA), chitosan and polyesters such as poly-L-arginine (PLA), poly(glycolic acid)(PGA), and poly(ε-caprolactone) (PCL), poly(trimethylene carbonate) (PTMC), and poly(propyleneoxide) (PPO)

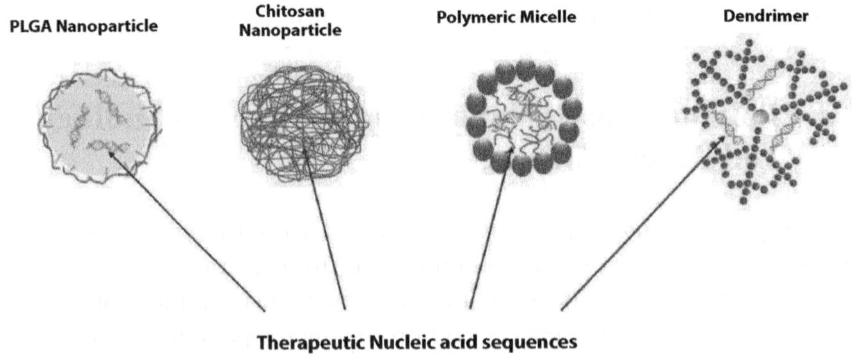

Figure 9.1 Polymeric materials used for gene delivery purposes towards cancer suppression.

[19–22]. Amphiphilic mixed polymers of the family of the triblock-copolymer Pluronic have also been regularly utilized towards the formulation of chimeric polymeric nano-vectors that exhibit a level of stimuli-responsiveness and microenvironment recognition, usually for the formation of biocompatible polymeric micellar [16].

Polymeric micelles exhibit a characteristic core-shell structure that enable the entrapment and complexation of different nucleic acid segments onto individual compartments, without them interacting and thus risk of losing their therapeutic properties [16]. Vehicles of high complexity, low polydispersity and of small sizes, polymeric micelles represent a non-toxic, high cargo loading vector that can surpass many extra- and intra-cellular biological barriers, with the polymers used for the construction of the hydrophobic core playing a crucial role in terms of the drug encapsulation and pharmacokinetics of the vehicle. Generally, the existence of hydrophobic interactions above the electrostatic nucleic acid complexation is preferred since it results in a more stable vector. Moreover, through the incorporation of different targeting moieties in the micellar outer shell (such as monoclonal antibodies), active targeting can be achieved, as well as a control over the cellular internalization process (receptor mediated delivery, caveolae or clathrin-mediated endocytosis), rather than follow a non-specific internalization pattern. Lysosomes and endosomal escape are significant obstacles for non-viral gene vectors to overcome. Polymer-based vectors and polyplexes are vulnerable to degradation in low pH environments like lysosomes and can be easily broken down by intracellular enzymes. This results in the conversion of the vectors into smaller polymers, making them unable to protect the genetic material [23].

Dendrimers are synthetic branched molecules of highly defined properties and structure, with a synthesis process that allows for high control over characteristics such as the number of functional groups (and generations), size, shape, charge and solubility that offer the possibility of active targeting through the incorporation of a number of moieties onto their outer shell (incorporation of PEG molecules for increased blood plasma circulation). These molecules are able to be synthesized having very specific sizes, by carefully choosing the monomer used to make them [19, 20]. These systems are capable of loading high amounts of nucleic acids, while they are non-immunogenic. Dendrimers derived from PLL have shown an excellent oligonucleic acid condensation capacity and thus are amongst the most researched.

Other well studied vectors include mesoporous silica nanoparticles, which show high biodegradability and can formulate stable bonds with different types of RNA molecules, such as siRNA, achieving targeted nucleic acid delivery [24]. A key aspect of their use in pre-clinical studies is their high tolerance that makes the possibility for rapid clinical translation imminent, while a high control of the porous structure and diameter, along with large surface areas are amongst the many advantages that offer for gene delivery [25].

Several pre-clinical studies have used aliphatic polyesters for gene delivery, since they represent a group of polymers that can be subjected to a great amount of structural and property engineering and design. They are highly biodegradable, able to formulate stable bonds with DNA molecules and amongst the most commonly utilized are PHP, PAGA, PBAE, PVL, aminated PAHA, PPE, and PLA [15].

It is worth mentioning that a great deal of prominent research when dealing with polymeric vectors, aims at the fabrication of chimeric, innovative carriers that either exhibit both therapeutic and diagnostic capabilities (theranostics) or try to accommodate both nucleic acids and various chemotherapeutic drug compounds and deliver them simultaneously to the affected pathogenic tissues, with their work representing a breakthrough of today's molecular biology.

Lastly, gelatin nanoparticles have been examined in terms of their nucleotide binding and condensation properties as gene delivery carriers, in terms of possible per os administration. These nanocarriers can be externally modified with protective molecules (such as PEG) with rather ease and once in the blood compartment, protect the therapeutic genes from rapid clearance and degradation [26].

When compared with viral vectors, the above delivery platforms hold a number of advantages in terms of their tunable architecture and morphology, chemical versatility and modifiable physicochemical characteristics (charge, ζ-potential, hydrodynamic diameter etc.) and most importantly in terms of immune system recognition and possible adverse effects that the repeated administration of viral vectors can have (Figure 9.2), and which as a result more often than not require the administration of immunosuppressive compounds (cytotoxicity, cytokine, chemokine release, systemic immune responses). Also, a notable risk that involves the use of viruses as vehicles for gene delivery is the possibility of their interaction with endogenous viruses or other pathogens that have infected the organism over the years and reside in tissues or the blood compartment [27–29].

The first polymeric based Phase I clinical trial that involved the use of non-viral vectors was conducted in 2010, using a nanoparticulate system namely CALAA-01 that involved the transfection of malignant melanoma cells (solid tumor) for the delivery of therapeutic siRNA against the M2 tumor gene subunit suppression (sequence siR2B+5). While the trial was

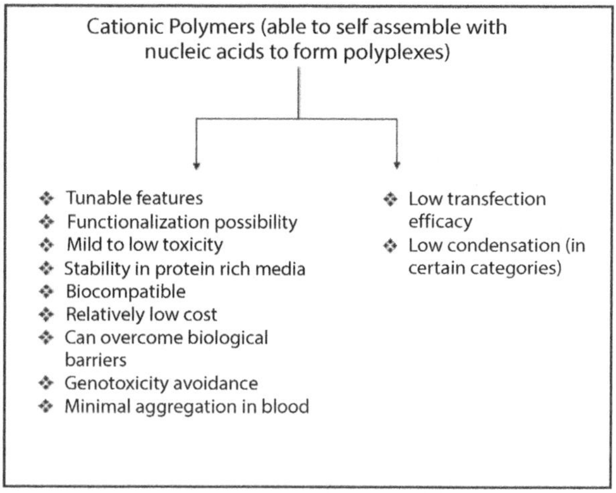

Figure 9.2 Advantages and limitations of non-viral polymer-based gene delivery vectors compared with viral vectors.

officially terminated after the Phase I due to dose-limiting toxicity issues, never the less it succeeded in exhibiting the successful polymeric siRNA transfectance of mammalian cells, building on the work of the first clinical trials involving the intraocular administration of therapeutic synthetic siRNAs in 2004 [17, 30].

Since then, a great deal of progress has been made in terms of siRNA transfectance towards cancer suppression. Thiolated gelatin nanoparticles were used for the encapsulation of polymerized anti-sense siRNA (poly-siRNA/tGel nanoparticles) aiming to down-regulate the expression of specific genes that played a crucial role in melanoma formation. The final formulation achieved higher tumor accumulation and tissue-specific distribution when compared with a naked siRNA tumor bearing model in mice, while also exhibiting limited cytotoxicity and high tissue tolerability [15]. To that day, polymers exhibit the greatest advantages in terms of siRNA delivery since they represent formulations that are versatile, of minimal cost, easy to produce and show high transfection rates [31].

The shortest nucleic acid sequences transferred for gene therapy purposes, siRNAs (along with miRNAs) are non-coding, about 20bp to 25bp long, gene silencing sequences that after their expression through the RNAi pathway, act by silencing a specific mRNA sequence that is responsible to a certain degree for the production of defective protein that enables malignancy growth [17].

An siRNA sequence responsible for hindering TNF-α gene-expression, were loaded in gelatin nanoparticles which were in turn enclosed by poly(ε-caprolactone) (PCL) microspheres is order to examine the potential for an oral gene delivery platform [15]. TNF-α is classified as a tumor promoting gene, specifically overexpressed in colon, as well as ovarian cancer, and is mostly related with advanced cancer growth. The previously mentioned nano-vectors (mean particle diameter of 279.2 ±3.2 nm), were orally administered in a SCID mice model and concluded that successful delivery and expression of the silencing sequences to the affected colon cells was achieved through the down regulation of the tumor necrosis factor-α in the pathological cells [15].

In another study, the transfection of human malignant cells of Ewing's sarcoma in a murine model with silencing siRNA was conducted against the EWS-FLI1 gene product protein, which is found in 90% of all Ewing sarcoma cases (with the individual fused genes residing in chromosomes 22 and 11 respectively – EWSR1, FLI1). The formulated multi-compartmentalized vectors utilized a cyclodextrin containing polycation offering protection to the encapsulated siRNA, along with transferrin used as a targeting moiety of the transferrin receptor of the pathogenic cells. The vectors were formulated by essentially mixing the individual components together and allow for self-assembly to occur, while the siRNA utilized (siGL3) was chosen based on the siEFBP2 which is the break point of the mutated chimeric gene. The formulation of the EWS-FLI1 chimeric gene (Figure 9.3), is mainly responsible for soft tissue and bone tissue tumorigenesis and can lead to the development of drug-resistant metastasis from which point the majority of the patients succumb. The results exhibited a drastic biochemical and molecular improvement and inhibition of malignant cell growth, while proliferation was hindered to a great degree but not stopped (either *in vitro* or in xenograft model – EWS-FLI1 cultured TC71 EFT cells) [32].

It is worth mentioning that in several cases the safety profile of an siRNA gene therapy application to mammalian cells has been questioned based on the immunogenic profile that this type of cell transfection has [32].

Amongst the most recent procedures in non-viral gene therapy is the transportation of genes aimed to amplify the activity of the host's innate immune cells at the cancerous microenvironment. Thus, components of the immune system are modulated to recognize and fight the malignant cells more efficiently promoting an overall better clinical outcome [33]. Various interleukin producing genes, which act mainly by stimulating the immune system, reduce inflammation and promote tissue regeneration, such as IL-12, IL-22 and IL-1Ra have been studied in terms of non-viral

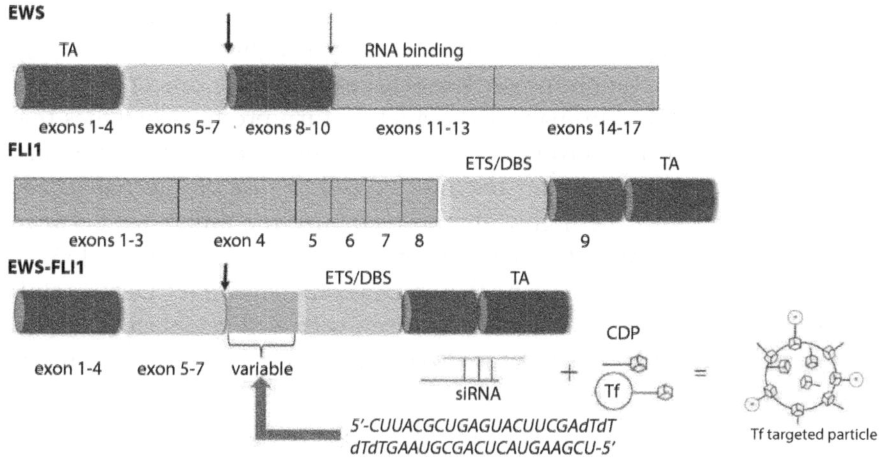

Figure 9.3 Chimeric EWS-FLI1 gene product siEFBP2 siRNA non-viral encapsulation and targeted transfection of Ewing's sarcoma malignant cells.

encapsulation and targeted delivery aiming for tumor regression (mostly in models resulting from malignant colonic cell lines) [33, 34].

In a recent study, DMP hybrid cationic micelles (DOTAP modified mPEG–PCL micelles) were developed for the complexation with the IL-12 interleukin gene to study its anticancer effect on various murine cancer models through either local administration or via systemic administration through the blood compartment. The complex achieved significant tumor regression in both a C26 colon carcinoma and also a LL/2 lung cancer *in vivo*. The successful targeted delivery and internalization of the encapsulated gene (no healthy tissues were affected) resulted in a magnified lymphocyte activity which inhibited cancer cell growth and metastasis [33].

Besides the strong complexation with interleukin related nucleic acids and adequate amounts of expression in the cells of interest (ELISA assay), an advantage of the formulated DMP micelles is that of low toxicity and high tolerance which is of great importance as previously mentioned, especially in therapeutic protocols that require repeated administration. The mean DMP particle size was 46 ± 5.6 nm and had a ζ-potential of 41.8 ± 0.5 mV [33].

Another area of interest is that of suicide-gene therapy of which the main principle lies in malignant cells transfection with apoptosis inducing genes in order to trigger cellular death, inhibiting proliferation [33, 34]. A critical aspect of this approach is that, in several cases, in order for the triggering signals to have any effect on cancer cell death they need to be "activated" during a specific phase of the cellular cycle.

The basic idea on how to induce apoptosis is based on the ability to convert an otherwise harmless substance into a highly toxic one for the malignant cells, by combining nucleic acids expressed in cancer cells with other genes not found in mammals, which in turn produce enzymes that are responsible for the conversance of the prodrug to its toxic counterpart. Due to the severe outcome for the affected cell in case of internalization, the vector used must be highly able to recognize the difference between malignant and healthy cells in order to avoid healthy cell apoptosis [35].

A novel non-viral suicide gene therapy was evaluated in terms of its ability to suppress tumor growth, in comparison with well-established clinical drugs such as paclitaxel (PTX) and irinotecan (CPT11). The polymeric nanocarrier that was used was poly{*N*-[2-(acryloyloxy)ethyl]-*N*-[*p*-acetyloxyphenyl]-*N*,*N*-diethylammonium chloride} (PQDEA), originating from the use of Poly (2-(diethylamino)ethyl methacrylate) or PDEA after DEA polymerization. PDEA, undergoes a self-catalyzed hydrolysis into polyacrylic acid resulting in a reversal of its charge from cationic to anionic and as a result can formulate stable bonds with deoxyribonucleic acids. Once internalized in the cancerous microenvironment, due to the abundance of esterases (responsible for the splitting of esters into alcohols and acids), the carrier dissociates and thus the encapsulated cargo is released [36]. The formulated polyplexes contained the genetic sequence responsible for the production of a major inflammatory cytokine, namely the tumor necrosis factor or TNF, while conjugated with DC-Chol/DOPE lipids, ultimately forming lipidic polyplexes with DNA. The use of pTRAIL plasmids used for the delivery and transfection of the malignant cells with the nucleic acid sequence inducing apoptosis was of vital importance. Despite related research, where the administration of such a gene in contradiction achieved the promotion of cancer growth, in this case the internalization achieved the apoptosis of HeLa tumors in nude mice and along with the administration of chemotherapeutic drugs managed to halt tumor growth and an extension of the life expectancy of the mice (especially in treatment-free periods). Lastly, the side effects occurring from the gene therapy treatment were minimal, when compared with the effects of the chemotherapeutics alone [36].

In another example, a suicide gene treatment protocol was developed against the proliferation of malignant gliomas, a direction towards gene transfection is focused when dealing with brain tumors, both in preclinical and in clinical studies (Figure 9.4).

Angiopep-2 is an amino acid sequence that has shown great promise in terms of its use as a ligand, for the transcendence of the blood brain barrier (BBB), a biological formation of specialized brain microvascular

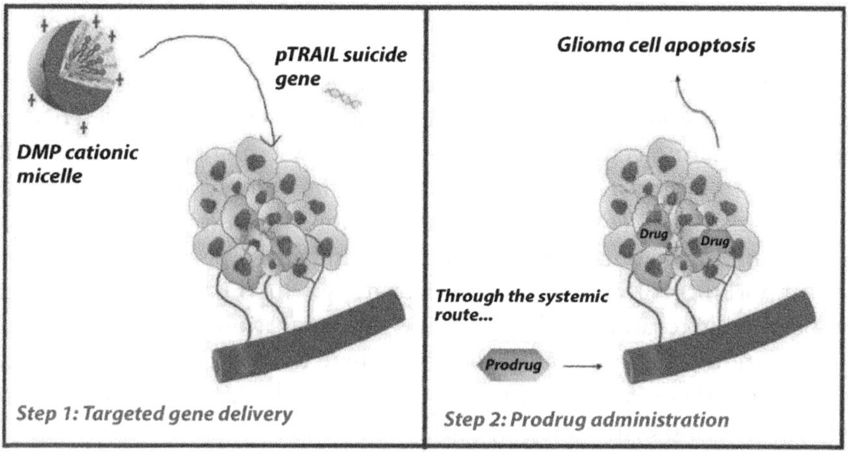

Figure 9.4 DMP cationic micelles used as non-vectors for suicide gene therapy against malignant glioma cells.

endothelial cells which allows only a small portion (≤2%) of low-molecular weight molecules to reach the brain. The formulated nanocarriers used poly (L-lysine)-grafted polyethylenimine (PEI-PLL) as a backbone via polyethyleneglycol (PEG) and designated as PPA, and exhibited an ability to penetrate the BBB and furthermore accumulate in certain key regions of the brain (use of the genetically modified mouse model BALB/c) showing promising anti-glioma activity. In the experiment, Cy5-labeled DNA along with the plasmid of PLOX-GFP-iresTK (TK) and I thymidine kinase gene (HSV-TK, suicide gene naturally found on the herpes simplex virus I) were used with the aim to convert the administered prodrug GCV into its toxic counterpart. The study, showed that the administration of the PPA/HSV-TK NPs combined with the administration of GCV, achieved a statistically significant increase in the survival rate of the test subjects induced with U87MG orthotopic glioma [35].

Several types of solid tumors can occur when problems in gene expression and regulation occur at a posttranscriptional level or more specifically at a microRNA (miRNA) level. These non-coding molecules are responsible for the regulation of various important genes, the function of which is associated with normal cell function. In some instances, a disorder may occur from the upregulation of such miRNA function (miR-21,19,20 miR-210,21,22 and miR-22123,24) resulting in an abnormal gene expression pattern, or in others from the downregulation of miRNA molecules (miR-34,12,13 miR-200,14,15 let-7, and miR-15516–18). In each instance, it has been exhibited that the successful delivery and transcription of nucleic acid sequences promoting or suppressing miRNA function can suppress tumorigenesis [20].

In terms of polymer-based delivery, PEI stands as the mostly utilized polycations designated for miRNA encapsulation and transference. A branched PEI molecule was used for the entrapment of miR-33a and miR-145 tumor suppressive sequences in a murine colorectal model in nude mice, bearing LS174T tumors. The intraperitoneal administration achieved a silencing of the oncogenic genes c-Myc and ERK5, while also stressing the importance of clever nanoparticle design in the formulation stage, by being able to address the issue of PEI aggregation-related toxicity, by understanding the different endothelial interactions of the same substance, obtaining different architectures (linear vs. branched, and long-chained or with smaller molecular weight) [20].

It is of high importance to stress in this point, that in terms of colorectal cancer, a promising approach is that of siRNA gene delivery, which is studied in many instances concurrently with the administration of various chemotherapeutic agents and genes in a single vector (with notable example that of poly (ethylene glycol)-ε-poly(caprolactone) block copolymer for co-loading of 5-fluorouracil (5-FU) and pEGFP (DNA) as DFNC). By transfected the malignant cells with the appropriate sequences, genes whose overexpression is related with tumor growth and high drug resistance such as P-gp and Bcl-2 can be successfully silenced. This approach, is even more promising in cases where a specific genetic mutation can be found in the majority of cases and as such the administration of the same product can act beneficial for multiple patients (estrogen receptor alpha (ERα), a steroid hormone receptor expressed in about 70% of cases of invasive breast cancers) [37–40].

Lastly, the recent work of several scientists has brought to light possible uses, in terms of non-viral miRNA gene delivery, of less common polymers, such as N-(2-Hydroxypropyl) meth acrylamide or HPMA, the polymerization of which gives the polymer poly(*N*-(2-hydroxypropyl) meth acrylamide). Polyplexes of HPMA with AMD3465, a known CXCR4 antagonist - the overexpression of which is associated with more than 23 different types of tumor growth and is associated with metastasis and drug resistance – were formulated to transfect human osteosarcoma U20S cells with miR-200c which acts by preventing tumor initiation. The results showed a mitigation of the ZEB1 gene regulation, a known oncogene, indicating adequate transfection ability *in vitro* [20].

In terms of formulations aimed both for gene delivery and the imaging/diagnosis of solid tumors, an interesting approach used folic acid (FA) polymeric nanoparticles with active targeting capabilities, complexed with

Au nanoparticles in order to achieve theranostic dual modality. Biodegradable 'smart' cationic poly(aspartic acid)s-grafted chitosan–cyclodextrin conjugates (CCPEs), mixed with FA were able to formulate stable pDNA bonds for the delivery of a suicide gene and exhibited high transfection on n FR-positive HeLa cells, stressing once more the importance of various polysaccharides on nucleic acid delivery. By using multiple cell lines, the FA-chitosan conjugates, functionalized with aspartic acid exhibited greater gene transfection rates in comparison with CD (chitosan) based complexes alone in folate receptor-positive tumor cells. In the study, the plasmid nucleic acid used was pRL-CMV, encoding Renilla luciferase, while it also stressed the possible use of other B vitamins that could be used as targeting moieties (since they are overexpressed in most cancer cells), such as biotin [41–43]. In general terms of active targeting, in the non-viral vector development amongst the most prominent are a variety of peptides such as transferrin, epidermal growth factor (EGF), RGD, HA and specific carbohydrates, and finally, specific antibodies and antibody fragments are able to target antigens related with tumor-genesis such as the HER2 receptor and are currently used in polymer vector-based transfection [44].

Finally, after complexion of CS with Au nanoparticles, bioimaging was achieved, with the final formulations achieving a mean hydrodynamic diameter of 92.2 nm and a surface charge of +24 mV. In this case, the suicide genes used, could successfully convert the prodrug 5-fluorocytosine into 5-fluorouracil (5-FU) which is cytotoxic and once more, the importance of chitosan in theranostic non-viral applications was stressed. After flow cytometry analysis, the apoptotic effect on tumor cells was evaluated and concluded the successful induced cell death [41].

9.2.1 Lipid-Based Nanocarriers: Their Technology and Recent Advances

Lipid-based nanocarriers of nucleic acids are very common in gene therapy due to their benefits over viral vectors. It is well known that the dominant lipid nanoplatforms in cancer gene therapy are lipoplexes – the organized structure of the combination of cationic liposomes and nucleic acids, followed by Lipid Nanoparticles (LNPs) [45–51]. In general, these formulations comprised of specific types of lipids – each with a pronounced purpose [48, 52, 53]. Namely, there is a lipid with positive charge incorporating and condensing the negatively charged nucleic acid via electrostatic interactions. This cationic charge could be either permanent (cationic lipids-DOTAP is one of their major representatives) or temporary (ionizable lipids) and is usually – but not necessarily – owed to a quaternary ammonium group [54, 55]. There are plenty of cationic and ionizable lipids, some of them can be seen in Figure 9.5. They can influence not only platform's cargo but also other properties, such as transfection and gene silencing efficacy depending on their chemical moieties [49, 54–57]. Even though cationic lipids are extensively used, nanoparticles with high positive charge are cleared out by systemic circulation immediately due to interactions with plasma proteins and macrophages uptake. They can even cause non-selective toxicity to blood cells and activation of multiple immune pathways [46, 48, 49, 53, 58]. Therefore, novel biodegradable ionizable lipids are preferred over cationic, although there are still toxicity issues that should be considered [48, 53, 58]. Referring to ionizable lipids, their charge depends on their pKa, which is one of the most critical selection parameters [49, 53, 55, 56]. Furthermore, there is at least another lipid with a neutral overall charge which helps the vehicle to be stable in storage and in use and/or for efficient endosomal escape via rearrangement of the molecules and thus rupture of the endosome membrane. Zwitterionic phospholipids with long acyl chains such as DSPC, unsaturated phospholipids with fusogenic characteristics such as DOPE, or cholesterol which has an extraordinary effect on lipid bilayer fine-tuning the fluidity of the membrane, pertain to helper lipids – as they are usually called [49, 50, 52, 55, 56]. Cholesterol's existence and proportion in the formulation are crucial factors regarding encapsulation and transfection efficacy [48, 55, 56, 58]. In most cases, there is a lipid incorporated to the structure with stealth properties, such as PEGylated lipids. PEGylation is a common way to mask on positive charge and its cytotoxic consequences. Besides steric stabilization, PEGylation has also an impact on physicochemical and morphological characteristics of the nanostructures. The proportion and the sort of PEG moiety are both critical for formulation's properties. However, PEGylation could be challenging as well [46, 48, 49, 53, 55, 56]. Lastly, some innovative formulations include materials with targeting properties such as ligand-associated lipids which have an affinity to receptors of specific cancer cell types or tumor-microenvironment favoring effective treatment over systemic side effects [47, 49, 52, 56].

Lipid-based gene vectors prevail for their ability to be customized due to intended needs [46, 47, 53]. In this manner, it is important their good clinical translation and the deep knowledge of the elements affecting their performance *in vitro* and especially *in vivo*. Except from their lipid composition – which is of great importance, their physicochemical and morphological characteristics are equally essential [46, 48]. Particle size influence *in vivo* behavior: extremely short or large nanoparticles are not preferred [46]. Even though Enhanced Permeability and Retention (EPR)

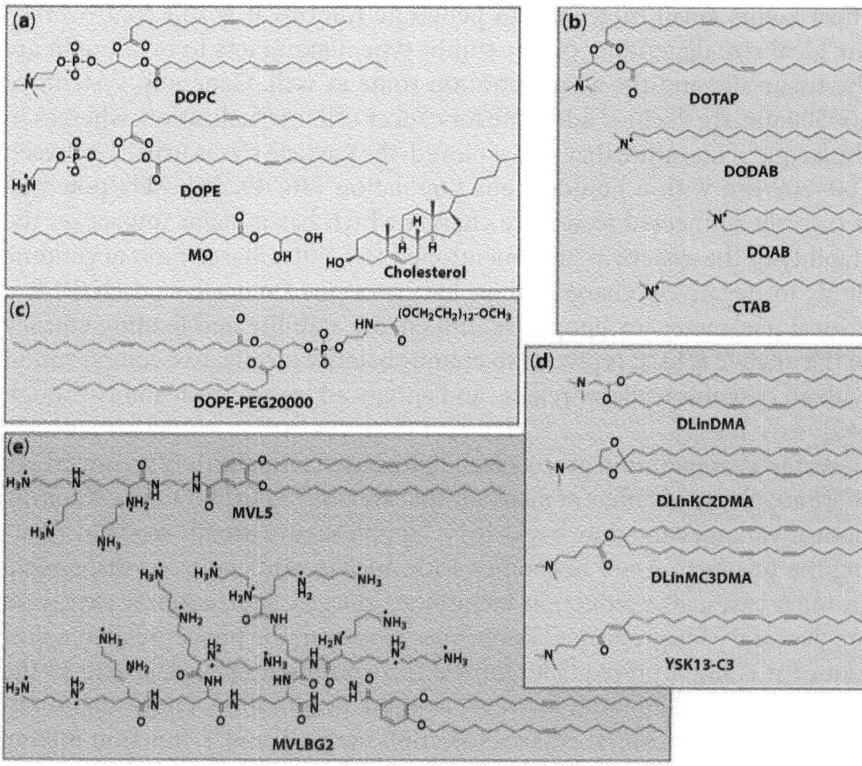

Figure 9.5 Molecular structures of lipids used in gene therapy formulations and discussed throughout the text. (a) Typical zwitterionic (1,2-dioleoyl-sn-glycero-3-phosphocholine (DOPC) and 1,2-dioleoyl-sn-glycero-3-phosphoethanolamine (DOPE)) and neutral lipids (monoolein (MO) and Cholesterol). (b) Typical cationic lipids (1,2-dioleoyl-3-trimethylammonium propane (DOTAP), dimethyldioctadecylammonium bromide (DODAB), didodecyldimethylammonium bromide (DDAB) and cetyltrimethylammonium bromide (CTAB)). (c) PEGylated lipid with DOPE backbone and PEG Mw of ca. 2000 Da. (d) Ionizable cationic lipids (1,2-dilinoleyloxy-3-dimethylaminopropane (DLinDMA), 2,2-dilinoleyl-4-(2-dimethylaminoethyl)-[1,3]-dioxolane (DLinKC2DMA), O-(Z,Z,Z,Zheptatriaconta-6,9,26,29-tetraen-19-yl)-4-(N,N-dimethylamino)butanoate (DLinMC3DMA) and 2- (dimethylamino)propyl-(12Z,15Z)-3-((9Z,12Z)-octadeca-9,12-dien-1-yl) henicosa-2,12,15-trienoate (YSK13-C3). (e) Multivalent cationic lipid N1-[2-((1S)-1-[(3-aminopropyl)amino]-4-[di(3-aminopropyl)amino]butylcarboxamido) ethyl]-3,4-di[oleyloxy]-benzamide (MVL5, nominal charge of +5) and dendritic headgroup cationic lipid N1-2-[((1S)-1,4-di[(1S)-1,4-di((1S)-1,4-di[(3- aminopropyl) amino]-butylcarboxamido)butyl]carboxamidobutyl)carboxamido]ethyl-3,4-di[(Z)-9-octadecenyloxy]benzamide (MVLBG2, nominal charge of +16). Adapted from [49].

effect favors small molecules to penetrate tumors, it is still controversial the ideal size depending on the tumor type. Parameters to be thought are the tissue site and the administration route as well. Generally, systems of 50–500 nm are deemed adequate for cancer cell internalization, whereas in the frequent literature that was explored, the particle size is usually between 100–200 nm with a homogenous population [46, 49, 50]. Zeta potential is directly connected to surface charge, which has a major impact on the stability of the system as aforementioned [59, 60]. Charge ratio of cationic lipids to negatively charged phosphate groups of nucleic acid (N/P) is a critical parameter for lipoplexes influencing stability and loading efficacy of the nucleic acid. It is pivotal to counterbalance an effective complexation with an easy intracellular release and enhanced transfection ability though [49].

As far as preparation protocols of lipid-based nanocarriers is concerned, there are plenty of different methods that are utilized in literature exhibiting both advantages and drawbacks [61]. Lipoplexes are usually made incubating the pre-formulated liposomes with nucleic acid's aqueous dispersion. In most cases, size reduction techniques such as sonication or extrusion are necessary for preformed liposomes. However, the preformulated nanoplatform is not a prerequisite for the incorporation of nucleic acid in the system. Other techniques are reverse phase evaporation, encapsulation in de-stabilized liposomes, ethanol injection method, and T-Junction mixing [45, 49, 61]. Popular protocols for liposome preparation include thin film hydration – which is the selected method for the majority of the publications that were used, solvent evaporation, and ethanol injection method [62, 63]. The ethanol injection technique – and similar approaches, are the most preferred methods for LNPs production [61]. Microfluidics is a scalable and robust technique which is based on solvent-exchange method, shows variations, and could be utilized for liposomes, lipoplexes, or LNPs preparation [45, 49, 61, 62].

Lipid based nanoparticles for cancer gene therapy are deemed promising carriers due to their biodegradability, versatility and easy-to-prepared [46–49]. However, there are still a lot of physiological obstacles to overcome for effective tumor penetration especially via systemic administration and overall stability and safety concerns [46, 53]. Active targeting and the usage of novel biocompatible and biodegradable materials are potential solutions. Certainly, there is no master key nanocarrier. In fact, the therapeutic efficacy of them is a multifactorial concept; the sort of nucleic acid, the cancer type, the lipid composition and relative proportion of the components, surface modifications, physicochemical parameters, and the preparation method contribute to a lipid gene vector with favorable physicochemical and morphological characteristics which is biocompatible and effective in cancer treatment. Hence, optimization of lipid components as well as the conduction of further and more specific preclinical studies are essential for efficacy and safety reasons [53]. Some of the most recent lipid-based gene vectors for cancer treatment are going to be discussed below.

Hattori's scientific group investigated extensively the influence of different kind of lipids into lipoplexes for siRNA delivery and their efficiency for lung cancer. Namely, in 2019 they provoked the impact of different type of helper lipid (cholesterol, DOPE, or both) on the effectiveness of the lipid gene vector in lung accumulation. However, helper's lipid combination with different type of cationic lipid-in terms of alkyl chain's saturation and length, was the major parameter for effective gene silencing due to fluidization effect. It is noteworthy that *in vitro* GL3 luciferase gene knockdown in breast cancer cell lines displayed different results than for Tie2 gene *in vivo*. They also declared that cationic lipids: DC-1-16 and DC-1-18, have great potentials as excipients for siRNA lipoplexes in this cancer type, while in 2022 they showed respective results for the importance of the chemical moieties of different sort of lipids and their amalgamation. In this case, they used a phospholipid and a cationic lipid in 1:1 molar ratio and the most promising lipoplexes comprised of DDAB or LP-TC-1-12 with DOPE and LP-TC-1-12 with POPE. In all formulations they used a charge ratio of 4:1 (molar) as the optimal for total RNA binding [57, 64].

Another preformulation study of Es *et al.* (2018) demonstrated the magnitude of molar charge ratio and its influence in other physicochemical characteristics. For this purpose, they created lipoplexes of EPC, DOTAP, and DOPE with siRNA for GL3 luciferase gene silencing in cancer cell lines trying different charge ratios. There is a critical point of charge ratio (3.27) where the physicochemical properties reach their optimum and no further changes can take place. They also observed differences in morphology between liposomes and lipoplexes and parameters affecting these [62].

Sanchez-Arribas and colleagues (2020) synthesized a cationic amino acid-modified lipid (C12ANHC18) and used it with MOG lipid for siRNA lipoplex's development. A model protein was encoded to investigate simultaneously transfection efficiency and safety in cancer cell lines. The system is deemed as a potential gene nanovector, while their results emphasized in lipoplex's microstructure and complexation with plasma proteins [65].

Ravula *et al.* (2021) composed an innovative peptide-lipid molecule and thus lipoplexes with redox characteristics. This lipid was mixed with DOTAP, DOPE, and a model or TRAIL plasmid DNA in different molar and charge ratios. Afterwards, they find the prominent lipoplex with an ideal charge ratio of five considering their physicochemical and internalization properties as well as their superiority over marketed available and neat DOTAP:DOPE lipoplexes. The system showed positive results in different cancer types at a minimum 50 μg/mL concentration and safety *in vitro* [66].

Diego-González *et al.* (2022) developed cationic liposomes of DODAB:MO (1:2 molar ratio) integrating different sort of siRNAs in various charge ratios against pancreatic cancer. The nucleic acids were encoding YAP and FOSL-1 genes and the lipoplexes consisted of two siRNAs for each target or four siRNAs for both targets. Preformulation studies demonstrated a preferable charge ratio of 15 for completely entrapment of RNAs, while all of the lipoplexes had a zeta-potential of 40–50 mV, an average size of 130 nm, and homogeneity. *In vitro* and *in vivo* experiments indicated that lipoplexes silencing both factors had the best performance, which was not improved by a dose augmentation. Despite biocompatibility, an entirely tumor disappearance was not achieved. Hence, it is suggested as a supplementary therapy [67].

Vaidya *et al.* (2022) composed novel cationic lipids (3,4-DMA) of 10, 12, or 14 C and mixed them with DOTAP, cholesterol, or PEG in different ratios and siRNA searching for the best formulation for selective insertion in breast cancer cell lines. The gene knockdown target was survivin and charge ratio was constant (5:1). They declared that C12-DMA:PEG (1:0.5) lipid gene carrier exhibits a good protein silencing efficacy *in vitro* compared to commercially available lipoplex, neat siRNA, and the other formulations, while it was no harmful for physiological human cell lines – 15% toxicity over 70–80% for cancer cell lines [68].

Survivin-T34A is a protein with antitumor properties encoded to mRNA into Zhang *et al.* (2019) nanoplatform against colon cancer. This novel lipoplex had about 180 nm size and peculiar shape due to protamine which incorporated completely the nucleic acid into the DOTAP:cholesterol shell. The vector was compared to plenty other relevant nanostructures proving to be superior both in safety and anticancer efficacy in different tumor models even by systemic administration (80% reduction of metastatic incidents). Besides, it seems to enhance immune responses in tumor microenvironment [69].

Rao and his scientific group (2018) proposed a DOTAP:cholesterol DNA lipoplex comprising of a bi-shRNA encoding a multi variant for KRAS gene against pancreatic ductal adenocarcinoma, while exceptional interest was given to the preparation method. In fact, differences in physicochemical characteristics, robustness, and effectiveness were noticed, whereas ethanol injection method stood out. Due to its exceptional results the novel carrier is at an early clinical stage [70].

Jiang *et al.* (2019) suggested another cationic DOTAP:cholesterol lipoplex, which is accumulated in the lungs, has a good toxicological profile and integrate a pDNA encoding a micro RNA (miR-143) acting through multiple mechanisms. The system has 200 nm size and 30 mV z-potential, and the optimum charge ratio is 4:1 (weight). *In vivo* experiments were performed by systemic and local administration and compared to cis-platin. The results were optimistic for the metastasis battle either at early or progressive lung tumors [71].

Barzegari Firouzabadi and colleagues (2019) encapsulated miR-143 in an optimal PEGylated liposome after the investigation of different formulations of DPPC, DOTAP, cholesterol, and/or mPEG 2000-DSPE. A homogenous population of 140 nm, a positive zeta potential of 14 mV, reduced toxicity, and increased efficacy against osteosarcoma *in vitro* were observed over free miR-143 [72].

Andey *et al.* (2019) accomplished to create a lipid gene vector for lung tumors with potentials in supporting therapy. Namely, the DPPC, DOTAP, and mPEG 2000-DSPE lipoplex of SOX2 siRNA exhibited a particle size of 90 nm, a zeta potential of 15 mV, 90% RNA encapsulation and a good anti-cancer activity *in vivo* especially in case of co-administration to cis-platin. Of paramount importance is also its positive influence on other endogenous substances that hamper chemotherapy and favor metastatic events [73].

PEGylation's prominence was highlighted by Lee and Ahn (2018). DC-cholesterol: DOPE lipoplexes with or without the presence of mPEG2000-DSPE (optimum quantity 2.5% mol) were compared *in vivo* against solid ovarian tumors by systemic administration. The liposomes complexed KSP siRNA and their charge ratio was 10:1 (weight), while the outcomes of this work signified superiority over non-PEGylated lipoplexes regarding efficiency, excretion, and biocompatibility [74].

Hu and co-workers (2022) formulated a gene carrier for glioblastoma via intranasal administration that internalized actively into the cancer cells. Namely, lipoplexes with siRNA encoding c-Myc were developed by incorporating two different types of peptides – an octaarginine (R8) and a penetratin. The former one condenses and protects siRNA due to attractive forces in the core and the last one is at the surface in an optimum quantity enhancing absorption by nasal mucous membrane. DOTAP, DOPE, and cholesterol were the rest of the excipients with or without the presence of mPEG2000-DSPE. Additionally, they find out that charge ratio of R8 peptide/siRNA and PEGylation affect particle size, whereas they proved the nanovector to be selective, effective, and non-toxic compared to corresponding marketed available lipoplexes [75].

Another recent example of a peptide-ligand lipid nanocarrier was that of Bao and colleagues (2020) forming DNA lipoplexes of PEDF for metastatic colorectal cancer to lungs. The lipoplex consisted of DOTAP, cholesterol, mPEG2000-cholesterol, and iRGD-PEG2000-DSPE was characterized by 200 nm particle size, 6:1 (weight) charge ratio and 20 mV zeta potential. The peptide was used for active targeting due to its affinity to specific proteins that are in a plethora in cancer cells. The protein expression and the tumor inhibition were successful via multiple mechanisms with no toxic effects in other organs [76].

iRGD peptide was used for targeting purposes from Xiao *et al.* (2020) as well. They proposed the synthesis of a shRNA cationic lipoplex translating eIF3i for the treatment of metastatic melanoma in lungs. Lipoplexes were formulated by DOTAP, cholesterol, and iRGD-PEG2000-DSPE in a volume ratio of liposomes: shRNA 6:1, whereas particle size was around 200 nm, zeta-potential about 15 mV, and knock down efficacy 60%. They correlated them *in vitro* and *in vivo* to non-targeting lipoplexes achieving limitation of tumor growth or progression to metastatic situations at a rate of 96%. Intravenous injection to mice was also examined for safety assessment and the preliminary results were encouraging [77].

Folate-conjugated lipoplexes for active targeting was composed by Tie and his scientific group (2020) due to their establishment of the association between folate receptors and tumor environment. Plasmid encoding BIM-S gene was complexed with DOTAP, cholesterol, mPEG-succinyl-cholesterol, and folate-PEG-succinyl-cholesterol liposomes – in a charge ratio of 6:1 (weight), for lung cancer therapy. Not only positive results concerning efficacy and safety were provoked *in vitro* and *in vivo* but also prevailed over non-targeted relevant lipoplexes. The nanoplatforms were characterized by an average size of 150 nm and a slightly positive charge (15 mV) [78].

He *et al.* (2018) also utilized folate-modified pDNA lipoplexes but in this case for gene co-translation of Cas9 and DNMT against ovarian cancer. DOTAP, cholesterol, methoxy- and folate- PEG-succinyl-cholesterol were used as lipid components. Lipoplexes were effectively complexed at a charge ratio of 12:1 having 150 nm size and 35 mV zeta potential and revealed anti-tumor efficacy and a better toxicological profile compared to anticancer drug [79].

CRISPR Cas9 is a useful method to moderate genome with great possibilities for anticancer therapy. Upon comprehensive preformulation studies, Sadat Hosseini *et al.* (2019) composed a stable, efficient, and non-toxic lipid gene vector for Cas9 delivery comprising of DOTAP, DOPE, cholesterol, and PEGylated cholesterol. Moreover, they highlighted the magnitude of charge ratio (optimum 3:1 for 39% cell internalization and gene silencing), cholesterol rafts, DOPE's arrangement, and PEGylation. Referring to the last one, they dealt with its combined lipid and the appropriate time of its addition during preparation [51].

Rosenblum and his fellow-workers (2020) investigated the use of some new ionizable lipids in LNPs formation for gene therapy exploiting CRISPR Cas9 technology. They proposed two novel nanocarriers of Lipid 8, cholesterol, DSPC, DMG–PEG, and DSPE -PEG targeting simultaneously different pathways against glioblastoma and efficiently entrapping (over 90%) an mRNA and a sgRNA – encoding Cas9 and PLK1 gene respectively. Targeting properties were provided to the system by adding EGFR antibodies against ovarian cancer. Both therapies were utilized by local administration and exhibited good physicochemical characteristics and promising results especially the EGFR-associated – approaching 80% gene silencing [80].

The mixture of a lipoplex and a linear peptide which has an affinity for proteins that are mainly located in skin cancer cells was introduced by Mohammed-Saeid's *et al.* (2018). The liposomes were prepared by DOPE and a 12-7N(GK)-12 Gemini surfactant and complexed with a model pDNA in 10:1 charge ratio with an average size of 150 nm and a heterogenous population. The gene delivery and the specificity of this system against melanoma was confirmed *in vitro* over other formulations [81].

The combination of gene and drug therapy favors effective cancer treatment with less adverse effects. Zhang C. *et al.* (2019) formulated lipid nanoparticles composed of DOPE, cholesterol, DMG-PEG2000, PEG, and an innovative paclitaxel – amino lipid (PAL) limiting triple negative breast cancer *in vitro* and *in vivo* compared to clinically used formulations. P53 mRNA was encapsulated in the lipid-based vector at 90% and drug's entrapment was equally satisfying [82].

Referring to anticancer drug paclitaxel, it was also combined with cationic lipoplexes of DOTAP: cholesterol and siRNA encoding Plk-1 gene by Bulbake and colleagues (2018). The co-delivery system was optimized in terms of lipid and charge ratio, loading cargo, and sonication time. The optimum formulation of 230 nm size and 12 mV potential showed effective gene silencing and overall better performance compared to separated therapies [83].

Xu *et al.* (2019) suggested the combinational therapy of pDNA lipoplexes with anticancer drug resveratrol. The lipoplex was synthesized by a novel cationic lipid with a peptide-moiety called CDO14, DOPE and pDNA for p53 gene expression, whereas the preferred charge ratio for effective gene editing was 3:1 (weight). The system was used against cervical and breast cancer concluding to efficient *in vitro* anti-tumor activity in correlation to other delivery platforms [84].

A niosome for synergistic therapy was also formed by Hemati *et al.* (2018) with encouraging results *in vitro* against variant cancer types. Upon preformulation screening, they designed a stimuli-responsive niosome with positive charge and stealth properties of Tween-60, cholesterol, DPPC, DOTAP, and DSPE-PEG2000 loading two drug substances and a siRNA encoded CDC20 protein. The desired charge ratio was 40:1 (weight) [85].

Another example of a multi-purpose lipid nanovector was introduced by Swami *et al.* (2021). They successfully composed pH-sensitive lipoplexes consisting of lecithin, cholesterol, DOTAP, DOPE (3:3:2:1 molar ratio) with the conjugation of SIRT1 shRNA and drug encapsulation of docetaxel. Lipoplexes were optimized with an average size of 200 nm, 20mV zeta potential, 40:1 charge ratio for utter binding, and an adequate drug entrapment. Comprehensive experiments were conducted not only for their superiority and effectiveness over available therapies against breast cancer but also for safety reasons [47].

The aforementioned molar ratio of this lipid mixture was also confirmed as an optimum by Kumar *et al.* (2020). The authors performed preformulation studies in pDNA lipoplexes encoding an imaging protein as model gene and probed effectiveness and viability in different cancer and normal cell lines as well as *in vivo*. Their prominent formulation was characterized by a charge ratio 10:1 (weight) with pH dependent characteristics [50].

Immunotherapy is very common in cancer treatment focusing on targeting genes that induce immune responses. Typical examples are Li and colleagues (2020) as well as Lei *et al.* (2020), who were interested in interleukin (IL) translation-particularly IL-12 and IL-15 respectively, by mRNA lipoplexes. The former one group used DOTAP, DSPC, cholesterol and PEG2000-DSPE or TT3 (ionizable lipid), DOPE, cholesterol, and C14-PEG2000 against multiple tumors via vaccination. Their findings indicate excessive immune reactions by multiple pathways *in vitro*, while *in vivo* they combined the mRNA with a protein before the lipoplex formation with even greater results [86]. The last one research suggested the usage of protamine with DOTAP:cholesterol liposomes against metastatic colorectal cancer, while an interesting outcome was the alteration in system's morphology after the incorporation of the nucleic acid [87]. Although there are a lot of recent publications highlighting the anti-cancer properties of lipid-based gene vaccines [88–91], it is believed that they exceed the purpose of this context and would be no further discussed.

Finally, Hajdu *et al.* (2018) amalgamated pDNA lipoplex (N/P 5:1) with radiolabeled 89Zr as a theranostic vector and proved that the imaging element did not affect the properties of lipid nanocarrier – particles having a homogenous population of around 120 nm and a zeta potential of 30 mV. DOPE lipid, gemini surfactants adjusted with 12-7NGK-12 and deferoxamine as well as interferon gamma gene were mixed in a specific order and used against melanoma *in vitro* [92].

9.2.2 Inorganic Nanocarriers: Their Technology and Recent Advances

Apart from polymer and lipid-based nanocarriers, also inorganic nanomaterials, are used as carriers for cancer gene therapy [93]. The inorganic nanomaterials don't possess the common limitations that viral vectors have and provide some desirable properties for biomedical applications [94]. Their chemical and thermal stability, facile functionalization, low toxicity, the variety on synthesis protocols and their potential for scalability are among those properties that make them promising candidates as carriers for cancer gene therapy [95]. Metallic NPs (e.g., gold NPs), iron oxide NPs, quantum dots (QDs) are among the common NPs that are chosen as gene delivery vehicles [96]. The main categories of nanomaterials that are usually utilized in gene therapy, are presented below.

Carbon Nanotubes (CNTs) have thoroughly studied for their potential use as drug carriers, particularly in gene therapy, due to their structure. CNTs are graphene layers that have been rolled either as single-walled carbon nanotubes (SWNTs) consisting of one layer, or as multi-walled carbon nanotubes (MWNTs) including multiple SWNTs. Their surface provides a friendly environment for binding and carrying drugs, contrast agents, etc. [97]. According to Pantarotto *et al.*, the common protocol that is applied to functionalize its surface, includes oxidation in acidic solution and 1,3-dipolar cycloaddition reaction, to achieve a functionality based on covalent bounds. Alternatively, a link between CNTs and non-polar rings (e.g., DNA skeleton) leads to non-covalent bounds. Depending on the application needs, the produced water-soluble CNT-based system can be used for gene therapy [98]. CNTs form complexes with plasmid DNA, and as they provide good cellular uptake, they allow gene delivery. Furthermore, if fluorescent compounds are co-bound, then *in vitro* and *in vivo* studies can track the cellular uptake [99].

Mesoporous silica NPs can be utilized in siRNA or DNA delivering to the target cells. Their porous structure can be perfect for the encapsulation of drugs and gene. They can be prepared through easy low-cost efficient techniques; they are biocompatible and have large surface area for molecules loading [100]. If they are positive charged functionalized, then are suitable for siRNA delivery. According to Zhou *et al.* if the porous of silica NPs are quite small (e.g., 3–4 nm) [100], or according to Di Gioia *et al.* if the surface is negatively charged post Polyethylenimine (PEI) modification [101], then plasmid DNA delivery is achievable. Also, dendrimers or cationic lipids can be used for surface modification of silica NPs allowing DNA adsorption [102]. DNA encapsulated on silica NPs pores, can be protected against degradation by nucleases, until its release in the target area.

Magnetic NPs have small size, allowing cellular penetration and thus are ideal as gene nanocarriers. Magnetic NPs have been widely used in various biomedical applications (e.g., as carriers for bio-separation, contrast agents, protein immobilization [103]. Coating with silica or other biocompatible materials, decreases their toxicity. Imaging of magnetic NPs within cells is feasible through transmission electron microscopy (TEM) and magnetic resonance imaging (MRI) [104]. Gene delivery and precise targeting is also achievable via magnetic NPs. Core–shell magnetic NPs with their self-assembly can be used for nucleic acid delivery with adeno- and lentivirus vectors [105] and iron oxide NPs is applied in siRNA delivery and transfection [106]. Coated magnetic NPs present optimized properties.

Gold Nanoparticles AuNPs are considered as multifunctional carriers, due to their easy preparation, and their biocompatibility. Surface modification of AuNPs with positively charged molecules such as amino acids or cationic peptides leads to the development of a gene delivery vector [107]. DNA or siRNA can be bound on this positive charged form, entering cellular membrane. According to Poletaeva *et al.* PEI-coated AuNPs prepared with controlled size and surface charge can be used for siRNA delivery [108]. These PEI-coated AuNPs were non-toxic and showed good gene silencing effect in cancer cells. Combination of AuNPs with polymers allowed single-stranded and double-stranded DNA functionalization for gene therapy with high binding affinity, low toxicity and immunity to nuclease [109]. Polyethylene glycol (PEG)-modified cationic AuNPs conjugated with nucleic acids have also been exploited for gene delivery [110]. DNA loaded AuNPs embedded in sharp carbonaceous carriers were used for efficient DNA delivery into plants with minimal damage [111]. The multifunctionality of AuNPs is depicted in Figure 9.6.

QDs are considered efficient siRNA carriers since they can be utilized as fluorescence probes *in vivo* for theragnostic applications. Organic-hydride modification of QDs eliminated their toxicity [112]. Positively charged CdTe QDs can carry plasmid DNA according to Banerjee *et al.* [113]. Another study focused on the development of l-arginine-functional-modified CdSe/ZnSe QDs, used as siRNA carriers, silencing HPV18 E6 gene in HeLa cells [114]. QDs complexes can be used as nanocrystal

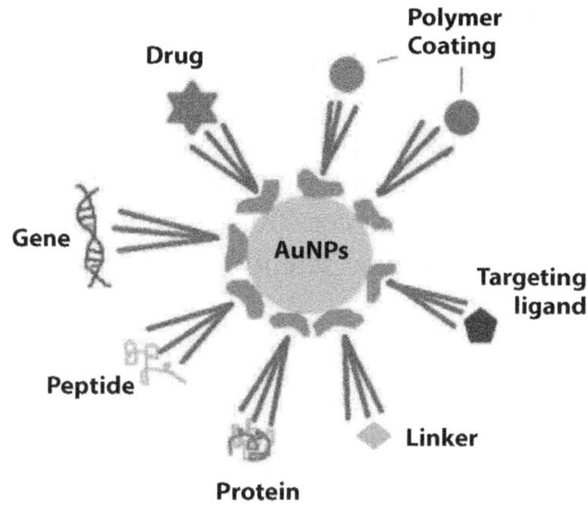

Figure 9.6 Schematic representation of AuNPs multifunctionality.

probing agents for gene delivery, transfection and real-time tracking. Amino acid modified QDs are used as siRNA carriers for gene silencing in parallel with fluorescent tracking [115].

Calcium phosphate can be also used for gene therapy due to its excellent biocompatibility. Calcium phosphate coprecipitated with DNA is used as a non-viral vector for efficient transfection into nucleus of transformed human endothelial cells. The aggregation of these nanohybrids is a limitation and usually this complex is replaced by another of the available choices [116].

9.3 Conclusions

The successful encapsulation and intracellular delivery of a wide range of therapeutic nucleic acid molecules, aiming for the treatment, at their most fundamental, i.e., genetic level, of a wide range of disorders, with the use of novel, tailor-made non - viral vectors has exhibited great promise during the last decade at a pre-clinical level. As with any gene delivery system, viral and non - viral, three components are essential for a successful gene transfection, a plasmid-based gene expression system, the correct therapeutic nucleic acid sequence, responsible for the most parts for the translation of the therapeutic protein, and finally the proper vector for the delivery of the plasmid to the specific pathogenic region inside the organism, we wish to treat (specific organ or cell type). Since, the research area of vectors for gene delivery applications is a very wide field of study, and the nanoparticles (polymeric, lipidic and inorganic) seem to be promising carriers in gene therapy, soon optimization of the available nanoplatforms will be probably feasible. Functionalization of nanoparticles will allow effective gene delivery. Gene therapy has paved the way towards the effective treatment of various diseased that are associated with defective genes, such as neurodegenerative disease or cancer. Thus, effective delivery of therapeutic genetic material into cells is still a scientific challenge.

References

1. Khan, M., Ong, Z.Y., Wiradharma, N., Attia, A.B., Yang, Y.Y., Advanced materials for co-delivery of drugs and genes in cancer therapy. *Adv. Healthc. Mater.*, 1, 4, 373–92, 2012.
2. Xu, H., Li, Z., Si, J., Nanocarriers in gene therapy: A review. *J. Biomed. Nanotechnol.*, 10, 12, 3483–507, 2014.
3. Li, Y., Tang, K., Zhang, X., Pan, W., Li, N., Tang, B., Tumor microenvironment responsive nanocarriers for gene therapy. *Chem. Commun. (Camb.)*, 58, 63, 8754–8765, 2022.
4. Zeng, X., Li, Z., Zhu, C., Xu, L., Sun, Y., Han, S., Research progress of nanocarriers for gene therapy targeting abnormal glucose and lipid metabolism in tumors. *Drug Deliv.*, 28, 1, 2329–2347, 2021.
5. Portney, N.G. and Ozkan, M., Nano-oncology: Drug delivery, imaging, and sensing. *Anal. Bioanal. Chem.*, 384, 3, 620–30, 2006.
6. Sharma, A., Jha, N.K., Dahiya, K., Singh, V.K., Chaurasiya, K., Jha, A.N., Jha, S.K., Mishra, P.C., Dholpuria, S., Astya, R., Nand, P., Kumar, A., Ruokolainen, J., Kesari, K.K., Nanoparticulate RNA delivery systems in cancer. *Cancer Rep.* 3, 5, e1271, 2020.
7. Revia, R.A., Stephen, Z.R., Zhang, M., Theranostic nanoparticles for RNA-based cancer treatment. *Acc. Chem. Res.*, 52, 6, 1496–1506, 2019.
8. Mainini, F. and Eccles, M.R., Lipid and polymer-based nanoparticle siRNA delivery systems for cancer therapy. *Molecules*, 25, 11, 2692, 2020.
9. Ambrosio, N., Voci, S., Gagliardi, A., Palma, E., Fresta, M., Cosco, D., Application of biocompatible drug delivery nanosystems for the treatment of naturally occurring cancer in dogs. *J. Funct. Biomater.*, 13, 3, 116, 2022.
10. de Brito E Cunha, D., Frederico, A.B.T., Azamor, T., Melgaço, J.G., da Costa Neves, P.C., Bom, A.P.D.A., Tilli, T.M., Missailidis, S., Biotechnological evolution of siRNA molecules: From bench tool to the refined drug. *Pharmaceuticals*, 15, 5, 575, 2022.
11. Kontogiannis, O. and Karalis, V., On the *in vivo* kinetics of gene delivery vectors. *bioRxiv*, 2022, http://dx.doi.org/10.1101/2022.02.11.22269834.
12. Sung, Y.K. and Kim, S.W., Recent advances in the development of gene delivery systems. *Biomater. Res.*, 23, 1, 8, 2019.

13. Bono, N., Ponti, F., Mantovani, D., Candiani, G., Non-viral *in vitro* gene delivery: It is now time to set the bar! *Pharmaceutics*, 12, 2, 183, 2020.
14. Xia, W., Tao, Z., Zhu, B., Zhang, W., Liu, C., Chen, S. *et al.*, Targeted delivery of drugs and genes using polymer nanocarriers for cancer therapy. *Int. J. Mol. Sci.*, 22, 17, 9118, 2021.
15. Chen, C.-K., Huang, P.-K., Law, W.-C., Chu, C.-H., Chen, N.-T., Lo, L.-W., Biodegradable polymers for gene-delivery applications. *Int. J. Nanomed.*, 15, 2131–50, 2020.
16. Yousefpour Marzbali, M. and Yari Khosroushahi, A., Polymeric micelles as mighty nanocarriers for cancer gene therapy: A review. *Cancer Chemother. Pharmacol.*, 79, 4, 637–649, 2017.
17. Piotrowski-Daspit, A.S., Kauffman, A.C., Bracaglia, L.G., Saltzman, W.M., Polymeric vehicles for nucleic acid delivery. *Adv. Drug Deliv. Rev.*, 156, 119–132, 2020.
18. Amreddy, N., Babu, A., Muralidharan, R., Munshi, A., Ramesh, R., Polymeric nanoparticle-mediated gene delivery for lung cancer treatment. *Top. Curr. Chem.*, 375, 2, 35, 2017.
19. Palmerston Mendes, L., Pan, J., Torchilin, V., Dendrimers as nanocarriers for nucleic acid and drug delivery in cancer therapy. *Molecules*, 22, 9, 1401, 2017.
20. Kapadia, C.H., Luo, B., Dang, M.N., Irvin-Choy, N., Valcourt, D.M., Day, E.S., Polymer nanocarriers for MicroRNA delivery. *J. Appl. Polym. Sci.*, 137, 25, 48651, 2020.
21. Mirón-Barroso, S., Correia, J.S., Frampton, A.E., Lythgoe, M.P., Clark, J., Tookman, L. *et al.*, Polymeric carriers for delivery of RNA cancer therapeutics. *Noncoding RNA*, 8, 4, 58, 2022.
22. Ke, L., Cai, P., Wu, Y.-L., Chen, X., Polymeric nonviral gene delivery systems for cancer immunotherapy. *Adv. Ther. (Weinh.)*, 3, 6, 1900213, 2020.
23. Medina-Kauwe, L.K., Xie, J., Hamm-Alvarez, S., Intracellular trafficking of nonviral vectors. *Gene Ther.*, 12, 24, 1734–1751, 2005.
24. Zhang, B., Liu, Q., Liu, M., Shi, P., Zhu, L., Zhang, L. *et al.*, Biodegradable hybrid mesoporous silica nanoparticles for gene/chemo-synergetic therapy of breast cancer. *J. Biomater. Appl.*, 33, 10, 1382–1393, 2019.
25. Rejeeth, C. and Vivek, R., Comparison of two silica based nonviral gene therapy vectors for breast carcinoma: Evaluation of the p53 delivery system in Balb/c mice. *Artif. Cells Nanomed. Biotechnol.*, 45, 3, 489–494, 2017.
26. Xu, J., Ganesh, S., Amiji, M., Non-condensing polymeric nanoparticles for targeted gene and siRNA delivery. *Int. J. Pharm.*, 427, 1, 21–34, 2012.
27. Jones, C.H., Chen, C.-K., Ravikrishnan, A., Rane, S., Pfeifer, B.A., Overcoming nonviral gene delivery barriers: Perspective and future. *Mol. Pharm.*, 10, 11, 4082–4098, 2013.
28. Zhou, Q.-H., Wu, C., Manickam, D.S., Oupický, D., Evaluation of pharmacokinetics of bioreducible gene delivery vectors by real-time PCR. *Pharm. Res.*, 26, 7, 1581–1589, 2009.

29. Belmadi, N., Midoux, P., Loyer, P., Passirani, C., Pichon, C., Le Gall, T. *et al.*, Synthetic vectors for gene delivery: An overview of their evolution depending on routes of administration. *Biotechnol. J.*, 10, 9, 1370–1389, 2015.
30. Davis, M.E., Zuckerman, J.E., Choi, C.H.J., Seligson, D., Tolcher, A., Alabi, C.A. *et al.*, Evidence of RNAi in humans from systemically administered siRNA via targeted nanoparticles. *Nature*, 464, 7291, 1067–1070, 2010.
31. Du, L., Zhou, J., Meng, L., Wang, X., Wang, C., Huang, Y. *et al.*, The pH-triggered triblock nanocarrier enabled highly efficient siRNA delivery for cancer therapy. *Theranostics*, 7, 14, 3432–3445, 2017.
32. Hu-Lieskovan, S., Heidel, J.D., Bartlett, D.W., Davis, M.E., Triche, T.J., Sequence-specific knockdown of EWS-FLI1 by targeted, nonviral delivery of small interfering RNA inhibits tumor growth in a murine model of metastatic Ewing's sarcoma. *Cancer Res.*, 65, 19, 8984–8992, 2005.
33. Men, K., Huang, R., Zhang, X., Zhang, R., Zhang, Y., He, M. *et al.*, Local and systemic delivery of interleukin-12 gene by cationic micelles for cancer immunogene therapy. *J. Biomed. Nanotechnol.*, 14, 10, 1719–30, 2018.
34. Men, K., Huang, R., Zhang, X., Zhang, R., Zhang, Y., Peng, Y. *et al.*, Delivery of interleukin-22 binding protein (IL-22BP) gene by cationic micelle for colon cancer gene therapy. *RSC Adv.*, 8, 30, 16537–16548, 2018.
35. Gao, S., Tian, H., Xing, Z., Zhang, D., Guo, Y., Guo, Z. *et al.*, A non-viral suicide gene delivery system traversing the blood brain barrier for non-invasive glioma targeting treatment. *J. Control. Release*, 243, 357–369, 2016.
36. Qiu, N., Gao, J., Liu, Q., Wang, J., Shen, Y., Enzyme-responsive charge-reversal polymer-mediated effective gene therapy for intraperitoneal tumors. *Biomacromolecules*, 19, 6, 2308–2319, 2018.
37. Li, Y., Thambi, T., Lee, D.S., Co-delivery of drugs and genes using polymeric nanoparticles for synergistic cancer therapeutic effects. *Adv. Healthc. Mater.*, 7, 1, 1700886, 2018.
38. Mohammadinejad, R., Dehshahri, A., Sagar Madamsetty, V., Zahmatkeshan, M., Tavakol, S., Makvandi, P. *et al.*, *In vivo* gene delivery mediated by non-viral vectors for cancer therapy. *J. Control. Release*, 325, 249–275, 2020.
39. Li, L., Hu, S., Chen, X., Non-viral delivery systems for CRISPR/Cas9-based genome editing: Challenges and opportunities. *Biomaterials*, 171, 207–218, 2018.
40. Grewal, I.K., Singh, S., Arora, S., Sharma, N., Polymeric nanoparticles for breast cancer therapy: A comprehensive review. *Biointerface Res. Appl. Chem.*, 11, 4, 11151–11171, 2020.
41. Chuan, D., Jin, T., Fan, R., Zhou, L., Guo, G., Chitosan for gene delivery: Methods for improvement and applications. *Adv. Colloid Interface Sci.*, 268, 25–38, 2019.
42. Lu, Y.-J., Lan, Y.-H., Chuang, C.-C., Lu, W.-T., Chan, L.-Y., Hsu, P.-W. *et al.*, Injectable thermo-sensitive chitosan hydrogel containing CPT-11-loaded EGFR-targeted graphene oxide and SLP2 shRNA for localized drug/gene delivery in glioblastoma therapy. *Int. J. Mol. Sci.*, 21, 19, 7111, 2020.

43. Balan, V., Dodi, G., Mihai, C.T., Serban, A.M., Ursachi, V.C., Biotinylated chitosan macromolecule based nanosystems: A review from chemical design to biological targets. *Int. J. Biol. Macromol.*, 188, 82–93, 2021.
44. Kim, J., Wilson, D.R., Zamboni, C.G., Green, J.J., Targeted polymeric nanoparticles for cancer gene therapy. *J. Drug Targeting*, 23, 7–8, 627–641, 2015.
45. Amoabediny, G., Haghiralsadat, F., Naderinezhad, S., Helder, M.N., Kharanaghi, E.A., Arough, J.M. *et al.*, Overview of preparation methods of polymeric and lipid-based (niosome, solid lipid, liposome) nanoparticles: A comprehensive review. *Int. J. Polym. Mater.*, 67, 6, 383–400, 2018.
46. Liu, C., Zhang, L., Zhu, W., Guo, R., Sun, H., Chen, X. *et al.*, Barriers and strategies of cationic liposomes for cancer gene therapy. *Mol. Ther. Methods Clin. Dev.*, 18, 751–764, 2020.
47. Swami, R., Kumar, Y., Chaudhari, D., Katiyar, S.S., Kuche, K., Katare, P.B. *et al.*, pH sensitive liposomes assisted specific and improved breast cancer therapy using co-delivery of SIRT1 shRNA and docetaxel. *Mater. Sci. Eng. C*, 120, 111664, 2021.
48. Kiaie, S.H., Majidi Zolbanin, N., Ahmadi, A., Bagherifar, R., Valizadeh, H., Kashanchi, F. *et al.*, Recent advances in mRNA-LNP therapeutics: Immunological and pharmacological aspects. *J. Nanobiotechnol.*, 20, 276, 2022.
49. Gaspar, R., Coelho, F., Silva, B.F.B., Lipid-nucleic acid complexes: Physicochemical aspects and prospects for cancer treatment. *Molecules*, 25, 21, 5006, 2020.
50. Kumar, Y., Kuche, K., Swami, R., Katiyar, S.S., Chaudhari, D., Katare, P.B. *et al.*, Exploring the potential of novel pH sensitive lipoplexes for tumor targeted gene delivery with reduced toxicity. *Int. J. Pharm.*, 573, 118889, 2020.
51. Hosseini, E.S., Nikkhah, M., Hosseinkhani, S., Cholesterol-rich lipid-mediated nanoparticles boost of transfection efficiency, utilized for gene editing by CRISPR-Cas9. *Int. J. Nanomed.*, 14, 4353–4366, 2019.
52. Zhu, Y., Shen, R., Vuong, I., Reynolds, R.A., Shears, M.J., Yao, Z.C. *et al.*, Multi-step screening of DNA/lipid nanoparticles and co-delivery with siRNA to enhance and prolong gene expression. *Nat. Commun.*, 13, 1, 4282, 2022.
53. Vlatkovic, I., Non-Immunotherapy Application of LNP-mRNA: Maximizing efficacy and safety. *Biomedicines*, 9, 5, 530, 2021.
54. Luiz, M.T., Dutra, J.A.P., Tofani, L.B., De Araújo, J.T.C., Di Filippo, L.D., Marchetti, J.M. *et al.*, Targeted liposomes: A nonviral gene delivery system for cancer therapy. *Pharmaceutics*, 14, 4, 821, 2022.
55. Hou, X., Zaks, T., Langer, R., Dong Y., Lipid nanoparticles for mRNA delivery. *Nat. Rev. Mater.*, 6, 1078–1094, 2021.
56. Hald Albertsen, C., Kulkarni, J.A., Witzigmann, D., Lind, M., Petersson, K., Simonsen, J.B., The role of lipid components in lipid nanoparticles for vaccines and gene therapy. *Adv. Drug Deliv. Rev.*, 188, 114416, 2022.

57. Hattori, Y., Tamaki, K., Ozaki, K., Kawano, K., Onishi, H., Optimized combination of cationic lipids and neutral helper lipids in cationic liposomes for siRNA delivery into the lung by intravenous injection of siRNA lipoplexes. *J. Drug Deliv. Sci. Technol.*, 52, 1042–1050, 2019.
58. Fang, H. and Chen, Q., Applications and challenges of biomaterial mediated mRNA delivery. *Explor. Targeting Antitumor Ther.*, 3, 4, 428–444, 2022.
59. Pippa, N., Pispas, S., Demetzos, C., Physico-chemical characterization and basic research principles of advanced drug delivery nanosystems, in: *Intelligent Nanomaterials, 2nd Edition*, A. Ashutosh Tiwari, Y.K. Mishra, H. Kobayashi, A.P.F. Turner (Eds.), pp. 107–126, Beverly, MA (USA): Scrivener Publishing LLC, 2017.
60. Triantafyllopoulou, E., Pippa, N., Demetzos, C., Protein-liposome interactions: The impact of surface charge and fluidisation effect on protein binding. *J. Liposome Res.*, 33, 1, 77–88, 2023.
61. Evers, M.J.W., Kulkarni, J.A., van der Meel, R., Cullis, P.R., Vader, P., Schiffelers, R.M., State-of-the-art design and rapid-mixing production techniques of lipid nanoparticles for nucleic acid delivery. *Small Methods*, 2, 9, 1700375, 2018.
62. Eş, I., Ok, M.T., Puentes-Martinez, X.E., De Toledo, M.A.S., Favaro, M.T.P., Cavalcanti, L.P. *et al.*, Evaluation of siRNA and cationic liposomes complexes as a model for *in vitro* siRNA delivery to cancer cells. *Colloids Surf. A Physicochem. Eng. Asp.*, 555, 280–289, 2018.
63. Pattni, B.S., Chupin, V.V., Torchilin, V.P., New developments in liposomal drug delivery. *Chem. Rev.*, 115, 19, 10938–10966, 2015.
64. Hattori, Y., Tang, M., Torii, S., Tomita, K., Sagawa, A., Inoue, N. *et al.*, Optimal combination of cationic lipid and phospholipid in cationic liposomes for gene knockdown in breast cancer cells and mouse lung using siRNA lipoplexes. *Mol. Med. Rep.*, 26, 2, 253, 2022.
65. Sánchez-Arribas, N., Martínez-Negro, M., Villar, E.M., Pérez, L., Aicart, E., Taboada, P. *et al.*, Biocompatible nanovector of siRNA consisting of arginine-based cationic lipid for gene knockdown in cancer cells. *ACS Appl. Mater. Interfaces*, 12, 31, 34536–34547, 2020.
66. Ravula, V., Lo, Y.L., Wang, L.F., Patri, S.V., Gemini lipopeptide bearing an ultrashort peptide for enhanced transfection efficiency and cancer-cell-specific cytotoxicity. *ACS Omega*, 6, 35, 22955–22968, 2021.
67. Diego-González, L., Fernández-Carrera, A., Igea, A., Martínez-Pérez, A., Real Oliveira, M.E.C.D., Gomes, A.C. *et al.*, Combined inhibition of FOSL-1 and YAP Using siRNA-lipoplexes reduces the growth of pancreatic tumor. *Cancers*, 14, 13, 3102, 2022.
68. Vaidya, S., Jeengar, M.K., Wadaan, M.A., Mahboob, S., Kumar, P., Reece, L.M. *et al.*, Design and *in vitro* evaluation of novel cationic lipids for siRNA delivery in breast cancer cell lines. *Evid. Based Complementary Altern. Med.*, 2022, 9231641, 2022.

69. Zhang, X., Men, K., Zhang, Y., Zhang, R., Yang, L., Duan, X., Local and systemic delivery of mRNA encoding survivin-T34A by lipoplex for efficient colon cancer gene therapy. *Int. J. Nanomed.*, 14, 2733–2751, 2019.
70. Rao, D.D., Luo, X., Wang, Z., Jay, C.M., Brunicardi, F.C., Maltese, W. *et al.*, KRAS mutant allele-specific expression knockdown in pancreatic cancer model with systemically delivered bi-shRNA KRAS lipoplex. *PLoS One*, 13, 5, e0193644, 2018.
71. Jiang, Q., Yuan, Y., Gong, Y., Luo, X., Su, X., Hu, X. *et al.*, Therapeutic delivery of microRNA-143 by cationic lipoplexes for non-small cell lung cancer treatment *in vivo*. *J. Cancer Res. Clin. Oncol.*, 145, 12, 2951–2967, 2019.
72. Barzegari Firouzabadi, F., Oryan, S.H., Sheikhha, M.H., Kalantar, S.M., Javed, A., Preparation and evaluation of a novel liposomal nano-formulation in metastatic cancer treatment studies. *Cell J.*, 21, 2, 135–142, 2019.
73. Andey, T., Bora-Singhal, N., Chellappan, S.P., Singh, M., Cationic lipoplexes for treatment of cancer stem cell-derived murine lung tumors. *Nanomedicine*, 18, 31–43, 2019.
74. Lee, J. and Ahn, H.J., PEGylated DC-Chol/DOPE cationic liposomes containing KSP siRNA as a systemic siRNA delivery carrier for ovarian cancer therapy. *Biochem. Biophys. Res. Commun.*, 503, 3, 1716–1722, 2018.
75. Hu, Y., Jiang, K., Wang, D., Yao, S., Lu, L., Wang, H. *et al.*, Core-shell lipoplexes inducing active macropinocytosis promote intranasal delivery of c-Myc siRNA for treatment of glioblastoma. *Acta Biomater.*, 138, 478–490, 2022.
76. Bao, X., Zeng, J., Huang, H., Ma, C., Wang, L., Wang, F. *et al.*, Cancer-targeted PEDF-DNA therapy for metastatic colorectal cancer. *Int. J. Pharm.*, 576, 118999, 2020.
77. Xiao, W., Zhang, W., Huang, H., Xie, Y., Zhang, Y., Guo, X. *et al.*, Cancer targeted gene therapy for inhibition of melanoma lung metastasis with eIF3i shRNA loaded liposomes. *Mol. Pharm.*, 17, 1, 229–238, 2020.
78. Tie, Y., Zheng, H., He, Z., Yang, J., Shao, B., Liu, L. *et al.*, Targeting folate receptor β positive tumor-associated macrophages in lung cancer with a folate-modified liposomal complex. *Signal Transduction Targeting Ther.*, 5, 1, 6, 2020.
79. He, Z.Y., Zhang, Y.G., Yang, Y.H., Ma, C.C., Wang, P., Du, W. *et al.*, *In vivo* ovarian cancer gene therapy using CRISPR-Cas9. *Hum. Gene Ther.*, 29, 2, 223–233, 2018.
80. Rosenblum, D., Gutkin, A., Kedmi, R., Ramishetti, S., Veiga, N., Jacobi, A.M. *et al.*, CRISPR-Cas9 genome editing using targeted lipid nanoparticles for cancer therapy. *Sci. Adv.*, 6, 47, eabc9450, 2020.
81. Mohammed-Saeid, W., Soudy, R., Tikoo, R., Kaur, K., Verrall, R.E., Badea., I., Design and evaluation of gemini surfactant-based lipoplexes modified with cell-binding peptide for targeted gene therapy in melanoma model. *J. Pharm. Pharm. Sci.*, 21, 1, 363–375, 2018.

82. Zhang, C., Zhang, X., Zhao, W., Zeng, C., Li, W., Li, B. *et al.*, Chemotherapy drugs derived nanoparticles encapsulating mRNA encoding tumor suppressor proteins to treat triple-negative breast cancer. *Nano Res.*, 12, 4, 855–861, 2019.
83. Bulbake, U., Kommineni, N., Bryszewska, M., Ionov, M., Khan, W., Cationic liposomes for co-delivery of paclitaxel and anti-Plk1 siRNA to achieve enhanced efficacy in breast cancer. *J. Drug Deliv. Sci. Technol.*, 48, 253–265, 2018.
84. Xu, X., Liu, A., Bai, Y., Li, Y., Zhang, C., Cui, S. *et al.*, Co-delivery of resveratrol and p53 gene via peptide cationic liposomal nanocarrier for the synergistic treatment of cervical cancer and breast cancer cells. *J. Drug Deliv. Sci. Technol.*, 51, 746–753, 2019.
85. Hemati, M., Haghiralsadat, F., Yazdian, F., Jafari, F., Moradi, A., Malekpour-Dehkordi, Z., Development and characterization of a novel cationic PEGylated niosome-encapsulated forms of doxorubicin, quercetin and siRNA for the treatment of cancer by using combination therapy. *Artif. Cells Nanomed. Biotechnol.*, 47, 1, 1295–1311, 2019.
86. Li, Y., Su, Z., Zhao, W., Zhang, W., Momin, N., Zhang, C. *et al.*, Multifunctional oncolytic nanoparticles deliver self-replicating IL-12 RNA to eliminate established tumors and prime systemic immunity. *Nat. Cancer*, 1, 9, 882–893, 2020.
87. Lei, S., Zhang, X., Men, K., Gao, Y., Yang, X., Wu, S. *et al.*, Efficient colorectal cancer gene therapy with IL-15 mRNA nanoformulation. *Mol. Pharm.*, 17, 9, 3378–3391, 2020.
88. Miao, L., Li, L., Huang, Y., Delcassian, D., Chahal, J., Han, J., Shi, Y. *et al.*, Delivery of mRNA vaccines with heterocyclic lipids increases anti-tumor efficacy by STING-mediated immune cell activation. *Nat. Biotechnol.*, 37, 10, 1174–1185, 2019.
89. Rybakova, Y., Kowalski, P.S., Huang, Y., Gonzalez, J.T., Heartlein, M.W., DeRosa, F. *et al.*, mRNA delivery for therapeutic anti-HER2 antibody expression *in vivo*. *Mol. Ther.*, 27, 8, 1415–1423, 2019.
90. Lai, I., Swaminathan, S., Baylot, V., Mosley, A., Dhanasekaran, R., Gabay, M. *et al.*, Lipid nanoparticles that deliver IL-12 messenger RNA suppress tumorigenesis in MYC oncogene-driven hepatocellular carcinoma. *J. Immunother. Cancer*, 6, 1, 125, 2018.
91. Zhang, R., Men, K., Zhang, X., Huang, R., Tian, Y., Zhou, B. *et al.*, Delivery of a modified mRNA encoding IL-22 binding protein (IL-22BP) for colon cancer gene therapy. *J. Biomed. Nanotechnol.*, 14, 7, 1239–1251, 2018.
92. Hajdu, I., Makhlouf, A., Solomon, V.R., Michel, D., Al-Dulaymi, M., Wasan, K.M. *et al.*, A 89Zr-labeled lipoplex nanosystem for image-guided gene delivery: Design, evaluation of stability and *in vivo* behavior. *Int. J. Nanomed.*, 13, 7801–7818, 2018.
93. Sheoran, S., Arora, S., Samsonraj, R., Govindaiah, P., Vuree, S., Lipid-based nanoparticles for treatment of cancer. *Heliyon*, 8, 5, e09403, 2022.

94. Zu, H. and Gao, D., Non-viral vectors in gene therapy: Recent development, challenges, and prospects. *AAPS J.*, 23, 4, 78, 2021.
95. Montané, X., Bajek, A., Roszkowski, K., Montornés, J.M., Giamberini, M., Roszkowski, S., Kowalczyk, O., Garcia-Valls, R., Tylkowski, B., Encapsulation for cancer therapy. *Molecules*, 25, 7, 1605, 2020.
96. Loh, X.J. and Lee, T.-C., Gene delivery by functional inorganic nanocarriers. *Recent Pat. DNA Gene Seq.*, 6, 108–114, 2012.
97. Zare, H., Ahmadi, S., Ghasemi, A., Ghanbari, M., Rabiee, N., Bagherzadeh, M., Karimi, M., Webster, T.J., Hamblin, M.R., Mostafavi, E., Carbon nanotubes: Smart drug/gene delivery carriers. *Int. J. Nanomed.*, 16, 1681–1706, 2021.
98. Pantarotto, D., Singh, R., McCarthy, D., Erhardt, M., Briand, J.P., Prato, M., Kostarelos, K., Bianco, A., Functionalized carbon nanotubes for plasmid DNA gene delivery. *Angew. Chem. (Int. Ed. Engl.)*, 43, 39, 5242–5246, 2004.
99. Tang, L., Xiao, Q., Mei, Y., He, S., Zhang, Z., Wang, R., Wang, W., Insights on functionalized carbon nanotubes for cancer theranostics. *J. Nanobiotechnology*, 19, 1, 423, 2021.
100. Zhou, Y., Quan, G., Wu, Q., Zhang, X., Niu, B., Wu, B., Huang, Y., Pan, X., Wu, C., Mesoporous silica nanoparticles for drug and gene delivery. *Acta Pharm. Sin. B*, 8, 2, 165–177, 2018.
101. Di Gioia, S. and Conese, M., Polyethylenimine-mediated gene delivery to the lung and therapeutic applications. *Drug Des. Devel. Ther.*, 2, 163–88, 2009.
102. Carvalho, A.M., Cordeiro, R.A., Faneca, H., Silica-based gene delivery systems: From design to therapeutic applications. *Pharmaceutics*, 12, 7, 649, 2020.
103. Tran, H.V., Ngo, N.M., Medhi, R., Srinoi, P., Liu, T., Rittikulsittichai, S., Lee, T.R., Multifunctional iron oxide magnetic nanoparticles for biomedical applications: A review. *Mater. (Basel)*, 15, 2, 503, 2022.
104. Busato, A., Bonafede, R., Bontempi, P., Scambi, I., Schiaffino, L., Benati, D., Malatesta, M., Sbarbati, A., Marzola, P., Mariotti, R., Magnetic resonance imaging of ultrasmall superparamagnetic iron oxide-labeled exosomes from stem cells: A new method to obtain labeled exosomes. *Int. J. Nanomed.*, 11, 2481–90, 2016.
105. Christou, E., Pearson, J.R., Beltrán, A.M., Fernández-Afonso, Y., Gutiérrez, L., de la Fuente, J.M., Gámez, F., García-Martín, M.L., Caro, C., Iron–gold nanoflowers: A promising tool for multimodal imaging and hyperthermia therapy. *Pharmaceutics*, 14, 636, 2022.
106. Yang, Z., Duan, J., Wang, J., Liu, Q., Shang, R., Yang, X., Lu, P., Xia, C., Wang, L., Dou, K., Superparamagnetic iron oxide nanoparticles modified with polyethylenimine and galactose for siRNA targeted delivery in hepatocellular carcinoma therapy. *Int. J. Nanomed.*, 13, 1851–1865, 2018.
107. Hu, X., Zhang, Y., Ding, T., Liu, J., Zhao, H., Multifunctional gold nanoparticles: A novel nanomaterial for various medical applications and biological activities. *Front. Bioeng. Biotechnol.*, 8, 990, 2020.

108. Poletaeva, J., Dovydenko, I., Epanchintseva, A., Korchagina, K., Pyshnyi, D., Apartsin, E., Ryabchikova, E., Pyshnaya, I., Non-covalent associates of siRNAs and AuNPs enveloped with lipid layer and doped with amphiphilic peptide for efficient siRNA delivery. *Int. J. Mol. Sci.*, 19, 7, 2096, 2018.
109. Graczyk, A., Pawlowska, R., Jedrzejczyk, D., Chworos, A., Gold nanoparticles in conjunction with nucleic acids as a modern molecular system for cellular delivery. *Molecules*, 25, 1, 204, 2020.
110. Torres-Vanegas, J.D., Cruz, J.C., Reyes, L.H., Delivery systems for nucleic acids and proteins: Barriers, cell capture pathways and nanocarriers. *Pharmaceutics*, 13, 3, 428, 2021.
111. Roell, M.S., A novel trojan horse for molecule delivery into plants. *Plant Physiol.*, 184, 2, 548–549, 2020.
112. Wagner, A.M., Knipe, J.M., Orive, G., Peppas, N.A., Quantum dots in biomedical applications. *Acta Biomater.*, 94, 44–63, 2019.
113. Banerjee, A., Pons, T., Lequeux, N., Dubertret, B., Quantum dots-DNA bioconjugates: Synthesis to applications. *Interface Focus*, 6, 6, 20160064, 2016.
114. Li, J.M., Zhao, M.X., Su, H., Wang, Y.Y., Tan, C.P., Ji, L.N., Mao, Z.W., Multifunctional quantum-dot-based siRNA delivery for HPV18 E6 gene silence and intracellular imaging. *Biomaterials*, 32, 31, 7978–87, 2011.
115. Ghafary, S.M., Nikkhah, M., Hatamie, S., Hosseinkhani, S., Simultaneous gene delivery and tracking through preparation of photo-luminescent nanoparticles based on graphene quantum dots and chimeric peptides. *Sci. Rep.*, 7, 1, 9552, 2017.
116. Fu, H., Hu, Y., McNelis, T., Hollinger, J.O., A calcium phosphate-based gene delivery system. *J. Biomed. Mater. Res. A*, 74, 1, 40–48, 2005.

10

Progress in Non-Viral Delivery of Nucleic Acid: Advancement in Biomedical Technology

Anil Kumar Mavi[1*§], Manmohan Kumar[2§], Amarjeet Singh[3§], Mahendra Kumar Prajapati[4], Rakhi Khabiya[5], Saurabh Maru[6†] and Dhruv Kumar[7‡]

[1]*Department of Pulmonary Medicine, Vallabhbhai Patel Chest Institute, University of Delhi, Delhi, India*
[2]*Immunobiology Laboratory, Department of Zoology, University of Delhi, Delhi, India*
[3]*Department of Zoology, Sri Venkateswara College, University of Delhi, Delhi, India*
[4]*Department of Pharmaceutics, School of Pharmacy and Technology Management, SVKM's NMIMS, Shirpur, Maharashtra, India*
[5]*Acropolis Institute of Pharmaceutical Education and Research, Indore, Madhya Pradesh, India*
[6]*Department of Pharmacology, School of Pharmacy and Technology Management, SVKM's NMIMS, Babulde, Banks of Tapi River, Shirpur, Maharashtra, India*
[7]*School of Health Sciences and Technology (SoHST), UPES University, Dehradun, Uttarakhand, India*

Abstract
Biomedical technology is a multidisciplinary field which involves application of technology and engineering principles in better understanding of biological systems at molecular level, and advances innovative methods to advance medical and bio-technological applications with an emphasis on human diseases and health.

Biomedical technology targets for improvement of malfunctioning genes accountable for disease *via* gene-therapy. Gene-therapy employs nucleic acids as functional therapeutic agents for biological treatment of a wide variety of diseases.

**Corresponding author*: anilmavi100@gmail.com
†*Corresponding author*: skmaru@gmail.com
‡*Corresponding author*: dhruvbhu@gmail.com, dhruv.kumar@ddn.upes.ac.in
§*Equal contributors*

Rishabha Malviya, Sonali Sundram and Neelam Jain (eds.) Integration of Biomaterials for Gene Therapy, (281–322) © 2023 Scrivener Publishing LLC

Delivery of nucleic acids is a highly studied approach for the cure of various ailments, and its accomplishment entirely relies on delivery. Although, there is growing need for a safe, effective, non-pathogenic means for nucleic acid delivery. Several types of delivery system exist, such as viral, non-viral, and combined-hybrid deliver system. Non-viral nucleic acid delivery system is two categories, physical and chemical non-viral delivery system. Physical methods refer to the introduction of nucleic acid by application of the physical force. Chemical methods utilize natural or synthetic carriers for delivery of nucleic acid into cells. The current chapter will put an emphasis on an in-depth insight on ways of non-viral nucleic acid delivery systems, its advantages and disadvantages, how it exhibits low immunogenicity, and its use in current biomedical technology.

Keywords: Non-viral delivery, nucleic acids, DNA, RNA, oligonucleotides, siRNA, miRNA, physical transfection, chemical transfection

10.1 Introduction

Over the past 20 years, the clinical application of biomedical technology such as gene therapy has been explored for treating or averting a wide array of genetic human diseases and cancer. The usage of gene therapy as a therapeutic approach is considerable due to its imperative capabilities for treatment of acquired and inherited lethal diseases caused due to genetic abnormalities, e.g., genetic deficiencies, cancer, cardiovascular abnormalities, autoimmune conditions, and AIDS. Gene therapy employs transfer of a gene, which is a fundamental molecular biological tool employed to maneuver cells, and/or study the function of a gene/genes, and the expression of encoded protein in *in-vitro* studies. In case of treatment of above-mentioned diseases, gene transfer is aimed at curbing genetic anomalies *via* introduction of specific genes encoding functional proteins. The ideal gene transfer delivery systems should have no or less cytotoxicity, high transfection efficiency, and target cell specificity. Besides this, gene transfer is also used for the production of bulk production of proteins as vaccines for direct therapeutic application. Though, the achievement of clinical trials of gene therapy is restricted due to numerous technical constraints. The fundamental bio-engineering demur of gene therapy is to innovate a safe and effective nucleic acid delivery system. Both viral and non-viral vectors are currently employed in clinical trials for systemic delivery of the genes. The clinical trials of gene therapy done so far relied on use of altered viruses, like adenoviruses, lentiviruses, retroviruses, and adeno-associated viruses for targeted gene delivery. The viral vector-based nucleic acid delivery system has advanced the field of gene therapy significantly, however numerous limitations are allied with the viral vector-based gene delivery system ranging from inadequate nucleic acid packaging capacity, carcinogenesis, immunogenicity, broad tropism, and difficulty in production of the viral vectors. Table 10.1 enlists the immune response triggered by viral vectors. These limitations can be addressed by using the non-viral gene delivery system, particularly with respect to bio-safety, immunogenicity, and biocompatibility. Synthetic non-viral delivery vehicles exhibit reduced immunogenicity, deliver large amount of genetic material, and are quite easy to produce in comparison to viral vectors. Synthetic (or artificial) nucleic acid delivery vehicles are able to form spatio-temporal complex with nucleic acids, and thus, shields them from extra- and intra-cellular nucleases, and hence, are considered as a substitute to viral vectors. Engineering of non-viral nucleic acid delivery systems requires great effort to reasonably design nucleic acid nano-formulations to overawe the similar cellular barriers which are possess innately by viruses to overcome cellular barricades.

Various types of nucleic acids are transfected into the cells for gene therapy in biomedical technology. The types of nucleic acids used are following (Figure 10.1):

- Deoxyribonucleic Acid (DNA)
  DNA is transfected into the cells *via* non-viral vector includes plasmid DNA (pDNA) [2]. The linear form or supercoiled form of pDNA can be transfected, and transfection of supercoiled pDNA yield higher transfection efficiency as compared to linear pDNA, however, linearized DNA integrates more efficiently into the host genome [3]. Transfection of pDNA shows reduced immunogenicity in comparison to viral DNA transfection.
- Ribonucleic Acid (RNA)
  RNA can be transfected into the cells *via* non-viral delivery system [4]. RNA transfection yields higher transfection efficiency because it doesn't require transport into the nucleus. However, in case of RNA transfection, protein expression is transient and RNA is less stable than DNA, thus, making transfected RNA prone to degradation [5].
- Small RNA and Oligonucleotides
  Small RNA is 18–200 base pairs (bp) long, and these minute nucleic acids have the capability to control post-transcriptional gene regulation and RNA alteration [6]. Examples of small RNA include miRNA (micro-RNA), (siRNA) small interfering RNA, (piRNA) piwi-interacting RNA, and (shRNA) short hairpin RNA.

Table 10.1 Immune response generated against viral vectors.

Viral vector type used	Adenoviral viral vector	Adenovirus-associated viral vector	Lentivirus	References
Innate immune response	Activation of vascular endothelial cells and platelets, production of inflammatory cytokines, cell death of macrophage	$CD8^+$ T-cell response, Activation of complement system	Enhanced IFN-α/β response triggers adaptive responses	[1]
Adaptive immune response to transgene product	Efficient $CD8^+$ T-cell response, Possible antibody formation against transgene	Less efficient $CD8^+$ T-cell response, Variable antibody response	Efficient T- and B-cell response	
Adaptive immune responses to vector	Formation of neutralization antibody, $CD8^+$ T-cells directed to viral proteins	Formation of neutralization antibody, $CD8^+$ T-cells directed to capsid	Neutralization antibody formation, T-cell response against viral proteins	

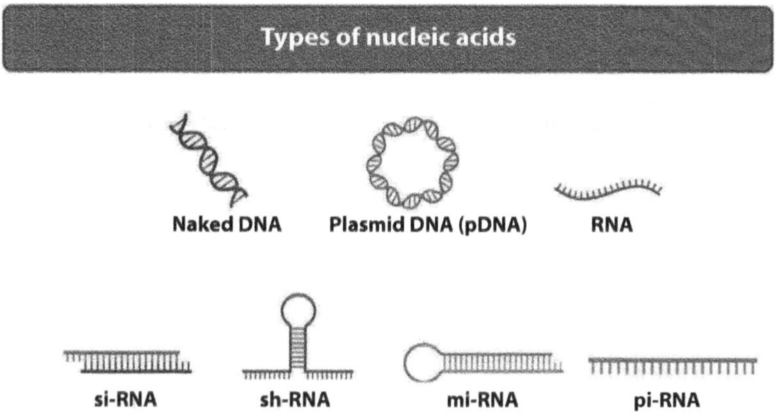

Figure 10.1 Types of nucleic acid used for transfection.

At present times, short-length oligonucleotides are also employed to study the knock-in/knock-down effects of specific gene. These can be synthesized artificially as well, and frequently utilized oligonucleotides are categorized as mimic, and antagonist. Mimic binds to a targeted mRNA which lead to transcriptional repression [7], whereas antagonist binds with the complementary small RNA (e.g., miRNA) and inhibits activity of this RNA, and hence increase expression of the gene which is targeted by that particular miRNA [7]. Recently, modified oligonucleotides are also employed such as agomirs and antagomirs, which are chemically altered in order to advance binding affinity with the target [8]. Another type of modified oligonucleotides used for transfection studies is Locked Nucleic Acid (LNA), which comprises an additional methylene bridge to enhance stability of the ribose ring. Even in few cases, transfection of LNA does not require transfection reagent [9], thus, minimizing the additional secondary effects that occurs due to transfection reagent at the time of transfection.

Non-viral nucleic acid delivery vehicles are classified into two groups – Delivery of nucleic acids *via* chemical means and nucleic acid delivery *via* physical methods (Figure 10.2).

10.2 Physical Methods of Non-Viral Nucleic Acid Delivery System

The physical means of delivery systems of non-viral nucleic acid have shown potential for transfecting the cells that are quite problematic to transfect. Delivery of nucleic acids by physical methods includes transient

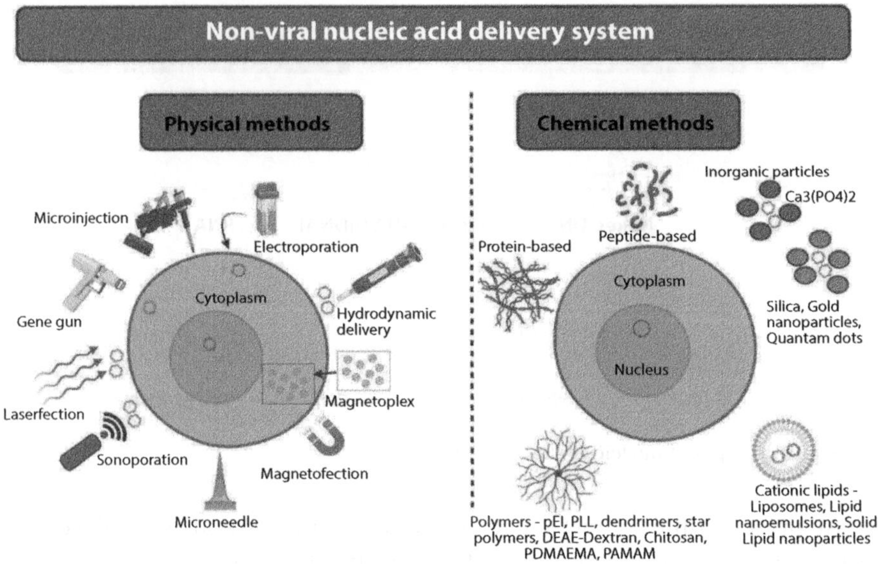

Figure 10.2 Physical and chemical methods of non-viral nucleic acid delivery.

pore development on cell's plasma membrane via mechanical disruption for delivery of naked nucleic acids. The genetic molecule i.e., nucleic acids are inserted into the cells without requirement for complex formation with any type of auxiliary chemical agent. Thus, the physical methods of nucleic acid transfer aim to deliver nucleic acids into the cells directly and with a possible attempt to circumvent problems linked with targeting, endocytic degradation pathways and immunogenicity.

10.2.1 Electroporation

Different nucleic acids are inserted into cells by employing high-voltage electric shocks. This method is known as electroporation. Electroporation was used to deliver pDNA and linear DNA expressing thymidine kinase of herpes simplex into lyoma cells of mouse in 1982 [10]. The application of an exterior electric field in electroporation can outperform the capacitance of cell's plasma membrane as well as may induce disruption (momentary but revocable) in the permeability of cellular membrane. The transient disruption of the plasma membrane enables the movement of various particles into cell including nucleic acids. This translocation comprised of simple diffusion or electrophoretic channel over cell's plasma membrane. Initially, this technique was devised for the purpose of gene transfer but now it is routinely employed with diversity of molecules such as drugs, antibodies, dyes, oligonucleotides, etc. [11].

The technical simplicity of electroporation is the ability to treat entire populations of cells, the extensive variety of applications for transfer of any macromolecule, the high rate of transfection efficiency for some cells, and the capacity to transiently or stably transfect primary and adherent or suspension cell lines are some of its benefits. Additionally, electroporation is quite helpful in transfecting the cells which are usually problematic to transfect by other modes of delivery systems, and it is better suited to meet experimental needs like cell sorting or analysis of transfectants, or the precise control of agents that cells are exposed to. The need for specialized equipment, the need to empirically determine the conditions for effective transfection as well as its efficacy in comparison to other methods of transfection, the sizeable amounts of DNA needed, and the low viability of treated cells were all historically considered to be disadvantages of electroporation. Although recent research has offered solutions for the latter two issues, it is feasible that standard electroporation technique can be altered to some extent so that it affects membranes of the cellular organelles with negligible recognizable consequence on cell's plasma membrane by using narrow brief pulses of quite high-voltage electric currents. Different pulse amplitudes can be used in electroporation to permeabilize cell's plasma membrane (the increase in the amplitude, the more area permeabilization will be there), whereas the duration of the pulse or pulse number affect the grade of permeabilization. The up-take of minor molecule isn't affected by charge since diffusion is used to carry such molecules into the cell. The uptake of heavy molecules, however, is dependent on their size, despite the fact that electroporation has been used to successfully load molecules like dextran [12]. Electrophoretic forces have been demonstrated to be crucial for the effective introduction of DNA [13]. Now, electroporation has been used extensively in the clinical trials and proved to be an effective means of nucleic acid delivery. In the year, 2014, the first electroporation-enhanced DNA vaccine Phase II trials were announced. Furthermore, DNA vaccination via intradermal and intramuscular route introduced by electroporation against HIV proved to be safe, tolerated by immune system and acceptable by the patients [14]. The delivery of Ebola and Zika DNA vaccines via intradermal route introduced by electroporation were demonstrated to be safe in healthy volunteers [15, 16].

10.2.2 Laserfection/Optical Transfection/Optotransfection

Cellular membranes can be momentarily permeabilized by laser light, which enables the introduction of almost any molecule existing in the surrounding liquid. Numerous techniques for using various types of light for laserfection have been devised based on this phenomenon. Laserfection can be divided into two groups: only laser light is used for delivering the nucleic acids; and chemical agents are coupled with optical transfection for the delivery of genes (nucleic acids).

Uncertainty still exists regarding the process through which chemicals gain entry into cells at the time of laserfection. Though, it seems that laser light or laser beam-induced shock wave induces a minute temporary pore in the cell membrane, and causes mechanical tension to be induced which facilitates the pore formation into the cell [17].

The membranes of endocytic vesicles are used to enclose photosensitive compounds for photochemical internalization, a subsequent optical transfection technique. Laser light is used to activate photosensitizers when a targeted molecule is pulled into an endosome through endocytosis. This causes the generation of ROS, and damages endosomes' membrane, releasing molecule's contents (such as nucleic acids) into the cytoplasm.

Optical transfection appears to be as effective as conventional transfection techniques. One of the main benefits of laserfection is that it can transfect single cell, and is easier to use than other techniques that can achieve a similar result [18]. Studies contrasting the effects of injecting mRNA into the dendrite and neuronal cell bodies have demonstrated the efficacy of optical transfection [19]. It has been demonstrated that the whole transcriptome may be transferred into other cell types via laserfection, which reprograms the receiving cell [20].

10.2.3 Gene Injection (or Biolistic Methods)

A mechanical technique called biolistic transfection has probable uses in widespread array of cellular and tissues types. Since it can pass through cell barriers, it was initially employed to transfect plant cells. The nucleic acid (such as DNA or RNA) immobilized on the sub-cellular sized particles is used in biolistic transfection. Then, using a "gene gun," these immobilized particles are augmented to a high velocity and fired into the cells. Therefore, it is independent of the biochemical characteristics or cell growth rate. The biolistic approach has immense potential in gene therapy applications, because it can be employed for both the studies such as *in-vitro*, and *in-vivo* cellular transfections experiments [21]. Only biolistic method of delivery of nucleic acid permits transfection of desired nucleic acids into deeper tissues, which enables the entry of nucleic acids into cells underlying into deeper layer of the tissues.

The ability of the biolistic methods to penetrate physical barriers like the epidermis, the possibility of multiple uses within a single sample, transfection of huge numbers of cells in single use, the potential for co-transfection of simultaneously two or more nucleic acid such as DNAs in a single injection via gene gun, as well as its speed and ease of use are its greatest advantages [22]. The gene cannon itself is unfortunately the method's weak point, even though costs related to its use are minimal [23]. Nano-biolistic is a revolutionary variation of the biolistic method that uses particles that are about 40 nm smaller in size (normal biolistic uses particles that are about 1 m in diameter). When evaluating cellular architecture in small cells like HEK293 cells and in situations where tissue damage is a concern, this approach is just as successful as traditional biolistic transfection [24].

Jet injection is one of the physical pressures employed in the biolistic procedures to introduce genetic material into a cell. Jet injection is rapid entry of the nucleic acids into differentiated cells or tissues for transfection using a compressed column of fluid (jet). The force that allows the nucleic acids to enter into the cells in this method is provided by the jet itself. The force of injection is connected to the efficacy of the transfer of nucleic acids by jet injection. The transfected cells are located in a constrained area surrounding the jet injection's route. It is possible to find expressing cells up to 2 cm away from injection site. For instance, the skin surface can be treated with the jet injection to affect underlying muscle or fat cells. The capacity, jet's launching force, and tissue being injected all affect how long the jet injection route is. The force of the jet decreases as it passes through tissue because of resistance from the tissue. The solution pools when the force is lost. Near this gathered material, there are not many transfected cells to be found. Both explanted tissue cultures and live animals can be used for the transfection of nucleic acids by jet injection [25–27]. It has been employed to deliver nucleic acids in pigs, mice, sheep, chickens, and rabbits with significant success, and barely caused any tissue injury.

There are many research uses for temporary introduction of pDNA via jet injection. First, using reporter genes like luciferase, 3-galactosidase, or chloramphenicol acetyl transferase (CAT), ideal situations for jet injection have been recognized [28]. The method can then be used to assess parameters specific to a tissue or a differentiation that cannot be reached in cells grown in tissue culture. For instance, certain groups of genes, such as those encoding milk proteins, express differently depending on a cell's developmental stage or extracellular matrix. Reporter genes coupled to the promoters of the protein encoding gene are utilized to identify promoter binding elements in differentiated cells such as epithelial cells of mammary glands afterward jet injection or to assess tissue-specific promoter activity [25], examine the expression of RNA and proteins [27]. This method could be employed to research tissue-specific protein processing or alternative splicing. For instance, when sheep that are breastfeeding have their mammary epithelial cells transfected, the produced protein is secreted into the milk [27].

The method's affordability, versatility, and simplicity are its main benefits. There is no need for additional equipment besides the injector. It makes use of inexpensive, quickly-preparable plasmid DNA. The kind of answer is not crucial. Injectors that are manufactured are sold for under a thousand dollars. These gadgets are incredibly portable. Animals and tissues from a range of species can be treated using this method. The jet-injection method is currently particularly beneficial for three research applications: genetic investigations where tissue-culture is not an option, genetic immunization, and research studies that use subtle methods to assess the expression of the genes. The main drawback of this procedure is transfected cells travel in confined track. The restricted channel of the transfected cells must segregated among neighboring tissue for investigating the expression of the genes, or several injections should be employed to transfect the huge portions of the tissues.

10.2.4 Magnetic Nanoparticles (MgNPs)

The process of magnetofection is quite similar with usage of transfection reagents such as Lipofectamine, pEI, FuGENE HD. In this process, the resultant complex formed between nucleic acid with cationic polymer-coated magnetic nanoparticles is referred as magnetoplex. Magnetic force is employed to control the behavior of magnetoplex, and then it enters into the cells *via* endocytosis, thus, increasing transfection efficiency. The magnetic nanoparticles mostly used in this process include iron oxide nanoparticles. Various surface modifications have been done in iron oxide nanoparticles in order to reduce cytotoxicity and increase transfection efficiency, such as coating agents (lauroyl sarcosinate, oleic acid, Pluronic F-127, Chitosan, TAT peptide, MCM48, LEG, PLL, PEI etc.) [29]. The delivery of hVEGF-siRNA bound to iron oxide nanoparticles into the nude mice, significantly reduced the growth of hepatocellular carcinoma [30]. PEI-coated magnetic nanoparticles efficiently bind with the nucleic acids, remain stable in solution along with high transfection efficiency. However, the major drawback of magnetofection is the need to design magnetic nanoparticles complexed with nucleic acid, and as well as strong external magnetic force is also required, and besides this, diminutive particle size

(<50 nm) is not suitable for magnetofection, and too large particle size (>5 µm) inhibit entry into the cells.

In the biomedical technology nowadays, magnetic nanoparticles (MNPs) are frequently utilized for diagnostic analysis, medication delivery, and bioimaging. Pure metal nanoparticles like Iron (Fe), Cobalt (Co), and Nickel (Ni) are among the most common types of magnetic nanoparticles, along with that nanoparticle of metal alloys, e.g., iron-platinum nanoparticles and cobalt-platinum, nano-ferrites like magnetite or magnetic oxide (Fe_3O_4) and hematite (Fe_2O_3), and metal-doped iron oxides like manganese ferrite ($MnFe_2O_4$), cobalt ferrite ($CoFe_2O_4$), and zinc ferrite ($ZnFe_2O_4$). Ferromagnetism is required for magnetic nanoparticles to be effective in magnetofection. $CoFe_2O_4$ and $MnFe_2O_4$ nanoparticles exhibited much sturdier magnetism as compared with other magnetic materials; however the toxicity of these two nanoparticles restricts their current usage in biomedical technology applications. On the other hand, frequently used Fe_3O_4 and γ-Fe_2O_3 nanoparticles showed excellent super-paramagnetic features, and these nanoparticles demonstrated biocompatibility, and thus are safe for biological use. The preparation and synthesis of magnetic nanoparticles includes several steps such as co-precipitation, decomposition via thermal methods, hydro- or solvent-thermal and formation of micro-emulsions. Several distinct physical assets, especially super-paramagnetism and high magnetic receptiveness are responsible for potential application of MgNPs in biomedical technology [31]. Super-paramagnetic nanoparticles only exhibit magnetism in existence of applied external magnetic field. Magnetic nanoparticles are used in other biological applications as well such as separation via magnetic ways, mostly for proteins, DNA, and another biological particle extraction [32]. The super-paramagnetic nanoparticles can produce huge quantities of heat instigating due to relaxation of an alternating magnetic field. Hence, magnetic nanoparticles can also be exploited as hyperthermia agents, and thus utilized in certain disease treatments [33]. The MgNPs are also employed as contrast materials for magnetic resonance imaging through fluctuating relaxation duration of tissues in order to affect the strength of tissues' signal, and recover contrast of miscellaneous biological samples in MRI [34]. Furthermore, magnetic nanoparticles can also be utilized as nucleic acid carrier system to improve transfection efficiency of nucleic acids [35]. This technique can augment the efficiency of transfection and precisely transfer nucleic acids into targeted site. The application of MgNPs could encourage the gene transfection efficiency by upholding the delivery agents at target site via the applied external magnetic field. Plank *et al.* and Mah were the first to introduce the magnetofection for nucleic acid delivery [36, 37].

10.2.5 Sonoporation

In order to quickly permeabilize cell membranes and introduce nucleic acids into the cells, ultrasonic waves are used and this process is known as sonoporation. Microbubbles, which are small (1–10 μm) gas-filled spheres stabilized by shells, help with the scattering of the sound and lower the threshold for the cavitation of the membranes. Ultrasound waves cause the breakdown of the microbubbles which are travelling with the desired nucleic acids via a particular area in the blood circulation, and thus localization of gene delivery is made possible. According to a published report, the usage of needles of successively narrower gauge for injecting microbubbles via intravenous (i.v.) route enhances transfection efficiency [38]. Diabetes, cardiovascular conditions, and some malignancies have all been treated with sonoporation in *in-vitro* and *in-vivo* models. Additionally, it has been demonstrated to stimulate bone tissue regeneration and to increase blood-brain barrier permeability [39, 40]. Transfection efficiency is improved by altering microbubbles through polymer coating. In this way, Un *et al.* performed DNA vaccination against metastatic melanoma using lipoplex bubbles with mannose [41]. Microbubbles containing location-specific ligands that guarantee nucleic acid transport to specified tissues including the nuclear import can be used to administer targeted gene therapy [42].

10.2.6 Hydrodynamic Delivery

The delivery of the nucleic acids via hydrodynamics is one of physical non-viral ways established initially for intracellular transfer of gene but later on, it is employed to transfer the various types of nucleic acids and macromolecules, e.g., oligonucleotides, ribonucleic acid, polypeptides, and certain molecules which are impermeable to transfer into the cellular membrane. This process comprises a speedy injection of a massive amount of a solution (i.e., controlled hydrodynamic pressure) into a vasculature such as capillaries in order to enable the transference of molecules such as nucleic acids into the parenchyma cells. Earlier reports demonstrated successful introduction of nucleic acids using hydrodynamics, such as introduction of pDNA into the skeletal muscle of rats [43], and systemic nucleic acid delivery in a mice model [44, 45]. The delivery of nucleic acids via hydrodynamics was developed on the basis of advances in the understanding of structural and functional properties of capillaries. Sole reason that parenchyma cells are targeted via hydrodynamic delivery is because of close association between endothelium of blood capillary and parenchyma cells, which allows instantaneous access of the nucleic acid into parenchyma cells on the disruption of endothelial layer. It produces pores in membrane of parenchyma cells from which nucleic acid gain entry into cell. The overall efficiency of gene delivery via hydrodynamics depends upon the structure of capillaries, cellular architecture surrounding the capillary and the applied hydrodynamic force [46]. Sinusoids or fenestrated capillaries are much more sensitive to the hydrodynamic nucleic acid delivery as compared to the continuous capillaries because sinusoids exhibit large breaches between adjacent endothelial cells, and have thin or no basal lamina. Nucleic acid delivery via hydrodynamics is used as an experimental research technique for transferring nucleic acids into the cells to investigate their physiological, biochemical, and therapeutic action in context of an animal model. Introduction of nucleic acid by hydrodynamic injection of pDNA is used in evaluating connection of particular genes responsible for the onset or suppression of pathophysiological conditions. This means of non-viral nucleic acid transfer provides a significant transfection efficacy, as evident in some studies, whereby long-lasting gene expression was observed in transfected animals and reaches at the therapeutic levels [47–49].

The most successful application of gene delivery using hydrodynamics is seen in hepatocytes of rodents. Additionally, delivery of nucleic acids via hydrodynamics is also successful in liver, skeletal muscle, tumor cells, myocardium, and kidney. However, the transfer of nucleic acid via hydrodynamics is well endured by animal models, but the influences of this process on animal models for tail vein injection are quite apparent. The complications associated with hydrodynamic delivery included immobility and labored breathing for a short duration immediately after the injection into the mice. Furthermore, rats reacted more severely and observed they stopped breathing after the injection [50]. The breathing ability was stimulated by a tender massage of the abdominal area of the animal and facilitated the animal recovery. And these critical signs of nucleic acid delivery via hydrodynamics on cardiovascular functioning ensue and recede speedily [51, 52]. Hydrodynamic gene delivery is utilized in some gene therapy studies as a possible therapeutics such as Fabry disease, hemophilia, phenylketonuria, myocarditis, organophosphate toxicity, liver regeneration, and different types of cancer.

10.2.7 Microneedle

Microneedles can transfer both the low-molecular weight and high-molecular weight molecules into the systemic circulation including nucleic acids via penetration of the stratum corneum layer of the skin. It is a minimally-invasive delivery system. Microneedles are shaped in diverse needle sizes ranging from 25–2000 μm in length [53] with various shapes and intervals for attachment with the support to the base. The delivery of nucleic acids via microneedles is easy to use, patient-friendly, and pain-free, and this technique has the potential for self-administration. The designing of microneedles involves various molecules like glass, stainless steel, ceramics, polymers, and dextrin. Nowadays, microneedles have been employed for delivery of nucleic acids via transdermal route. Microneedles can reproducibly and proficiently transfer the nucleic acids into the skin for treating various genetic skin diseases, hyper proliferative diseases, and cancers [54]. Various types of microneedles are available such as coated microneedles, solid microneedles, hollow microneedles, and dissolved microneedles which are in use for the treatment of skin cancers including squamous cell carcinoma, melanoma, and cell carcinoma [55]. In case of the delivery of nucleic acid via microneedles; nucleic acids are encapsulated using liposomes, polymers, cell-penetrating peptides and nanoparticles for effective nucleic acid transfer into cell. Although microneedles proved effective in delivering nucleic acids including siRNA into the skin, but various complications still need to addressed. For example, first - the motorized solid microneedles are costly and requires two-step application process, and besides this, skin need to be punctured for quite a long-time duration in order to achieve high gene expression which may cause objectionable damage to the skin [56]. And the collapse of the microneedles in or under the skin is additional safety concern. Second – the microneedles are composed of biodegradable components which almost liquefy completely after the application into the skin, thus biocompatibility issue can be there because the degradation of such microneedles solely depends upon the interior assembly of the polymer material used for designing. Moreover, the benefit of pain-free application makes it an optimal nucleic acid delivery system and improvisation in fabrication of microneedles, usage of biocompatible materials in designing microneedles can greatly increase the potential of microneedles in biomedical technology.

10.2.8 Microinjection

Microinjection is a direct method for the introduction of the nucleic acid either into cell's cytoplasm or nucleus. This microsurgical process is usually conducted on a single cell, by utilizing a glass needle (glass microcapillary pipette), micro-manipulator, and a microinjector. Microinjection is a widely used method that has been used to investigate a number of experimental problems in all kinds of cells, and it is widely used to develop transgenic animals. It is a quintessential approach in the study of mammalian oocytes and early embryos. In addition to nuclear microinjection of nucleic acid (DNA), it is used to transfer the nucleic acid into mitochondria also. It is especially useful for transfecting cells like mesenchymal stem cells or smooth muscle cells that are challenging to handle using other techniques [57, 58]. A technique for cloning and the creation of transgenic animals called nuclear transfer was developed from a single-cell microinjection. In nuclear manipulation, the nucleus of somatic cell is separated, and anticipated nucleus is transferred at the level of the single cell. External nucleus may be microinjected or fused into the host oocyte. The donor cells' stage of differentiation has a significant impact on this process' effectiveness [59]. Nuclear transplantation creates an autologous ESC line which is derived primarily from a cloned embryo, as opposed to reproductive cloning. In order to create useful ESCs for cell replacement, therapeutic cloning is done. Theoretically, single-cell microinjection has a transduction effectiveness of approximately 100% in healthy, successfully transfected cells [60]. As a result, the only independent variable in a properly executed microinjection technique is the injected substance.

Precision dosing of the injected material is a benefit of the microinjection approach, which is crucial for introducing many foreign substances, such as peptides or polypeptides, into a single cell [61]. Additionally, when compared to chemical or viral transfection, microinjection has lesser cytotoxicity and enables the discriminatory transfer of nucleic acids into cytosol or nucleus, exclusively in 1° neurons of human [58]. More than 100–200 cells cannot be injected using microinjection for each therapy. However, this quantity is typically sufficient to be statistically significant even in a 1×10^6 cell culture. Despite the possibility of transcription analysis, the modest number of transfected cells may restrict more extensive protein research. Since the injection itself puts physical stress on cells, appropriate controls are necessary to confirm that injection doesn't damage viability of cell. Additionally, the non-integrated genome material that is delivered into dividing cells during transfection is diluted with individual cell division. Development of transgenic animals, IVF, transduction-challenged cells studies, and the effects of injected molecules in mixed cell cultures are only a few of the research and clinical sectors where microinjection has been widely used [58]. Reports suggest delivery of adenoviral vectors in nude mice effectually in a regulated method in xenotransplants [62].

10.3 Advantages and Disadvantages of Physical Transfection

Certain advantages associated with physical transfection are – i) No requirement of carriers, simple and safe methods for nucleic acid deliver; as nucleic acid can be directly used for the delivery thus there is no need of additional chemicals and transport carriers to deliver nucleic acid inside the cell, ii) good efficiency and reproducibility; as already stated that no additional chemicals are required for transfer of nucleic acid into the cells, thus nucleic acids aren't altered and yields good reproducibility and efficiency iii) precise delivery of nucleic acids; nucleic acids delivered via physical methods are delivered at the targeted site inside the cells or tissues, along with minimum loss in amount of delivered nucleic acid.

Disadvantages in using chemical transfection as non-viral nucleic acid delivery systems are – i) Degradation of nucleic acid may usually results in reduced transfection efficiency; this is major problem in delivering nucleic acid via physical methods. The nucleic acid which is to be delivered isn't complexed with any chemicals or carriers which renders them susceptible to the action of nucleases inside the cells and outside the cells. This lead to degradation of the nucleic acids and thus negatively affects the transfection efficiency, ii) inapplicable for deep-tissue targets; although physical methods precisely targets transfer of nucleic acids into the tissues or cells but it can't transfer nucleic acids into deeper layer of tissues that is another major disadvantage of this method because the cancerous cells, and mutant cells which are present in the deeper layer of the tissues will not respond to the gene therapy even if the nucleic acids are introduced by physical methods, iii) incorrect usage may lead to tissue damage; delivery of nucleic acids via physical methods need precision and accuracy because if these methods are used incorrectly, it may damage the tissues which may be irreparable in certain cases or requires surgical treatments.

10.4 Chemical Methods of Non-Viral Nucleic Acid Delivery System

Chemical means of non-viral nucleic acid delivery systems have been proved quite effective in transfecting the cells. Delivery of nucleic acids by chemical methods include coating of the nucleic acids with inorganic particles, nanoparticles, polymers, peptides or lipids in order to gain entry into desired target cell by fusing with cellular plasma membrane or through endocytosis Thus, the genetic material that is to be inserted the into cells is required to form a complex with the chemical agents prior to the introduction into the cell. Thus, the chemical means of delivery of nucleic acid delivery aim to deliver nucleic acids into cells via chemical agent-nucleic acid complex, and with a possible attempt to circumvent problems linked with endocytic degradation pathways, digestion by enzymes in the cytoplasm and immunogenicity.

10.4.1 Inorganic Particles

Inorganic materials are much steadier as compared to organic materials, and have been utilized as non-viral nucleic acid carriers. The first reported non-viral delivery of a gene was done using calcium phosphate around 1960s. But now, commonly used inorganic particles are – gold nanoparticles, silica-based systems (like mesoporous silica nanoparticles), and quantum dots [63]. Inorganic particles incorporated *via* nanomaterials, and certain features of nanoparticles such as high mono-dispersity, and stable suspension in aq. solutions allow them to be functionalized on the surface with amino groups (NH_3 group) of nucleic acid such as DNA [64]. Inorganic nanomaterials have become an attractive therapeutic nucleic acid delivery vehicle due to numerous advantages – accurate control of the size, tuneable surface properties, and large loading efficiency [65].

a) Calcium Phosphate

Calcium phosphate is a positively charged molecule, particularly used for the introduction of the nucleic acid into mammalian cells. Originally, transfer of nucleic acid into the cells using this chemical was developed by Vander Eb and Graham [66]. Calcium phosphate forms a precipitation complex with nucleic acid. The resultant complex of nucleic acid and calcium phosphate is mixed with phosphate buffer in a dropwise manner in order to provide proper aeration for avoiding the formation of clumping of the nucleic acid. The use of calcium phosphate as a non-viral transfection material is accepted by fact that it is non-toxic, cheap, easy to synthesize, biocompatible and easy to degrade [67].

The complex of calcium phosphate and nucleic acid gain entry into cells *via* caveolae- and clathrin-mediated endocytosis [68]. After internalization, calcium phosphate- nucleic acid complex remain stable till late endosome, however, calcium phosphate dissolves due in the lysosome. The release of component ions inside lysosome raises osmotic pressure, leading to burst of lysosome, and release of nucleic acid into cytosol [67, 68]. Calcium phosphate mediated transfection is suitable for certain cell lines; but efficacy of such transfection depends on cell type. The disadvantages include that some cells (e.g., HeLa cells) are supersensitive to calcium phosphate transfection procedure, and efficiency of DNA uptake in some cells are very poor. Several other factors such as consistency of the reagent and minute pH variations can reduce the transfection efficiency and is critical for reproducibility. The raising concern is that it also engenders DNA mutation at very high rate which occurs in quick duration of time. Such mutations are principally substitutions and deletions of bases [69]. In order to resolve these issues, numerous modifications have been done such as nanocomposite particles of calcium phosphate efficiently delivered pDNA and thus, showed high transfection efficiency and also exhibited high biocompatibility.

b) Silica Nanomaterials

Mesoporous compounds are attractive targets for precise loading of the cargo, its release due to well-defined nanoscale-sized pores and chemical stability [70]. Mesoporous silica nanomaterials gained much attention for transfer of siRNA and DNA due to easy binding of nucleic acids on pre-adsorbed cationic polymers (e.g., pEI), easy endosomal escape based on presence of charged moieties that lead to membrane destabilization [71]. Synthesis of ultra-large pores by incorporating mesoporous silica nanoparticles with certain agents at higher temperature increased genetic material loading and transfection efficiency [72]. Li *et al.* reported reduced cytotoxicity and efficient siRNA transfection when siRNA incorporated under dehydration conditions [73]. pEI-coated mesoporous materials delivered siRNA along with chemotherapeutic agent in drug-sturdy tumor cell [74]. Mesoporous materials have been also used to deliver peptides, therapeutic drugs and protein-based vaccine [70] as well but its role in nucleic acid delivery has been discussed here only. One such non-viral vectors is ORMOSIL nanoparticles which allows effective nucleic acid transfer, and it showed a prodigious potential for efficient therapeutic operation of neural stem cells [75].

c) Gold Nanoparticles (AuNPs)

Gold nanoparticles serve as the striking target for nucleic acid delivery due to various advantages. Gold nanoparticles cab be modulated in scalable manner with less-size dispersity, cytotoxicity, biodistribution, highly adjustable optical properties, and removal from the host system can be regulated by controlling size of particle, and functionality of surface [76–78], and thus, gold nanoparticles are the foremost metal nanoparticles used for transfer of nucleic acids in biomedical applications. Straightforward formulation and soft surface chemistry of gold nanoparticles permits it to tailor with different biomolecules which is quite impossible with another metal nanoparticles like magnetic nanoparticles. Metallic core of gold nanoparticles provides a solid support system to the nucleic acids, and also allows it to withstand degradation even after inestimable dilutions. Gold nanoparticles are currently utilized for gene therapy for *in vitro* studies, and in pre-clinical animal model studies because of reduced cytotoxicity, high genetic material encapsulation, effective uptake and enhanced half-life [79]. The immunogenicity of gold nanoparticles is 25-fold lesser than lipid-based non-viral delivery vehicles. The contributions of gold nanoparticles-mediated nucleic acid transfer have enhanced the treatment of cancer drastically. For example, gold nanoparticles efficiently delivered PLK1-siRNA in the cancerous cells for blocking miotic progression of cell cycle [80].

d) Quantum Dots
These are semiconductor particles of nanometer-size, which are produced from heavy metals such as Lead selenide (PbSe), Cadmium selenide (CdSe), and Greenockite (CdS). These traditional quantum dots usually result in high cytotoxicity, and poor biocompatibility. Photoluminescent carbon quantum dots (CQDs) gained much appreciation due to their reduced cytotoxicity, biocompatibility, surface modifications and strong fluorescence. CQDs showed efficient capacity for condensation of pDNA, high transfection efficiency and tracking of the cellular trafficking pathway [81]. The coating of quantum dots with natural biopolymer i.e., chitosan lowered toxicity of quantum dots. PEGylated quantum dots efficiently delivered F3-siRNA to the HeLa cells [82].

10.4.2 Lipid-Based Nucleic Acid Delivery System

a) Cationic Lipids
Transfection of cells with cationic (positively charged molecule) lipids is known as lipofection or liposome transfection [83]. Liposome transfection employs a cationic lipid (or liposomes) that are amphiphilic in nature, interacts with nucleic acids such as DNA/RNA which contain negatively charged (anionic) phosphate moieties, and with the cell membranes *via* electro-statistic interaction. It permits to transfer RNA, DNA, and protein molecules of a varied array of molecular weight, and this method is routinely used for transient and stable transfection [83]. In case of transient transfection, nucleic acid doesn't integrate into the host cell genome [84]; the transfected nucleic acid in this case can be a plasmid DNA or oligonucleotides. And due to non-integration into genome of host, expression

of transgene is lost after replication of the host cell. Transient transfection is normally employed for short-term research work in order to study the effects of a particular knock-in or knock-down gene. In case of stable transfection, introduced nucleic acid integrates into the host genome or maintain as an extra-chromosomal element. The transgene expresses constitutively even in the replicated cells. Stable transfection is usually employed for long-term genetic and pharmacological studies for the production of a particular protein at a large-scale. The cationic lipid/liposome-nucleic acid complex arrives *via* endocytosis into cell, followed by release of complex into the cytoplasm. When mRNA is used for liposome transfection as the nucleic acid component, it remains in the cytoplasm and when DNA is used as the nucleic acid component for liposome transfection, it needs to be moved into the nucleus. Worldwide, cationic liposomes are utilized as non-viral nucleic acid transfer vehicles for human clinical trials [85], and cationic liposomes have the potential of carrying a large of piece of nucleic acid approximately up to 1×10^6 base pairs [85]. The most regularly used cationic lipid for liposome transfection is Lipofectamine/Lipofectamine 2000, which consists of a 3:1 mixture of DOSPA, and DOPE. Lipofectamine/Lipofectamine 2000 form complex with nucleic acids in order to permit them to overawe electrostatic repulsion of cellular plasma membrane. In addition, the molecules of cationic lipid of Lipofectamine are designed with the help of helper/neutral co-lipid, and this construction upsurges the transfection efficacy of pDNA, mRNA, and siRNA in both dividing and non-dividing cells [86, 87].

Another central constituent of cationic lipid is linker that is responsible for its stability, cytotoxicity, transfection efficiency of liposomes, and biodegradability of lipid carriers at the target sites. The different types of linkers are used in liposomes preparation are: ether, ester, amide, carbonyl, ester, disulfide, acylhydrazone, urea and carbamate etc. [83]. Ether- and amide-based linkers are more stable as compared to ester-based linkers. The amide-linked lipids aids in improved combination of liposomal membrane, and increases uptake of nucleic acid by cell in comparison to the ester-based linkers [88]. Interestingly, monodisperse liposomes display stronger encapsulation efficiency which allows to synthesize nano-liposomes for the encapsulation of siRNA [89]; thus, making the monodisperse liposomes a fascinating approach for non-viral delivery of nucleic acids. Liposome transfection offer various advantages due to the usage of extensive variety of cells, ranging from 1° cells to numerous cell lines, and exhibits high efficiency as well.

However, the transfection efficiency of lipofection relies on several constraints such as the quantity and quality of nucleic acid (DNA/RNA/

siRNA/pDNA), the ratio of transfection reagent and the amount of nucleic acid, incubation period of lipid-nucleic acid complex and the density of cells at the time of lipid-nucleic acid complex addition [83, 90].

b) Lipid Nano-Emulsions
Nano-emulsions are a colloidal particulate system, solid spheres ranging from 10-1000 nm. So, a lipid nano-emulsion is composed of oil, water and an emulsifying agent (i.e., surfactant). It is considered superior in terms of reduced cytotoxicity and stability as compared to liposomes [91].

c) Solid Lipid Nanoparticles (SLNPs)
Solid lipid particles (SLPs) are synthesized from lipids which are solid at both room temperature (RT), and body temperature. SLNPs are quite effective in shielding nucleic acid from nuclease degradation [92]. SLNPs are used for targeted transfer of therapeutics into macrophages. And for the study, a mannan-based phosphatidylethanolamine (PE)-associated ligand is manufactured which was used to formulate Man-SLNPs–DNA *via* surface adjustments of DNA-loaded cationic SLNPs. The resultant complex served as an effective transfer system for targeted delivery of nucleic acid. The marvelous accomplishment of the vaccines from Moderna (also known as Spikevax®) and Pfizer-BioNTech (also known as Comirnaty®) [93] in battling SARS-CoV-2 virus has confirmed the worth and swift translational potential of lipid nanoparticles (LNPs).

10.4.3 Polymer-Based Nucleic Acid Delivery System

Many diverse types of polymers were efficaciously investigated for non-viral nucleic acid transfer in biomedical technology. One of the frequently used methods for transfection is the usage of cationic polymers, which are verified as the non-viral vehicle for the nucleic acid transfer [94]. Because of their positive charge, cationic polymers served as another substitute for viral carriers for transfer of pDNA, mRNA, siRNA into the cells. The negatively-charged nucleic acid forms complex with cationic polymers *via* electrostatic interactions. Cationic polymers neutralize the negative charges of nucleic acids to be delivered resulting in the formation of nanoscale polymer-nucleic acid complex (referred as polyplex). The direct introduction of naked nucleic acids into the cells is rapidly digested by cellular nucleases. And the complex of nucleic acid with cationic polymers prevents nucleic acid against such type of degradation [64]. The cationic polymers are synthetic non-viral nucleic acid delivery vectors that trigger efficient DNA compaction as compared with the cationic lipids. Thus, these advantages make them an appropriate nucleic acid delivery vehicle.

Polymers are synthetic non-viral nucleic acid delivery vehicles which have capability to transfer nucleic acids into cells [95]. Most common cationic polymers are linear or branched depending upon the structure. The most generally used polymers for non-viral nucleic acid delivery are polyethylene imines (PEIs), poly L-lysine (PLL), dendritic polyamidoamine (PAMAM), PDMAEMA, Diethylaminoethyl-dextran (DEAE-dextran), and chitosan etc. [96].

a) Polyethylene Imine (pEI)

It was first announced by Behr [97], and regarded as exemplar for polymer-mediated transfer of the transgene into the cells, since it was initial primitive polymers which exhibited high transfection efficiency [98]. pEI is an organic polymer, and categorized into two forms – linear pEI and branched pEI, and both are frequently used for transfection for *in vitro* and *in vivo* research work [99]. Linear pEI owns 1° and 2° NH_3 groups, and branched pEI contains 3° NH_3 groups. Because of high charge concentration, it forms complex with delivering genes, and it also serve as a proton buffer for buffering low endosomal pH, and resulting in osmotic imbalance, facilitating rupturing of endosomal membrane, and escape of polyplex into cytoplasm [100]. Though, biosafety of pEI remains a matter of concern for its clinical trials. The high charge density of pEI could induce necrosis of the cells both *in vivo* and *in vitro* studies [100, 101].

b) DEAE-Dextran

It is a branched cationic polymer that forms intricate complex with negatively charged groups present on the nucleic acids. DEAE-dextran is utilized as conjugating mediator for nucleic acids and as coating agent of liposomes as well [102]. DEAE-dextran contains tertiary amines. The mixture of nucleic acid such as DNA with DEAE-dextran forms a positively charged complex *via* electrostatic interactions that sticks with negatively charged cellular plasma membrane. Then, resulting complex gains entry into cytoplasm *via* osmotic shock persuaded by glycerol, DMSO, or endocytosis. Introduction of the transgene using DEAE-dextran is easy to perform, and inexpensive. Several studies reported that transfection using DEAE-dextran as a complexing agent is capable in delivering both the nucleic acids, i.e., RNA and DNA [102, 103]. However, the foremost shortcoming of DEAE-dextran mediated transfection is toxicity, i.e., the enhanced volume of DEAE-dextran is lethal in transfected cells [64]. In addition to this, delivery of nucleic acid using this method is suitable only for transient transfection and even the transfection effectiveness varies between cell types used.

c) Dendrimers

Dendrimers are nano-sized, synthetic, branched positively charged hydrophilic polymers containing groups at terminal ends which form complex with nucleic acid (DNA or antisense oligonucleotides), and characterized by spherical symmetry [104]. The complex formed in this case is known as dendriplexes [104]. The terminal groups of exterior phases of dendrimers can be chemically modified. Dendrimers possess reduced toxicity to the cells and dendrimer derivatives aids in overcoming the cellular obstacles through nanoscale hole formation in plasma membranes of the cell [105, 106]. Poly(amido amine) (PAMAM) dendrimers and Poly(propylene imine) (PPI) dendrimers are the two most well-studied dendrimers, and poly(L-lysine) and phosphorus dendrimers (P-dendrimers), and carbosilane dendrimers are well-exploited for drug and nucleic acid delivery systems, and even for antibacterial and antiviral agents. PLL dendrimers triggers apoptosis and thereby hinders in the growth of cancerous cells, and healthy cells remain harmless to the action of PLL dendrimers [107]. Carbosilane are compounds which contains silicon and carbon. Numerous types of carbosilane dendrimers have been developed using silicone chemistry. They can be anionic and cationic depending upon the molecules present at the periphery. Carbosilane dendrimer 2G-03NN24 complexed with Nef-siRNA effectively inhibited HIV replication [108], Carbosilane dendrimers 2G-03NN24 and 2G-NN16 can employed for transferring nucleic acids for the treatment of HIV infection. Carbosilane dendrimer 2G-03NN24 decreased M2 polarization of macrophages and thus, could be used for antitumor therapy [109].

d) Star Polymer

Star polymers are simple branched polymers containing numerous linear chains joined to a central core, thus, exhibiting a 3D hyperbranched structure. Star polymers gained attention for widespread studies due to their exclusive topological structures and striking physico-chemical properties of star polymers. Star-shaped polymers are of mounting attention for applications in nucleic acid transfer systems, chiefly because of extremely competent delivery apparatus. Exceptional globular star-shape imparts a elevated charge concentration to star polymers, which advantageous for the interaction with the nucleic acids. In addition, peripheral and internal active chemical groups provide an appropriate pathway for various functionalities. Also, studies demonstrated that vital features for delivery vehicles are molar mass, branching gradation, architecture and density of charges of star polymers [110], and these properties also affect the cytotoxicity [111]. Star polymers are usually characterized into two groups – regular star polymers, and miktoarm star polymers.

Regular star polymers contain equal arm sections, such as block copolymer and homopolymer segments. The polymers of miktoarm star contain two or more arm segments with different chemical compositions, molecular weight and different peripheral groups, thus, exhibit asymmetrical branching, and are also known as heteroarm star polymers are subcategorized into different types on the basis of their chemical structures, molecular weight, functional groups, and topologies. Numerous studies have established that these are excellent vehicles for pDNA, siRNA, and mRNA [64, 112, 113, 114]. The conjugation of these polymers with DMAEMA arms; DEGMA and DMAEMA co-polymer arms efficiently form DNA complex and successfully transfected the complex into human cells [115]. The amalgamation of stars polymers with P(DMAEMA-co-POEGMA-OH) as non-viral transporters of mRNA, and pDNA were established [64]. α-CD-OEI star polymers demonstrated exceptional efficiency of gene transfection, and reduced cellular toxicity in Cos7 and HEK293 cells [116].

e) Poly-L-Lysine (PLL)

PLL is an FDA approved natural homopolymer, is an adroitly cation polymer which has been extensively investigated for non-viral transfer of nucleic acid [117]. It is composed of L-Lysine (an amino acid) in a polymerized form. Every single repeating element of PLL exhibits a positive charge on its side chain (i.e., on ε-amine) at physiological pH, and this property has been exploited for the compaction of pDNA by PLL [117]. However, the increase in the length of PLL is directly correlated with its cytotoxicity [117]. Thus, cytotoxicity of the delivery vehicles employed in transfection possess a major issue. The ideal nucleic acid delivery system should be non-toxic for the transfected cells and at the same time, should yield high transfection efficiency as well. The structural amendments such as PEGylation permits to reduce the cytotoxicity of the polymer *via* incorporation of poly(ethylene) glycol into the non-viral delivery vehicles [112]. Additionally, PEGylation also averts non-specific connections between the non-viral delivery vehicle and the plasma membrane of the cells [118]. Similarly, PEGylation of PLL exhibited reduced cytotoxicity, and immunogenicity [118]. PLL dendrimers exhibits good biocompatibility, water solubility, resistance against proteolytic digestion due to their flexible edifice and protein-like features.

f) Chitosan

Chitosan is a natural linear cationic polysaccharide, containing erratically dispersed β–(1-4)-linked N-acetyl-D-glucosamine and D-glucosamine molecules which is found in cuttlefish, crabs, lobsters, shrimps, and fungi. It is synthesized by enzymatic N-deacetylation of chitin or by using alkaline methods. It is used as a suitable candidate for non-viral nucleic acid transfer because of biodegradability, biocompatibility and reduced cytotoxicity [119]. The complex formation occurs between negatively charged nucleotides in pDNA with 1° NH_3 groups of chitosan *via* electrostatic interactions. However, several problems such as reduced transfection efficacy, poor solubility and poor specificity confines usage of chitosan in transferring nucleic acid [120]. Modification of chitosan by PEGylation solved the poor solubility and transfection efficiency problem [121]. In addition, chitosan shows poor buffering capacity which restricts endosomal escape of nucleic acid cargoes, ultimately leading to poor gene expression. Various strategies have been employed such as use of fusogenic peptides and protonable groups, conjugation with urocanic acid, and modification of histidine in chitosan.

h) PAMAM

PAMAM is an extremely branched, mono-disperse synthetic (or artificial) macromolecules. The core of PAMAM is composed of ethylenediamine, and a series of methyl acrylate and ethylenediamine (EDTA) are supplemented [122]. PAMAM contains internal cavities for encapsulation of nucleic acid and the terminal branches of PAMAM are improvised with diverse functional groups, e.g., amino, carboxyl, hydroxyl, methyl groups [122]. Amine-group is exploited for transfer of nucleic acids into cells because of which PAMAM-nucleic acid complex is established by electrostatic interaction preventing degradation of the cargo i.e., nucleic acids from the nucleases. Certain feature such as spherical structure, solubility, biocompatibility, controlled release of the cargo, and non-immunogenicity makes it a suitable choice for non-viral nucleic acid delivery [122].

i) PDMAEMA

PDMAEMA contains tertiary amines and yields approx. 90% transfection efficiency as compared to pEI [123]. PDMAEMA of molecular weight (>112 kDa) make smaller polyplexes and shows high transfection efficiency but at the cost of cytotoxicity [123]. Cellular toxicity and transfection proficiency are unswervingly correlated with the molecular weight; therefore, a need of optimum molecular weight arises. The mixing of short

PDMAEMA with biocompatible cyclodextrin and hydroxypropyl cellulose showed increased transfection of the nucleic acid along with reduced cytotoxicity [124]. PDMAEMA-PCL-PDAEMA triblock polymers complexed with chemotherapeutic drug, paclitaxel efficiently transfected VEGF-siRNA yielding a significant higher gene knockdown efficiency and showed a putative therapeutic agent against cancer.

10.4.4 Protein-Based Nucleic Acid Delivery System

The importance of bio-degradable polymers has been widely exploited in biomedical technology ranging from drug delivery, tissue engineering, bio-nanotechnology and gene therapy. In case of protein-based non-viral vectors, albumin and gelatin are the frequently used proteins, both are characterized by their high bio-compatibility, reduced antigenicity, and accessible for modification. Gelatin is a natural polymer, extracted from animal collagen by patrial acid or alkaline hydrolysis [125]. According to FDA, gelatin has been regarded to be "generally regarded as safe (GRAS)", and thus it has been approved for cosmetic, pharmaceutical and food products. Gelatin is a polyelectrolyte which has less-charge density which needs to be modified *via* cationization by interacting with nucleic acids that are negatively charged. The cationization of gelatin is done by mixing it with pEI, ethylenediamine (EDA), cholamine, spermidine, putrescine etc. The delivery of STAT6-siRNA using gelatin nanocarriers effectively silenced STAT6 gene-expression, leading to death of A549 cells [126].

Albumin is most plentiful blood protein found in human, produced in liver. It exhibits negatively charged surface, and it is highly stable because of presence of inner disulfide bonds formed *via* interaction of 34 cysteine. Albumin-based nanocarriers offer numerous advantages like gelatin polymers, thus, making it a suitable choice in using it as a carrier for transfer of nucleic acids. Albumin nanoparticles shield probity of condensed nucleic acids in a capsule, and thus preventing its enzymatic digestion. hMDA-7 plasmid encapsulated with albumin nanoparticles induced apoptosis of human pancreatic cancer cells (BxPC3 cells). The effective transfer of Akt1 oligonucleotides encapsulated in albumin nanoparticles in A549 cells, and human epithelial carcinoma cells (KB cells) [127] implicated its high transfection efficiency.

10.4.5 Peptide-Based Nucleic Acid Delivery System

Peptides are bio-compatible and bio-degradable building chunks of amino acids accessibly regulated to augment different functions including as a nucleic acid delivery system [128]. Peptides that 10-30 amino acids in length which are positively charged are referred as cell-penetrating peptides (CPPs) [128], and these are also referred as protein transduction domains. They can be synthesized artificially or can be naturally occurring peptides [129]. Cell-penetrating peptides can be categorized into cell-specific peptides and non-cell-specific peptides depending upon the delivery of the cargo for the targeted site. They can be conjugated with lipidic or polymeric compounds, or these can formulate intricate complex with nucleic acids by electrostatic interactions in the nucleic acid carriers. Cell-penetrating peptides have the exclusive capability in bypassing the plasma membrane of target host cells through endocytosis (which is an energy-dependent mechanism) or direct penetration (which is an energy-independent mechanism). Cell-penetrating peptides can migrate into the host or target cells and help to deliver nucleic acid into the live animal cells by competently internalizing across the cellular plasma membrane [130]. Examples of peptide-based polymer for non-viral nucleic acid delivery are – Tat, Ant, 8-Lysine, 8-Arginine, MAP, TP10, Pep-7, Penetratin, and CADY [129]. Cell-penetrating peptides are categorized into several groups on the basis of sequence homology – Hydrophilic CPPs, Amphiphilic CPPs, Targeting ligand-CPP conjugates, and activatable CPPs.

Hydrophilic CPPs comprise one or more segments of positive charged groups which are important for condensation of nucleic acid and infiltration of the surface membrane. Arginine plays a vital role in membrane penetration due to the presence of guanidine group which has strong affinity for negatively charged phospholipids, among the basic amino acids of hydrophilic CPPs. As the pores close immediately formed due to penetration of the membrane, thus cytotoxicity of hydrophilic CPPs is quite low. An arginine oligomer referred as R12 due to presence of 12 arginine amino acids has been studied for delivering anti-Her2 siRNA in mouse and lead to significant drop in the growth of tumor cells [131].

Amphiphilic CPPs share assets with the phospholipids, can embed into lipid plasma membrane of cells and bypass the cell membrane *via* formation of transient channels or lipid rafts [132]. Amphiphilic CPPs are usually classified into bipartite CPPs and α-helical CPPs on the basis of primary sequence and motifs formed in the secondary structures respectively. Example of amphiphilic CPPs is penetratin that effectively delivered exogenous siRNA into the cells.

Targeting ligand CPPs are currently exploited for transfer of nucleic acids into a particular type of cells. The cell-specific targeting capability of CPPs can be increased by conjugating these molecules to a widespread array of ligands, e.g., aptamers, antibodies, carbohydrates and homing peptides. Furthermore, the choice of particular ligands can be directed *via* biological and chemical approaches for various targeting studies related with tumor cells, and β-pancreatic cells. Examples of targeting ligand CPPs include Lyp-1 (responsible for inhibition of metastasis in MDA-MB-435 cells), and SP5-52 (responsible for inhibition of angiogenesis) [133].

Activatable CPPs are hairpin-structure peptides which consist a poly-cationic motif and a polyanionic motif bridged through cleavable peptide loop [134]. The cleavage of peptide loop by a disease-associated proteolytic enzyme triggers activation of catalytic domain which facilitates penetrating function of the CPP [135].

10.5 Advantages and Disadvantages of Chemical Transfection

Certain advantages associated with chemical transfection are – i) No use of viral vectors; as already mentioned viral vectors triggers immune response that may result in deterioration of nucleic acid which is delivered using viral vectors, and even it can induce formation of antibody response against viral vectors, and nucleic acid which will make the host unresponsive to the gene therapy treatment with the same nucleic acid in future as well, therefore chemical treatment efficiently mask the nucleic acid and don't let it recognize as foreign agent to the host immune system, ii) High transfection efficiency; delivery of nucleic acids yield high transfection efficiency because it enhance the endocytic uptake or fusion with plasma membrane of host cells for delivery of nucleic acids, and chemical complex shields delivered nucleic acid from activity of hydrolytic enzymes in endolysosomes or phagolysosomes and in cytoplasm from the action of nucleases, iii) Easy to use and effective on *in-vitro* cultured cells; delivery of nucleic acids specially pDNA, miRNA, and siRNA are easily and efficiently transferred into the cells in order to study effect of a particular gene of its function. For e.g., Lipofectamine, HiPerFect has been widely used and routinely used in research studies for easy and efficient nucleic acid delivery, iv) No packaging size constraint; another major advantage is that the amount and size of nucleic acid that has to be delivered present no problem when it comes to deliver them via chemical methods. The packaging size is quite large, and high molecular weight pDNA, DNA, RNA or huge amount of siRNA can be packaged and delivered easily. The disadvantages in using chemical transfection as non-viral nucleic acid delivery carriers are – i) May cause toxicity due to the types of chemicals used in some cells;

a major constraint in using chemical methods for transfer of nucleic acids is that chemicals that are used for complex formation with the nucleic acid or chemical used as the carriers may be toxic for the host cells, and may trigger cytotoxicity which nullify the use of these chemical methods for delivering nucleic acids. However, a thorough research on the usage the types of chemicals can solve this problem and thus can offer numerous biocompatible and non-cytotoxic chemicals for using them as carriers or complexing agents, ii) Variation in transfection efficiency depending upon the cell types; the transfection efficiency varies quite a bit due to the need of the complexing of the nucleic acids with the different chemicals and different cell types, iv) Not easy to precisely target the specific cells; this is in contrast with the physical methods of nucleic acids delivery system. The delivery of nucleic acids can't be precisely controlled and targeted at the specific site inside the cells and tissues which poses a problem in using chemical methods for delivering nucleic acids into the cells or tissues. However, the target specific delivery can be solved by using certain biological agents such as amino acids which are specific for a particular cell or subcellular location inside the cells.

10.6 Cellular Barriers for Nucleic Acid Delivery Faced by Non-Viral Vectors

The successful transportation of a nucleic acid into the cell isn't a simple mission as there are numerous hurdles faced by non-viral vectors, and thus makes this process difficult. These barriers can be classified into two major categories and shown in Figure 10.3.

a) Extracellular barriers – Extracellular barriers may result in dilapidation of nucleic acids or rapid clearance of vectors containing nucleic acids, thereby rendering them futile for the envisioned therapeutic purpose. This includes several organs, e.g., kidney, spleen, liver, and blood-brain barriers, and these barriers are also referred as physiological barriers because they display removal mechanisms that filter or uptake vector-containing nucleic acids out from the host body [136]. Other impediments like endo- and exo-nuclease function of blood constituents, activation of immune response. These barriers need to be overcome by non-viral vectors in order to reach the target cell.

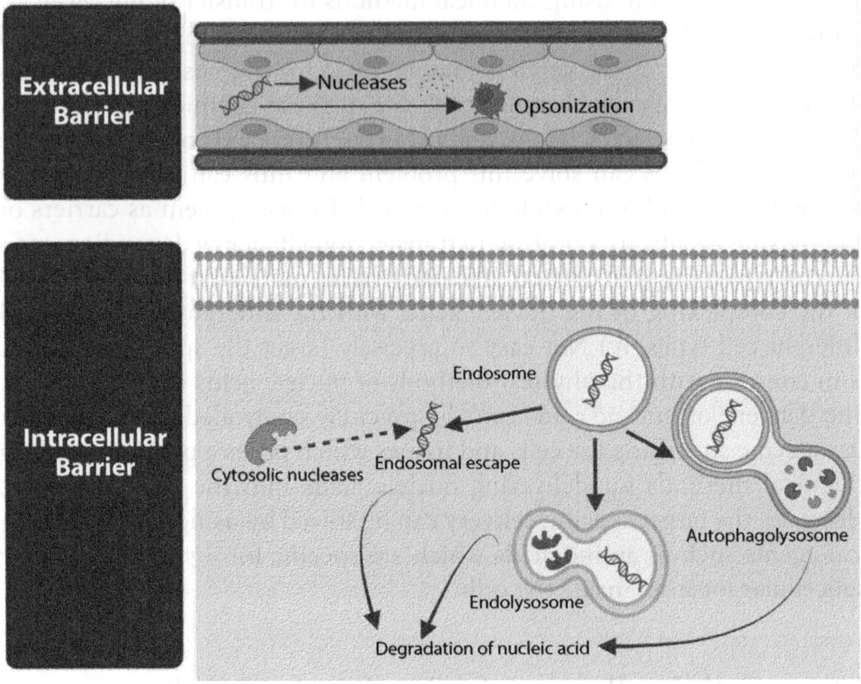

Figure 10.3 Intracellular and extracellular barriers for nucleic acid delivery.

b) Intracellular barriers – Surpassing of the extracellular barriers by the vectors allow them to access to the intracellular barriers. And intracellular barriers impede the unrestricted passage of the non-viral delivery systems to accomplish an efficient gene delivery. Intracellular barriers include plasma membrane of the cell, acidic environment of endosome, autophagy and cytoplasmic nucleases [69].

Therefore, a comprehensive knowledge about these barriers is crucial in using and designing non-viral vectors for nucleic acid delivery, and hence they must be altered in certain ways that these carriers can easily surpass extracellular and intracellular barriers. However, non-viral vectors need less modifications in comparison to viral vectors, because non-viral vectors do not activate host immune system or if they activate, it is not activated at a similar scale while using viral vectors. The non-viral carriers can transport nucleic acids into the host cells *via* endocytosis, phagocytosis or fusion with the host cell's plasma membrane whereas viral vectors rely on the endocytosis and phagocytosis.

10.7 Challenges and Limitations of Non-Viral Nucleic Acid Delivery System

The major technical limitations in attaining a nucleic acid delivery via non-viral vectors are categorized into following steps –

i) Efficiency of the delivery system and unloading of transgene into the target cells
ii) Perseverance
iii) Immune response and biocompatibility
iv) Regulatory issues, ethical concerns, and commercialization

However, much of the problems associated with non-viral nucleic acid delivery systems have been resolved by surface modifications, and using biopolymers. But still, a major concern needs to be addressed i.e., ethical concerns, regulatory issues and commercialization. Table 10.2 enlists the applications of non-viral vectors in *in-vivo* and *in-vitro* model system.

10.8 Conclusion

During past years, significant improvement has been achieved in scientific field in diverse research field allied with designing of innovative nucleic acid transfer systems, along with advantage of information attained in structural biology, genomics and molecular biology, has upstretched up rational optimism to deliberate systematic application of delivery system via non-viral ways into biomedical technology to deal with numerous distressing diseases. The non-viral vector delivery system displayed relevant advantages as compared to the viral vector counterparts. For instance, they are tolerated by the host immune system and thus, are non-immunogenic, no limitation for genetic packing capacity, and the production of these delivery systems is quite effortless, and inexpensive due to reduced restrictions linked with the biological molecules. The non-viral vector delivery systems are also regarded as drugs, rather as the biological molecules. These factors augment clinical translation of non-viral nucleic acid delivery carriers; though, reality is that these vectors have been unsuccessful at translational level at clinical trials. Various apprehensions that non-viral vectors embrace reduced transfection efficiency, transient gene expression, complications in targeting nucleic acids to gain entry into targeted host cells after *in-vivo* administration.

Numerous approaches have been successfully employed to address these difficulties such as mixing of fusogenic agents for buffering capacity,

Table 10.2 Various non-viral vectors and their gene delivery target used in specific *in-vivo/in-vitro* models.

Diseases	Non-viral vector	Nucleic acid delivered	Target cells	References
Parkinson's disease	PEGylated liposomes	TH gene	Nigral-striatial neurons	[137]
Parkinson's disease	PEGylated PLL	hGDNF gene	Dopaminergic neurons	[138]
Parkinson's disease	PEGylated PLL	BDNF gene	Dopaminergic neurons	[139]
Glioblastoma cancer	siRNA-coated gold nanoparticles	Bcl-2-like 12	Glioblastoma multiforme	[140]
Glioblastoma cancer	pEI-PLL	HSV-TK	Glioblastoma cells	[141]
Pancreatic cancer	Liposomes	Anti-PKN3 siRNA (Atu027)	HeLa, HUVEC, EOMA cells	[142]
Breast cancer	Thioaptamer (PAMAM)	miRNA against CD44	Breast cancer tumor cells	[143]
Glioblastoma cancer	RGD	anti-siRNA $\alpha v\beta 3$	Glioma cells	[144]

modifications of surface molecules. The profound understanding of interplay among elaborate physico-chemical characteristics of non-viral delivery systems (such as morphology, size, outer surface chemistry, charge, and solubility) offers evidences for adapting its next generation with better and significant enhanced chances of clinical approaches. Future developments in this area will propel the progress of novel bioinstrumentation to certify the automated and regulated nucleic acid release into the host cell.

References

1. Shirley, J.L., de Jong, Y.P., Terhorst, C., Herzog, R.W., Immune responses to viral gene therapy vectors. *Mol. Ther.*, 28, 3, 709–22, 2020.

2. Balak, J.R., de Graaf, N., Zaldumbide, A., Rabelink, T.J., Hoeben, R.C., de Koning, E.J. *et al.*, Highly efficient *ex vivo* lentiviral transduction of primary human pancreatic exocrine cells. *Sci. Rep.*, 9, 1, 1–10, 2019.
3. Hardee, C.L., Arévalo-Soliz, L.M., Hornstein, B.D., Zechiedrich, L., Advances in non-viral DNA vectors for gene therapy. *Genes*, 8, 2, 65, 2017.
4. Ylösmäki, L., Polini, B., Carpi, S., Martins, B., Smertina, E., Feola, S. *et al.*, Harnessing therapeutic viruses as a delivery vehicle for RNA-based therapy. *PLoS One*, 14, 10, e0224072, 2019.
5. Oh, S. and Kessler, J.A., Design, assembly, production, and transfection of synthetic modified mRNA. *Methods*, 133, 29–43, 2018.
6. Watson, C.N., Belli, A., Di Pietro, V., Small non-coding RNAs: New class of biomarkers and potential therapeutic targets in neurodegenerative disease. *Front. Genet.*, 10, 364, 2019.
7. Smith, C.E. and Zain, R., Therapeutic oligonucleotides: State of the art. *Annu. Rev. Pharmacol. Toxicol.*, 59, 605–30, 2019.
8. Hu, Y.H., Zhang, B.C., Zhou, H.Z., Guan, X.L., Sun, L., *Edwardsiella tarda*-induced miRNAs in a teleost host: Global profile and role in bacterial infection as revealed by integrative miRNA–mRNA analysis. *Virulence*, 8, 7, 1457–64, 2017.
9. Hillebrand, F., Ostermann, P.N., Müller, L., Degrandi, D., Erkelenz, S., Widera, M. *et al.*, Gymnotic delivery of LNA mixmers targeting viral SREs induces HIV-1 mRNA degradation. *Int. J. Mol. Sci.*, 20, 5, 1088, 2019.
10. Neumann, E., Schaefer-Ridder, M., Wang, Y., Hofschneider, P., Gene transfer into mouse lyoma cells by electroporation in high electric fields. *EMBO J.*, 1, 7, 841–5, 1982.
11. Young, J.L. and Dean, D.A., Electroporation-mediated gene delivery. *Adv. Genet.*, 89, 49–88, 2015.
12. Luft, C. and Ketteler, R., Electroporation knows no boundaries: The use of electrostimulation for siRNA delivery in cells and tissues. *SLAS Discovery*, 20, 8, 932–42, 2015.
13. Furuya, M., Yasuchika, K., Mizutani, K., Yoshimura, Y., Nakatsuji, N., Suemori, H., Electroporation of cynomolgus monkey embryonic stem cells. *Genesis*, 37, 4, 180–7, 2003.
14. Edupuganti S, C., De Rosa, S., Elizaga, M., Lu, Y., Han, X., Huang, Y. *et al.*, Intramuscular and intradermal electroporation of HIV-1 PENNVAX-GP® DNA vaccine and IL-12 is safe, tolerable, acceptable in healthy adults. *Vaccines*, 8, 4, 741, 2020.
15. Tebas, P., Roberts, C.C., Muthumani, K., Reuschel, E., White, S., Khan, A.S. *et al.*, ZIKA-001: Safety and immunogenicity of an engineered DNA vaccine against ZIKA virus infection, in: *Open Forum Infectious Diseases*, vol. 4, pp. S300–S301, Oxford University Press US, New York, USA, 2017.
16. Tebas, P., Kraynyak, K.A., Patel, A., Maslow, J.N., Morrow, M.P., Sylvester, A.J. *et al.*, Intradermal SynCon® Ebola GP DNA vaccine is temperature stable and safely demonstrates cellular and humoral immunogenicity advantages in healthy volunteers. *J. Infect. Dis.*, 220, 3, 400–10, 2019.

17. Kumar, P., Nagarajan, A., Uchil, P.D., Optical transfection. *Cold Spring Harb. Protoc.*, 2018, 12, pdb–top096222, 2018.
18. Stevenson, D.J., Gunn-Moore, F.J., Campbell, P., Dholakia, K., Single cell optical transfection. *J. R. Soc. Interface*, 7, 47, 863–71, 2010.
19. Barrett, L.E., Sul, J.Y., Takano, H., Van Bockstaele, E.J., Haydon, P.G., Eberwine, J.H., Region-directed phototransfection reveals the functional significance of a dendritically synthesized transcription factor. *Nat. Methods*, 3, 6, 455–60, 2006.
20. Antkowiak, M., Torres-Mapa, M.L., Stevenson, D.J., Dholakia, K., Gunn-Moore, F., Femtosecond optical transfection of individual mammalian cells. *Nature*, 8, 6, 1216–33, 2013.
21. Ettinger, M., Peckl-Schmid, D., Gruber, C., Laimer, M., Thalhamer, J., Hintner, H. *et al.*, Transcutaneous gene gun delivery of hNC16A Induces BPAG2-specific tolerance. *J. Invest. Dermatol.*, 132, 6, 1665–71, 2012.
22. Bergmann-Leitner, E.S. and Leitner, W.W., Vaccination using gene-gun technology, in: *Malaria Vaccines*, pp. 289–302, Humana Press, New York, Springer; 2015.
23. O'brien, J.A. and Lummis, S.C., Biolistic transfection of neuronal cultures using a hand-held gene gun. *Nat. Protoc.*, 1, 2, 977–81, 2006.
24. O'Brien, J.A. and Lummis, S.C., Nano-biolistics: A method of biolistic transfection of cells and tissues using a gene gun with novel nanometer-sized projectiles. *BMC Biotech.*, 11, 1, 1–6, 2011.
25. Furth, P.A., Shamay, A., Wall, R.J., Hennighausen, L., Gene transfer into somatic tissues by jet injection. *Anal. Biochem.*, 205, 2, 365–8, 1992.
26. Furth, P.A., Shamay, A., Hennighausen, L., Gene transfer into mammalian cells by jet injection. *Hybridoma*, 14, 2, 149–52, 1995.
27. Kerr, D., Furth, P., Powell, A., Wall, R., Expression of gene-gun injected plasmid DNA in the ovine mammary gland and in lymph nodes draining the injection site. *Anim. Biotechnol.*, 7, 1, 33–45, 1996.
28. Furth, P.A., Kerr, D., Wall, R., Gene transfer by jet injection into differentiated tissues of living animals and in organ culture. *Mol. Biotechnol.*, 4, 2, 121–7, 1995.
29. Kami, D., Takeda, S., Itakura, Y., Gojo, S., Watanabe, M., Toyoda, M., Application of magnetic nanoparticles to gene delivery. *Int. J. Mol. Sci.*, 12, 6, 3705–22, 2011.
30. Chen, J., Zhu, S., Tong, L., Li, J., Chen, F., Han, Y. *et al.*, Superparamagnetic iron oxide nanoparticles mediated 131I-hVEGF siRNA inhibits hepatocellular carcinoma tumor growth in nude mice. *BMC Cancer*, 14, 1, 1–8, 2014.
31. Akbarzadeh, A., Samiei, M., Davaran, S., Magnetic nanoparticles: Preparation, physical properties, and applications in biomedicine. *Nanoscale Res. Lett.*, 7, 1, 1–13, 2012.
32. Wierucka, M. and Biziuk, M., Application of magnetic nanoparticles for magnetic solid-phase extraction in preparing biological, environmental and food samples. *TrAC Trends Anal. Chem.*, 59, 50–8, 2014.

33. Liu, X.L. and Fan, H.M., Innovative magnetic nanoparticle platform for magnetic resonance imaging and magnetic fluid hyperthermia applications. *Curr. Opin. Chem. Eng.*, 4, 38–46, 2014.
34. Avasthi, A., Caro, C., Pozo-Torres, E., Leal, M.P., García-Martín, M.L., Magnetic nanoparticles as MRI contrast agents, in: *Surface-Modified Nanobiomaterials for Electrochemical and Biomedicine Application*, pp. 49–91, 2020.
35. Bi, Q., Song, X., Hu, A., Luo, T., Jin, R., Ai, H. *et al.*, Magnetofection: Magic magnetic nanoparticles for efficient gene delivery. *Chin. Chem. Lett.*, 31, 12, 3041–6, 2020.
36. Plank, C., Zelphati, O., Mykhaylyk, O., Magnetically enhanced nucleic acid delivery. Ten years of magnetofection—Progress and prospects. *Adv. Drug Deliv. Rev.*, 63, 14-15, 1300–31, 2011.
37. Mah, C., Zolotukhin, I., Fraites, T., Dobson, J., Batich, C., Byrne, B., Microsphere-mediated delivery of recombinant AAV vectors *in vitro* and *in vivo*. *Mol. Ther.*, 1, 5, S293, 2000.
38. Browning, R.J., Mulvana, H., Tang, M., Hajnal, J.V., Wells, D.J., Eckersley, R. *et al.*, Influence of needle gauge on *in vivo* ultrasound and microbubble-mediated gene transfection. *Ultrasound Med. Biol.*, 37, 9, 1531–7, 2011.
39. Bekeredjian, R., Kroll, R.D., Fein, E., Tinkov, S., Coester, C., Winter, G. *et al.*, Ultrasound targeted microbubble destruction increases capillary permeability in hepatomas. *Ultrasound Med. Biol.*, 33, 10, 1592–8, 2007.
40. Sheyn, D., Kimelman-Bleich, N., Pelled, G., Zilberman, Y., Gazit, D., Gazit, Z., Ultrasound-based nonviral gene delivery induces bone formation *in vivo*. *Gene Ther.*, 15, 4, 257–66, 2008.
41. Un, K., Kawakami, S., Suzuki, R., Maruyama, K., Yamashita, F., Hashida, M., Suppression of melanoma growth and metastasis by DNA vaccination using an ultrasound-responsive and mannose-modified gene carrier. *Mol. Pharmaceutics*, 8, 2, 543–54, 2011.
42. Deng, Q., Chen, J.-L., Zhou, Q., Hu, B., Chen, Q., Huang, J. *et al.*, Ultrasound microbubbles combined with the NFκB binding motif increase transfection efficiency by enhancing the cytoplasmic and nuclear import of plasmid DNA. *Mol. Med. Rep.*, 8, 5, 1439–45, 2013.
43. Budker, V., Zhang, G., Danko, I., Williams, P., Wolff, J., The efficient expression of intravascularly delivered DNA in rat muscle. *Gene Ther.*, 5, 2, 272–6, 1998.
44. Liu, F., Song, Y.K., Liu, D., Hydrodynamics-based transfection in animals by systemic administration of plasmid DNA. *Gene Ther.*, 6, 7, 1258–66, 1999.
45. Zhang, H., Zhu, J., Bergougnou, M.A., Hydrodynamics in downflow fluidized beds (1): Solids concentration profiles and pressure gradient distributions. *Chem. Eng. Sci.*, 54, 22, 5461–70, 1999.
46. Yoshino, H., Hashizume, K., Kobayashi, E., Naked plasmid DNA transfer to the porcine liver using rapid injection with large volume. *Gene Ther.*, 13, 24, 1696–702, 2006.

47. Miao, Z.Q. and Xie, Y., Effects of water-depth on hydrodynamic force of artificial reef. *J. Hydrodynam., Ser. B*, 19, 3, 372–7, 2007.
48. Aliño, S.F., Crespo, A., Dasi, F., Long-term therapeutic levels of human alpha-1 antitrypsin in plasma after hydrodynamic injection of nonviral DNA. *Gene Ther.*, 10, 19, 1672–9, 2003.
49. Yew, N.S., Przybylska, M., Ziegler, R.J., Liu, D., Cheng, S.H., High and sustained transgene expression *in vivo* from plasmid vectors containing a hybrid ubiquitin promoter. *Mol. Ther.*, 4, 1, 75–82, 2001.
50. Hodges, B.L. and Scheule, R.K., Hydrodynamic delivery of DNA. *Expert Opin. Biol. Ther.*, 3, 6, 911–8, 2003.
51. Suda, T. and Liu, D., Hydrodynamic gene delivery: Its principles and applications. *Mol. Ther.*, 15, 12, 2063–9, 2007.
52. Zhang, X., Dong, X., Sawyer, G.J., Collins, L., Fabre, J.W., Regional hydrodynamic gene delivery to the rat liver with physiological volumes of DNA solution. *J. Gene Med.: A Cross-Disciplinary Journal for Research on the Science of Gene Transfer and Its Clinical Applications*, 6, 6, 693–703, 2004.
53. Donnelly, R.F., Singh, T.R.R., Woolfson, A.D., Microneedle-based drug delivery systems: Microfabrication, drug delivery, and safety. *Drug Delivery*, 17, 4, 187–207, 2010.
54. McLean, W.I. and Moore, C.T., Keratin disorders: From gene to therapy. *Hum. Mol. Genet.*, 20, R2, R189–R97, 2011.
55. Ita, K., Dissolving microneedles for transdermal drug delivery: Advances and challenges. *Biomed. Pharmacother.*, 93, 1116–27, 2017.
56. Birchall, J., Coulman, S., Pearton, M., Allender, C., Brain, K., Anstey, A. *et al.*, Cutaneous DNA delivery and gene expression in *ex vivo* human skin explants via wet-etch microfabricated microneedles. *J. Drug Targeting*, 13, 7, 415–21, 2005.
57. Stewart, M.P., Sharei, A., Ding, X., Sahay, G., Langer, R., Jensen, K.F., *In vitro* and *ex vivo* strategies for intracellular delivery. *Nature*, 538, 7624, 183–92, 2016.
58. Tiefenboeck, P., Kim, J.A., Leroux, J.C., Intracellular delivery of colloids: Past and future contributions from microinjection. *Adv. Drug Delivery Rev.*, 132, 3–15, 2018.
59. Stein, P. and Schindler, K., Mouse oocyte microinjection, maturation and ploidy assessment. *JoVE*, 53, e2851, 2011.
60. Zhang, Y. and Yu, L.C., Microinjection as a tool of mechanical delivery. *Curr. Opin. Biotechnol.*, 19, 5, 506–10, 2008.
61. Chow, Y.T., Chen, S., Wang, R., Liu, C., Kong, C.W., Li, R.A. *et al.*, Single cell transfection through precise microinjection with quantitatively controlled injection volumes. *Sci. Rep.*, 6, 1, 1–9, 2016.
62. Aerts, J.M., Martinez-Madrid, B., Leroy, J.L., Van Aelst, S., Bols, P.E., Xenotransplantation by injection of a suspension of isolated preantral ovarian follicles and stroma cells under the kidney capsule of nude mice. *Fertil. Steril.*, 94, 2, 708–14, 2010.

63. Loh, X.J., Lee, T.C., Dou, Q., Deen, G.R., Utilising inorganic nanocarriers for gene delivery. *Biomater. Sci.*, 4, 1, 70–86, 2016.
64. Fus-Kujawa, A., Prus, P., Bajdak-Rusinek, K., Teper, P., Gawron, K., Kowalczuk, A. *et al.*, An overview of methods and tools for transfection of eukaryotic cells *in vitro*. *Front. Bioeng. Biotechnol.*, 9, 634, 2021.
65. Malmsten, M., Inorganic nanomaterials as delivery systems for proteins, peptides, DNA, and siRNA. *Curr. Opin. Colloid Interface Sci.*, 18, 5, 468–80, 2013.
66. Graham, F.L. and Van Der Eb, A., A new technique for the assay of infectivity of human adenovirus 5 DNA. *Virology*, 52, 2, 456–67, 1973.
67. Lee, D., Upadhye, K., Kumta, P.N., Nano-sized calcium phosphate (CaP) carriers for non-viral gene deilvery. *Mater. Sci. Eng.:B*, 177, 3, 289–302, 2012.
68. Olton, D.Y., Close, J.M., Sfeir, C.S., Kumta, P.N., Intracellular trafficking pathways involved in the gene transfer of nano-structured calcium phosphate-DNA particles. *Biomaterials*, 32, 30, 7662–70, 2011.
69. Kumar, P., Nagarajan, A., Uchil, P.D.J., Calcium phosphate-mediated transfection of eukaryotic cells with plasmid DNAs. *Cold Spring Harb. Protoc.*, 2019, 10, pdb–rot095430, 2019.
70. Torres-Vanegas, J.D., Cruz, J.C., Reyes, L.H., Delivery systems for nucleic acids and proteins: Barriers, cell capture pathways and nanocarriers. *Pharmaceutics*, 13, 3, 428, 2021.
71. Rueda-Gensini, L., Cifuentes, J., Castellanos, M.C., Puentes, P.R., Serna, J.A., Muñoz-Camargo, C. *et al.*, Tailoring iron oxide nanoparticles for efficient cellular internalization and endosomal escape. *Nanomaterials*, 10, 9, 1816, 2020.
72. Kim, C., Yoon, S., Lee, J.H., Facile large-scale synthesis of mesoporous silica nanoparticles at room temperature in a monophasic system with fine size control. *Microporous Mesoporous Mater.*, 288, 109595, 2019.
73. Li, X., Chen, Y., Wang, M., Ma, Y., Xia, W., Gu, H., A mesoporous silica nanoparticle–PEI–fusogenic peptide system for siRNA delivery in cancer therapy. *Biomaterials*, 34, 4, 1391–401, 2013.
74. Meng, H., Liong, M., Xia, T., Li, Z., Ji, Z., Zink, J.I. *et al.*, Engineered design of mesoporous silica nanoparticles to deliver doxorubicin and P-glycoprotein siRNA to overcome drug resistance in a cancer cell line. *ACS Nano*, 4, 8, 4539–50, 2010.
75. Bharali, D.J., Klejbor, I., Stachowiak, E.K., Dutta, P., Roy, I., Kaur, N. *et al.*, Organically modified silica nanoparticles: A nonviral vector for *in vivo* gene delivery and expression in the brain. *Proc. Natl. Acad. Sci.*, 102, 32, 11539–44, 2005.
76. Kim, S.T., Saha, K., Kim, C., Rotello, V.M., The role of surface functionality in determining nanoparticle cytotoxicity. *Acc. Chem. Res.*, 46, 3, 681–91, 2013.
77. Takeuchi, I., Onaka, H., Makino, K., Biodistribution of colloidal gold nanoparticles after intravenous injection: Effects of PEGylation at the same particle size. *Biomed. Mater. Eng.*, 29, 2, 205–15, 2018.

78. Zhu, M., Perrett, S., Nie, G., Understanding the particokinetics of engineered nanomaterials for safe and effective therapeutic applications. *Small*, 9, 9-10, 1619–34, 2013.
79. Wong, J.K., Mohseni, R., Hamidieh, A.A., MacLaren, R.E., Habib, N., Seifalian, A.M., Will nanotechnology bring new hope for gene delivery? *Trends Biotechnol.*, 35, 5, 434–51, 2017.
80. Xue, C., Hu, S., Gao, Z.-H., Wang, L., Luo, M.-X., Yu, X. *et al.*, Programmably tiling rigidified DNA brick on gold nanoparticle as multi-functional shell for cancer-targeted delivery of siRNAs. *Nat. Commun.*, 12, 1, 1–12, 2021.
81. Zhou, J., Deng, W., Wang, Y., Cao, X., Chen, J., Wang, Q. *et al.*, Cationic carbon quantum dots derived from alginate for gene delivery: One-step synthesis and cellular uptake. *Acta Biomater.*, 42, 209–19, 2016.
82. Pranatharthiharan, S., Patel, M.D., D'Souza, A.A., Devarajan, P.V., Inorganic nanovectors for nucleic acid delivery. *Drug Deliv. Transl. Res.*, 3, 5, 446–70, 2013.
83. Zhi, D., Bai, Y., Yang, J., Cui, S., Zhao, Y., Chen, H. *et al.*, A review on cationic lipids with different linkers for gene delivery. *Adv. Colloid Interface Sci.*, 253, 117–40, 2018.
84. Riedl, S.A., Kaiser, P., Raup, A., Synatschke, C.V., Jérôme, V., Freitag, R., Non-viral transfection of human T lymphocytes. *Processes*, 6, 10, 188, 2018.
85. Ewert, K., Slack, N.L., Ahmad, A., Evans, H.M., Lin, A.J., Samuel, C.E. *et al.*, Cationic lipid-DNA complexes for gene therapy: Understanding the relationship between complex structure and gene delivery pathways at the molecular level. *Curr. Med. Chem.*, 11, 2, 133–49, 2004.
86. Zylberberg, C. and Matosevic, S., Bioengineered liposome–scaffold composites as therapeutic delivery systems. *Ther. Deliv.*, 8, 6, 425–45, 2017.
87. Šimčíková, M., Prather, K.L., Prazeres, D.M., Monteiro, G.A.J.B., Reviews, G.E., Towards effective non-viral gene delivery vector. *Biotechnol. Genet. Eng. Rev.*, 31, 1-2, 82–107, 2015.
88. Zylberberg, C., Gaskill, K., Pasley, S., Matosevic, S., Engineering liposomal nanoparticles for targeted gene therapy. *Gene Ther.*, 24, 8, 441–52, 2017.
89. Belliveau, N.M., Huft, J., Lin, P.J., Chen, S., Leung, A.K., Leaver, T.J. *et al.*, Microfluidic synthesis of highly potent limit-size lipid nanoparticles for *in vivo* delivery of siRNA. *Mol. Ther.-Nucleic Acids*, 1, e37, 2012.
90. Paecharoenchai, O., Niyomtham, N., Apirakaramwong, A., Ngawhirunpat, T., Rojanarata, T., Yingyongnarongkul, B.E. *et al.*, Structure relationship of cationic lipids on gene transfection mediated by cationic liposomes. *AAPS PharmSciTech*, 13, 4, 1302–8, 2012.
91. Al-Dosari, M.S. and Gao, X., Nonviral gene delivery: Principle, limitations, and recent progress. *AAPS J.*, 11, 4, 671–81, 2009.
92. Ramamoorth, M. and Narvekar, A., Non viral vectors in gene therapy-an overview. *J. Clin. Diagn. Res.*, 9, 1, GE01, 2015.
93. Thomas, S.J., Moreira Jr., E.D., Kitchin, N., Absalon, J., Gurtman, A., Lockhart, S. *et al.*, Safety and efficacy of the BNT162b2 mRNA Covid-19 vaccine through 6 months. *N. Engl. J. Med.*, 385, 19, 1761–73, 2021.

94. Schallon, A., Synatschke, C.V., Jérôme, V., Müller, A.H., Freitag, R., Nanoparticulate nonviral agent for the effective delivery of pDNA and siRNA to differentiated cells and primary human T lymphocytes. *Biomacromolecules*, 13, 11, 3463–74, 2012.
95. Bartlett, D.W. and Davis, M.E., Impact of tumor-specific targeting and dosing schedule on tumor growth inhibition after intravenous administration of siRNA-containing nanoparticles. *Biotechnol. Bioeng.*, 99, 4, 975–85, 2008.
96. Englert, C., Brendel, J.C., Majdanski, T.C., Yildirim, T., Schubert, S., Gottschaldt, M. *et al.*, Pharmapolymers in the 21st century: Synthetic polymers in drug delivery applications. *Prog. Polym. Sci.*, 87, 107–64, 2018.
97. Boussif, O., Lezoualc'h, F., Zanta, M.A., Mergny, M.D., Scherman, D., Demeneix, B. *et al.*, A versatile vector for gene and oligonucleotide transfer into cells in culture and *in vivo*: Polyethylenimine. *Proc. Natl. Acad. Sci.*, 92, 16, 7297–301, 1995.
98. Sonawane, N., Szoka, F.C., Verkman, A.S., Chloride accumulation and swelling in endosomes enhances DNA transfer by polyamine-DNA polyplexes. *J. Biol. Chem.*, 278, 45, 44826–31, 2003.
99. Eliyahu, H., Barenholz, Y., Domb, A., Polymers for DNA delivery. *Molecules*, 10, 1, 34–64, 2005.
100. Yuan, W. and Li, H., Polymer-based nanocarriers for therapeutic nucleic acids delivery, in: *Nanostructures for Drug Delivery*, pp. 445–60, Elsevier, Amsterdam, Netherlands, 2017.
101. Andersen, M.Ø., Lichawska, A., Arpanaei, A., Jensen, S.M.R., Kaur, H., Oupicky, D. *et al.*, Surface functionalisation of PLGA nanoparticles for gene silencing. *Biomaterials*, 31, 21, 5671–7, 2010.
102. Siewert, C., Haas, H., Nawroth, T., Ziller, A., Nogueira, S., Schroer, M. *et al.*, Investigation of charge ratio variation in mRNA–DEAE-dextran polyplex delivery systems. *Biomaterials*, 192, 612–20, 2019.
103. Kumar, P., Nagarajan, A., Uchil, P.D., DEAE-dextran transfection. *Cold Spring Harb. Protoc.*, 2018, 7, pdb–top096263, 2018.
104. Shcharbin, D., Pedziwiatr, E., Bryszewska, M.J., How to study dendriplexes I: Characterization. *J. Controlled Release*, 135, 3, 186–97, 2009.
105. Majoros, I.J., Williams, C.R., Becker, A.C., Baker, J.R., Surface interaction and behavior of poly (amidoamine) dendrimers: Deformability and lipid bilayer disruption. *J. Comput. Theor. Nanosci.*, 6, 7, 1430–6, 2009.
106. Hong, S., Rattan, R., Majoros, I.J., Mullen, D.G., Peters, J.L., Shi, X. *et al.*, The role of ganglioside GM1 in cellular internalization mechanisms of poly (amidoamine) dendrimers. *Bioconjugate Chem.*, 20, 8, 1503–13, 2009.
107. Gorain, B., Choudhury, H., Pandey, M., Amin, M.C.I.M., Singh, B., Gupta, U. *et al.*, Dendrimers as effective carriers for the treatment of brain tumor, in: *Nanotechnology-Based Targeted Drug Delivery Systems for Brain Tumors*, pp. 267–305, Academic Press, Massachusetts, USA, 2018.
108. Perisé-Barrios, A.J., Jiménez, J.L., Domínguez-Soto, A., de la Mata, F.J., Corbí, A.L., Gomez, R. *et al.*, Carbosilane dendrimers as gene delivery agents for the treatment of HIV infection. *J. Controlled Release*, 184, 51–7, 2014.

109. Perisé-Barrios, A.J., Gómez, R., Corbí, A.L., de La Mata, J., Domínguez-Soto, A., Muñoz-Fernandez, M.A., Use of carbosilane dendrimer to switch macrophage polarization for the acquisition of antitumor functions. *Nanoscale*, 7, 9, 3857–66, 2015.
110. Wu, W., Wang, W., Li, J., Star polymers: Advances in biomedical applications. *Prog. Polym. Sci.*, 46, 55–85, 2015.
111. Xiu, K., Yang, J., Zhao, N., Li, J., Xu, F., Multiarm cationic star polymers by atom transfer radical polymerization from β-cyclodextrin cores: Influence of arm number and length on gene delivery. *Acta Biomater.*, 9, 1, 4726–33, 2013.
112. Gibson, T.J., Smyth, P., Semsarilar, M., McCann, A.P., McDaid, W.J., Johnston, M.C. *et al.*, Star polymers with acid-labile diacetal-based cores synthesized by aqueous RAFT polymerization for intracellular DNA delivery. *Polym. Chem.*, 11, 2, 344–57, 2020.
113. Dearnley, M., Reynolds, N.P., Cass, P., Wei, X., Shi, S., Mohammed, A.A. *et al.*, Comparing gene silencing and physiochemical properties in siRNA bound cationic star-polymer complexes. *Biomacromolecules*, 17, 11, 3532–46, 2016.
114. Mendrek, B., Sieroń, Ł, Zymełka-Miara, I., Binkiewicz, P., Libera, M., Smet, M. *et al.*, Nonviral plasmid DNA carriers based on N, N′-dimethylaminoethyl methacrylate and di (ethylene glycol) methyl ether methacrylate star copolymers. *Biomacromolecules*, 16, 10, 3275–85, 2015.
115. Yang, C., Li, H., Goh, S.H., Li, J., Cationic star polymers consisting of α-cyclodextrin core and oligoethylenimine arms as nonviral gene delivery vectors. *Biomaterials*, 28, 21, 3245–54, 2007.
116. Godbey, W.T., Gene delivery, in: *An Introduction to Biotechnology*, W.T. Godbey, (Ed.), pp. 275–312, Academic Press Inc., London, UK, 2014.
117. Jiang, Y., Li, Y., Richard, C., Scherman, D., Liu, Y., Hemocompatibility investigation and improvement of near-infrared persistent luminescent nanoparticle ZnGa 2 O 4: Cr 3+ by surface PEGylation. *J. Mater. Chem. B*, 7, 24, 3796–803, 2019.
118. Nayerossadat, N., Maedeh, T., Ali, P.A., Viral and nonviral delivery systems for gene delivery. *Adv. Biomed. Res.*, 1, 1, 27, 2012.
119. Lee, M.-K., Chun, S.-K., Choi, W.-J., Kim, J.-K., Choi, S.-H., Kim, A. *et al.*, The use of chitosan as a condensing agent to enhance emulsion-mediated gene transfer. *Biomaterials*, 26, 14, 2147–56, 2005.
120. Barua, S., Ramos, J., Potta, T., Taylor, D., Huang, H.-C., Montanez, G. *et al.*, Discovery of cationic polymers for non-viral gene delivery using combinatorial approaches. *Comb. Chem. High Throughput Screen.*, 14, 10, 908–24, 2011.
121. Zhang, Y., Chen, J., Zhang, Y., Pan, Y., Zhao, J., Ren, L. *et al.*, A novel PEGylation of chitosan nanoparticles for gene delivery. *Biotechnol. Appl. Biochem.*, 46, 4, 197–204, 2007.
122. Abedi-Gaballu, F., Dehghan, G., Ghaffari, M., Yekta, R., Abbaspour-Ravasjani, S., Baradaran, B. *et al.*, PAMAM dendrimers as efficient drug and

gene delivery nanosystems for cancer therapy. *Appl. Mater. Today*, 12, 177–90, 2018.

123. Agarwal, S., Zhang, Y., Maji, S., Greiner, A., PDMAEMA based gene delivery materials. *Mater. Today*, 15, 9, 388–93, 2012.
124. Xu, F., Ping, Y., Ma, J., Tang, G., Yang, W., Li, J. *et al.*, Comb-shaped copolymers composed of hydroxypropyl cellulose backbones and cationic poly ((2-dimethyl amino) ethyl methacrylate) side chains for gene delivery. *Bioconjugate Chem.*, 20, 8, 1449–58, 2009.
125. Madkhali, O., Mekhail, G., Wettig, S.D., Modified gelatin nanoparticles for gene delivery. *Int. J. Pharm.*, 554, 224–34, 2019.
126. Youngren, S.R., Tekade, R.K., Gustilo, B., Hoffmann, P.R., Chougule, M.B., STAT6 siRNA matrix-loaded gelatin nanocarriers: Formulation, characterization, and *ex vivo* proof of concept using adenocarcinoma cells. *Biomed. Res. Int.*, 2013, 858946, 2013.
127. Li, H., Liu, Y., Chen, L., Liu, Q., Qi, S., Cheng, X. *et al.*, Folate receptor-targeted lipid-albumin nanoparticles (F-LAN) for therapeutic delivery of an Akt1 antisense oligonucleotide. *J. Drug Targeting*, 26, 5-6, 466–73, 2018.
128. Thapa, R.K. and Sullivan, M.O., Gene delivery by peptide-assisted transport. *Curr. Opin. Biomed. Eng.*, 7, 71–82, 2018.
129. Taylor, R.E. and Zahid, M., Cell penetrating peptides, novel vectors for gene therapy. *Pharmaceutics*, 12, 3, 225, 2020.
130. Alhakamy, N.A., Nigatu, A.S., Berkland, C.J., Ramsey, J.D., Noncovalently associated cell-penetrating peptides for gene delivery applications. *Ther. Deliv.*, 4, 6, 741–57, 2013.
131. Kim, S.W., Kim, N.Y., Choi, Y.B., Park, S.H., Yang, J.M., Shin, S., RNA interference *in vitro* and *in vivo* using an arginine peptide/siRNA complex system. *J. Controlled Release*, 143, 3, 335–43, 2010.
132. Tai, W. and Gao, X., Functional peptides for siRNA delivery. *Adv. Drug Delivery Rev.*, 110, 157–68, 2017.
133. Tripathi, P.P., Arami, H., Banga, I., Gupta, J., Gandhi, S., Cell penetrating peptides in preclinical and clinical cancer diagnosis and therapy. *Oncotarget*, 9, 98, 37252, 2018.
134. Shi, N.Q., Gao, W., Xiang, B., Qi, X.R., Enhancing cellular uptake of activable cell-penetrating peptide–doxorubicin conjugate by enzymatic cleavage. *Int. J. Nanomed.*, 7, 1613, 2012.
135. Li, H., He, J., Yi, H., Xiang, G., Chen, K., Fu, B. *et al.*, siRNA suppression of hTERT using activatable cell-penetrating peptides in hepatoma cells. *Biosci. Rep.*, 35, 2, e00181, 2015.
136. Okholm, A.H. and Kjems, J., DNA nanovehicles and the biological barriers. *Adv. Drug Delivery Rev.*, 106, 183–91, 2016.
137. Tasset, A., Bellamkonda, A., Wang, W., Pyatnitskiy, I., Ward, D., Peppas, N., Wang, H., Overcoming barriers in non-viral gene delivery for neurological applications. *Nanoscale*, 14, 10, 3698–719, 2022.

138. Yue, P., Miao, W., Gao, L., Zhao, X., Teng, J., Ultrasound-triggered effects of the microbubbles coupled to GDNF plasmid-loaded PEGylated liposomes in a rat model of Parkinson's disease. *Front. Neurosci.*, 12, 222, 2018.
139. Lin, C.Y., Lin, Y.C., Huang, C.Y., Wu, S.R., Chen, C.M., Liu, H.L., Ultrasound-responsive neurotrophic factor-loaded microbubble-liposome complex: Preclinical investigation for Parkinson's disease treatment. *J. Controlled Release*, 321, 519–28, 2020.
140. Kong, L., Wu, Y., Alves, C.S., Shi, X., Efficient delivery of therapeutic siRNA into glioblastoma cells using multifunctional dendrimer-entrapped gold nanoparticles. *Nanomedicine*, 12, 23, 3103–15, 2016.
141. Malik, Y.S., Sheikh, M.A., Xing, Z., Guo, Z., Zhu, X., Tian, H., Chen, X., Polylysine-modified polyethylenimine polymer can generate genetically engineered mesenchymal stem cells for combinational suicidal gene therapy in glioblastoma. *Acta Biomater.*, 80, 144–53, 2018.
142. Kokkinos, J., Ignacio, R.M., Sharbeen, G., Boyer, C., Gonzales-Aloy, E., Goldstein, D., McCarroll, J.A., Phillips, P.A., Australian pancreatic cancer genome initiative. Targeting the undruggable in pancreatic cancer using nano-based gene silencing drugs. *Biomaterials*, 240, 119742, 2020.
143. Fan, W., Wang, X., Ding, B., Cai, H., Wang, X., Fan, Y. *et al.*, Thioaptamer-conjugated CD44-targeted delivery system for the treatment of breast cancer *in vitro* and *in vivo*. *J. Drug Targeting*, 24, 4, 359–71, 2016.
144. Mirzaei, S., Mahabady, M.K., Zabolian, A., Abbaspour, A., Fallahzadeh, P., Noori, M. *et al.*, Small interfering RNA (siRNA) to target genes and molecular pathways in glioblastoma therapy: Current status with an emphasis on delivery systems. *Life Sci.*, 275, 119368, 2021.

11

The Junction of Biomaterials and Gene Therapy – Current Strategies and Future Directions

Ranjan Kumar Singh[1*], Sunita Panchawat[2], Chennu M.M. Prasada Rao[1], Joohee Pradhan[2], Rajeswari Tanniru[1], Deepika Bairagee[2] and Ajay Kumar Garg[1]

[1]School of Pharmacy, Raffles University, Neemrana, Rajasthan, India
[2]Department of Pharmaceutical Sciences, Mohanlal Sukhadia University, Udaipur, Rajasthan, India

Abstract

Biomaterials are a key part of technologies that are used in clinical settings to deliver therapeutic agents. We now know a lot more about how cells use nucleic acids, and this has sparked interest in developing both older (like antisense oligonucleotides, plasmid DNA, and transposons) and newer (like short interfering RNA, messenger RNA, and non-coding RNAs) nucleic acid agents to treat a wide variety of diseases. This review will look at the progress made in biomaterials to make it easier to use nucleic acids for therapeutic purposes. First, we look at the different kinds of nucleic acids and how each one works in a clinical setting. Using recent improvements in T-cell-based therapy as an example, we list the different ways biomaterials could be used to make a difference in this exciting therapeutic intervention technology. We believe that this modality will soon become a model for other types of cellular therapies. Then, we look at how biomaterials can help with new nucleic acid technologies. In particular, we look at the design of intelligent nanoparticles, the use of mRNA instead of plasmid DNA, long-acting (integrating) expression systems, and the *in vitro*/*in vivo* expansion of engineered T-cells. We explain the role of biomaterials in these new nucleic acid technologies so that in the near future, nucleic acids can have a bigger impact in the clinical world.

Keywords: Biomaterials, gene therapy, gene transfer technology, and delivery systems

**Corresponding author*: rxsingh8@gmail.com

Rishabha Malviya, Sonali Sundram and Neelam Jain (eds.) Integration of Biomaterials for Gene Therapy, (323–348) © 2023 Scrivener Publishing LLC

11.1 Introduction

11.1.1 Gene Therapy

Gene therapy is a medical strategy that addresses the underlying genetic issue to treat or prevent disease. Instead of employing drugs or surgery, gene therapy (GT) approaches let doctors change a person's genetic composition to treat a problem. Diseases that are untreatable by conventional methods can now be treated thanks to gene therapy. Since 2016, two chimeric antigen receptor T-cell products for B-cell cancers and four additional products for serious monogenic disorders, such as -thalassemia, a rare form of vision loss, spinal muscular atrophy, and a rare form of primary immunodeficiency, have been approved by the European Medicines Agency (EMA) and the U.S. Food and Drug Administration (FDA). The goal of gene therapy for genetic disorders is to establish the long-lasting expression of the therapeutic gene, or transgene, at a level adequate to reduce or eliminate the symptoms of the condition with the fewest possible side effects. There are two main approaches: either the gene is delivered in a nonintegrating vector to a long-lived post-mitotic or slowly dividing cell, ensuring the expression of that gene for the life of the cell, or the gene is delivered in an integrating vector to a precursor or stem cell, ensuring the gene is passed to every daughter cell (the vector is designed to integrate at one or more loci in the patient's chromosomes [1]. In its broadest definition, GT is a technique for giving a patient's somatic cells the genetic information they need to produce certain therapeutic proteins to treat or control the disease. Gene therapy has made treatments for illnesses that are beyond the scope of conventional methods possible. The goal of gene therapy for genetic disorders is to achieve the therapeutic gene's or transgene's permanent expression at a level that can reduce or eliminate disease symptoms while causing the fewest possible side effects. There are two main approaches: either the gene is delivered in a non-integrating vector to a long-lived post-mitotic or slowly dividing cell, ensuring the expression of that gene for the life of the cell, or the gene is delivered in an integrating vector to a precursor or stem cell, ensuring the gene is passed to every daughter cell (the vector is designed to integrate at one or more loci in the patient's chromosomes). The second scenario keeps the transferred DNA outside of the chromosomes, avoiding the need to integrate the therapeutic DNA into the patient's cells. Transduction of stem cells normally occurs *ex vivo* and requires an integrating vector, in contrast to *in vivo* gene delivery for long-lived post-mitotic cells. GT can be summed up as a way to provide a patient's somatic cells with the genetic information they need to produce particular therapeutic proteins to treat or control their disease. The goal of somatic gene therapy is to overcome the drawbacks of administering therapeutic proteins directly, such as limited bioavailability, systemic toxicity, *in vivo* instability, high rates of hepatic and renal clearance, and expensive manufacturing costs. Some of the restrictions related to using recombinant therapeutic proteins may be overcome by giving a patient a therapeutic gene as a predrug [2]. Gene therapy currently consists of three parts: a gene that encodes a specific therapeutic protein, a plasmid-based gene expression system that regulates a gene's activity inside a target cell, and a gene delivery system that manages the transfer of the gene expression plasmids to particular parts of the body. In addition to the therapeutic gene, gene expression plasmids also contain several other genetic components such as transcript stabilizers, polyadenylation sequences, and introns that control the host cells' ability to produce stable proteins and secrete them. The distribution and access of a gene expression system to the target cell, as well as recognition by a cell-surface receptor followed by intracellular trafficking and nuclear translocation, are all aspects of how gene delivery systems are intended to control the placement of a gene within the body. Contrarily, biomaterials serve a crucial role in modern medicine by helping patients recover from illness or damage and restoring function. Biomaterials are used in medical applications to sustain, improve, or replace damaged tissue or a biological function. They can be natural or manufactured. The goal of gene therapy is to regulate or correct the expression of a gene that is expressing improperly by inserting genetic material, such as DNA or RNA, into the cellular gene regulatory system.

Despite the therapeutic promise of nucleic acids (NAs), their usage is restricted by issues such as short cellular absorption, short *in vivo* half-lives, hasty clearance, and fast enzymatic degradation. To evade this issue and offer targeted and efficient gene delivery, various vectors – roughly split into viral and non-viral vectors – have been developed. Viral vectors were previously proposed as feasible carriers due to their site-specificity and naturally formed transduction capacities; however, due to immunogenicity and pathogenicity, their extensive clinical application is hindered. These limitations and advancements in nanotechnology have led to the formation of nanoparticle DNA delivery techniques. There have been numerous distribution strategies devised, each with its benefits and drawbacks. Modifications to the liberation methods, the addition of substances like PEG to extend residence times, as well as the introduction of cell-targeting peptides have all significantly increased these delivery systems' efficacy. Despite advancements, there are still issues and unanswered questions in this field of study. Some of these contain the fate of nanomaterials in living things, their durable detrimental effects on gene expression and toxicity, as well as the conventional ADMETox methods. The need for further fundamental research into the principles behind gene activity and gene delivery is essentially limiting the development of efficient nanoparticle-DNA formulations, and this research must also proceed. Our understanding of biochemical processes and phenotype/genotype relationships have greatly improved with the discovery of gene delivery, which offers advantages over currently used pharmaceutical treatments. Gene therapy's efficacy depends on several factors, including DNA protection, high transfection efficacy, reduced toxicity, and nonspecific effects, high potency even at DNA dosages, adaptation to numerous treatment regimens and diseases, and productive vectors to overcome extracellular and intracellular barriers. Despite considerable research into DNA delivery vectors, it is still difficult to safely and efficiently transfer DNA into specific cells or organs [3]. There are two main ways to do this. A non-integrating variable is employed to distribute a gene to a long-lived post-mitotic or steadily dividing cell to make sure that the gene is conveyed throughout the cell's lifetime. An integrating variable is employed to deliver a gene to a reference or stem cell to make sure that the gene is passed on to each daughter cell (the vector is made to combine at one or more loci on the patient's genetic material). In the second case, extrachromosomal stabilization stabilizes the transferred DNA without the therapeutic DNA having to be put into the chromosomes of the patient's cells. Gene distribution to long-lived post-mitotic cells is usually done *in vivo*. Stem cell transduction, on the other hand, is usually done outside of the body and requires an integrating vector.

11.1.2 *Ex Vivo*

In situ in *ex vivo* transmission, the patient's cells are removed, injected with the desirable qualities, and then return to the patient using results similar to those used in hematopoietic juvenile microbial grafts (although in this case, the transfer includes a lot of autologous transference cells). This procedure is referred to as *ex vivo* transduction (Figure 11.1). This method necessitates the utilization of a facility with a high level of technological sophistication for the processing of the cells, as well as a genomic vehicle (or vector). The quality of the drug delivery *in vivo*, on the other side, is comparable to that of other drug specialists. It is common practice to provide the vector and gene build as an outpatient treatment once it has been defrosted and prepared by a pharmacist. The vector and gene build is kept in a frozen state.

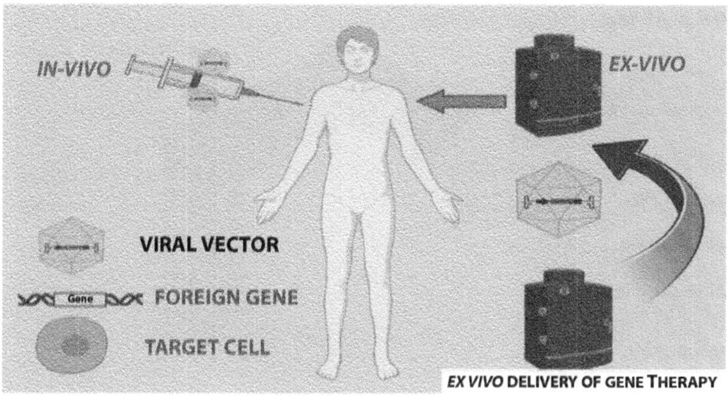

Figure 11.1 A diagram depicts both *in vivo* and *ex vivo* gene therapy. *In vivo* gene therapy (left) uses viral vectors that are injected directly into the patient's body, whereas *ex vivo* gene therapy (right) uses genetically altered cells that have been removed from the patient's body before being reintroduced.

11.1.3 *In Vivo* Gene Therapy

In vivo gene transfer studies, also known as studies in which the carrier is required to fill into the patient, utilize a variety of gene-delivery vehicles in monogenic illnesses; however, the majority of the research being conducted right now uses AAV (Adeno-associated virus) vectors. To generate a transgenic AAV vector, a harmless, non-enveloped parvovirus was utilized. This enabled the creation of the vector. Since the AAV vector DNA is not merged into the patient's genome, the majority of it is kept in the cell as a permanent episome. This is the case even though it is transferred. As a consequence of this, the occurrence of insertional mutagenesis is quite improbable when employing AAV vectors. In 2012, the European Medicines Agency (EMA) issued a license for the use of Alipogene tiparvovec (Glybera, uniQure) as a treatment for recurring or severe pancreatitis in patients who have a rare inherited lipid condition known as lipoprotein lipase deficiency. This was the first commercial AAV vector product and the first gene therapy for a genetic disease.

11.2 Viral Gene Therapy

The process of viral gene therapy involves the transfer of genes into a host organism by employing the highly developed and effective transfection mechanism of viruses. In the field of gene therapy research, viral carriers

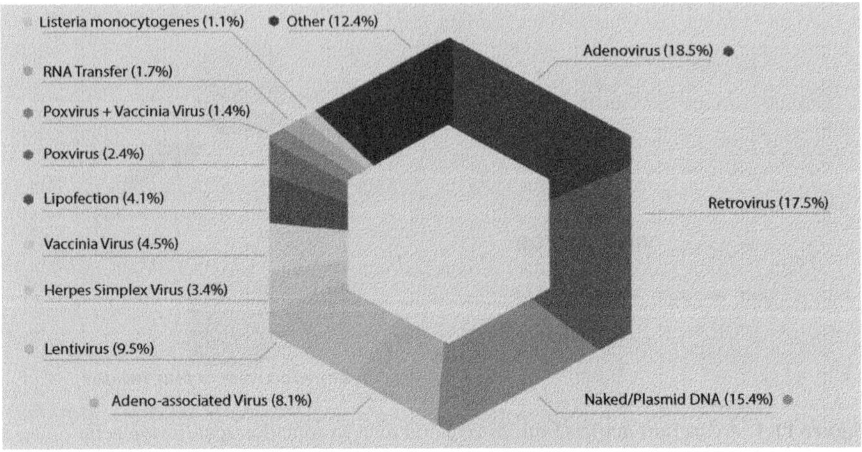

Figure 11.2 Clinical trials for gene therapy use many vectors [2].

have garnered a lot of attention due to the high transfection efficiency that they offer. To construct viral vectors, a portion of the viral DNA must be removed; this renders viral replication ineffective. After then, a gene that encodes the intended result is inserted into the generally very small gap left by the deletion of the viral gene section. This gap is normally caused by the removal of the viral gene segment. Therefore, there is a limit placed on the total quantity of genetic information that can be transferred into a viral vector. The price of production, the difficulties of production, and the worries over safety are further restraints. Mammalian cells, like those of viruses, have developed the ability to resist the process of viral transfection. Because of their considerable potential to provoke an immune response, viral vectors cannot be used routinely as delivery methods for genes. In addition, there is a possibility that viral vectors will come back together in the body at some point and set off the immunological response of the complement system. Research into non-viral techniques of gene transfer has been inspired by the hazards and restrictions of viral vectors as a means of gene transfer. Despite this, Figure 11.2 illustrates the vectors that were applied in the clinical research of gene therapy.

11.2.1 Retroviral Vectors

Several RNA viral vectors in use today are of retroviral origin. The genome size of these viruses ranges from 8 to 11 kilobases, and their outside diameter is 80 to 130 nm (kb). The viral Genome, integrase, and reverse transcriptase are all contained inside the capsid. The desired gene is easily substituted for replication components in retroviral vectors, but all other crucial RNA sequences are kept. The long terminal repetition (LTR) is particularly important because it is required in wrapping, reverse transcription, inclusion, and transcription control. The packaging cells responsible for making the viral proteins needed to create retroviral vectors are also responsible for making the vectors themselves. Since the vectors are free of viral genes, they can absorb exogenous gene sequences up to 7–10 kb in length. To date, retroviral vectors have been the most often employed gene transfer vehicle for *ex vivo* treatment for severe combined immunodeficiency (SCID). Retroviruses, viral vectors, and spuma viruses, all of which contain extra viral protein, are some examples of RNA viral vectors that are now undergoing research and development. These viruses encode the structural genes gag, pol, and env [3].

11.3 DNA Viral Vectors

11.3.1 Adenoviral Vectors

Upgraded adenoviruses, which were first identified as the causal agents of the cold virus in 1953, are the most thoroughly researched DNA viral vectors. Adenoviruses have an open, icosahedral structure with a size of 70–90 nm. A double-stranded DNA molecule of 36–38 kilobase pairs makes up the massive viral genome. The replication and assembly of viruses, as well as the more complicated processes of DNA replication and transcription, only take place in the nucleus of infected cells. There are nine primary complex transcription units (PCUs) that are further subdivided into initial (E1A, E1B, E2-E4) and later (L1-L5) transcripts through inverted terminal repeats to every side (ITR). Gene transfer using adenoviral vectors is effective in many different animals outside humans, including rodents, pigs, rabbits, and even nonhuman primates. Transduction of various cell types is possible using various delivery methods. Transgene expression is induced by intraperitoneal (intraperitoneal) infusion into the peritoneal, kidneys, pancreas, cerebrospinal fluid, skeletal muscle, brain, cardiac muscle, or coronary artery. However, the transgenic expression inside the spleen and liver is the predominant impact of intravenous treatment in animals. Regardless of injection method or cell type, transgenic expression generated by the first series in adenoviral vectors is often short-lived. The potential of adenoviral vectors as carriers for gene-therapy treatments for cancer has been extensively researched. There has been a lot of interest in oncolytic adenoviral vectors recently because of their ability to replicate in tumor cells. Adenoviral vectors replicate in cancer cells, increasing the viral inoculum and killing off the tumor cells. Chinese regulators have greenlit the use of anti-tumor adenovirus H101 (Shanghai Sunway Biotech, Shanghai, China), in which the E1b55K gene was deleted [4].

11.4 Adeno-Associated Viral Vectors

Paramyxovirus, a living being's adenovirus satellite viruses, and herpes simplex virus were used to create what is now known as an adeno-associated virus (Adenoviral) vector (HSV). About 100 distinct serotypes of adeno-associated virus have been identified in non-human primates, and another 12 have been identified in humans, each with its unique targeting capacity. AAV serotype 2 was the first serotype to have a clone discovered and used for transgene transmission between serotypes (AAV2). The icosahedral virion has a dimension of 18–26 nm and contains 4–5 kb of DNA in a single strand. The virion appears to have no preference for which DNA strand it transports, being able to carry either sense or antisense strand. Seven main transcription units (Rep40, Rep52, Rep68, Rep78, VP1, VP2, and VP3) are encoded by Rep and Cap genes in the viral genome DNA. Inverted terminal repeats (ITRs) are palindromic sequences found at both the beginning and the end of the viral genome. Although AAV is normally found in an episomal form, these ITRs are necessary for the location incorporation of AAV DNA into a static location on chromosome 19. AAV in its natural state appeared appealing as a vector because of its ability to specifically integrate into chromosome 19. Most of the AAV genomes are also still in the nucleus of the host cell in the episomal state since AAV vectors lack the rep gene. The risk posed by AAV vectors is far lower than that posed by other viral vectors. Intramuscular administration of AAV2 vectors was followed by intraportal delivery, in Phase I investigations of hemophilia B gene therapy. The temporary gene expression is probably controlled by the activation of memory CD8+ T lymphocytes against the AAV capsid protein instead of the Factor IX protein. Due to its tiny size (4681 nt) or poor delivery of nucleotides bigger than its genome, the AAV requires further research to mitigate the immune reaction and enhance its packaging capabilities.

11.4.1 Herpes Simplex Viral Vectors

The total diameter of the living beings' neurotropic virus HSV is between 180 and 200 nm. It has a double-stranded genome that is about 150 kilobases in size and encodes for roughly 80 proteins. Before infecting central nervous system cells, the replication occurs in mucosal surfaces or on the skin through direct contact. Both dormant and active processes are readily apparent. Both viral replication and death of cells occur throughout the lytic cycle. Thanks to the latent infection, the virus can stay dormant in the host for an extremely long time. Both viral glycoproteins gB and gD link to the heparin sulfate moiety of the cell surface protein, initiating a fusion event that allows the virus to enter the cell. Recent research has shown that chronic pain can be prevented or even reversed *in vivo* through gene delivery mediated by HSV vectors [5, 6].

11.4.2 Other Types of Viral Vectors

In addition to the most well-known viral vectors discussed in the previous sections, a great number of other viruses have been taken into consideration for gene-delivery vehicles. Although the fact that most of these systems have not been utilized in clinical trials, there is a plethora of research proving the numerous benefits associated with their development and application in preclinical and developmental settings. Some members of the alphavirus genus, including the Sindbis virus, the Semliki Forest virus, and the Venezuelan equine encephalitis virus, have been engineered so that they can act as delivery vehicles for genes. Alternatively, you could insert an AAV vector in gutless adenoviruses and then modify the vector to transiently generate Rep protein. This would allow you to achieve site-specific integration and steer clear of insertional mutagenesis [7]. Two further hybrid viral vectors are HSV/AAV and HSV/Epstein-Barr virus [8]. It's interesting how numerous groups have attempted to encase the viral vectors in lipids or polymers [9]. Whenever a virus is administered *in vivo*, the major function of the coating is to disguise the virus so that it cannot be detected by the immune system of the host. This delays the breakdown of the virus, which in turn stimulates tissue transduction and lengthens the time that vectors can circulate across an area. The capacity to retarget viral vectors is another benefit of viral encapsulation. This is accomplished by first removing the viral vectors' natural tropism, and afterward retargeting the virus by combining targeting ligands with polymers.

11.5 Non-Viral Gene Therapy

Over the last 10 years, there has been a significant amount of interest in producing a non-viral vector that is capable of matching the degree of

Table 11.1 Types of non-viral gene delivery methods.

Physical non-viral gene delivery methods	Chemical-based nonviral vectors
• Needle and jet injection	• Cationic lipids
• Hydrodynamic gene transfer	• Cationic polymers
• Gene gun	
• Electroporation	
• Sonoporation	

specificity and gene expression afforded by viral vectors while enhancing the safety, gene inserts capability and immunogenicity. Recently, several suggestions made on ways of physicochemical distribution for non-viral-based systems. Examples of biomaterials that are less likely to provoke an immune response in the body include lipids, exposed DNA, chromatids, plasmids, cationic polymers, and conjugate complexes. Non-viral vectors that are used as delivery systems for gene therapy are another example of this type of material [10]. The following are two examples of popular non-viral gene delivery techniques (Table 11.1) [11].

11.5.1 Advances in Gene Delivery

The purpose of gene therapy is to regulate and rectify the expression of the gene that is expressed abnormally by inserting genetic material into the cellular gene regulatory system. This genetic material can be in the form of DNA or RNA. Nucleic acids (NAs) have a significant amount of promise for use in therapeutic applications; nevertheless, the applications of NAs are restricted due to factors including low cellular uptake, short *in vivo* half-lives, rapid clearance, and rapid enzymatic destruction. Numerous vectors, which can be generally classified as either viral or non-viral vectors, have been designed to get around this problem and provide efficient and focused gene delivery. In the beginning, it was thought that viral vectors may serve as viable carriers due to the inherent transduction capabilities and site specificity of viruses. Immunogenicity and pathogenicity, on the other hand, prevent the extensive clinical application of these agents. Because of their size selectivity and their naturally occurring transduction capacity, viral vectors were initially presented as potential candidates for suitable carriers. However, because of the immunogenicity and pathogenicity of these agents, their clinical application is restricted. These restrictions, in conjunction with recent developments in nanotechnology, have resulted in the creation of systems for the transport of DNA in the form of nanoparticles. There have been numerous distribution systems devised, each of which comes with its own set of benefits as well as drawbacks. The efficacy of these delivery methods has been greatly improved as a result of several recent changes to the delivery mechanisms, the introduction of cell-targeting peptides, and the addition of chemicals such as PEG, which are designed to extend the residence durations. There are still problems and questions that need to be resolved in this area of research, despite the progress that has been made. These include the nanomaterials' long-term impacts on gene expression and toxicities, as well as the traditional ADMETox testing procedures. Additionally, the nanomaterials' fate *in vivo* must be determined. To produce DNA nanoparticle compositions that are risk-free and reliable, preclinical and clinical research need to be well designed. As a direct consequence of this, this field has to progress as well. The requirement for additional fundamental study into the mechanics of gene activation and transport will invariably function as a barrier to the creation of effective nanoparticle-DNA formulations.

Our knowledge of molecular pathways and the relationships between phenotype and genotype has been significantly expanded as a result of the development of gene therapy, which offers the possibility of improvements in comparison to the pharmaceutical treatments that are now in use. The efficacy of gene therapy as a treatment is contingent upon several factors. These factors include i) DNA preservation, ii) high transfection effectiveness, iii) diminished toxic effects and lack of nonspecific impacts, iv) high potency even at quantities of DNAs, v) adaptation to various treatment regimens as well as illnesses, and (vi) effective vectors to avoid intracellular and extracellular barriers.

Even though a huge number of DNA delivery vectors have been created as a result of significant research, it is still an extremely difficult task to transport DNA to the cells or organs of interest in a way that is both safe and effective. This method of DNA delivery requires the elimination of several barriers, both extracellular and intracellular before it can be used successfully; these barriers can be eliminated with the use of effective vectors. To carry out its role as a delivery system, a vector needs to be composed of three distinct functional moieties, which are as follows: i) a ligand that is specific to the targeted cell or tissue for improved target capacity to impart more efficacious cellular uptake by receptor-mediated endocytosis, and (ii) a water-insoluble component to facilitate efficient unpacking of the compounds in the cytoplasmic compartment and endosomal escape following cellular intake. a ligand that is specific to the target cell or tissue for improved target ability to impart extra effective cellular uptake by receptor-mediated endocytosis In addition, it is possible to generate target-specific vectors by utilizing the variations in angiogenesis and metastasis between malignant and normal cells. This can be accomplished by building nanoparticle-DNA complexes that target tumor cells. Polymeric nanoparticles are quickly establishing themselves as DNA delivery methods both *in vitro* and *in vivo* as a result of the increasing prevalence of the utilization of nanoparticle-DNA complexes. According to an analysis of the relevant published research, polymeric nanoparticles have the potential to be utilized in the delivery of functional DNA to the target cells. In addition, anti-tumor therapy research has been conducted using cancers of varying types to test the efficacy of transporting DNA in nanoparticles. Even though systemic injection into nonhuman primates has been used in several successful animal model experiments, there are still a lot of hurdles that need to be jumped before DNA delivery can be utilized as a cutting-edge therapeutic strategy.

Long circulation durations, minimal immunogenicity, strong biocompatibility, selective targeting, successful penetration through barriers such as the blood-brain barrier and vascular endothelium, and a self-regulating release without detrimental clinical effects are all properties that should be present in the ideal DNA carrier system for nanoparticles. It is necessary to have in-depth knowledge of the efficiency and kinetics of DNA-vector systems to incorporate DNA complexes in therapeutic uses in the most effective manner possible. Processing DNA and finding improved methods for encasing or complexing it with polymers are both areas in which there is still much work to be done. In this day and age of manufacturing genetic therapy complexes, educational establishments, and companies need to work together to develop processes for the manufacturing of more secure polymeric nanoparticles and DNA, as well as dependable analytical methods to characterize formulations while they are in the process of forming and being store [12].

11.6 Recent Advances in the Development of Gene Delivery Systems

Gene therapy is a novel approach that can be used to treat and even prevent a wide variety of illnesses. To address a patient's ailment without resorting to medications or surgery, physicians may be able to use the gene therapy technique to implant a gene into a cell: i. replacing a disease-causing mutated gene with a healthy gene, ii. knocking out or deactivating a mutated gene that is functioning improperly, and iii. introducing new genes into the cells to prevent any diseases are all gene therapy strategies being researched by some scientists and medical professionals.

Gene therapy has the potential to treat a variety of ailments, like cancer, genetic abnormalities, and some forms of viral infection; however, the treatment is currently too dangerous to be widely employed. Only diseases for which no other effective treatments exist have had gene therapy studies. To initiate gene expression, the genetic material must get into the nucleus of the host cells [13]. Gene therapy has progressed to the point that it can now give genetic information to a patient's somatic cells, allowing those cells to produce precise therapeutic proteins that can treat inherited diseases. It is necessary to have a comprehensive awareness of the interactions that take place between the cell that is being targeted and the method that will be used to deliver the gene to correctly build the system. Gene delivery systems are comprised of three components: the plasmid-based gene expression scheme which regulates how a gene works within the target cell; a gene that encodes a specific therapeutic protein; and a gene delivery system that regulates how the gene expression plasmid is transported to a specific location inside the body. These three components work together to control the delivery of the therapeutic gene. For the system that delivers genes to work properly, the external genetic material must be able to maintain its integrity within the host cells [14, 15]. The process of transgenic expression began in the 1980s with the development of virus-based vectors as a technique. In 1984, chimpanzees were vaccinated against Hepatitis B using the vaccinia virus as the vector for the immunization. On the other side, non-viral gene delivery technologies were the first to describe how DNA exposures inside the body may modify cellular phenotypic characteristics.

11.7 Development of Gene Delivery Systems

11.7.1 Viral Vector Gene Delivery Systems

Viruses need to be capable of injecting their DNA into the cells they infect for them to be able to transmit genes. These profit from the capacity of a virus to multiply the genetic material that it carries. As their design prevents DNA liposome breakdown, viruses are an excellent medium for the transfer of genetic material. When it comes to gene delivery methods, the genes open up intending to be crammed into the replication. The infectious agents that spread viruses are called viral vectors. Retroviruses, adenoviruses, adeno-associated viruses, and herpes simplex viruses are some examples of viruses that have been employed as delivery methods for genes. Germline and somatic gene delivery methods are the two distinct types of these kinds of mechanisms. Despite the advantages of germline gene delivery technology, germline gene therapy cannot be employed ethically. The only method of human gene transfer so far has been the modification of somatic cells. There are three distinct methods for delivering somatic genes into cells: *in vivo* delivery, *ex vivo* delivery, and *in situ* delivery. The genetic components are transported directly into the tissue of interest through the use of *in vivo* delivery techniques. Before being implanted into the bone marrow of the desired tissue, genetic materials that have been produced and changed *in vitro* as part of *ex vivo* delivery systems must first undergo this process. The hereditary material is transferred to the tissues that were designated in advance. There are not any immunologic reactions; rather, there are just tactics that are used in a range of situations. Only an extremely minute fraction of the cells that were transplanted are still viable today. Using *in situ* delivery mechanism is utilized to transport the genetic material in a manner that is not only efficient but also precise. Because most of today's distribution methods don't ask for any kind of exact targeting, the tactic is perfect for the situation. Even though it is the least developed method available right now, it nevertheless has a chance of being useful.

11.8 Viral Vectors Based on DNA for Gene Delivery Systems

DNA-based viral vectors, when utilized for the transport of genes, are frequently more robust and integrate into the genomes of their hosts. Lentiviruses, poxviruses, adenoviruses, adeno-associated viruses, retroviruses, the human foamy virus (HFV), and herpes viruses are all examples of viral vectors that are based on DNA. The transmission of genetic information to the host cells is accomplished by the use of viral vectors in gene delivery systems that are based on DNA. The genetic material is effectively transferred to the host cells by the viral vectors. Plasmids that include transgenes for gene therapy are included in the category of DNA-based viral vectors that are used for gene delivery. Despite being in early clinical trials, Genetic material viral vectors have shown promise as gene delivery methods for cancer, AIDS, neurological illnesses like Parkinson's and Alzheimer's, and cardiovascular problems [16].

11.9 Viral Vectors Based on RNA for Gene Delivery Systems

At this time, it is possible for viral vectors based on RNA for gene delivery to directly convert infectious RNA transcripts. When RNA is used, the delivery of genes is typically temporary and does not become permanent. Gene therapy often makes use of a human foamy virus, onco-retroviral vectors, and lentiviral vectors. These are all examples of RNA-based gene delivery methods. This complex mechanism supplies RNA-dependent polymerase complexes as well as negative-strand RNA template molecules. Patients undergoing transplants for AIDS-related lymphoma have benefited from the use of RNA-based HIV gene delivery approaches. These procedures involve the use of CD 34+ cells that have been transformed with lentiviral vectors.

11.10 Oncolytic Viral Vectors for Gene Delivery Systems

Oncolytic viruses (OVs) represent a fresh approach to the treatment of cancer and diseases associated with cancer. Recently, they have shifted their attention to research on cancer to make their medications more effective. It has been reviewed both the benefits and the drawbacks of the multiple modifications that have been made to them to increase their contagiousness, anti-tumor resistance, and therapeutic efficacy for the contact between OVs and cancer cells. The oncolytic adenovirus which co-expresses IL-12 and IL-18 increases tumor-specific immunity by increasing the development of T cells that express IL-12R2 or IL-18R. This occurs as a result of the oncolytic adenovirus co-expressing both IL-12 and IL-18. The end goal is to create a virus that is capable of successfully replicating within the host, locating a specific target, and killing cancer cells. The expression of decorin that is produced by an adenovirus leads to the activation of the p53 gene and the apoptosis of mitochondria, which both lead to the death of cancerous cells. Through the application of gene therapy, oncolytic adenovirus (Ad) vectors present the possibility of a different approach to the treatment of cancer. T cells that express IL-23 and p35 and generate IFN and TNF are the ones that need to be activated to generate antitumor immunity. According to some reports, cytokine immune-gene therapy may be one of the most effective methods now available for treating cancer.

11.10.1 Non-Viral Vector Gene Delivery System

In recent times, physical-chemical techniques for the delivery of non-viral payloads have been developed. Biocompatible materials, such as lipids, naked DNA, chromosomal, plasmids, cationic polymers, and conjugate complexes, are less likely to cause an immune response in the body when in comparison to non-viral vectors or gene therapy delivery systems. The following (Table 11.2) is a list of biomaterial-based yields that are currently available for purchase and include some of them notwithstanding all of them.

The process of moving genetic material through the cellular membrane is done on purpose via a physical mechanism that is referred to as non-viral gene delivery. There are a variety of different physical methods that can be used to transfer genes, including hydroboration, magneto-fiction, sonoporation, photoporation, and ballistic DNA injection. The direct delivery of genetic material is what is meant when people talk about needle injections. In contrast, the technique known as ballistic DNA injection involves injecting gold-coated DNA fragments inside cells. The technique of producing pores inside a cell membrane using electroporation, which allows for the passage of genetic material, is known as electroporation. In the procedure known as sonoporation, a sound wave is used to pierce the membrane of a cell to provide room for the transfer of genetic material. During the process known as photo-poration, a laser pulse is used to penetrate the cell membrane, which then makes it possible for genetic material to get inside the cell. Magneto-fiction is a technique that concentrates nucleic acid particles inside of target cells by employing magnetic particles which have complexed with DNA and an external magnetic field. The hydrodynamic capillary effect is referred to as hydroboration, and it is responsible for modifying the permeability of cells. The non-viral chemical approaches for gene delivery systems generate particles that help in delivering genes into the cells by utilizing either artificial or organic components. These particles can be made from either organic or artificial materials. Synthetic vectors can compress vast amounts of genetic information to facilitate greater genetic transfers. This is accomplished by electrostatically contacting and connecting with RNA or DNA. Endocytosis is the process that allows non-viral chemical vectors to enter cells after this point. The two most common forms of non-viral vectors are called polymers and liposomes respectively. Lipoplexes, which are generated when non-viral vectors based on liposomes combine, simplify the process of gene delivery. When positively charged liposomes and opposite charges DNA come into touch with one another, the result is the spontaneous formation of

Table 11.2 Some commercially available biomaterial-based products [17].

S. no.	Biomaterials	Marketed name	License year	Indication	Treatment
1.	PLGA microspheres	**Leupron Depot®**	1989	Advanced prostate cancer	Hormone therapy
2.	6 silicone rubber tubes, cross-linked polydimethylsiloxane	**Norplant® implant**	1990	Contraceptive	Implant (Levonorgestrel)
3.	Lipid nanoparticles	**Doxil®**	1995	Breast neoplasms; multiple myeloma; ovarian neoplasms; Kaposi's sarcoma	Chemotherapy (Doxorubicin)
4.	PCPP-SA	**Gliadel®**	1996	Treatment to surgery for recurrent glioblastoma multiforme	Intracranial wafer
5.	PLGA	**Neutropin subcutaneous Depot®**	1999	Growth hormone deficiency	Microspheres
6.	Virus-like particle	**Gardasil®**	2006	Cervical, vulvar, and vaginal cancers	Vaccine delivery
7.	Poly (L-lactide)	**Absorb GT1**	2016	Coronary artery disease	Vascular scaffold system

(Continued)

Table 11.2 Some commercially available biomaterial-based products [17]. (*Continued*)

S. no.	Biomaterials	Marketed name	License year	Indication	Treatment
8.	Lipid nanoparticles	**Onpattro®**	2018	Polyneuropathy of hereditary transthyretin-mediated amyloidosis in adults	Gene therapy (siRNA)
9.	Polyethylene glycol-based Hydrogel conjugated with fluorescein	**Dextenza®**	2018	Ocular inflammation and pain following ophthalmic surgery	Insert (Dexamethasone)
10.	Lipid nanoparticles	**Comirnaty®**	2020	SARS-CoV-2 infection	Vaccine delivery

Table 11.3 A broad comparison of viral and non-viral transfection vectors [18].

Vectors	Different criteria				
	Immunogenicity, Inflammation	**Mutagenesis**	**Vector production**	**Transfection efficiency**	**Duration of gene expression**
Viral	High-medium risk	High-medium risk	Laborious, batch-to-batch variability	High efficiency	Long term
Nonviral	Medium-low risk	Medium-low risk	Ranges in difficulty	Medium-low efficiency	Medium, transient

Table 11.4 Some examples of mixed hydrogels for biomedical applications [19].

Hydrogel system	Comment
Collagen, HAp, Alendronate	Scaffolds, bone regeneration
Alginate, hydroxyapatite nanowire	0.2–0.5 mm HAp wire, Ca as cross-linker; bone defect repair
NIPAM, mesoporous SiO2	Sustainable drug release
Poly(l-lysine), polypeptide, silica colloid	Miniemulsion; redox, pH-sensitive
Collagen/polyNIPAAm, MM	Stimuli-responsive hybrid, elastic
Hyaluronate, CSA-coated MNPs	Intra-articular injection
Chitosan, GPTMS, GP	Defect filler/solidifiable gel
ECH cross-linked chitin, nano-HAp	Scaffolds for tissue engineering
TiO2 NP, carboxymethyl cellulose	Amino-modified NPs, scaffold

NP nanoparticle, NIPAM, NIPAAm, N-isopropyl acrylamide, MM montmorillonite, nHAp nano-hydroxyapatite, CS chondroitin sulfate, PVA poly(vinyl alcohol), CSA chondroitin sulfate A, MNPs magnetic nanoparticles, ECM extracellular matrix, PEG poly(ethylene glycol), GP glycerophosphate, ECH epichlorohydrin.

lipoplexes. In the case of non-viral vectors based on polymers, the polymers produce polyplexes by interacting with the DNA.

Because of this, the production of viral vector cells for careful planning and can take anywhere from two weeks to six months. Furthermore, differences in batches or sources might make comparisons and replication difficult. In light of these constraints, non-viral vectors that are less costly, risky, and erratic in their production are continually being developed to compete with the efficacy of viral vectors. A comparison between viral and non viral transfection vectors are mentioned in Table 11.3, as well as in Table 11.4, we have included some examples of mixed hydrogels for biomedical applications.

11.10.2 Nano-Sized Organic-In-Organic Particles

11.10.2.1 Cerasomes

Proteins, genes, drug delivery vehicles, and cells all carry physiological components in varying amounts. Targeting is the most important part of the distribution technique. Vascular mobility follows endocytosis susceptibility.

Cationic phospholipids form liposomes, spherical bilayer-containing vesicles, in water. Liposome transfection (lipofection) can deliver nucleotides to cells. Cerasomes can transport DNA since they are as small as viral shells. HepG2 cells and HeLa were used to confirm the cerasome's DNA-distribution capabilities for its latest drug delivery status. In conjunction with supra-molecular investigations, another study reviewed organo-silanetriols and the production of caged compounds and cerasomes.

11.10.2.2 Magnetoliposomes

The nucleus of the cerasome in magnetoliposomes is replaced by a magnetic moiety in magnetoliposomes. They are efficient pharmaceutical carriers as a result of the magnetic guidance that is provided for the delivery. The numerous proposals have all taken into account the process of selecting the magnetic material, as well as the materials that will be used for the shell that will surround it.

11.10.2.3 Nanomachine

Targeted distribution of medicines, DNA, or proteins is the most crucial aspect of drug delivery; In 2016 Yang and Yu reviewed the use of mesoporous silica-based NPs for cancer therapy. The idea of a nanomachine has gained widespread acceptance recently.

11.11 Practical Application of Gene Delivery Methods

The development of the optimal method for delivering genes is necessary for the successful application of gene therapy. When used in therapeutic contexts, the gene delivery method makes use of non-immunogenic vectors with cells preferentially to provide sufficient levels of transgene expression to achieve the desired effect. Using analytical software that examines the patterns of gene expression, DNA or RNA microarrays, which are employed in a variety of approaches to gene sequencing, can identify hundreds of genes at once. The technology for delivering genes was utilized in the production of a hybrid biosynthetic vector, which has the potential to be used for the delivery of a possible vaccine. By mixing *Escherichia coli* with synthetic polymers, certain vectors can circumvent the conventional roadblocks that stand in the way of gene delivery. The system of gene therapy presents a great prospect for the treatment of a wide variety of diseases, including cancer and various infections that are caused by genetic defects. The non-viral vector provides several benefits that its viral counterpart lacks, including a low cost of replication and minimal immunogenicity. There is no limit on the amount of DNA that may be packed into a non-viral vector and altered with ligands to specifically target a certain cell.

In the context of animal gene transfer, research into and evaluation of the utilization of gene delivery systems constructed of artificial polymeric materials have been carried out. Plasmid DNA is one approach that has the potential to be used for safely transporting the gene into the nucleus of the cells that are being targeted. Gene delivery systems that are based on non-viral vectors are a relatively new therapeutic strategy that has lately come to light as a novel and possibly useful therapeutic strategy.

11.11.1 Gene Delivery Systems Based on Cationic Polymers

Researchers looking into gene delivery systems have looked at the possible advantages of polymeric carriers for genes in charged biopolymers. These polymeric carriers include liposomes and derivatives of chitosan, for example. Polymeric systems can conceal the negative DNA ions and condense big genes into small molecular structures when applied in combination with cationic polymers for gene delivery. Polyplex is the name of the material that contains cationic polymer-based nucleic acid. Lipoplexes, which are cationic non-viral lipid-based gene carriers, is currently undergoing testing on people [20]. In gene therapy, the gene complexes are an essential component of the targeted delivery process. The majority of study focuses on investigating how the DNA complex is affected by ligands that are already covalently linked to it. The cationic polymers can combine with the ligands that are being targeted. One of the rivals is poly (L-lactide), which has a long history of application as a polymer to attach specific ligands. Under a wide range of circumstances, the charged polymer condensation of plasmid DNA will run into difficulties with compacting. Condensing agents are most commonly utilized when working using multivalent cationic polymers. The tertiary structures of complexes that resemble non-condensing plasmid DNA have been investigated by researchers. These complexes contain minimal amounts of hydrophobic stearyl-poly (L-lactide). Despite this, DNA condensation has made it possible to link hydrophilic segments by utilizing hydrophobic stearyl chains, polyethylene glycol, dextran, or other hydrophilic components.

11.11.2 Gene Delivery Systems Based on Polysaccharides

The incorporation of grafted oligoamine residues into natural polysaccharides has resulted in the production of a great number of biodegradable cationic polymers. The notion of grafting involves attaching side chain oligomers to hydrophilic polysaccharides that are either linear or branched to permit three-dimensional contact with the anionic surfaces of double-strand DNA strands. According to the polysaccharide-based gene delivery technology, the structure of the cationic polysaccharide is thought to play a crucial role in the transfection activity for gene transport. This is because cationic polysaccharides are ionic. The mucosal transport of complexes such as peptides, proteins, oligonucleotides, and plasmids has been examined with the use of colloidal polysaccharide particles. Nanoparticles composed of natural polysaccharides have also been used in research looking at the delivery of drugs and genes. The formation of polysaccharide-based nanoparticles required several different stages, including electrolyte complexing, covalent and ionic cross-linking, and hydrophobic polysaccharide self-assembly. Pectin and chitosan are two examples of naturally occurring polysaccharides that have been investigated for their potential use as drug or gene transporters. Chitosan families have strongly suggested polycationic derivatives of poly-N-acetyl-D-glucosamine. This is because poly-N-acetyl-D-glucosamine is a positively charged molecule that has considerable mucoadhesion. The structural portions of pectin known as rhamnogalacturonan (RG-I) and homo-galacturonan (HG) have both smooth and hairy regions in their composition.

11.11.3 Poly(Ethylenimine)-Based Gene Delivery

In most cases, the cationic polymers that are employed for gene complex formation are polyamines that, when exposed to physiological circumstances, change into cationic forms. The chemical compound known as poly(ethylenimine) has numerous nitrogen atoms, as well as 1°, 2°, and 3° amine groups (PEI). As a consequence of the presence of these amine groups, they might become more efficient as buffers. These trigger a breach in either the endosome or the lysosome, which makes it possible for protons to be released into the cytoplasm via the proton sponge action [20]. PEI has the potential to contribute, albeit in a limited way, to the buffering capacity of other functional molecules. Poly(cystamine bis (acrylamide)-diamino hexane) [poly(CBA-DAH)] linked with PEI was designed to increase the effectiveness of transfection while simultaneously decreasing the weight ratio. Poly(CBA-DAH) is made up of several disulfides, and an intracellular reductant like glutathione can break it down into its parts in the cytoplasm (GSH). The precursor amino acids were changed further into triple-peptide that is GSH in the cytosol. To validate the successful production of vectors and polyplexes containing pDNA, research has been conducted on the synthesis of a bio-reducible polymer that is coupled with PEI for effective gene transport. Other investigations have demonstrated that other compounds, like imidazole, histidine, and PEI, possess greater buffering capabilities than the previously mentioned chemicals. There have been investigations into the use of biodegradable poly(ethylenimine) s for the delivery of plasmid DNA. In addition, water-soluble branching poly(Ethylenimine)-cholesterol lipo-polymers were produced for gene transfer. Gene delivery was made possible by the use of improved linear polyethyleneimine/cholesterol conjugates for DNA complexation. A gene delivery technique that makes use of linear polyethyleneimine-cholesterol/ DNA complexes has had its efficacy and bio-distribution investigated. A biodegradable gene carrier composed of polyethyleneimine with acid-labile connections had been designed and put through its paces in the testing phase before it was ready for use. Techniques for the delivery of genes based on the reducible polymer poly (amido ethylenimine) have been created and tested.

11.11.4 Polymer-Based Gene Delivery (L-lysine)

Poly(L-lysine) is a cationic polymer that is used in gene transfers (PLL). At a pH of 7, the PLL molecule is a type of homo-polypeptide that belongs to the class of cationic polymers. There is a possibility that PLL contains an amino group that is hydrophilic and positively charged. To assist in the co-adsorption of plasmid DNA, Replica Particles (RP) based upon PLL cross-linked through a homo-bi-functional linker have been designed for use in gene delivery systems. The *in-vivo* transport of genes to the liver has been improved by stabilizing poly-L-lysine/DNA polyplexes. The use of poly-L-lysine in the creation of gene delivery systems that have been characterized has taken place. In mouse lung endothelial cells, an N-terminally modified poly(L-lysine)-antibody conjugate had been used previously as a carrier for targeted gene delivery. Non-viral methods of delivering genes are successful when using biodegradable poly(ethylene glycol)-co-poly (L-lysine)-g-histidine multi-block copolymers. In the section under polymeric gene delivery, both poly(-L-lysine) and copolymers for gene delivery were discussed and analyzed. In addition, gene delivery methods utilizing poly-L-lysine-graft-PEG-comb polycation copolymers of this type have been developed. Initially, researchers worked on developing and studying Nano-particle DNA carriers utilizing poly(L-lysine) grafted polysaccharide copolymers and triplex DNA delivery systems as a gene carrier. Gene delivery systems also saw the development of novel non-viral DNA delivery vectors of the triplex systems and poly(L-lysine) grafted polyethylene glycol. Recent years have seen a rise in the number of clinical trials done in the field of biological applications.

11.12 Conclusion

While not exhaustive, the three gene delivery methods mentioned in this chapter are commonly utilized in clinical and preclinical research or advanced development; 67% of clinical trials employed viral vectors, whereas 24% used non-viral techniques. Even though the methodologies presented in this chapter have contributed to research and gene therapies and will continue to do so, a range of viral and non-viral vectors, a few of which are yet to be fully exploited or even discovered, are anticipated to enrich the current toolset. One gene delivery method may not suit all gene delivery needs. Three gene delivery technologies have improved. If gene therapy is to succeed, much work remains to build a human gene therapy method. Despite viral vectors' success and clinical trials, viral proteins and DNA encourage carcinogenesis and immune response. A method to incorporate transgenes into such a targeted chromosome without repeated transportation and insertion mutagenesis may solve this challenge. Non-viral vector-based gene delivery research has used organic and synthetic materials, but they transfer genes orders of magnitude worse than viral vectors. Thus, current non-viral vectors are not considered beneficial. Given this article's strategies' pros and cons, no single method will work for every application. Current physical techniques can be rectified with modern technology, computers, and a better knowledge of biological systems. Gene delivery for gene therapy may depend on our ability to integrate engineering, cell biology, and computer science.

References

1. High, K.A. and Roncarolo, M.G., Gene therapy. *N. Engl. J. Med.*, 10, 4, 455-64, 2019.
2. Ginn, S.L., Amaya, A.K., Alexander, I.E., Edelstein, M., Abedi, M.R., Gene therapy clinical trials worldwide to 2017: An update. *J. Gene Med.*, 20, 5, 1–16, 2018.
3. Kamimura, K., Suda, T., Zhang, G., Liu, D., Advances in gene delivery systems. *Pharm. Med.*, 25, 5, 293–306, 2011.
4. Piedra, P.A., Oka, K., Chan, L., Kim, I., Jo, A., Lifetime correction of genetic deficiency in mice with a single injection of the helper-dependent adenoviral vector. *Proc. Natl. Acad. Sci. U. S. A.*, 98, 23, 13282–13287, 2001.

5. Williams, D.F., On the nature of biomaterials. *Biomaterials* [Internet], 30, 30, 5897–909, 2009, Available from: http://dx.doi.org/10.1016/j.biomaterials.2009.07.027.
6. Glorioso, J.C. and Fink, D.J., Herpes vector – mediated gene transfer in the treatment of chronic pain. *Molecular Therapy*, 17, 1, 13–8, 2009.
7. Recchia, A., Parks, R.J., Lamartina, S., Toniatti, C., Pieroni, L., Palombo, F. *et al.*, Site-specific integration mediated by a hybrid adenovirus/adenoassociated virus vector. *Proc. Natl. Acad. Sci. U. S. A.*, 96, 6, 2615–20, 1999.
8. Oehmig, A., Fraefel, C., Breakefield XO, A.M., Herpes simplex virus type 1 amplicons and their hybrid virus partners. *Curr. Gene Ther.*, 4, 4, 385–408, 2004.
9. Fisher, K.D., Stallwood, Y., Green, N.K., Ulbrich, K., Mautner, V., Seymour, L.W., Polymer-coated adenovirus permits efficient retargeting and evades neutralizing antibodies. *Gene Ther.*, 8, 5, 341–8, 2001.
10. Yang, R., Chen, F., Guo, J., Zhou, D., Luan, S., Recent advances in polymeric biomaterials-based gene delivery for cartilage repair. *Bioact. Mater.* [Internet], 5, 4, 990–1003, 2020, Available from: https://doi.org/10.1016/j.bioactmat.2020.06.004.
11. Al-Dosari, M.S. and Gao, X., Nonviral gene delivery: Principle, limitations, and recent progress. *AAPS J.*, 11, 4, 671–81, 2009.
12. Jones, C.H., Chen, C.K., Ravikrishnan, A., Rane, S., Pfeifer, B.A., Overcoming nonviral gene delivery barriers: Perspective and future. *Mol. Pharm.*, 10, 11, 4082–98, 2013.
13. Neufeld, E.F., Sweeley, C.C., Rogers, S., Friedmann, T., Roblin, R., Gene therapy for human genetic disease? *Science (80-)*, 178, 4061, 648–9, 1972.
14. Pack, D.W., Hoffman, A.S., Pun, S., Stayton, P.S., Design and development of polymers for gene delivery. *Nat. Rev. Drug Discovery*, 4, 7, 581–93, 2005.
15. Nayerossadat, N., Ali, P., Maedeh, T., Viral and nonviral delivery systems for gene delivery. *Adv. Biomed. Res.*, 1, 1, 27, 2012.
16. Patil, S.D., Rhodes, D.G., Burgess, D.J., DNA-based therapeutics and DNA delivery systems: A comprehensive review. *AAPS J.*, 7, 1, E61–77, 2005.
17. Han, X., Alu, A., Liu, H., Shi, Y., Wei, X., Cai, L. *et al.*, Biomaterial-assisted biotherapy: A brief review of biomaterials used in drug delivery, vaccine development, gene therapy, and stem cell therapy. *Bioact. Mater.*, [Internet], 17, 2022, 29–48, 2022, Available from: https://doi.org/10.1016/j.bioactmat.2022.01.011.
18. Uludag, H., Ubeda, A., Ansari, A., At the intersection of biomaterials and gene therapy: Progress in non-viral delivery of nucleic acids. *Front. Bioeng. Biotechnol.*, 7, JUN, 1–21, 2019.
19. Stone, D., Novel viral vector systems for gene therapy. *Viruses*, 2, 4, 1002–7, 2010.
20. Benjaminsen, R.V., Mattebjerg, M.A., Henriksen, J.R., Moghimi, S.M., Andresen, T.L., The possible proton sponge effect of polyethyleneimine (PEI) does not include a change in lysosomal pH. *Mol. Ther.*, 21, 1, 149–57, 2013.

12

Utilization of Silk for Gene Delivery

Swatantra K. S. Kushwaha[1]*, Shruti Khare[2] and Neelottama Kushwaha[1]

[1]Krishna Institute of Pharmacy and Sciences, Kanpur, India
[2]Advance Institute of Biotech and Paramedical Sciences, Kanpur, India

Abstract

Numerous studies on the application of gene delivery have been conducted; however, because gene-based formulations require viruses as gene vectors, they have not been effective in attaining commercial adoption. Such vectors often raise questions about immune response and systemic toxicity. As a result, the chapter provides an overview of recent research that has moved its emphasis to non-viral vectors made of biocompatible polymeric formulations. Because they may be genetically modified to have certain chemical properties, silks are self-assembling, biodegradable, biocompatible proteins that are valuable for the transportation of medications and DNA. Silk fibroin has been used by numerous researchers to create medicine delivery systems. Silk proteins have the ability to self-assemble into mechanically robust, biodegradable, and non-cytotoxic material structures, which may be employed in the transfer of genes. Because silk fibers can be genetically modified to have certain chemistry, molecular weight, and other design characteristics. The chapter provided a summary of the various developments using silk fibroin as a delivery vehicle. Due to its biocompatibility, predictable degradability, and customizable drug release qualities, silk has been employed frequently in the development of drug delivery systems. A variety of pharmaceuticals and important biological drugs like proteins and DNA can be stabilized and encapsulated by silk-based drug delivery methods to increase their stability and duration in the bloodstream. The feasibility of these methods for delivering small molecule pharmaceuticals, proteins, and genes will be discussed in this chapter.

Keywords: Gene delivery, biocompatible proteins, self-assemble, silk-based drug delivery methods, gene vector, customizable drug release

**Corresponding author*: swatantrakushwaha15@gmail.com

Rishabha Malviya, Sonali Sundram and Neelam Jain (eds.) Integration of Biomaterials for Gene Therapy, (349–366) © 2023 Scrivener Publishing LLC

12.1 Introduction

To increase the effectiveness of therapy and the life expectancy of patients, polymer-based drug delivery systems have been used in place of conventional formulations as a more practical alternative. These systems, which also modify their chemical and physical properties and address the main difficulties associated with administrations, such as accurate drug release, intracellular transport, and biocompatibility, contained in the active drug moiety. An optimal drug delivery system should maintain the loaded API, allow for the modification of its release kinetics, and reduce side effects by tissue-specific targeting, especially in the case of highly targeted therapies like anticancer drugs. Despite being valued as a natural resource for the textile industry for millennia, silk has most recently garnered a lot of attention as a possible copolymer for use in biological and pharmacological applications [1, 2]. Structure and properties of silk is shown in Figures 12.1 and 12.2 respectively.

12.2 Dimensional Structure of Silk

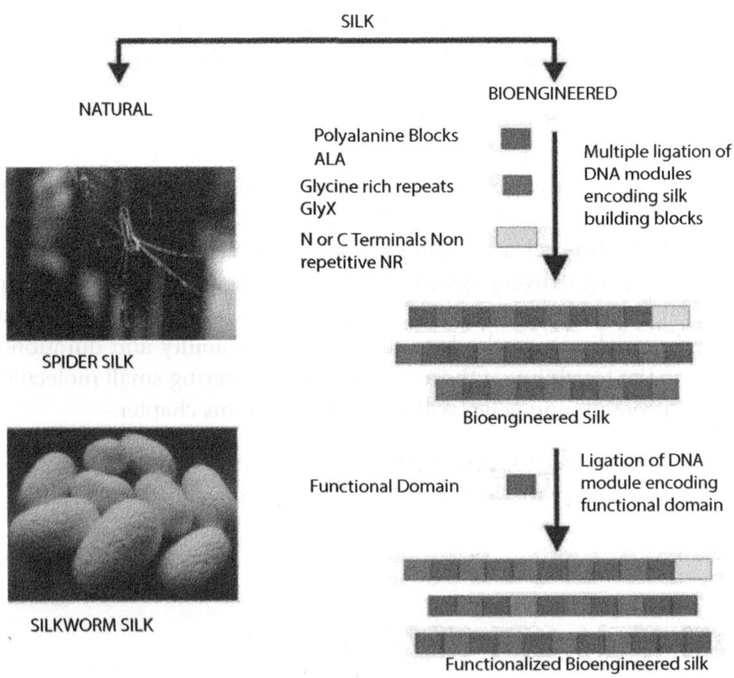

Figure 12.1 Dimensional structure of silk [3].

12.3 Properties of Silk

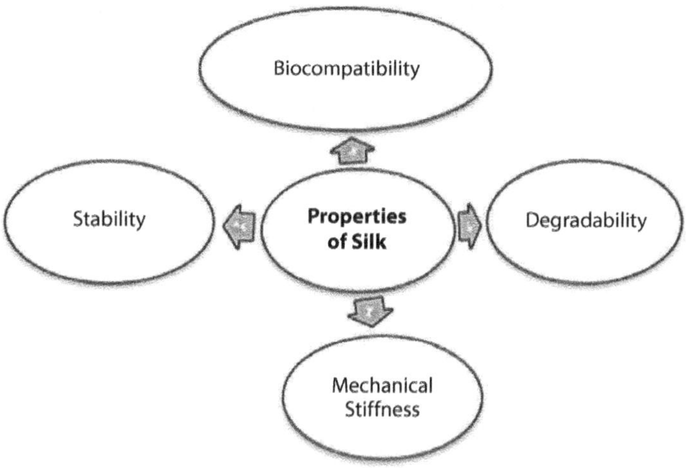

Figure 12.2 Different properties of silk.

12.3.1 Biocompatibility

As per FDA, Silk fibroin is a biocompatible substance with little immunogenic effects. Hepatocytes, osteoblasts, and mesenchymal cells are only a few examples of the diverse cells with which the silk fibroin has proven to be very compatible [4]. The various studies results showed that the cross-linking of silk fibroin with substances like hexafluoroisopropanol or methanol results in formulations that are efficient against inflammatory responses. The use of calcium polyphosphate and silk fibroin in conjunction for various bone deficiencies is another illustration of biocompatibility [5].

12.3.2 Mechanical Strength

Silk fibroin has enough mechanical strength, but it must match the target tissue's rigidity in order to be effective. Mechanical rigidity is a key component of silk fibroin based drug delivery system for pharmacological and biological purposes. In order to increase their mechanical strength, polymers like polylactic glycolic acid and collagen are cross-linked, but this causes cellular toxicity or immunogenicity [6].

12.3.3 Stability

One of the most essential components in pharmaceutical formulations is the stability of the silk based polymeric substances. The possibility that silk fibroin will gel when kept over an extended period of time. As a result, the various climatic variables are preserved because the less stable helical structure of a silk fibroin sheet is created in very humid environments [7]. The glass transition temperature must be maintained because, when compared to other proteins, silk fibroin has the maximum robustness under thermal stress. Tg also affects the crystallinity and porosity of silk fibroin [8, 9].

12.3.4 Degradability

Silk fibroin needs to be properly degradable in order to be susceptible to proteolytic enzymes [10, 11]. The alteration in the particle size, cross-linking, and crystalline structure of silk fibroin alters its breaks down. Previous studies revealed that the rat fibroblast cells' phagocytic activity is directly influenced by the breakdown of silk fibroin [12]. Raw silk degrades more quickly than pure silk fibroin because the protein sericin promotes cellular breakdown [13].

12.4 Extraction of Fibroin from Silk Worm

A core of fibroin, which makes up 75% of the weight of a silk fiber, and sericin, which makes up 25% of the weight of a silk fiber, surround the core. Sericin can be eliminated by the thermochemical process known as degumming, which employs sodium bicarbonate as the degumming agent and chaotropic salt solution to convert silk II into silk I, according to Aramwit and Melke *et al.* Sarin is a mixture of several water-soluble globular glycoproteins with the potential to elicit an immune response. The viscous liquid is then centrifuged for 20 minutes at 9000 rpm and 40°C after being dialyzed for 48–72 hours against ultrapure water. The soluble regenerated fibroin can be kept in the supernatant for at least a month before it turns into an irreversible gel. Figure 12.3 has summarized different steps of fibroin formation as it is composed of 75% of total silk content. Fibroin solutions should be lyophilized for long-term storage; the powder that results will be stable for many years at room temperature [14, 15].

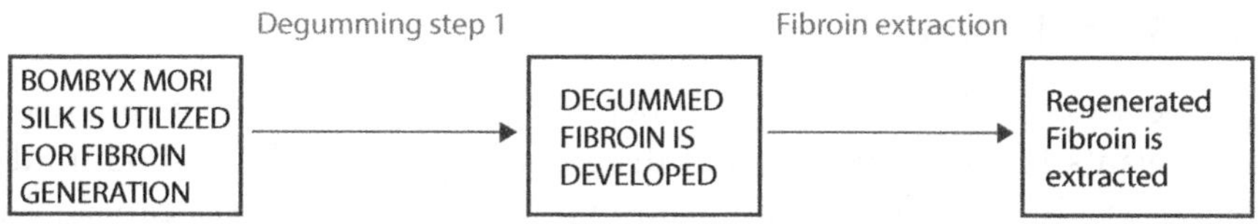

Figure 12.3 Steps of fibroin formation.

12.5 Fabrication of Silk in Different Therapeutics Carriers

Several methods are used to generate silk fibroin-based drug delivery systems. Each delivery method generates a special delivery with special characteristics, such as altered release kinetics, stability, and other qualities that can be useful in certain applications. Hydrogels, films, micro- and nanoparticles, nanofibers, lyophilized sponges, and polymeric particles coated with silk fibroin are just a few of the many different types of drug delivery methods that have been developed. The chapter reviews the silk-based drug delivery techniques that have received the most in-depth research.

12.5.1 Drug Delivery System Based on Hydrogels

Silk fibroin aqueous solution is utilized to create hydrogels. The transformation of sol to gel uses both natural and synthetic polymeric solutions. Shearing or water exclusion, followed by an evaporation technique, is the main method used to prepare the gel form. The stability of the gel form, which can previously be destroyed by oxidative or enzyme reactions, is due to the -sheets of silk fibroin [16].

12.5.2 Silk Nanoparticles

Silk-based nanoparticulate systems are utilized to modify drug delivery and to encapsulate active medicinal ingredients, primarily anticancer therapeutics. The regulated drug delivery is carried out through a silk-based nanoparticulate system, which can be created by shaping polyvinyl alcohol into the right shapes and sizes. Their charge and lipophilicity are the primary variables impacting the effectiveness of medication distribution and encapsulation in such systems. Salting out is one of the most used methods for producing nanoparticles. Salting out agent made of potassium phosphate [17].

12.5.3 Silk Films

Covalently coated with therapeutic peptides, silk films have been employed as implants for bone growth and drug delivery. Using the carbodiimide chemistry, BMP-2, RGD, and parathyroid hormone (PTH) can all be directly immobilized on silk films for bone regeneration. Human bone marrow-derived stem cells were cultivated on decorative silk sheets, and immobilized BMP-2 was used to encourage differentiation. The ability of adenosine kinase-deficient embryonic stem cells to synthesize adenosine over an extended period of time has been studied using silk films as a promoter [18].

12.5.4 Silk Nanofibers

It's possible for scaffolds used in tissue engineering to mimic the extracellular matrix's structural and biological characteristics. Numerous fibrous collagens that are enclosed in proteoglycans make up the extracellular matrix. The hierarchical structures created by collagen fibers grouped in a three-dimensional porous network range in size from many fibers with nanometer-scale diameters to macroscopic tissue architectures. With nanoscale threads and microscale interconnecting holes, electrospinning produces structures that resemble the extracellular matrix's topology. Electrospun scaffolds comprised of silk fibroin fibers are hence promising. For *in-vitro* bone production from hMSCs, electrospun silk fibroin fiber scaffolds containing BMP-2 and/or hydroxyapatite (HAP) nanoparticles have been studied [19]. he main factors affecting the efficiency of drug distribution and encapsulation in such systems are their charge and lipophilicity. One of the most popular methods for producing silk fibroin particles is the salting-out procedure. For instance, Lammel *et al.* produced silk fibroin particles with regulated diameters ranging from 500 nm to 2 m using potassium phosphate as the salting out agent [20].

12.6 Utilization of Silk for Gene Therapy

Gene therapy is used to treat acquired disorders like cancer and hemoglobinuria as well as hereditary diseases like cystic fibrosis. Due to the therapeutic genes' high molecular weight and negative charges, the successful delivery of therapeutic genes to the desired location depends on an efficient gene delivery system. The delivery of exogenous genes in many cell types has been studied using both viral and non-viral systems. Viral systems like lentivirus, adenovirus, and herpes simplex virus, are frequently used because of their simplicity and high efficacy. Now a days silk offers a special way to create well-defined gene carriers with excellent transfection efficiency and minimal toxicity, as well as being amenable to bioengineering for fine-tuning of chemistry and sequencing. Recent studies have shown that silk fibroin-based gene delivery approaches provide high transfection effectiveness, biocompatibility, and DNase resistance [21]. The most important aspect of gene delivery involves introduction of genes into the cells of a tissue for the therapeutic effect. RNA is degraded to treat various diseases in recent advances like RNA interference and gene silencing. But the most widely applied method of gene delivery has been plasmid DNA transfer into tumor cells. Oncogene activity is reduced in this instance, and the cell's tumor suppressor regulatory genes are activated [22]. Gene therapy entails giving the cells the essential genetic resources, such as genes, DNA, in order to replace, regulate, or express the genes/proteins required for biological function [23].

Gene therapy, which is often used to treat hereditary disorders, involves delivering genes or DNA to the cells in order to control or express the genes for cellular functions. The majority of the time, a carrier is necessary to get across different intracellular barriers, such as cell or endosome membranes, and to keep the genetic components from degrading. Fibroin has been taken into account as a potential gene transport pathway due to its adaptability, nontoxicity, high transfection effectiveness, and DNase resistance. In order to construct magnetic-silk/polyethyleneimine core-shell nanoparticles for accurate gene delivery into cells, researcher adjusted the amount of silk fibroin employed in particle creation, the size, and the zeta potential of the particles [24].

12.7 Properties of Silk Fibroin as Biomaterial

Following are the some specific properties of silk which are crucial to utilize the silk for gene as well as drug delivery.

12.7.1 Silk Fibroin's Amino Acid Composition and Derivatization

The cocoon of the domesticated silkworm *Bombyx mori* contains one of the most distinctive types of silk. The fibroin heavy chain and light chain are the two structural proteins necessary for the development of *Bombyx mori* silk (25 kDa). The protein sericin, which functions as glue, holds the chains together (20 kDa to 310 kDa). Two fibroin filaments in the silk thread employed in the cocoon are encased and held together by silk sericin, a naturally occurring polymer generated by the silkworm Bombyx mori. The amino acid composition of silk fibroin offers possibilities for chemical alterations [25]. Researchers have looked into a variety of side groups as potential reactive side groups for the chemical modification of silk fibroin, side groups like amines, alcohols, phenols, carboxyl groups, and thiols. For instance, to increase cell adhesion, the primary amines of peptides like the RGD sequence have been derivatized with carboxylic acid side groups using aspartic and glutamic acids, which together make about 2.3 percent of the total amino acid composition of silk fibroin [26].

12.7.2 Chain Length and Molecular Weight of Silk Fibroin

A polymer's mechanical and biodegradable characteristics are greatly influenced by its molecular weight, which also affects its potential for use in drug delivery. Because of this, tailored polyesters with different lactide: glycolide ratios, such as poly (lactide-co-glycolide) (PLGA), have been widely used for drug delivery applications [27].

12.7.3 Solubility of Silk Fibroin

Crystalline silk fibroin does not solubilize the solvents used to dissolve polymeric solutions in medication delivery. To dissolve silk fibroin, lithium bromide, lithium must be present in a solution for longer periods of time without aggregation. It is therefore possible to make silk fibroin matrices that differ from those made from solutions with lower silk fibroin concentration, such as denser pattern that thiocyanate, and other highly concentrated salt solutions are routinely utilized. The ability to regulate the solubility of silk fibroin enables higher concentrations of silk fibroin to may impact release kinetics [28].

12.7.4 Hydrophobic and Crystal Nature of Silk Fibroin

As the silk fibroin blocks are hydrophobic in nature so they are responsible for the crystal regions by the formation of intermolecular β-sheets. The interaction between medicinal molecules and silk fibroin are affected by the alterations in both crystallinity and hydrophobicity. The changes made to the crystallinity also affect how quickly silk fibroin breaks down, which could be a possible strategy for developing drug delivery systems with a variety of release kinetics [29].

12.7.5 Biocompatibility of Silk Fibroin

According to reports, silk fibroin implants *in vivo* cause a foreign body reaction that is on par with or even less severe than that caused by the most common biomaterials now in use. In a 1-year implantation research with rats, scaffolds made from HFIP and all-aqueous solutions were investigated. The host immune response to the implanted scaffolds, according to Wang *et al.* (2008), was modest and local, which is comparable to the outcomes of earlier research using silk fibroin films. The host animals tolerated these scaffolds well [30].

12.7.6 Thermal Stability of Silk Fibroin

Low sheets content in untreated silk fibroin matrix makes it more hygroscopic and sensitive to dampness. The storage stability of silk fibroin matrices has not been the subject of any comprehensive research. Up to a temperature of 140 °C, silk fibroin has outstanding thermal stability and is unaffected. The transitional glass temperature (Tg) is considered to be a key factor in protein self-assembly [31]. Researchers discovered that the dried silk fibroin silk films that were made from Bombyx Mori silk's tail at Tg of 175°c. Free molecular movement occurs above this temperature and transforms into a stable β sheet shape [32].

12.8 Summary of Silk-Based Formulations for Gene Delivery [33]

S. no.	Different formulations based approach	Genes utilized
1.	Recombinant elastic based polymeric hydrogels	Adenovirus [34]
2.	3D porous scaffolds	Renilla luciferase reported gene [35, 36]
3.	Bioengineered silk film	Human embryonic kidney cells
4.	SF-Coated PEI/DNA Complexes	pDNA
5.	Bioengineeredsilk–polylysine–ppTG1 nanoparticles	pDNA [37]

12.9 Examples of Some Delivery Approaches which Utilizes Silk as a Biomaterial for Gene Delivery

12.9.1 Bioengineered Silk and Elastin-Based Materials for Gene and Delivery Medication

According to several results, genetic engineering techniques enable the creation of protein-based drug delivery systems with precisely calibrated physicochemical properties, providing a higher level of customization than is feasible with synthetic polymers [38]. Elastin-like proteins, silk-like proteins, and silk-elastin-like proteins all have unique chemical and physical properties that make them suitable for use as drug delivery vehicles. Recombinant DNA technology also makes these proteins simple to manufacture [39, 40].

12.9.2 Multifunctional Spider Silk Polymers for Gene Delivery

The important transport methods that can be secure and efficient substitutes for the current viral based systems are multifunctional spider silk polymers [41– 43]. As a result of bioengineering spider silk human mesenchymal cells, multifunctional gene carriers can interact with human mesenchymal stem cells [44, 46]. These carriers have been successfully used to carry DNA to the nuclei of mammalian cells. The presence of specific functional sequences in the recombinant proteins, along with a nuclear localization sequence of the large tumor antigen of the Simian virus, a high affinity binding peptide for human mesenchymal cells, and a translocation motif of the hepatitis-B virus specific protein, as well as their roles in reducing and increasing gene transfection efficiency toward human mesenchymal cells, were investigated [45, 47]. The results demonstrate that these spider silk proteins function like an efficient carrier without the well-known drawbacks of viral delivery techniques [48, 49].

12.9.3 Insertable Silk Fabricated with Silk Fiber for Drug Delivery

In a ground-breaking piece of work, Yavuz *et al.* transformed silk fibroin into a disc-based material that could carry either immunoglobin G antibody or the HIV inhibitor chemokine analogue. Through the use of silk fibroin stacking, water vapor annealing, and methanol treatment, three distinct formulations were created. The protein cargo was able to be stabilized and the release profile was altered by these formulations. Due to its relatively porous structure, the methanol-treated formulation released a significant amount of Immunoglobin G, which provides a slow and control release. In human colorectal and blood samples, the protein created following the water vapor annealing for potent HIV inhibitor established a prolonged release for 31 days and was successful in preventing HIV infection [50, 51].

12.9.4 Gene Carriers Based on Spider Silk that Target Specific Tumor Cells

The current studies have developed and done the analysis of brand-new nanoscale silk-based ionic compounds with pDNA that can target tumor cells precisely [52, 53]. Bioengineered poly(L-lysine) domains were added to silk-based block copolymers to interact with plasmid DNA as cationic liposomes, which binds to specific tumor cells to deliver target-specific plasmid DNA [54, 55]. *Escherichia coli* was used to create genetically altered silk proteins that contained poly L-lysine and the tumor homing peptides. The highly metastatic epithelial human breast cells from a human breast tumor and melanoma cells were then transfected *in vitro* and *in vivo*. As a control cell line for *in vitro* tumor-specific drug delivery, breast epithelial cells from the non-tumorigenic MCF-10A cell line were employed.

12.9.5 Nuclear Targeting Using Bioengineered Silk Gene Delivery System

The design and analysis of the impacts of different structures and their positions in recombinant silk proteins for nuclear specific drug incorporated genetic delivery were the main objectives of the current work. We demonstrated that the endosome escape unit blocked endosomal degradation, prolonging the cargo's rapid disintegration using intracellular tracking tests. Nuclear localization signal's effectiveness was enhanced by nuclear targeting, and in its lack, endosomal escape increased the cargo's breakdown to take a bit longer [56].

12.9.6 Layer-by-Layer Microcapsules Made of Silk Fibroin for Localized Gene Delivery

The scientist found that silk fibroin (SF) microcapsules constructed layer by layer can transport plasmid DNA. When silk fibroin with a fluorescent label was applied to polystyrene template particles, the quantity of layers increased fluorescence intensity while decreasing surface charge.

The researchers discovered that layer-by-layer constructed microcapsules made of silk fibroin were effective for plasmid DNA delivery. The number of silk fibroin layers was correlated with the fluorescence intensity and surface charge when fluorescently tagged silk fibroin was deposited onto polystyrene template particles. After the removal of polystyrene, hollow, monodisperse, and good mechanical strength silk fibroin microcapsules with a range of shell thickness were produced. *In vitro* tests were then conducted to assess plasmid DNA release, cytotoxicity, and eGFP expression. Both loading procedures resulted in sustained pDNA release over the course of three days, which was enhanced in the presence of protease [57, 58].

12.10 Some Highlights of Silk Fibroin

- ✓ Silk is a promising fabric material for the optimized administration of active pharmaceutical ingredient because of its biocompatibility, regulated kinetics of release and degradation to maintain the drug's bioactive properties and compatibility with sterilization techniques [59].
- ✓ Various types of delivery systems are prepared from the silk fibroin like injections, implantation, or transdermal routes.
- ✓ By prolonging the duration of release, plasma concentration circulation time, and accumulating in the target location, silk-based nanoparticles can be easily developed to decrease dosage frequency as well as systemic toxicity.
- ✓ Silk hydrogels are employed as implant systems and injectable sustained release depots when paired with a solid supported substance [60].
- ✓ Silk based implanted drug delivery system, including films, wafers, foams, and reservoirs, have been studied for focused administration of chemotherapeutic drugs [61].
- ✓ Silk based microfabricated needles are a promising transdermal delivering approach because they are simple to manufacture, offer regulated release, superior mechanical, and effective skin penetration [62, 63].

12.11 Conclusion

Due to its numerous differentiating benefits, such as adequate mechanical strength, process variability in an aqueous environment, biocompatibility, and a controlled disintegration time, silk fibroin is the ideal solution for the delivery of drugs. Because of this, silk fibroin nanoparticles have been successfully developed with high degree of stability while controlling the release rate of biomolecules. In biomedical research, the therapeutic regulated delivery of genetic elements such as silencing RNA, bidirectional RNA interface, or plasmid DNA to diseased cells or tissues is evolving into a workable technique with the potential to transfer the advances into helpful clinical applications. The idea of developing the appropriate drug delivery system has recently drawn a lot of interest. The current study aims to develop a platform based on proteins for non-viral DNA delivery without cytotoxicity or immunological reaction. The chapter has explored various silk-based gene delivery systems that have been identified as viable candidates for use as gene carriers [64]. These systems have the unique capacity to be quickly and readily customized and genetically altered to improve targeting and specificity. Simple surface adjustments may have an impact on drug release and binding, and they may even permit tailored drug delivery. A virtually infinite variety of drug delivery devices with unique release kinetics are possible when various silk fibroin drug delivery systems are combined into a single design.

References

1. Halpern, S.D., Ubel, P.A., Caplan, A.L., Solid-organ transplantation in HIV-infected patients. *N. Engl. J. Med.*, 347, 7, 284–7, 2002.
2. Numata, K. and Kaplan, D.L., Silk-based gene carriers with cell membrane destabilizing peptides. *Biomacromolecules*, 11, 3189–1397, 2010.
3. Philipp Seib, F. and Kaplan, D.L., Silk for drug delivery applications: Opportunities and challenges. *Isr. J. Chem.*, 53, 756–769, 2013.
4. Tokareva, O.S., Glettig, D.L., Abbott, R.D., Kaplan, D.L., Multifunctional spider silk polymers for gene delivery to human mesenchymal stem cells. *J. Biomed. Mater. Res. Part B*, 103, 7, 1390–401, 2014.
5. Dutta, D., Hossain, C.M., Biswas, A., Silk proteins in drug delivery: An overview. *Res. Pharm. Health Sci.*, 4, 4, 514–518, 2018.
6. Vepari, C. and Kaplan, D.L., Silk as a biomaterial. *Prog. Polym. Sci.*, 32, 8-9, 991–1007, 2007.
7. Zhao, Z., Li, Y., Xie, M.-B., Silk fibroin-based nanoparticles for drug delivery. *Int. J. Mol. Sci.*, 16, 3, 4880–4903, 2015.
8. Tokareva, O., Michalczechen-Lacerda, V.A., Rech, E.L., Kaplan, D.L., Recombinant DNA production of spider silk proteins. *Thematic Issue Biomater.*, 6, 6, 651–663, 2013.
9. Tomeh, M.A., Hadianamrei, R., Zhao, X., Silk fibroin as a functional biomaterial for drug and gene delivery. *Pharmaceutics*, 11, 49, 2019.

10. Bessa, P.C., Balmayor, E.R., Azevedo, H.S., Nürnberger, S., Casal, M., Van Griensven, M., Reis, R.L., Redl, H., Silk fibroin microparticles as carriers for delivery of human recombinant BMPs. Physical characterization and drug release. *J. Tissue Eng. Regen. Med.*, 4, 5, 349, 2010.
11. Bhowmik, D., Gopinath, H., Kumar, B.P., Duraivel, S., Kumar, K.P.S., Controlled release drug delivery systems. *Pharma Innov.*, 1, 10, 24–32, 2012.
12. Bhrany, A.D., Lien, C.J., Beckstead, B.L., Futran, N.D., Muni, N.H., Giachelli, C.M., Ratner, B.D., Cross-linking of an oesophagus acellular matrix tissue scaffold. *J. Tissue Eng. Regen. Med.*, 2, 6, 365–372, 2008.
13. Bhowmik, D., Gopinath, H., Duraivel, S., Sampath Kumar, K.P., Silk-based drug delivery systems. *Pharm. J.*, 1, 11, 2012.
14. Rahat, I. and Sharma, S.K., Use of silk fibroin-based nanoparticles as a novel drug delivery system. *Eur. J. Mol. Clin. Med.*, 7, 6, 2020.
15. Han, J., Wang, Q., Zhang, Z., Gong, T., Sun, X., Cationic bovine serum albumin based self-assembled nanoparticles as siRNA delivery vector for treating lung metastatic cancer. *Small*, 10, 524–535, 2014.
16. Wu, D., Zhang, Y., Xu, X., Guo, T., Xie, D., Zhu, R., Chen, S., Ramakrishna, S., He, L., RGD/TAT-functionalized chitosan-graft-PEIPEG gene nanovector for sustained delivery of NT-3 for potential application in neural regeneration. *Acta Biomater.*, 72, 266–277, 2018.
17. Pang, J., Xing, H., Sun, Y., Feng, S., Wang, S., Non-small cell lung cancer combination therapy: Hyaluronic acid modified, epidermal growth factor receptor targeted, pH sensitive lipid-polymer hybrid nanoparticles for the delivery of erlotinib plus bevacizumab. *Biomed. Pharmacother.*, 125, 1–10, 2020.
18. Dutta, D., Hossain, C.M., Biswa, A., Silk proteins in drug delivery: An overview. *Res. Pharm. Health Sci.*, 4, 4, 514–518, 2018.
19. Grigs, C.L. and Leong, K.W., Balancing protection and release of DNA: Tools to address a bottleneck of non-viral gene delivery. *J. R. Soc. Interface*, 20107, S67–S82, 2010.
20. Chambre, L., Moldes, Z.M., Parker, R.N., Kaplan, D.L., Bioengineered elastin- and silk-biomaterials for drug and gene delivery. *Adv. Drug Deliv. Rev.*, 160, 186–198, 2020.
21. Wu, J., Sahoo, J.K., Li, Y., Xu, Q., Kaplan, D.L., Challenges in delivering therapeutic peptides and proteins: A silk-based solution. *J. Controlled Release*, 345, 176–189, 2022.
22. Tokareva, O.S., Glettig, D.L., Abbott, R.D., Kaplan, D.L., Multifunctional spider silk polymers for gene delivery to human mesenchymal stem cells. *J. Biomed. Mater. Res. Part B, Appl. Biomater.*, 00B, 1–12, 2014.
23. Kim, S.Y., Naskar, D., Kundu, S.C., Bishop, D.P., Doble, P.A., Boddy, A.V., Chan, H.-K., Wall, I.B., Chrzanowski, W., Formulation of biologically-inspired silk-based drug carriers for pulmonary delivery targeted for lung cancer. *Sci. Rep.*, 5, 11878, 2015.

24. Yavuza, B., Morgan, J.L., Herrerac, C., Harrington, K., Perez-Ramirez, B., LiWang, P.J., Kaplan, D.L., Sustained release silk fibroin discs: Antibody and protein delivery for HIV prevention. *J. Controlled Release*, 301, 1–12, 2019.
25. Chakraborty, T., Sarma, A., Das, M.K., Silk fibroin: A smart biomaterial for long term and targeted nanotherapeutics. *Curr. Trends Pharm. Res.*, 3, 1, 45–82, 2016.
26. Altman, G.H., Diaz, F., Jakuba, C., Calabro, T., Calabro, T., Horan, R.L., Chen, J., Lu, H., Richmond, J., Kaplan, D.L., Silk-based biomaterials. *Biomaterials*, 24, 3, 401–416, 2003.
27. Numata, K., Reagan, M.R., Goldstein, R.H., Rosenblatt, M., Kaplan, D.L., Spider silk-based gene carriers for tumor cell-specific delivery. *Bioconjug. Chem.*, 22, 1605–1610, 2011.
28. Gobin, A.S., Froude, V.E., Mathur, A.B., Structural and mechanical characteristics of silk fibroin and chitosan blend scaffolds for tissue regeneration. *J. Biomed. Mater. Res. A*, 74, 3, 465–73, 2005.
29. Mottaghitalaba, F., Farokhib, M., Ali, M., Fatemeh, S., Hossein, A., Khanide, H., Silk fibroin nanoparticle as a novel drug delivery system. *J. Controlled Release*, 206, 161–176, 2015.
30. Yigit, S., Tokareva, O., Varone, A., Georgakoudi, I., Kaplan, D.L., Bioengineered silk gene delivery system for nuclear targeting. *Macromol. Biosci.*, 9, 1291–1298, 2014.
31. Liab, L., Puhla, S., Meinela, L., Germershausa, O., Silk fibroin layer-by-layer microcapsules for localized gene delivery. *Biomaterials*, 35, 27, 7929–7939, 2014.
32. Lucke, M., Mottas, I., Herbst, T., Hotz, C., Romer, L., Herold, M.S.H.M., Spinett, U.S.T., Engineered hybrid spider silk particles as delivery system for peptide vaccines. *Biomaterials*, 17, 105–115, 2018.
33. Jastrzebska, K., Kucharczyk, K., Florczak, A., Dondajewskaa, E., Mackiewicz, A., Dams-Kozlowskaa, H., Silk as an innovative biomaterial for cancer therapy. *Rep. Pract. Oncol. Radiother.*, 20, 2, 87–98, 2015.
34. Gottschalk, S., Sparrow, J.T., Hauer, J., Mims, M.P., Leland, F.E., Woo, S.L., Smith, L.C., A novel DNA-peptide complex for efficient gene transfer and expression in mammalian cells. *Gene Ther.*, 3, 448–457, 1996.
35. Lear, J.D. and DeGrado, W.F., Membrane binding and conformational properties of peptides representing the NH2 terminus of influenza HA-2. *J. Biol. Chem.*, 262, 6500–6505, 1987.
36. Plank, C., Oberhauser, B., Mechtler, K., Koch, C., Wagner, E., The influence of endosome-disruptive peptides on gene transfer using synthetic virus-like gene transfer systems. *J. Biol. Chem.*, 269, 12918–12924, 1994.
37. Subbarao, N.K., Parente, R.A., Szoka, F.C., Jr, Nadasdi, L., Pongracz, K., pH-dependent bilayer destabilization by an amphipathic peptide. *Biochemistry*, 26, 2964–2972, 1987.

38. Wyman, T.B., Nicol, F., Zelphati, O., Scaria, P.V., Plank, C., Szoka, F.C., Jr, Design, synthesis, and characterization of a cationic peptide that binds to nucleic acids and permeabilizes bilayers. *Biochemistry*, 36, 3008–3017, 1997.
39. Whittall, D.R., Baker, K.V., Breitling, R., Takano, E., Host systems for the production of recombinant spider silk. *Trends Biotechnol.*, 39, 6, 2021.
40. Ma, Y., Zeng, W., Ba, Y., Luo, Q., Ou, Y., Liu, R., Ma, J., Tang, Y., Hu, J., Wang, H., Tang, X., Mu, Y., Li, Q., Chen, Y., Ran, Y., Xiang, Z., Xu, H., A single-cell transcriptomic atlas characterizes the silk-producing organ in the silkworm. *Nat. Commun.*, 13, 3316, 2022.
41. Foong, C.P., Higuchi-Takeuchi, M., Malay, A.D., Oktaviani, N.A., Thagun, C., Numata, K., A marine photosynthetic microbial cell factory as a platform for spider silk production. *Commun. Biol.*, 3, 357, 2020.
42. Numata, K. and Kaplan, D.L., Silk-based delivery systems of bioactive molecules. *Adv. Drug Deliv. Rev.*, 62, 1497–1508, 2010.
43. Wang, Y., Kim, H.J., Vunjak Novakovic, G., Kaplan, D.L., Stem cell-based tissue engineering with silk biomaterials. *Biomaterials*, 27, 6064–6082, 2006.
44. Dewair, M., Baur, X., Ziegler, K., Use of immunoblot technique for detection of human ige and igg antibodies to individual silk proteins. *J. Allergy Clin. Immunol.*, 76, 537–542, 1985.
45. Zaoming, W., Codina, R., Fernandez Caldas, E., Lockey, F., Bukanz, S.C., Partial characterization of the silk allergens in mulberry silk extract. *J. Allergy Clin. Immunol.*, 97, 109–109, 1996.
46. Numata, K., Cebe, P., Kaplan, D.L., Mechanism of enzymatic degradation of beta-sheet crystals. *Biomaterials*, 31, 2926–2933, 2010.
47. Numata, K. and Kaplan, D.L., Silk-based gene carriers with cell membrane destabilizing peptides. *Biomacromolecules*, 11, 3189–3195, 2010.
48. Numata, K., Subramanian, B., Currie, H.A., Kaplan, D.L., Bioengineered silk protein-based gene delivery systems. *Biomaterials*, 30, 5775–5784, 2009.
49. Bini, E., Foo, C.W., Huang, J., Karageorgiou, V., Kitchel, B., Kaplan, D.L., RGD-functionalized bioengineered spider dragline silk biomaterial. *Biomacromolecules*, 7, 3139–3145, 20062006.
50. Kim, S.Y., Naskar, D., Kundu, S.C., Bishop, D.P., Doble, P.A., Boddy, *et al.*, Formulation of biologically-inspired silk-based drug carriers for pulmonary delivery targeted for lung cancer. *Sci. Rep.*, 5, 11878, 2015.
51. Choi, M., Choi, D., Hong, J., Multilayered controlled drug release silk fibroin nanofilm by manipulating secondary structure. *Biomacromolecules*, 19, 3096–3103, 2018.
52. Li, A.B., Kluge, J.A., Guziewicz, N.A., Omenetto, F.G., Kaplan, D.L., Silk-based stabilization of biomacromolecules. *J. Controlled Release*, 219, 416–430, 2015.
53. Jastrzebska, K., Kucharczyk, K., Florczak, A., Dondajewsk, E., Mackiewicz, A., Dams Kozlowsk, H., Silk as an innovative biomaterial for cancer therapy. *Rep. Pract. Oncol. Radiother.*, 20, 2, 87–98, 2015.

54. Florczak, A., Deptuch, T., Kucharczyk, K., Kozlowska, H.D., Systemic and local silk-based drug delivery systems for cancer therapy. *Cancers*, 13, 21, 5389, 2021.
55. Yavuz, B., Chambre, L., Kaplan, D.L., Extended release formulations using silk proteins for controlled delivery of therapeutics. *Expert Opin. Drug Delivery*, 16, 7, 741–756, 2019.
56. Das, S.K., Dey, T., Kundu, S.C., Fabrication of sericin nanoparticles for controlled gene delivery. *RSC Adv.*, 4, 2137, 2014.
57. Khan, M.R., Morikawa, H., Gotoh, Y., Miura, M., Ming, Z., Sato, Y., Iwasa, M., Structural characteristics and properties of Bombyx mori silk fiber obtained by different artificial forcibly silking speeds. *Int. J. Biol. Macromol.*, 42, 3, 264–270, 2008.
58. Keten, S., Xu, Z., Ihle, B., Buehler, M.J., Nanoconfinement controls stiffness, strength and mechanical toughness of β-sheet crystals in silk. *Nat. Mater.*, 9, 4, 359–367, 2010.
59. Xiao, S., Stacklies, W., Debes, C., Gräter, F., Force distribution determines optimal length of β-sheet crystals for mechanical robustness. *Soft Matter*, 7, 4, 1308–1311, 2011.
60. Sinsawat, A., Putthanarat, S., Magoshi, Y., Pachter, R., Eby, R.K., X-ray diffraction and computational studies of the modulus of silk (Bombyx mori). *Polymer*, 43, 4, 1323–1330, 2002.
61. Krasnov, I., Diddens, I., Hauptmann, N., Helms, G., Ogurreck, M., Seydel, T., Funari, S., Muller, M., Mechanical properties of silk: Interplay of deformation on macroscopic and molecular length scales. *Phys. Rev. Lett.*, 100, 4, 1–4, 2008. 048104/1–048104/4.
62. Sinsawat, A., Putthanarat, S., Magoshi, Y., Pachter, R., Eby, R.K., The crystal modulus of silk (Bombyx mori). *Polymer*, 44, 3, 909–910, 2003.
63. Salatin, S., Jelvehgari, M., Maleki-Dizaj, S., Adibkia, K.Z., A sight on protein-based nanoparticles as drug/gene delivery systems. *Ther. Deliv.*, 3, 1–13, 2015.
64. Niu, L., Chen, G., Feng, Y., Liu, X., Pan, P., Huang, L. *et al.*, Polyethylenimine-modified bombyx mori silk fibroin as adelivery carrier of the ING4-IL-24 coexpression plasmid. *Polymers*, 13, 3592, 2021.

13

Challenges and Emerging Problems in Nanomedicine Mediated Gene Therapy

Shalini Bhatt[1]*, Neha Faridi[2], Rakshit Pathak[1], Vinay Deep Punetha[1] and Mayank Punetha[1]

[1]PP Savani University, Surat, Gujarat, India
[2]Defence Institute of Bio-Energy Research, DRDO, Uttarakhand, India

Abstract
Nanomedicine merges nanotechnology and medical sciences for the diagnosis, management, and cure of fatal diseases for the healthy and disease-free living of mankind. Small dimensional structures along with the large surface-to-volume ratio enable nanoparticles to be potentially utilized for gene therapy applications. Intensive research in this field has resulted in the global deployment of various nanomedicine products. However, nanomedicines regardless of their wide therapeutic applications encounter a large number of noteworthy barriers to their clinical conversion, which are complex, challenging, as well as resource-intensive procedures that necessitate mobilization and earmarking of the resources, booming alliance and cooperation between the leading national and international collaborators. Therefore, for efficient utilization it is necessary to explore the challenges and drawbacks of nanomedicine and gene therapy from a variety of research and clinical study aspects. This chapter explores various challenges faced by nanomedicine-mediated gene therapy along with major problems faced by nanomedicines during drug therapy. Further, an attempt to showcase the various therapeutic approaches to gene delivery is also done. Hence, the chapter may guide innovative, adaptive, and socially responsible research in this field, so that nanomedicine in the future may play a progressively major as well as revolutionary role in medicine and public health.

Keywords: Biocompatibility, challenges, gene therapy, nanomedicines, nanoparticles

**Corresponding author*: shalini.bhatt@ppsu.ac.in

Rishabha Malviya, Sonali Sundram and Neelam Jain (eds.) Integration of Biomaterials for Gene Therapy, (367–416) © 2023 Scrivener Publishing LLC

13.1 Introduction

The non-specific target ability of the available medicaments forms a major problem in the health care system [1]. Worldwide application of nanotechnology on the way to medicine has led to the inception of a new class of science termed nanomedicine (NM), which amalgamates nanotechnology with diagnostic molecules or drugs to improve the ability to specifically target particular tissue or cells. NM involves the employment of nanotechnology as well as nanoscience in the treatment and prevention of lethal diseases of human [222]. Vast research in this area has headed the invention of new nano-biomaterials and nano-therapeutics as well as advancement in medical technology adding accurate diagnostics, contrast reagents, and medical instruments [2–5]. Nano-therapeutics have been investigated in clinical assessment and authorized for medical use presenting superior therapeutic efficiency than conventional molecular drugs [6–8] due to their small size (10–100 nm), huge specific surface area, and capability of surface functionalization [9]. These characteristics give nano-therapeutics minimal toxicity, superior pharmacokinetics, and enhanced bioavailability. As a result, nano-therapeutics circumvents the drawbacks of conventional medications and has the potential to satisfy the growing consumer demand for disease treatments [223]. Moreover, developing NM and confronting it with regular human use involves more challenges as compared to the production of other nanomaterial-based products. The development of a successful drug requires large-scale pre-clinical research, meticulous clinical trials, and adequate therapeutic indications. However, the existing status of the clinical transformation of NMs is disappointing. Various challenges connected with safety, regulation, and biological barriers have all led to the inefficient clinical translation of these products. The dominance of traditional drugs over nanomaterial-based medicine shows that the former still dominates the drug development processes. NM should use the coherent strategy of nanotechnology based on a complete grasp of biological processes along with the drug interactions rather than merely applying materials that are currently in high demand.

Gene therapy has the potential to cure various deadly diseases since most of the diseases are caused by defective copies of the genes. It holds a vast potential for treating a multitude of diseases by gene correction, gene silencing, or protein expression [10]. Therefore, inserting a healthy copy of the gene can sustain healthy living conditions. However, the introduction of the therapeutic gene to its appropriate cell forms a major challenging task. For systematic and resourceful delivery of nucleic acid pack to the target tissue, the deoxyribonucleic acid (DNA) needs to be combined with a delivery model. In the 1990s, during a gene therapy study in hollowed-out adenoviruses for packaging the gene of interest for the treatment of liver disorders, found that one of the patients under trial dies post four days of receiving gene therapy [11]. Soon, the research focus was shifted to developing safer means for gene delivery, involving non-viral vectors such as nanoparticles which provide a comparatively safer means of gene therapy.

A comprehensive accord among scientists and researchers is much needed such that NM together with personalized and regenerative medicine could transform the existing concept of human diseases and treatments. To carefully translate the laboratory innovation in NMs to the treatment center requires extensive global efforts. These efforts may include the development as well as assessment of relatively new and efficient methods that can identify potential short- and long-term health risks by generalizing the (co-relating) *in vitro* results for the prognosis of chronic *in vivo* effects. For better medical health care, it is necessary to explore the challenges and drawbacks of NM and gene therapy from a variety of research and clinical study aspects. Biocompatibility, safety, large-scale manufacturing, and cost-effectiveness, are some of the challenges to name a few. This chapter explores various challenges faced by NM-mediated gene therapy from design and development to clinical translation and release in the market.

13.2 Why Nanomedicine Over Traditional Drugs?

Nanotechnology has been used in the field of medicine in which progressive understanding of both the systems; biological as well nanomaterial; and the composite problems of pathophysiological conditions, emphasizes the demand for genetic as well as cellular level therapies. NM cures diseases by way of very small mass compounds at specified diseased areas to bring out the preferred therapeutic effect with less adverse effects. On the other hand, general medicine cures disease with lesser specificity and abroad range. Moreover, conventional drugs have inferior pharmacokinetics, lesser bioavailability, and a high level of toxicity, all of which limit their use [7]. Traditional small-molecule medicines may have substantial adverse effects and require careful attention to dosage and timing. Due to their poor solubility, they could not traverse the biological barriers thus lowering their effectiveness. NMs are targeted and have improved bioavailability. Table 13.1 represents FDA approved list of licensed products from the Office of Tissues and Advanced Therapies (OTAT) for cellular and gene therapy [12].

Table 13.1 List of FDA-approved cellular and gene therapy products.

Trade name	Product name	Disease	Manufacturer
ABECMA	Idecabtagene vicleucel	Relapsed or refractory multiple myeloma	Celgene Corporation
ALLOCORD	HPC, Cord Blood	Hematopoietic system diseases	SSM Cardinal Glennon Children's Medical Center
BREYANZI	Lisocabtagene maraleucel	Large B-cell lymphoma (LBCL), diffused large B-cell lymphoma (DLBCL)	Juno Therapeutics
CARVYKTI	Ciltacabtagene autoleucel	Relapsed or refractory multiple myeloma	Janssen Biotech
Clevecord	HPC, Cord Blood	Hematopoietic system diseases	Cleveland Cord Blood Center
Ducord	HPC, Cord Blood	Hematopoietic system diseases	Duke University School of Medicine
GINTUIT	Allogeneic Cultured Keratinocytes and Fibroblasts in Bovine Collagen	Surgically created vascular wound bed	Organogenesis Incorporated
HEMACORD	HPC, Cord blood	Hematopoietic system diseases	New York Blood Center
-	HPC, Cord Blood	Hematopoietic system diseases	Clinimmune Labs, University of Colorado Cord Blood Bank

(*Continued*)

Table 13.1 List of FDA-approved cellular and gene therapy products. (*Continued*)

Trade name	Product name	Disease	Manufacturer
-	HPC, Cord Blood	Hematopoietic system diseases	MD Anderson Cord Blood Bank
-	HPC, Cord Blood	Hematopoietic system diseases	LifeSouth Community Blood Centers, Inc.
-	HPC, Cord Blood	Hematopoietic system diseases	Bloodworks
IMLYGIC	Talimogene laherparepvec	Lesions in patients with melanoma recurrent after initial surgery	Amgen Inc.
LAVIV	Azficel-T	Nasolabial fold wrinkles	Fibrocell Technologies
LUXTURNA	Voretigene neparvovec-rzyl	Biallelic RPE65 mutation-associated retinal dystrophy	Spark Therapeutics
MACI	Autologous Cultured Chondrocytes on a Porcine Collagen Membrane	Cartilage defects	Vericel Corporation
PROVENGE	Sipuleucel-T	Prostate cancer	Dendreon Corporation
RETHYMIC	Allogeneic processed thymus tissue–agdc	Congenital athymia	Enzyvant Therapeutics GmbH

(*Continued*)

Table 13.1 List of FDA-approved cellular and gene therapy products. (*Continued*)

Trade name	Product name	Disease	Manufacturer
SKYSONA	Elivaldogene autotemcel	Cerebral adrenoleukodystrophy	Bluebird bio, Inc.
STRATAGRAFT	Allogeneic cultured keratinocytes and dermal fibroblasts in murine collagen-dsat	Thermal burns	Stratatech Corporation
TECARTUS	Brexucabtagene autoleucel	Mantle cell lymphoma (MCL)	Kite Pharma, Inc.
YESCARTA	Axicabtagene ciloleucel	Large B-cell lymphoma	Kite Pharma Inc.
ZYNTEGLO	Betibeglogene autotemcel	β-thalassemia	Bluebird bio Inc.
ZOLGENSMA	Onasemnogene abeparvovec	Spinal Muscular Atropy (Type I)	Novartis Gene Therapies, Inc.

Source: www.fda.gov

13.3 Nanomedicine for Gene Therapy

The introduction of NM was prompted by the imperative need to advance current medical treatment methods for improved diagnosis and therapy. They provide site-specific disease treatment with relatively low-mass compounds to manifest the desired clinical effects with lesser complications. Since cells and genes are already sub-micrometer-sized, a thorough understanding of cellular biology at the nanoscale is necessary to achieve the outstanding benefits of NM [224]. To supplement NM, genes are now potentially employed as a therapeutic agent in gene therapy, which evolved as a result of fusing with NM. Gene therapy involves three aspects namely, gene silencing using small interfering ribonucleic acid (siRNA), short hairpin RNAs (shRNAs), and microRNAs (miRNA); gene replacement in which plasmids and viral vectors are used to transfer the desired gene; and lastly gene editing which involve modification of mutations using specific nucleases [13]. The potential for gene therapy has advanced rapidly with the development of genomic technologies. Genomic databases have greatly aided in gene-mediated therapy such that the altered genetic material responsible for a particular disease could be manipulated to treat the particular disease. Gene therapy was initially used to replace the inherited defective genes in genetic disorders. However, with the advancement in the carrier system, gene therapy evolved rapidly and is being employed against genetic disorders as well as cancer, nervous disorders, and heart diseases. Nevertheless, the transport of therapeutically relevant genes concentrates on two key aspects of gene therapy: the efficient mechanism required to transfer and deliver the gene, and how the gene enters target cells and fulfills its intended purpose. All the nucleic acids are made up of repeating units of nucleotides which are made up of a five-carbon sugar coupled via phosphate backbone to the nitrogenous bases. Although RNA and DNA have different sizes and can include either ribose or deoxyribose sugar, all nucleic acids are negatively charged and hydrophilic. These vehicles consist of particles comprised of lipids, polymers, silica, gold, and other materials. The most frequently used materials for transporting nucleic acids are lipids and polymers. For efficient *in vitro* gene transfer such as DNA and RNAs, mRNA, microRNA (miRNA), and (siRNAs), the majority of commercially available transfection reagents employ cationic lipids. Exogenous nucleic acids (NA) have intrinsic pharmacokinetic limitations, such as electrostatic repulsions, enzymatic degradation, restricted cellular uptake, rapid clearance, etc., which must be overcome for exogenous NA to be internalized by cells and require highly efficient delivery vehicles to overcome

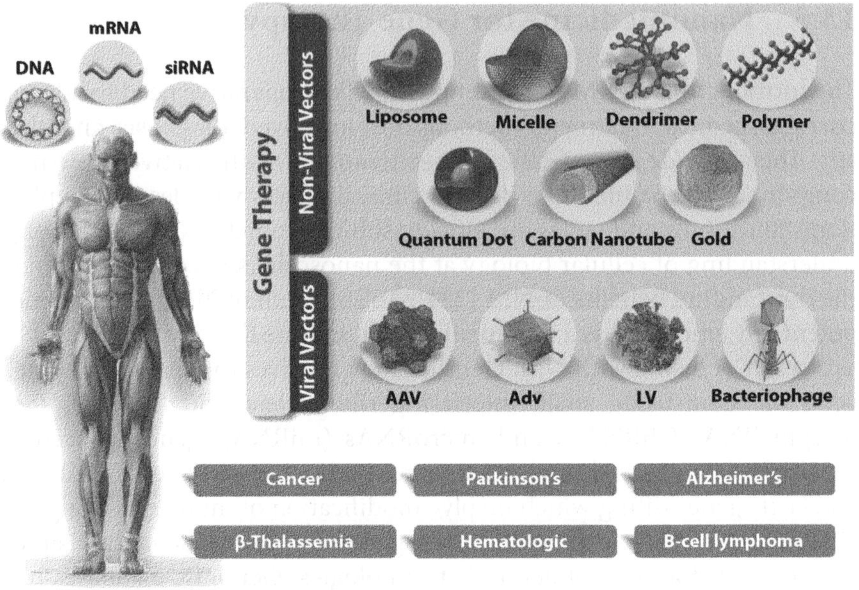

Figure 13.1 Viral and non-viral vectors used in nano-medicine mediated gene therapy (AAV: Adeno-associated virus; Adv: Adenovirus; LV: Lentivirus).

these limitations. Various systems available for gene carriage have been discussed in the upcoming sections. Figure 13.1 depicts representative images of various viral as well as non-viral vectors used in gene therapy.

13.3.1 Viral Vectors

Employing viral systems for carrying the gene, at first, gained great attention because they possess an inherent capability to swamp the biological barriers of the humans and interfere with the replication mechanism inside the nucleus. Gene delivery has been devised using viruses like the Adeno-Associated Virus (AAV), which has its virulent genes deleted, and Lentiviruses, which explicitly state their lack of virulence. Regardless of these advantages viruses possess significant limitations which restrict their use for gene therapy. Viral vectors activate the host's immune response thus lowering the effect of subsequent gene delivery. They often undergo random recombination events with the genome of the host and thereby upregulation of the oncogenic genes which later activate hypersensitive reactions and non-specific immunological responses. Furthermore, viral glycoproteins can trigger the adaptive immune response, which can develop immunological memory and decrease the efficiency of using that viral vector in the future. In addition to these issues, it was challenging to solve the problem of virally mediated gene delivery due to the size restriction of the transgene which could be packed into the viral vector, and the practical constraints encountered during the mass fabrication of viral vectors.

13.3.2 Nanoparticle-Based Vectors

Owing to the difficulties encountered using viral vectors, a lot of focus on developing non-viral vectors such as lipids, peptides, polymers, inorganic particles, and combinations of different types, for efficient and safe gene delivery has emerged. In comparison to viral vectors, non-viral vectors are simpler to construct, slightly immunogenic and carcinogenic, and do not have the possibility of virus recombination or size limitations for the genes that may be transferred [14]. Additionally, non-viral vectors can be structurally altered for easy transport of genetic material to target cells. Nanoparticles (NPs), a class of non-viral gene vectors, provide the best platform for combining all desirable properties into a single gene delivery device. Vast research by the scientific community has been done on its application for DNA therapy and delivery. NPs are among the various nano-constructs that make promising vectors for non-viral gene transfer. They have undergone successful testing, both *in vivo* and *in vitro*, for gene transfer [15–18]. Principally NPs are made to reach the target site by their small size and charge [19]. They possess a high surface area-to-volume ratio, thereby increased surface energy makes them highly reactive and a potent candidate for gene therapy [20, 21]. The small size of these materials is comparable to the size of that of protein and offers great potential for targeted delivery of genes and penetrating deep into the tissues with solid tumors with high specificity. Owing to their small sizes they can interact with the cellular biomolecules and deliver genetic materials such as DNA, RNA or siRNA into target cells or tissues for the expression of desired gene. When in contact with biological fluids, they can adsorb biomolecules especially proteins on its surface forming structures known as protein corona. Other molecules adsorbed include lipids, DNA, and, RNA. Figure 13.2 clearly represents illustration of nanomaterial mediated gene delivery. Moreover, the magnetic, electrical, as well as optical properties of the nanomaterials, offer benefits for gene therapy. They can be engineered to variable shapes, sizes, and chemical compositions, thus enabling their interaction with particular biological targets [22]. Table 13.2 provides brief information about the various classes of NPs used in gene delivery.

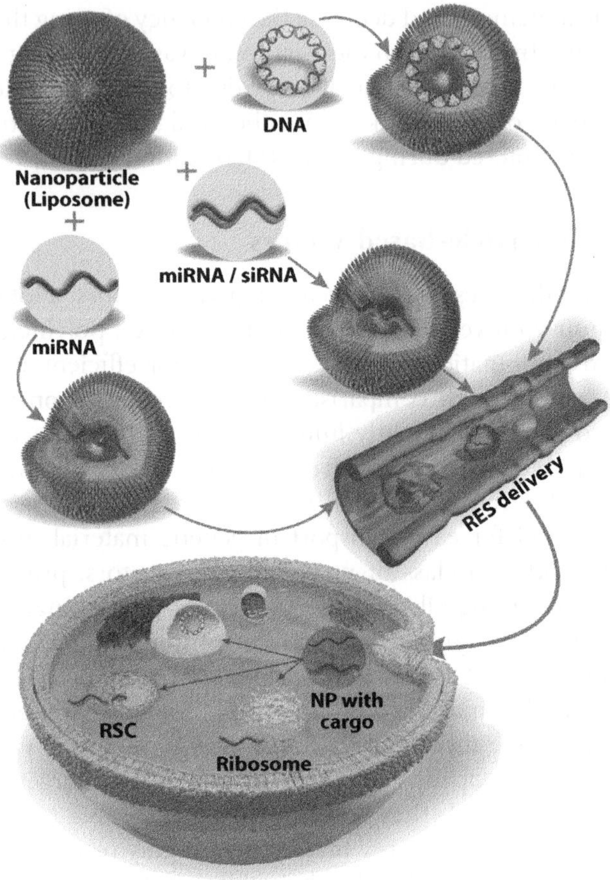

Figure 13.2 Schematic illustration of nanomaterial mediated gene delivery showing liposome as a non-viral vector carrying three different cargoes DNA, mRNA or siRNA and delivering to its target tissue or cells (NP: Nanoparticle; RES: Reticuloendothelial system; RSC: Remodels the structure of chromatin complex).

Lipid-Based Nanoparticles

Since early research using lipid compounds was conducted in the middle of the 1980s and 1990s, a lot of effort has been made to create NPs based on lipids. Lipid molecules constitute the architectural structure of the cellular as well as organelle membrane, which facilitated its use in gene therapy. Lipid-based formulations produces a vesicle-like shape with a double membrane layer that is hydrophobic on one side and hydrophilic on the other and have proved to be a highly efficient and effective delivery platform for DNA, mRNA, or siRNA [36–39]. These can enclose both lipids as well as lipophilic compounds. Several features including reproducibility

Table 13.2 Nanoparticles used in gene therapy.

Nanoparticle class	Nanoparticle	Advantages	Limitations	References
Organic Nanoparticles	Dendrimers	Uniform size, shape and branch length; tuneable pharmacokinetics and bio-distribution; high loading capacity and large surface area	Size and surface chemistry	[23, 24]
	Liposomes	Ligand conjugation, used for hydrophilic & hydrophobic cargo, biocompatible, longer duration in circulation	Cytotoxicity due to non-specific uptake	[25, 26]
	Polymers	Biodegradable, used for hydrophilic & hydrophobic cargo	Self-aggregation	[27, 28]
Inorganic	Metallic (Iron, Gold)	Tunable size, simple functionalization	Detailed information on uptake, cytotoxicity, biocompatibility is required, toxicity	[29, 30]
	Magnetic	Large loading capacity, controlled release ability, biocompatibility	Detailed assessment of acute and chronic toxicity, hyperthermiz	[31]
	Silica	Tunable pore size, large surface area, high loading capacity	Detailed information on cytotoxicity, uptake, biocompatibility is required	[32, 33]
	Carbon derived	High loading capacity, diverse functionalization, biocompatible	Very less *in vivo* studies, toxicity	[34, 35]

and scaling of the formulation process, physico-chemical properties of the particle size; pharmacokinetic properties, internalization, and release of nucleic acid, have been investigated and optimized to develop lipid NP formulation of clinical utility [40–44]. Further, these optimization approaches resulted in the creation of "Onpattro", a breakthrough in lipid NP-RNA interference therapy for liver gene regulation that is made up of an ionizable cationic lipid, a polyethylene glycol (PEG)-lipid, a phospholipid and cholesterol [40, 45, 46]. Several lipids NP-based therapeutic approaches using nucleic acids have been emerged for the treatment of various tumors, cancer, and genetic and infectious diseases by altering the nucleotide sequence of the encapsulated genetic material [47].

Cationic lipids are positively charged molecules having a cationic *polar head*, and hydrophobic *tail* consisting of saturated or unsaturated alkyl chains or cholesterol, greatly influencing the physical characteristics of the lipid bilayer; *a linker* connecting the polar head with the non-polar tail domain, determining the chemical stability, transfection efficiency and biodegradability of the lipid molecule; lastly, a backbone domain which separates the head from its non-polar domain [48]. Popular cationic lipids include dioleoyl trimethyl ammonium propane (DOTAP) and dioleylpropyl trimethyl ammonium chloride (DOTMA). Lipid-based NP complexed with adenoviral vectors are used in gene therapy to overcome the limitations of the very vector including the lack of adenoviral receptors on the target cells, hepatotoxicity, and formation of neutralizing antibodies [49]. Typically, cationic lipids in combination with neutral helper lipids namely cholesterol and DOPE, are used in transfection applications. These helper lipids reduce the interaction between the cationic surface and ions leading to the removal of counter ions from the surface of lipids and further facilitating the lipoplexes formation with the DNA. Moreover, they promote full as well as stable packaging of DNA molecules by the lipid along with lipoplex stability [50–52]. Thus, formed cationic liposomes protect the enclosed DNA from the actions of nucleases, successfully delivering it by interacting with the negatively charged cell membrane. Recently a cationic liposome consisting of DOTAP, DOPE, and cholesterol in conjunction with poly (aspartate-graft-PEI423) (PAE) (PEI: Polyethylenimine) and plasmid DNA was constructed to improve the transfection efficiency and lipopolyplex thus formed offering a promising platform for gene delivery [53]. Solid lipid nanoparticles (SLN) consisting of high melting point lipids have been created and shown to be effective in gene delivery [54]. They consist of triglycerides, fatty acids, partial glycerides, and steroids; considered as biocompatible and have been approved for pharmacological use in humans. Cationic SLN, like positively charged liposomes, bind via electrostatic forces with DNA.

Cationic lipid emulsions have too emerged and evidenced to act as lipid-based gene carriers. Its constituents include a cationic lipids shell surrounding the hydrophobic core of oil, and surfactants. Cationic surfactant components form electrostatic interactions with the phosphate groups on the nucleic acids, thus bestowing it with gene transfer [55]. This nano-sized lipid-based emulsion system is biodegradable, more stable, less toxic, and provides continued circulation in the blood, thus upholding extended therapeutic effects. However, a lot of work needs to go into creating an optimal CLE system that exhibits robust and long-lasting gene expression both *in vitro* and *in vivo* [56].

Polymer-Based Nanoparticles

Cationic polymers due to their simple production and adaptability, have shown to be a potent non-viral gene delivery method. At physiological pH, these polymers can be attached to the nucleic acids via electrostatic interactions, thereby enabling gene transfer [57]. Polymer-based NPs are used in DNA delivery because, these particles upon mixing with DNA, self-assemble, condense and neutralize the DNA molecule forming the nano-sized complexes termed polyplexes [58, 59]. These polyplexes offer wide adaptability, bioavailability, target specificity, and stability [60]. As compared to lipid -based systems, these polyplexes have relatively smaller sizes, constricted distribution, effective against nucleases, and controllable physical factors. Moreover, cationic residues present in these polyplexes may increase the interaction with the cell and can also trigger the transfer of nucleic acid to the cytoplasm by disrupting the vesicular membrane. Natural polymers like dextran, gelatin, chitosan, etc., and synthetic polymers such as poly (l-ornithine), poly (l-lysine) (PLL), diethylaminoethyl-dextran, linear as well as branched polyethyleneimine (PEI), poly-beta-amino esters (PBAE), etc., are being employed in these lipoplexes for gene therapy.

Dendrimers are a class of novel polymers with unique spherical architecture possessing mono-dispersibility, well-defined structures, polyvalency, and the ability for surface functionalization, all of which bestow it with great potential for gene therapy [61]. They get attached to the DNA, RNA and other anti-sense nucleotides through electrostatic interactions forming a dendrimer complex termed a dendriplex [62]. The most extensively researched dendrimer is poly-amido-amine (PAMAM), which exhibits great solubility and reactivity as a result of the existence of multiple functional groups at its periphery and vacant internal cavities. PAMAM dendrimers from the sixth generation are often utilized dendrimers as gene delivery vectors [63].

Polyion Complex (PIC) Micelles

Polyplexes are often modified by coated with PEG to form polyion complex micelle. Thus, formed micelle has increased transfection efficiency as well as prolonged circulating half-life. A highly specific gene transfer to cells at the diseased site will result from the use of physical or chemical sensitive vectors in combination with the genes expressed under the regulation of a particular promoter. One of the PIC micelles consists of PLL-PEG co-polymer, where PLL and pDNA form a hydrophobic core by electrostatic interactions and a hydrophilic PEG layer surrounding the core. PIC micelle possesses extraordinary properties such as colloidal stability, higher solubility, nuclease resistance, minimal cross-interaction with other cell components, and extended presence in blood circulation, as compared to lipoplexes and polyplexes vector systems.

Hybrid Nanoparticles

Cationic liposomes form large DNA-liposome complexes such that DNA molecules within the complex are not properly condensed. Therefore, liposome-polycation complexes were introduced in which DNA molecule is packed in a highly condensed form of about 30–100 nm in diameter. In the liposome-polycation-DNA (LPD) NPs, the DNA molecule is first pre-condensed with polycation which is further mixed up with cationic liposomes forming a virus-like structure with compacted DNA located within the lipid membrane [64, 65] PLL, PEI and protamine sulfate are usually used polycation and cationic liposomes are comprised of neutral lipids DOPE (1,2-dioleyl-sn-glycerol-3-phosphoethanolamine), cationic lipids like DOTAP, DOTMA, di-methyl amino ethane carbamoyl cholesterol (DC-chol) [66]. LPD consisting of anionic liposomes to entrap polycation-condensed DNA by Leaf Huang's Group is termed LPD-II. The surface of LPD may be functionalized with numerous ligands to enhance its cellular endocytosis by a specific receptor-mediated process [67].

Multi-functional enveloped nano-device (MEND) was constructed which was similar to LPD in structure except that the DNA core and lipid envelop exist as a separate structure [68–70]. Here, polycations used are PLL, stearyl octa-arginine (STR-R8), and protamine, which are used for DNA condensation. The lipid envelope is made up of neutral lipid DOPE and anionic lipid CHEMS (cholesteryl hemisuccinate) facilitating DNA delivery into the nucleus by increasing the internalization as well as fusogenic properties. MEND has certain ligands including antibodies, sugar, transferrin, and peptides, attached to the lipid envelope, forming a tetralamellar MEND (T-MEND), exhibiting an elevated level of transgene expression [71, 72].

Other hybrid NPs include multilayered-NPs which are formed by layer-by-layer (LbL) technique by deposition of polycations and polyanions. In multilayered gene vector, alternate layers of specific ligands, fusogenic lipids as well as nuclear localization signals are attached such that the constructed vector possess target specificity, endosomal escape, and greater nuclear delivery [73, 74].

Poly (D,L-lactide-co-glycolide) (PLGA) Nanoparticles

PLGA forms an efficient vector system for delivering sustained pDNA, and together with that has been accepted for human use by FDA in conjugation with poly lactic acid PLA. PLGA particles are nuclease resistant and biocompatible. Moreover, the polymer matrix provides a regulated release of the gene from the NPs [75]. Encapsulated pDNA in the PLGA tend to get inactivated owing to the use of harsh conditions including shear stress and organic solvents for encapsulation of DNA into the PLGA NPs. Moreover, it is observed that there is a slow release of DNA, as negative charge poses obstruction in DNA incorporation. PLGA can also be fabricated by surface adsorption of DNA via ionic interaction between the polycationic PLGA and polyanionic DNA molecule [76]. Cationic surfactant including cetyl trimethyl ammonium bromide (CTAB) and dimethyl di-dodecyl ammonium bromide (DMAB); polymers namely PEI or a combination of both cationic surfactant as well as polymers has been for the formation of cationic NPs which further enhance DNA loading and its intracellular uptake by interaction with the negatively charged cell membrane. Cationic PLGA particles are prepared by combining with the PEI.

Several limitations are faced by PLGA particles for their use in DNA delivery applications. Encapsulation of large-sized hydrophilic pDNA in the hydrophobic PLGA microsphere is a major challenge [77, 78]. Moreover, a decrease in the microsphere pH due to hydrolysis of PGLA, results in exposure of pDNA to the harsh environment, thereby leading to DNA degradation [79, 80]. Due to the slow release of DNA, antigen production is delayed in the case of DNA vaccination, which further inhibits an optimum immune response. There is a definite need for creating new kinds of biodegradable NPs due to the constraints of PLGA NPs for DNA delivery. The transfer of DNA was the main focus up until recently, but now sustained and controlled gene expression is becoming more and more significant. Due to their minimal cytotoxicity, the biodegradable PLGA NPs are also attractive candidates for the intracellular delivery of gene therapies.

Peptide-Based Nanoparticles

Peptide-based NPs form the peptide-DNA complexes employing short basic peptides consisting of positively charged amino acids including arginine, histidine, and lysine that interact with the negatively charged phosphate groups present in the DNA molecule [81, 82]. Thus, the formed positively charged complex of peptide-DNA interacts with the negatively charged cell membrane, thereby facilitating easy internalization into the cell. Moreover, the amphiphilic nature of the peptide facilitates conformational changes in acidic conditions that allow them to escape from the endosome and lysosome, enabling nuclear gene delivery and expression [83].

Inorganic Nanoparticles

Inorganic NPs comprising gold NPs, quantum dots, iron oxide NPs, have been extensively described as an alternative means for gene delivery, owing to their remarkable intracellular activities and potent cellular imaging abilities. Making NP competent for gene delivery application requires three strategies [84]. In the first method, negatively charged DNA is combined with positively charged inorganic NP to produce a complex. The second method uses a responsive linker to directly conjugate the genetic material onto the inorganic NP. In order, the third technique makes use of a cationic amphiphilic polymer that is generated from the NP to facilitate the complex formation between the inorganic NP and genetic material.

A. Quantum dots (QD) are crystalline, luminescent NPs that can be employed for *in vivo* cancer targeting and diagnostics as well as fluorescent probes for living cells [85, 86]. QD-based NPs can be used as both imaging probes and multifunctional delivery systems to track and monitor the siRNA transfection process *in situ* over an extended period [87]. This will enable a thorough investigation of the gene silencing process. QD-based NPs are synthesized by the double combination of different semiconductor materials like zinc sulfide (ZnS), Cadmium sufide (CdS), Cadmium telluride (CdTe), lead telluride (PbTe), lead sulfide (PbS), etc. Most of the reported QD-based vectors use cadmium to develop QD-based vector and their potential toxicity is still heavily disputed and unresolved, making it difficult for these vectors to be used for *in vivo* as well as clinical research. The breakdown of QD NPs in the biological system results in the release of cadmium ions from the particle's surface into the biological environment which may further induce severe acute toxicity [88–90].

B. Magnetic NPs are spherical and have the dimension of a nanometer scale. They have been employed as nanocarriers in gene delivery [91]. Strong magnetic fields are utilized to direct their accumulation in the target tissues, and they have been employed successfully to cure cancer [92]. The inherent magnetic properties of these particles have enabled their use in magnetic resonance imaging (MRI). Additionally, under static magnetic fields, MNP-based transfection, also known as magneto-fection, can lead to extremely quick transfection and enhanced transduction efficiency [93]. They usually have an organic covering, a magnetic core, and several functional molecules on their surface [94]. Different magnetic materials, such as pure iron, cobalt, and chromium metals; alloys like cobalt-platinum ($CoPt_3$), iron-platinum (FePt), and iron-zinc (FeZn), as well as iron oxides like magnetite (Fe_3O_4) and maghemite (γ-Fe_2O_3), can be used to create magnetic cores. These magnetic cores are often prone to agglomeration, oxidation, as well as corrosion by acid or base and hence, need to be protected from the environment to prevent their degradation. Therefore, proper coating with cationic polymers is done to isolate these cores from the environment preventing their further degradation [95]. These positively charges polymer such are PEI or PAMAM or chitosan coated on the surface of cores forms the surface for the adsorption of the negatively charged nucleic acids. Moreover, polymeric coatings have been customized to increase the pharmacokinetics as well as the release behavior of these magnetic NPs.

C. Gold nanoparticles (AuNPs) have been researched extensively as multifunctional gene carriers due to their simplicity in synthesis, great biocompatibility, well-defined surface chemistry, and simplicity in molecular imaging using fluorescence resonance energy transfer (FRET). AuNPS-based vectors are prepared by surface functionalization with the positively charged molecules including cationic peptides, amino acids, and tertiary amine-containing molecules [96, 97].

D. Mesoporous silica NPs are generated by precursor silica i.e., tetra ethyl ortho silicate (TEOS), a pore-creating agent such as cetyl tri methyl ammonium bromide (CTAB), and reducing conditions utilizing sodium hydroxide and heat [98, 99].

The silanol group on the surface of these NPs, imparts positive charge for functionalization with nucleic acids and medicines, and their high surface area due to channel development in their structure is their main advantages [100, 101].

E. Carbon-Derived NPs are nano-sized carbon elements formed by various methods like grinding, heating, carbonization, and activation [102, 225]. One such carbon NPs includes carbon nanotubes (CNT) in which one or more sheets of carbon atoms enclose a hollow tube-like structure where genetic material may be deposited due to their large surface: volume ratio and hence used in gene therapy applications [103]. However, the use of CNT is restricted by its toxicity [104, 105]. In recent years, numerous functionalization alternatives have been created to lessen the toxicity of CNTs, such as by adding an amino group to the structure of the NPs to lessen their hydrophobic nature [106]. Graphene NPs, which are based on yet another carbon configuration, are characterized by their hexagonal monolayer structure [107]. These NPs have a large surface area that can bind nucleic acid and may impart electrical and thermal properties [108].

13.4 Complications in Nanomedicine-Mediated Gene Therapy

Gene therapy is a potential method for treating several illnesses, including cancer, hereditary disorders, and chronic or viral infections. Exogenous NA have intrinsic pharmacokinetic limitations, such as electrostatic repulsions, enzymatic degradation, restricted cellular uptake, rapid clearance, etc., which must be overcome for exogenous NA to be internalized by cells and require highly efficient delivery vehicles to overcome these limitations [109]. Yet, in large part delivery of nucleic acids NA and targeted delivery. Moreover, gene delivery methods based on non-viral vectors have faced difficulties in translating to clinical success due to preclinical experimental studies which insufficiently co-relate the therapeutic efficiency. These problems as illustrated in Figure 13.3, form a major area of concern in gene therapy and prevent the execution of the desired therapeutic output. Furthermore, while building therapeutic

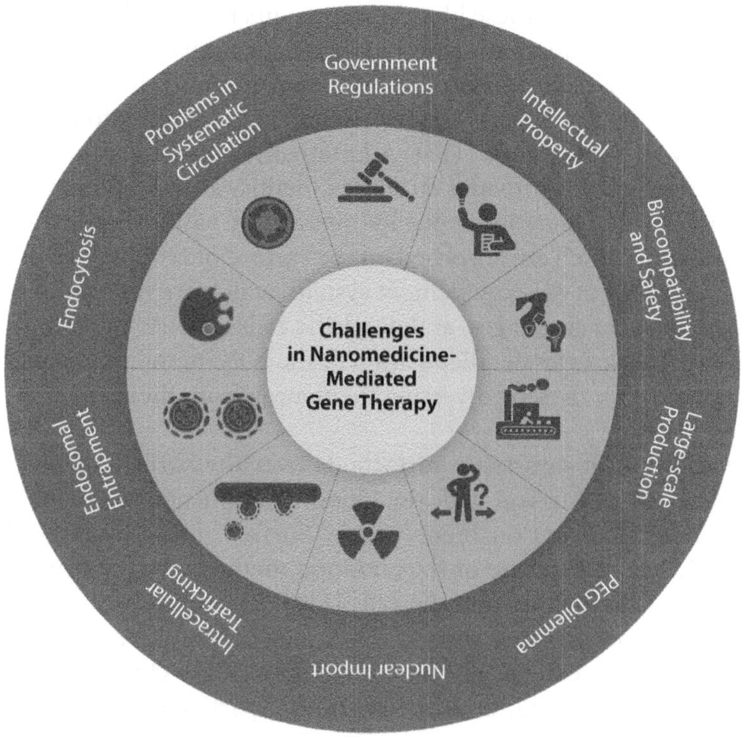

Figure 13.3 Challenges associated with the NM-mediated gene therapy.

platforms, the manufacturing and distribution procedures should be taken into account since they are important factors for clinical application [110]. Discussed below are some of the major hurdles faced during the gene therapy process.

13.4.1 Problems in the Systemic Circulation

Naked DNA when administered intravenously, might be quickly degraded and eliminated by the circulatory system's reticulo-endothelial cells [111]. The gene would be protected from this attack by polymeric delivery, which might also delay the gene's removal from the body, thereby allowing extended time for circulation as well as quick reach to the target site. Yet body fluids create obstructions when in contact with the NPs. Once inside the body, NPs encounter different components of the circulatory system, which is the fluid system in the body consisting of various plasma proteins, blood corpuscles, enzymes, factors, and the immune system cells [112].

The binding affinity of these blood components for the NPs exerts a negative impact on the functioning of the NP. Also, conformational changes are induced on the NPs, by adsorption of the serum opsonin proteins on the surface of NP by the process termed opsonization, leading to their phagocytosis as well as clearance [113]. Additionally, protein-corona formation on the surface of NPs provides an altered identity as well as function to the NP, which further may influence its interaction with other cell surfaces. These proteins which are adsorbed on the surface of NPs include globulins and components of the complement system, which are further recognized by monocytes, macrophages, and dendritic cells (DC) of the immune system [114]. This accelerated deposition of the NPs on the reticuloendothelial system (RES), thereby diverting them away from their intended target location [115].

The addition of non-ionic surfactants over the surface of NPs avoids the interaction of carriers with the blood components, as blood cells too have anionic elements on their surface. Surface modification with PEG called "PEGylation", is one of the common methods for masking the NPs from recognition by the immune cells [116]. PEG reduces the interaction of various blood proteins with the NPs and hence prevents its recognition by RES, thus extending blood circulation. PEG chains can be covalently linked, entrapped, or adsorbed onto a particle's surface to achieve PEGylation. PEG can, however, be immunogenic, and people who have been exposed to substances that have been PEGylated may develop antibodies that are directed against PEG. These Anti-PEG antibodies can accelerate drug clearance, leading to hypersensitivity, and adverse effects [117]. As a result, many PEG substitutes are being researched, including polysarcosine, a biodegradable, more biocompatible, and less immunogenic polypeptoid [118].

13.4.2 Problem in Selective Binding with Target Cell

An increase in the duration of circulating time of the NPs, increases their accessibility to the target site. Endothelial cells lining the blood vessels are present as porous and continuous or discontinuous depending upon the cell morphology of the organ. NPs can readily enter the endothelial cells at discontinuous and fenestrated area and get collected quickly in the liver and spleen. This may be favorable when the NPs are intended to cure the medical condition affecting these specific organs. The enhanced permeability and retention (EPR) effect is seen in the tumorous cell due to the leaky vascular characteristics of the lymphatic system around the carcinoma-affected cells [119]. These pathological features of the diseased cells promote the accumulation of NPs and are termed as passive targeting. Once the NP gains access in the circulatory system of the body, its main hurdle is the selective identification of the target cell. Specific identification of the target site requires binding of the NPs with the affinity ligands [120], termed as active targeting of receptor-mediated targeting or ligand-targeting.

13.4.3 Intracellular Barriers

NPs for non-viral gene delivery techniques along with the struggle to get to the target cell and survive the systemic circulation must be delivered to the nucleus of the specific cell for transcription by the cell or the cytoplasm in case the NP carries siRNA. Thereafter, various intracellular barriers may prevent its delivery consisting of cell membrane barriers, endosomal escape, the release of the nucleic acid cargo into the cytoplasm, and active transport to the nucleus followed by nuclear import.

Endocytosis

Large macromolecular anionic nucleic acids are often not able to pass through the lipophilic anionic structures that make up cell membranes [121]. To create a complete net cationic complex that can interact with anionic cell membranes, non-viral gene delivery techniques try to complex nucleic acid cargo and mask their inherent negative charge [122, 123]. Cellular internalization by endocytosis occurs for the uptake of nutrients and forms a striking route by which genes can be translocated inside the cell thereby reaching the nucleus and leading to gene expression. NPs first bind to the cell surface receptors and later invagination of the membrane forms membrane-bound vesicles that internalize and trap the NP in vesicles called as endosomes [124–126]. Moreover, it condenses the DNA near to a relatively smaller size which is appropriate for the uptake by the cell. For instance, cell-penetrating peptide TAT, which is derived from the human immunodeficiency-1 virus (HIV-1), penetrates the cell depending upon the size of the NP. Smaller NPs, such as peptides with less than 30–40 amino acids, are internalized with TAT via the clathrin pathway, while larger cargoes of proteins or quantum dots that surpass the size of 500 Daltons are internalized by caveolae or macro-pinocytosis [127]. Furthermore, target cell type, receptors on target, incubation time, temperature, the concentration of vector, and properties of cargo and vector, are some other factors that determine the cell membrane internalization route [128, 129]. However, with time these endosomes get fused with lysosomes present in the cells leading to the degradation of NPs [126, 130].

Endocytosis involves three pathways by which vector for gene delivery is internalized by the cell and each of them has been discussed below:

A. Clathrin-Mediated Endocytosis (CME)
   It comprises the internalization of cargo via receptors, like proteoglycans, present on the cell membrane into the vesicles identified as clathrin-coated pits, which measure between 100–150 nm in diameter [125]. These pits further move within the cell by the cytoskeleton's microtubules, where they develop into endosomes. The binding of the NP with the target cell membrane requires the presence of specific ligands on the NP that interacts with the complementary receptor molecule on the cell surface. Various ligands used include transferrin, epidermal growth factor (EGF), low-density lipoprotein (LDL), etc. For instance, the incorporation of ligands such as transferrin into the NPs has made it possible to attach with the cancerous cell which is overexpressing the transferrin receptor. These transferrin receptors on the cancerous cell bind with its ligand on the NPs, thereby facilitating CME-mediated internalization of the NP [126, 131].
B. Caveolae-Mediated Endocytosis (CvME)
   Caveolae mediated pathway involves a molecule namely caveolin (Cav-1) for endocytosis. These Cav-1 proteins trigger the invagination of the plasma membrane and hence vesicle formation [132]. This occurs with smaller NPs of less than 50 nm and a surface charge within +15 and -15 mV. Additionally, caveolae-mediated endocytosis plays a major role in the directed delivery of NPs based on albumin, folic acid, oligonucleotides, HER2 (human epidermal growth factor receptor 2), TAT peptide, and integrin [133]. The capacity of the caveolae-mediated route to bypass lysosomal degradation is by far its greatest benefit. Therefore, the caveosome retains its integrity during brief interactions with endosomes without rupturing the caveolar coat, avoiding the lysosomes and enabling effective peri-nuclear trafficking into the nucleus, Golgi bodies, and endoplasmic reticulum [134]. However, many frequently used cancer cell lines, such as prostate cancer cells (PC-3), have been found to lack the capacity to produce caveolae, which may be

significant for delivery systems depending on this pathway for internalization [135]. On the contrary, caveolae have been observed to be elevated in some cancer cells, making them a potential candidate for delivery systems.

C. Macropinocytosis

It functions in nutrition uptake, immunological monitoring, and diseases, including cancer and viral [136, 137]. It begins with the actin-driven expansion of plasma membrane folds, which produces huge cup-shaped endosomes called macropinosomes, which ultimately combine with the endosomal pathway [138, 139]. NPs are very well internalized by this pathway as they do not involve clathrin or caveolae-based vesicles and hence are devoid of lysosomal degradation. Nearly all cells, except macrophages and endothelial cells of brain, exhibits such nonspecific absorption mechanism [140].

D. Direct internalization

Direct internalization occurs for cationic vectors which are rich in arginine amino-acids facilitated by the interaction between the (-) charged cell membrane and (+) charged amino-acids, thereby inducing the straight delivery into the cytoplasm avoiding the endosomes [141]. Octa-arginine (R8) peptides have also been used for gene delivery due to the potent cell penetration capability [142].

Endosomal Entrapment

After endocytosis, the vector may get trapped in the endosomal pathway, thereby leading to its degradation. Thus, endosomal entrapment represents a significant barrier to effective gene therapy. Since the majority of endosomes have an acidic environment, constant ion channel pumping may harm NPs ability to stay inside the endosomes for an extended period [143]. Therefore, the rate-limiting step of NPs endosomal escape affects the transfection efficiency rate. That is why the DNA cargo in the NP must be released via membrane disruption or endo-osmolytic reactions. Polyplexes are sensitive to pH change, so functional modulations are combined to surpass the endosomal degradation [144]. When these polyplexes are coupled with highly dense amino groups, endosomal escape of NPs into the cytosol is seen to be facilitated [145–147]. The amine group causes a rise in the endosomal pH causing excessive protonation as well as buffering within the endosomes, eventually sequestering counter ions to enter inside the endosomes, causing destabilization of the cell membrane, and finally bursting of the endosomes.

Vectors for gene delivery are designed in such a way that they provide excessive positive charges and cause subsequent bursting of the endosomes. Polymers with amine groups, like PEI, or histidine-rich peptides, which are exploited in numerous gene therapy delivery systems, both make use of the "proton sponge" effect [148–150].

Intracellular Trafficking

After the endosomal escape, the NP vector and its payload must next be delivered to the appropriate cellular compartment, such as DNA delivery to the nucleus or siRNA assembly into RNA-induced silencing complexes (RISC) in the cytoplasm. Successful transgene expression and uptake of NP by the specific cell can be mediated only when the cargo DNA inside the vector NP is released to deliver free DNA to its specific cellular compartment. But this process faces various difficulties including the limited drive of the macromolecules inside the cell slows down the flow of vectors into the nucleus and the degradation of the naked DNA by endonucleases of the cell [97, 151, 152]. Movement of vectors in the cytoplasm is inhibited due to the presence of other organelles and proteins as well as the cell cytoskeleton. The network of microtubules and related motor proteins including kinesins and dyneins maintains proper positioning of the organelles and intracellular trafficking of the vesicles, endosomes, and lysosomes [153, 154]. Non-viral gene delivery vectors may achieve far higher transfection efficiencies if they were able to directly access the nucleus through the cell's microtubule network.

Nuclear Import

Nuclear barriers prevent the gene from entering the transcription machinery of the cell. A nuclear envelope restricts the movement of molecules transported through it via the nuclear pore complex (NPC). To overcome this situation, NPs are modified in such a way that they are attached to nuclear localization sequences (NLS). This NLS consists of short peptides which act as fragments of signals that mediate transport from cytoplasm to the nucleus through the nuclear pore complex [155]. Vectors for gene therapy are functionalized with these NLS to facilitate nuclear import [156, 157]. Integration of human-derived NLS peptide based on insulin growth factor binding proteins (IGFBP) enhances the transfection efficiency of the non-viral vectors [158]. Also, NLS from simian virus 40 (SV40), was found to enhance the nuclear delivery [159].

Short Interfering RNA (siRNA) Delivery

Gene therapy employs the inherent phenomenon of RNA interference (RNAi) of the cells which has a major regulatory role in gene silencing.

For gene therapy, a short oligonucleotide of siRNA complementary to the mRNA sequence of the gene to be silenced is attached to the vector molecule [160, 161]. These siRNAs further binds at the complementary sites in the mRNA and thus inhibit the transcription of the gene [162]. Co-polymer of chitosan and PEI, as well as diblock co-polymer of methoxy poly (ethylene glycol) (MPEG) and polycaprolactone (PCL), conjugated with TAT, a cell-penetrating peptide, via a disulfide linkage, was found to be potentially effective in siRNA delivery for gene silencing [163, 164].

13.5 Challenges in the Clinical Translation of Nanomedicines

Nanomedicine (NM) provides new options to combat illnesses. They have shown important therapeutic benefits for a wide range of medical applications. However, a worldwide effort is required to securely bring experimental discovery to the clinic. Their clinical translation has still not advanced as quickly as the abundance of encouraging preclinical data would have predicted [165–167]. The process of translating NM into the clinical setting is expensive and time-consuming. Moreover, when compared to traditional formulation technology, which uses free medication dispersion in a base, NM technology is typically far more sophisticated. Clinical translation of nanomaterials as NM necessitates that the complexity and variability in their design, as well as development, must be kept as low as possible such that it could further be manufactured, reproduced, and characterized [168, 169]. The following section discusses various challenges during the clinical translation of NPs to NM.

13.5.1 Biological Challenges

The successful translation of NMs depends in large part on our understanding of the interplay between the biotechnology and nanotechnology, including the impact of disease pathophysiology on the accumulation, distribution, retention, and efficacy of NMs, as well as the biological and medical connection between the properties of the delivery system and *in vivo* behavior in animals vs. humans [170]. Therefore, it has been proposed that using a disease-driven strategy as well as designing and creating NMs which can take advantage of pathophysiological alterations in disease biology will enhance clinical translation of NMs [171, 172].

Drugs must overcome several biological obstacles to effectively treat their target diseases. To obtain high systemic bioavailability, oral medications must be very stable in the gastrointestinal tract and able to permeate intestinal epithelium [173, 174]. Likewise, successful drug transport across the epithelium is necessary for cutaneous, nasal, and pulmonary drug delivery. Drugs for the management of central nervous system disorders and brain disorders have to cross the major barrier termed the blood-brain barrier (BBB). Many NP-based techniques are being developed, including nanosphere, liposome, and cationic albumin NP, for delivering drugs across the BBB [175–178].

Therapeutic delivery to solid tumors imposes a major challenge as a large part of tumorous sites are poorly infused as well as have impaired lymphatic drainage that leads to a rise in the interstitial fluid pressure (IFP) [179]. Despite the leaky microvasculature of the tumor, increased IFP has been identified as one of the major mechanisms limiting the extravasation and trans-vascular transport of macromolecules, thus preventing the movement of molecules in the tumor interstitial space [180–182]. Further, dense tumor stroma along with elevated tumor cell density forms a major hurdle in therapeutic transfer in tumors [183].

13.5.2 Large-Scale Production

The steady rate in the clinical conversion of NM remains primarily because of the physio-chemical and structural complexity of the NP formulation. Poor quality control, complexities in scalability, raised manufacturing costs, lack of infrastructure, expertise, and reproducibility of results, all forms major obstacles in the large-scale production of NP medicine [184–186]. Clinical translation requires accessibility to a preparation process that enables the creation of sizable amounts of NMs that are also routinely generated at a high degree of quality and batch-to-batch repeatability to set specifications [187, 188]. Challenges in NP production most importantly arises when their production system becomes more multifaceted and complex; such as the need for chemical and biological modifications, to name a few ligand additions, the inclusion of NLS, and multiple targeting site. Multiple production steps are required to incorporate multiple components into a single NP carrier, which inevitably causes issues for large-scale good manufacturing production, raises the cost of production, and makes it more challenging to evaluate the quality assurance and quality control (QA and QC) of such products [189–191]. Furthermore, batch-to-batch stability of the NP is needed such that it produces identical results (reproducible) every time it is used for gene therapeutics. NP must be stable after production during long-term storage and after clinical administration.

13.5.3 Immunological Response

Once vectors/NPs for gene delivery are delivered systematically, they initially face the immunological response by the host body and are cleared instantly from circulation by macrophages of the spleen and Kupffer cells of the liver via opsonization followed by phagocytosis [192]. In opsonization, opsonin proteins including complement system proteins, immunoglobulins, and blood serum proteins bind to foreign particles, here it is a gene delivery vector/NPs. Later, the opsonized particle binds directly to macrophages which are then engulfed and removed from the circulation via phagocytosis or lead to induction of the complement system [193, 194]. Thus by avoiding opsonization, clearance may be prevented. The introduction of shielding groups, called "stealth" molecules made of long hydrophilic polymeric chains, is a widely used technique to resist opsonization. Polymers are usually elastic and charge-neutral molecules that avoid electrostatic interactions between nucleic acid-vehicle complexes, thereby improving the stability of the vector/NPs in systemic circulation. PEG, polyacrylamide, poly (N-viny-2-pyrrolidone), and poly (vinyl alcohol) are different polymers used to avoid clearance.

13.5.4 Overcoming PEG Dilemma

In cancer gene therapy, using PEG can hamper gene expression by obstructing the entry of the delivery system into the tumor cells. Interaction between the cationic delivery system and the negatively charged cell membrane is mainly electrostatic, which is usually hindered when PEG polymers are attached to the delivery systems. Additionally, increased stability of the PEGylated particles may prevent membrane fusion and lessen the impact of fusogenic peptides during endosomal disruption or cell internalization. Therefore a balance must be achieved by a good vector design such that an adequate balance between the aspects that makes PEG a desirable add-on to a vector and those that limit its efficacy [195, 196]. Addition of ligands to PEG specific for the target cell surface receptors such that it limits the endocytosis of the PEGylated delivery system [197, 198]. As a result of the development of a detachable or reversible PEGylation, "smart" delivery systems have been created that can take advantage of various intracellular or extracellular tumor environments, such as pH, enzyme complement, or reduction, and also serve as a targeting tool to guide vectors to tumors

[70, 199]. PEGylation substantially reduces the ability of nano-vehicles to exit endosomes, which significantly reduces the effectiveness of the delivery mechanism. Endocytosis is the main route for the cellular transport of NMs. When endocytosed cargo is internalized by clathrin-coated vesicles at the plasma membrane, it is first transported into early endosomes, which have an internal pH range of 5.0 to 6.5 [200]. Then, the early endosomes develop into late-endosomes, which eventually join with lysosomes, intracellular organelles having lumen pH of around (4.5–5.0) which is ideal for the hydrolytic enzymes. Thus, facilitating endosomal escape and ensuring cytosolic distribution of the vector is a limiting step in obtaining an effective delivery [201].

13.5.5 Biocompatibility and Safety

Comprehensive toxicological profiling of the developed NMs is indispensable, such that human safety is ensured before it is introduced into the market. Predicting potential side effects or toxicities *in vivo* requires knowledge of the activity and toxicities of the free drug, the behavior of various NM delivery systems and their interactions with biological elements, and the influence of drug release rate on target as well as off-target concentrations of bioavailable drug [202, 203]. Therefore, to maximize the clinical translation potential of NMs, rational design from the early stages of material selection, production method optimization, and product purification is crucial [204, 205]. Even though the biocompatibility of phospholipids and biodegradable polymers has been evaluated before the NM formulation, these can induce severe adverse effects on biocompatibility and bio-distribution when administered *in vivo* [189, 206, 207]. For proper analysis of the nano-toxicology of NMs in the early stages of clinical development, highly sensitive as well as validated assays using *in vitro* and *in vivo* methodologies are required to be performed [204, 208]. Standardized *in vitro* procedures are used to evaluate the possible risk of cytotoxicity, immune-toxicity, and genotoxicity of NMs, utilizing various cell culture models including lung, liver, brain, blood, placenta, and gastrointestinal system [204]. Current approaches for nano-toxicological evaluation of the NMs are limited as well as insufficient. Consequently, highly potent as well as specific novel techniques for nano-toxicological evaluation are under development [209, 210]. These include alternative test protocols, high-throughput screening methods, high-content analysis, and computational modeling; as well as involving simultaneous comparative analysis of different NMs [204, 211, 212]. A comprehensive study on the metabolism, absorption, distribution, and excretion of developing NM *in vivo* is essential to identify toxicological response to NMs [213, 214].

13.5.6 Intellectual Property (IP)

Intellectual property (IP) rights and strategies need to be carefully considered in light of the expanding market and competition. NMs are devised of complex components consisting of as in the case of gene therapy, a NP with its functionalized components and the cargo genetic material. A patent position on the enclosed cargo, carrier technology, and properties of the drug and carrier taken together are often necessary for the management of NMs [215]. NMs involve novel carrier technologies to deliver drugs or genes to the specified targeted location. The IP situation becomes more complicated and confusing with the complex drug or gene delivery systems where various components are owned by different companies. In this, IP approaches may implicate multiple patents linked to any particular technology and cross-licensing agreements [216]. Thus, new IP procedures and standards are needed to streamline the process from development to commercialization such that time as well as cost required in establishing license and partnership agreements may be minimized. Other notable problems that require attention are delays in patent review, patent thickets, and the issuing of invalid patents due to the huge growth in the number of nanotechnology patent applications over the past few decades.

13.5.7 Government Regulations

NM's commercialization depends upon the regulatory government policies in the domain of safety, manufacturing practices, patent protection, as well as quality control [189, 217]. Advancement in the development of NMs products has been hampered by the absence of defined regulatory and safety criteria in the pursuit of efficient clinical translation. NMs are now subject to standard regulation, which is governed by each nation's key regulatory body namely Food and Drug Administration (FDA), Therapeutic Goods Administration (TGA), and European Medicines Agency (EMA). Although the use of NMs dates back a long past, the regulatory measures used are not appropriate to ensure the safety, quality, as well as efficacy of the designed NMs for clinical use [217, 218]. There is a need for regulatory guidelines and standards for NPs, which can bridge regulations of both medicine and medical devices. Challenges hampering the NM regulation include a lack of unified definition and classification of NPs and NMs, lack of agreed regulation, and stability issues after scale-up for manufacturing [219–221].

13.6 Conclusion

The global spread of deadly diseases in the pro-modern times has caused disastrous effects on mankind. Before the development of vaccines, common human diseases like smallpox, tuberculosis, diphtheria, and polio caused severe morbidity as well as mortality. Integration of nanotechnology with the biomedical field has opened new horizons for research on the application of nanomaterials as a carrier for drug and gene delivery. This in turn has generated a new hope in contemporary times for curing various lethal diseases. An ideal NP for the drug as well as gene delivery must overcome the barriers including short circulation time, immunogenicity, poor biocompatibility, non-specific binding, and inefficient penetration of barriers like the vascular endothelium and the blood-brain barrier. Moreover, NP-driven delivery encounters a problem in the translocation of cargo to the nucleus. Therefore, the efficiency of NP-medicated gene delivery is lower than as observed by viral vectors. However, ongoing clinical research shows how NM has the potential to enhance miRNA- and siRNA-based treatments. Poor study design makes it difficult to apply NMs effectively in clinical settings; therefore, more careful research with the accurate characterization of NP and their biological effects is needed. For better medical health, it is necessary to explore the challenges and drawbacks of NMs and gene therapy from a variety of research and clinical study perspectives. The problems that have been highlighted are merely a portion of the enormous problems that exist and needed to be addressed and resolved as early as possible such that the full potential of the NMs could be realized.

References

1. Nanotechnology in medicine: Who are the leading private companies?, Medical Device Network, 2019. Available from: https://www.medicaldevice-network.com/comment/nanotechnology-in-medicine-who-are-the-leading-private-companies/.
2. Pallotta, A., Clarot, I., Sobocinski, J., Fattal, E., Boudier, A., Nanotechnologies for medical devices: Potentialities and risks. *ACS Appl. Bio Mater.*, 2, 1, 1–13, 2019, Available from: https://pubs.acs.org/doi/abs/10.1021/acsabm.8b00612.

3. Chen, Q., Sun, T., Jiang, C., Recent advancements in nanomedicine for 'cold' tumor immunotherapy. *Nano-Micro Lett.*, 13, 1, 1–47, 2021, Available from: https://link.springer.com/article/10.1007/s40820-021-00622-6.
4. Gowd, V., Ahmad, A., Tarique, M., Suhail, M., Zughaibi, T.A., Tabrez, S. *et al.*, Advancement of cancer immunotherapy using nanoparticles-based nanomedicine. *Semin. Cancer Biol.*, 86, 624–44, 2022.
5. Detappe, A., Bustoros, M., Mouhieddine, T.H., Ghoroghchian, P.P., Advancements in nanomedicine for multiple myeloma. *Trends Mol. Med.*, 24, 6, 560–74, 2018.
6. Germain, M., Caputo, F., Metcalfe, S., Tosi, G., Spring, K., Åslund, A.K.O. *et al.*, Delivering the power of nanomedicine to patients today. *J. Controlled Release*, 326, 164–71, 2020.
7. Beltrán-Gracia, E., López-Camacho, A., Higuera-Ciapara, I., Velázquez-Fernández, J.B., Vallejo-Cardona, A.A., Nanomedicine review: Clinical developments in liposomal applications. *Cancer Nanotechnol.*, 10, 1, 1–40, 2019, Available from: https://link.springer.com/articles/10.1186/s12645-019-0055-y.
8. Wolfram, J. and Ferrari, M., Clinical cancer nanomedicine. *Nano Today*, 25, 85–98, 2019, Available from: https://pubmed.ncbi.nlm.nih.gov/31360214/.
9. Zhong, Z. and Liu, Z., Nanomedicine in China. *Adv. Healthc. Mater.*, 7, 20, 1801051, 2018, Available from: https://onlinelibrary.wiley.com/doi/full/10.1002/adhm.201801051.
10. Francia, V., Schiffelers, R.M., Cullis, P.R., Witzigmann, D., The biomolecular corona of lipid nanoparticles for gene therapy. *Bioconjug. Chem.* 31, 9, 2046–59, 2020, Available from: https://pubs.acs.org/doi/pdf/10.1021/acs.bioconjchem.0c00366.
11. The rise, fall and resurgence of gene therapy, American Chemical Society, 2019. Available from: https://www.acs.org/content/acs/en/pressroom/presspacs/2019/acs-presspac-september-18-2019/the-rise-fall-and-resurgence-of-gene-therapy.html.
12. Approved cellular and gene therapy products, FDA, 2023. Available from: https://www.fda.gov/vaccines-blood-biologics/cellular-gene-therapy-products/approved-cellular-and-gene-therapy-products.
13. Sayed, N., Allawadhi, P., Khurana, A., Singh, V., Navik, U., Pasumarthi, S.K. *et al.*, Gene therapy: Comprehensive overview and therapeutic applications. *Life Sci.*, 294, 120375, 2022.
14. Zu, H. and Gao, D., Non-viral vectors in gene therapy: Recent development, challenges, and prospects. *AAPS J.* 23, 4, 1–12, 2021, Available from: https://link.springer.com/article/10.1208/s12248-021-00608-7.
15. Zhang, Q., Zhang, C., Feng, L., Wang, L., Qi, J., Dong, Y. *et al.*, Effects of nanoparticle-mediated co-delivery of bFGF and VEGFA genes to deep burn wounds: An *in vivo* study. *Colloids Surf. B Biointerfaces*, 209, 112135, 2022.

16. Kalluru, P., Shanmugam, M., Vankayala, R., Chiang, C.S., Hwang, K.C., Conquering multidrug resistant lung cancer by upconversion nanoparticles-mediated photodynamic therapy and gene silencing. *J. Chin. Chem. Soc.* 69, 8, 1305–17, 2022, Available from: https://onlinelibrary.wiley.com/doi/full/10.1002/jccs.202200001.
17. Sago, C.D., Lokugamage, M.P., Loughrey, D., Lindsay, K.E., Hincapie, R., Krupczak, B.R. *et al.*, Augmented lipid-nanoparticle-mediated *in vivo* genome editing in the lungs and spleen by disrupting Cas9 activity in the liver. *Nat. Biomed. Eng.*, 6, 2, 157–67, 2022, Available from: https://www.nature.com/articles/s41551-022-00847-9.
18. Ye, Z., Chen, J., Zhao, X., Li, Y., Harmon, J., Huang, C. *et al.*, *In vitro* engineering chimeric antigen receptor macrophages and T cells by lipid nanoparticle-mediated mRNA delivery. *ACS Biomater. Sci. Eng.* 8, 2, 722–33, 2022, Available from: https://pubs.acs.org/doi/abs/10.1021/acsbiomaterials.1c01532.
19. Roma-rodrigues, C., Rivas-garcía, L., Baptista, P.V., Fernandes, A.R., Gene therapy in cancer treatment: Why go nano? *Pharmaceutics*, 12, 2020, MDPI AG.
20. Yetisgin, A.A., Cetinel, S., Zuvin, M., Kosar, A., Kutlu, O., Therapeutic nanoparticles and their targeted delivery applications. *Molecules*, 25, 9, 2020, Available from: /pmc/articles/PMC7248934/.
21. Khan, I., Saeed, K., Khan, I., Nanoparticles: Properties, applications and toxicities. *Arabian J. Chem.*, 12, 7, 908–31, 2019.
22. Soares, S., Sousa, J., Pais, A., Vitorino, C., Nanomedicine: Principles, properties, and regulatory issues. *Front. Chem. Front. Media S.A.*, 6, 2018.
23. Wang, J., Li, B., Qiu, L., Qiao, X., Yang, H., Dendrimer-based drug delivery systems: History, challenges, and latest developments. *J. Biol. Eng.*, 16, 1, 1–12, 2022, Available from: https://jbioleng.biomedcentral.com/articles/10.1186/s13036-022-00298-5.
24. Gibbens-Bandala, B., Trujillo-Nolasco, M., Cruz-Nova, P., Aranda-Lara, L., Ocampo-García, B., Dendrimers as targeted systems for selective gene and drug delivery, in: *Nanotechnology in the Life Sciences* pp. 361–97, 2022, Available from: https://link.springer.com/chapter/10.1007/978-3-031-12658-1_13.
25. dos Santos Rodrigues, B., Banerjee, A., Kanekiyo, T., Singh, J., Functionalized liposomal nanoparticles for efficient gene delivery system to neuronal cell transfection. *Int. J. Pharm.* 566, 717, 2019, Available from: /pmc/articles/PMC6671319/.
26. Tenchov, R., Bird, R., Curtze, A.E., Zhou, Q., Lipid nanoparticles from liposomes to mRNA vaccine delivery, a landscape of research diversity and advancement. *ACS Nano* 15, 11, 16982–7015, 2021, Available from: https://pubs.acs.org/doi/full/10.1021/acsnano.1c04996.
27. Begines, B., Ortiz, T., Pérez-Aranda, M., Martínez, G., Merinero, M., Argüelles-Arias, F. *et al.*, Polymeric nanoparticles for drug delivery: Recent

developments and future prospects. *Nanomaterials* 10, 7, 1–41, 2020, Available from: /pmc/articles/PMC7408012/.

28. Madej, M., Kurowska, N., Strzalka-Mrozik, B., Polymeric nanoparticles—Tools in a drug delivery system in selected cancer therapies. *Appl. Sci.* 12, 19, 9479, 2022, Available from: https://www.researchgate.net/publication/363742108_Polymeric_Nanoparticles-Tools_in_a_Drug_Delivery_System_in_Selected_Cancer_Therapies.
29. Hamimed, S., Jabberi, M., Chatti, A., Nanotechnology in drug and gene delivery. *Naunyn-Schmiedeberg's Arch. Pharmacol.*, 395, 7, 769–87, 2022, Available from: https://link.springer.com/article/10.1007/s00210-022-02245-z.
30. Chandrakala, V., Aruna, V., Angajala, G., Review on metal nanoparticles as nanocarriers: Current challenges and perspectives in drug delivery systems. *Emergent Mater.*, 5, 1593–1615, 2022, Available from: https://link.springer.com/article/10.1007/s42247-021-00335-x.
31. El-Sherbiny, I.M., Elbaz, N.M., Sedki, M., Elgammal, A., Yacoub, M.H., Magnetic nanoparticles-based drug and gene delivery systems for the treatment of pulmonary diseases. *Nanomedicine (Lond).*, 12, 4, 387–402, 2017, https://doi.org/102217/nnm-2016-0341, Available from: https://www.futuremedicine.com/doi/10.2217/nnm-2016-0341.
32. Zhou, Y., Quan, G., Wu, Q., Zhang, X., Niu, B., Wu, B. *et al.*, Mesoporous silica nanoparticles for drug and gene delivery. *Acta Pharm. Sin. B*, 8, 2, 165, 2018, Available from: /pmc/articles/PMC5926503/.
33. Carvalho, A.M., Cordeiro, R.A., Faneca, H., Silica-based gene delivery systems: From design to therapeutic applications. *Pharmaceutics,* 12, 7, 1–45, 2020, Available from: /pmc/articles/PMC7407166/.
34. Debnath, S.K. and Srivastava, R., Drug delivery with carbon-based nanomaterials as versatile nanocarriers: Progress and prospects. *Front. Nanotechnol.*, 3, 15, 2021.
35. Mostafavi, E. and Zare, H., Carbon-based nanomaterials in gene therapy. *OpenNano*, 7, 100062, 2022.
36. Leung, A.K.K., Tam, Y.Y.C., Cullis, P.R., Lipid nanoparticles for short interfering RNA delivery. *Adv. Genet.* 88, 71–110, 2014, Available from: https://pubmed.ncbi.nlm.nih.gov/25409604/.
37. Kulkarni, J.A., Witzigmann, D., Chen, S., Cullis, P.R., van der Meel, R., Lipid nanoparticle technology for clinical translation of siRNA therapeutics. *Acc. Chem. Res.* 52, 9, 2435–44, 2019, Available from: https://pubmed.ncbi.nlm.nih.gov/31397996/.
38. Qiu, M., Li, Y., Bloomer, H., Xu, Q., Developing biodegradable lipid nanoparticles for intracellular mRNA delivery and genome editing. *Acc. Chem. Res.* 54, 21, 4001–11, 2021, Available from: https://pubmed.ncbi.nlm.nih.gov/34668716/.
39. Mashima, R. and Takada, S., Lipid nanoparticles: A novel gene delivery technique for clinical application. *Curr. Issues Mol. Biol.* 44, 10, 5013–27, 2022, Available from: https://pubmed.ncbi.nlm.nih.gov/36286056/.

40. Akinc, A., Querbes, W., De, S., Qin, J., Frank-Kamenetsky, M., Jayaprakash, K.N. *et al.*, Targeted delivery of RNAi therapeutics with endogenous and exogenous ligand-based mechanisms. *Mol. Ther.* 18, 7, 1357–64, 2010, Available from: https://pubmed.ncbi.nlm.nih.gov/20461061/.
41. Basha, G., Novobrantseva, T.I., Rosin, N., Tam, Y.Y.C., Hafez, I.M., Wong, M.K. *et al.*, Influence of cationic lipid composition on gene silencing properties of lipid nanoparticle formulations of siRNA in antigen-presenting cells. *Mol. Ther.* 19, 12, 2186–200, 2011, Available from: https://pubmed.ncbi.nlm.nih.gov/21971424/.
42. B.L.M., Y.K.T., M.J., S.M.A., X.D., Y.Y.T. *et al.*, Influence of polyethylene glycol lipid desorption rates on pharmacokinetics and pharmacodynamics of siRNA lipid nanoparticles. *Mol. Ther. Nucleic Acids* 2, 12, 2013, Available from: https://pubmed.ncbi.nlm.nih.gov/24345865/.
43. Chen, S., Tam, Y.Y.C., Lin, P.J.C., Sung, M.M.H., Tam, Y.K., Cullis, P.R., Influence of particle size on the *in vivo* potency of lipid nanoparticle formulations of siRNA. *J. Controlled Release* 235, 236–44, 2016, Available from: https://pubmed.ncbi.nlm.nih.gov/27238441/.
44. Belliveau, N.M., Huft, J., Lin, P.J., Chen, S., Leung, A.K., Leaver, T.J. *et al.*, Microfluidic synthesis of highly potent limit-size lipid nanoparticles for *in vivo* delivery of siRNA. *Mol. Ther. Nucleic Acids* 1, 8, e37, 2012, Available from: /pmc/articles/PMC3442367/.
45. Akinc, A., Maier, M.A., Manoharan, M., Fitzgerald, K., Jayaraman, M., Barros, S. *et al.*, The Onpattro story and the clinical translation of nanomedicines containing nucleic acid-based drugs. *Nat. Nanotechnol.* 14, 12, 1084–7, 2019, Available from: https://pubmed.ncbi.nlm.nih.gov/31802031/.
46. Adams, D., Gonzalez-Duarte, A., O'Riordan, W.D., Yang, C.C., Ueda, M., Kristen, A.V. *et al.*, Patisiran, an RNAi therapeutic, for hereditary transthyretin amyloidosis. *N. Engl. J. Med.* 379, 1, 11–21, 2018, Available from: https://pubmed.ncbi.nlm.nih.gov/29972753/.
47. Witzigmann, D., Kulkarni, J.A., Leung, J., Chen, S., Cullis, P.R., van der Meel, R., Lipid nanoparticle technology for therapeutic gene regulation in the liver. *Adv. Drug Deliv. Rev.* 159, 344–63, 2020, Available from: https://pubmed.ncbi.nlm.nih.gov/32622021/.
48. Ponti, F., Campolungo, M., Melchiori, C., Bono, N., Candiani, G., Cationic lipids for gene delivery: Many players, one goal. *Chem. Phys. Lipids*, 235, 105032, 2021.
49. Kim, S.Y., Lee, M.K., Lim, S.J., Current advances in developing cationic lipid-based nanoparticles as a vehicle for improving adenoviral gene delivery. *J. Pharm. Investig.* 46, 4, 393–402, 2016, Available from: https://link.springer.com/article/10.1007/s40005-016-0261-0.
50. Dabkowska, A.P., Barlow, D.J., Campbell, R.A., Hughes, A.V., Quinn, P.J., Lawrence, M.J., Effect of helper lipids on the interaction of DNA with cationic lipid monolayers studied by specular neutron reflection. *Biomacromolecules*,

13, 8, 2391–401, 2012, Available from: https://pubmed.ncbi.nlm.nih.gov/22724519/.
51. Arruda, D.C., Hoffmann, C., Charrueau, C., Bigey, P., Escriou, V., Innovative nonviral vectors for small-interfering RNA delivery and therapy, in: *Nanostructures for Novel Therapy: Synthesis, Characterization and Applications*, pp. 713–40, 2017.
52. Karmacharya, P., Patil, B.R., Kim, J.O., Recent advancements in lipid–mRNA nanoparticles as a treatment option for cancer immunotherapy. *J. Pharm. Investig.* 52, 4, 415–26, 2022, Available from: https://link.springer.com/article/10.1007/s40005-022-00569-9.
53. Li, Y., Yu, T., Zhe, H.L., Li, J.L., Jin, Y., Shan, Q.J., Development of a non-viral gene vector for enhancing gene transfection efficiency. *J. Drug Deliv. Sci. Technol.*, 75, 103669, 2022.
54. Jitta, S.R. and Kumar, L., Solid lipid nanoparticles-based drug and gene delivery to macrophages, in: *Macrophage Targeted Delivery Systems* pp. 203–24, 2022, Available from: https://link.springer.com/chapter/10.1007/978-3-030-84164-5_10.
55. Nam, H.Y., Park, J.H., Kim, K., Kwon, I.C., Jeong, S.Y., Lipid-based emulsion system as non-viral gene carriers. *Arch. Pharm. Res.* 32, 5, 639–46, 2009, Available from: https://www.researchgate.net/publication/26243295_Lipid-based_emulsion_system_as_non-viral_gene_carriers.
56. Choi, W.J., Kim, J.K., Choi, S.H., Park, J.S., Ahn, W.S., Kim, C.K., Low toxicity of cationic lipid-based emulsion for gene transfer. *Biomaterials.*, 25, 27, 5893–903, 2004.
57. Rai, R., Alwani, S., Badea, I., Polymeric nanoparticles in gene therapy: New avenues of design and optimization for delivery applications. *Polym. (Basel)* 11, 4, 2019, Available from: https://pubmed.ncbi.nlm.nih.gov/31027272/.
58. Tros de Ilarduya, C., Sun, Y., Düzgüneş, N., Gene delivery by lipoplexes and polyplexes. *Eur. J. Pharm. Sci.* 40, 3, 159–70, 2010, Available from: https://pubmed.ncbi.nlm.nih.gov/20359532/.
59. Anis, H.A., Gene therapy in the era of nanotechnology/a review of current data. *J. Cancer Prev. Curr. Res.* 10, 1, 2019, Available from: https://medcrave-online.com/JCPCR/JCPCR-10-00380.php.
60. Amreddy, N., Babu, A., Muralidharan, R., Munshi, A., Ramesh, R., Polymeric nanoparticle-mediated gene delivery for lung cancer treatment. *Top. Curr. Chem.*, 375, 2, 1–23, 2017, Available from: https://link.springer.com/article/10.1007/s41061-017-0128-5.
61. Rai, D.B., Pooja, D., Kulhari, H., Dendrimers in gene delivery, in: *Pharmaceutical Applications of Dendrimers*, pp. 211–31, 2020.
62. Mendes, L.P., Pan, J., Torchilin, V.P., Dendrimers as nanocarriers for nucleic acid and drug delivery in cancer therapy. *Molecules*, 22, 9, 1401, 2017, Available from: https://www.mdpi.com/1420-3049/22/9/1401/htm.
63. Wang, J., Li, B., Qiu, L., Qiao, X., Yang, H., Dendrimer-based drug delivery systems: History, challenges, and latest developments. *J. Biol. Eng.*, 16, 1, 1–12,

2022, Available from: https://jbioleng.biomedcentral.com/articles/10.1186/s13036-022-00298-5.
64. Kaur, J., Taleuzzaman, M., Das, V., Singh, A.D., Lipopolyplex-based delivery system for cancer vaccine delivery, in: *Nanotherapeutics in Cancer Vaccination and Challenges*, pp. 193–216, 2022.
65. Zare, M., Pemmada, R., Madhavan, M., Shailaja, A., Ramakrishna, S., Kandiyil, S.P. *et al.*, Encapsulation of miRNA and siRNA into nanomaterials for cancer therapeutics. *Pharmaceutics*, 14, 8, 1620, 2022, Available from: https://www.mdpi.com/1999-4923/14/8/1620/htm.
66. Tan, Y., Whitmore, M., Li, S., Frederik, P., Huang, L., LPD nanoparticles–novel nonviral vector for efficient gene delivery. *Methods Mol. Med.* 69, 73–81, 2002, Available from: https://pubmed.ncbi.nlm.nih.gov/11987799/.
67. Sabir, F., Qindeel, M., Zeeshan, M., Ain, Q.U., Rahdar, A., Barani, M. *et al.*, Onco-receptors targeting in lung cancer via application of surface-modified and hybrid nanoparticles: A cross-disciplinary review. *Processes*, 9, 4, 621, 2021, Available from: https://www.mdpi.com/2227-9717/9/4/621/htm.
68. Moriguchi, R., Kogure, K., Akita, H., Futaki, S., Miyagishi, M., Taira, K. *et al.*, A multifunctional envelope-type nano device for novel gene delivery of siRNA plasmids. *Int. J. Pharm.* 301, 1–2, 277–85, 2005, Available from: https://pubmed.ncbi.nlm.nih.gov/16019173/.
69. Kogure, K., Akita, H., Yamada, Y., Harashima, H., Multifunctional envelope-type nano device (MEND) as a non-viral gene delivery system. *Adv. Drug Deliv. Rev.*, 60, 4–5, 559–71, 2008.
70. McErlean, E.M., McCrudden, C.M., McCarthy, H.O., McErlean, E.M., McCrudden, C.M., McCarthy, H.O., Multifunctional delivery systems for cancer gene therapy, in: *Gene Therapy - Principles and Challenges*, 2015, Available from: https://www.intechopen.com/state.item.id.
71. Yamada, Y., Akita, H., Harashima, H., Multifunctional envelope-type nano device (MEND) for organelle targeting via a stepwise membrane fusion process. *Methods Enzymol.*, 509, 301–26, 2012, Available from: https://pubmed.ncbi.nlm.nih.gov/22568912/.
72. Sanghani, A., Kafetzis, K.N., Sato, Y., Elboraie, S., Fajardo-Sanchez, J., Harashima, H. *et al.*, Novel PEGylated lipid nanoparticles have a high encapsulation efficiency and effectively deliver MRTF-B siRNA in conjunctival fibroblasts. *Pharmaceutics*, 13, 3, 382, 2021, Available from: https://www.mdpi.com/1999-4923/13/3/382/htm.
73. Li, P. and Zhang, N., Layer-by-layer self-assembly vectors for gene delivery. *Curr. Gene Ther.*, 11, 1, 58–73, 2011.
74. Linnik, D.S., Tarakanchikova, Y.V., Zyuzin, M.V., Lepik, K.V., Aerts, J.L., Sukhorukov, G. *et al.*, Layer-by-layer technique as a versatile tool for gene delivery applications. *Expert. Opin. Drug Deliv.*, 18, 8, 1047–66, 2021, Available from: https://www.tandfonline.com/doi/abs/10.1080/17425247.2021.1879790.

75. Panyam, J. and Labhasetwar, V., Biodegradable nanoparticles for drug and gene delivery to cells and tissue. *Adv. Drug Deliv. Rev.*, 55, 3, 329–47, 2003.
76. Ramezani, M., Ebrahimian, M., Hashemi, M., Current strategies in the modification of PLGA-based gene delivery system. *Curr. Med. Chem.* 24, 7, 728–39, 2017, Available from: https://pubmed.ncbi.nlm.nih.gov/27919215/.
77. Kim, S.M., Patel, M., Patel, R., PLGA core-shell nano/microparticle delivery system for biomedical application. *Polym. (Basel)* 13, 20, 2021, Available from: /pmc/articles/PMC8540999/.
78. Su, Y., Zhang, B., Sun, R., Liu, W., Zhu, Q., Zhang, X. *et al.*, PLGA-based biodegradable microspheres in drug delivery: Recent advances in research and application. *Drug Deliv.* 28, 1, 1397, 2021, Available from: /pmc/articles/PMC8248937/.
79. Alvi, M., Yaqoob, A., Rehman, K., Shoaib, S.M., Akash, M.S.H., PLGA-based nanoparticles for the treatment of cancer: Current strategies and perspectives. *AAPS Open*, 8, 1, 1–17, 2022, Available from: https://aapsopen.springeropen.com/articles/10.1186/s41120-022-00060-7.
80. Makadia, H.K. and Siegel, S.J., Poly lactic-co-glycolic acid (PLGA) as biodegradable controlled drug delivery carrier. *Polym.*, 3, 3, 1377, 2011, Available from: /pmc/articles/PMC3347861/.
81. Thapa, R.K. and Sullivan, M.O., Gene delivery by peptide-assisted transport. *Curr. Opin. Biomed. Eng.* 7, 71–82, 2018, Available from: /pmc/articles/PMC6428085/.
82. Tarvirdipour, S., Huang, X., Mihali, V., Schoenenberger, C.A., Palivan, C.G., Peptide-based nanoassemblies in gene therapy and diagnosis: Paving the way for clinical application. *Molecules,* 25, 15, 2020, Available from: /pmc/articles/PMC7435460/.
83. Tarvirdipour, S., Skowicki, M., Schoenenberger, C.A., Palivan, C.G., Peptide-assisted nucleic acid delivery systems on the rise. *Int. J. Mol. Sci.*, 22, 16, 2021, Available from: /pmc/articles/PMC8396486/.
84. Loh, X.J., Lee, T.C., Dou, Q., Deen, G.R., Utilising inorganic nanocarriers for gene delivery. *Biomater. Sci.* 4, 1, 70–86, 2015, Available from: https://pubs.rsc.org/en/content/articlehtml/2016/bm/c5bm00277j.
85. Jha, S., Mathur, P., Ramteke, S., Jain, N.K., Pharmaceutical potential of quantum dots. *Artif. Cells Nanomed. Biotechnol.*, 46, sup1, 57–65, 2017, https://doi.org/101080/2169140120171411932, Available from: https://www.tandfonline.com/doi/abs/10.1080/21691401.2017.1411932.
86. Wagner, A.M., Knipe, J.M., Orive, G., Peppas, N.A., Quantum dots in biomedical applications. *Acta Biomater.*, 94, 44, 2019, Available from: /pmc/articles/PMC6642839/.
87. Lin, G., Chen, T., Zou, J., Wang, Y., Wang, X., Li, J. *et al.*, Quantum dots-siRNA nanoplexes for gene silencing in central nervous system tumor cells. *Front. Pharmacol.*, 8, 182, 2017.

88. Derfus, A.M., Chan, W.C.W., Bhatia, S.N., Probing the cytotoxicity of semiconductor quantum dots. *Nano Lett.* 4, 1, 11–8, 2004, Available from: https://pubs.acs.org/doi/abs/10.1021/nl0347334.
89. Hardman, R., A toxicologic review of quantum dots: Toxicity depends on physicochemical and environmental factors. *Environ. Health Perspect.*, 114, 2, 165–72, 2006, Available from: https://pubmed.ncbi.nlm.nih.gov/16451849/.
90. Soo Choi, H., Liu, W., Misra, P., Tanaka, E., Zimmer, J.P., Itty Ipe, B. *et al.*, Renal clearance of quantum dots. *Nat. Biotechnol.*, 25, 10, 1165–70, 2007, Available from: https://pubmed.ncbi.nlm.nih.gov/17891134/.
91. Kami, D., Takeda, S., Itakura, Y., Gojo, S., Watanabe, M., Toyoda, M., Application of magnetic nanoparticles to gene delivery. *Int. J. Mol. Sci.*, 12, 6, 3705–22, 2011, Available from: https://www.mdpi.com/1422-0067/12/6/3705/htm.
92. Akbaba, H., Erel Akbaba, G., Kantarcı, A.G., Fabrication and evaluation of cationic charged magnetic nanoparticles for enhanced gene delivery. *Fabad J. Pharm. Sci.*, 47, 2, 201–12, 2022, Available from: https://dergipark.org.tr/en/pub/fabadeczacilik/issue/72155/1134568.
93. Blokpoel Ferreras, L.A., Chan, S.Y., Vazquez Reina, S., Dixon, J.E., Rapidly transducing and spatially localized magnetofection using peptide-mediated non-viral gene delivery based on iron oxide nanoparticles. *ACS Appl. Nano Mater.*, 4, 1, 167–81, 2021, Available from: https://pubs.acs.org/doi/full/10.1021/acsanm.0c02465.
94. Majidi, S., Zeinali Sehrig, F., Samiei, M., Milani, M., Abbasi, E., Dadashzadeh, K. *et al.*, Magnetic nanoparticles: Applications in gene delivery and gene therapy. *Artif. Cells Nanomed. Biotechnol.*, 44, 4, 1186–93, 2015, http://dx.doi.org/103109/2169140120151014093, Available from: https://cogentoa.tandfonline.com/doi/abs/10.3109/21691401.2015.1014093.
95. Kudr, J., Haddad, Y., Richtera, L., Heger, Z., Cernak, M., Adam, V. *et al.*, Magnetic nanoparticles: From design and synthesis to real world applications. *Nanomaterials*, 7, 9, 2017, Available from: /pmc/articles/PMC5618354/.
96. Prabu, S.L., Kuppusami Suriyaprakash, T.N., Thirumurugan, R., Prabu, S.L., Kuppusami Suriyaprakash, T.N., Thirumurugan, R., Medicated nanoparticle for gene delivery, in: *Advanced Technology for Delivering Therapeutics*, 2017, Available from: https://www.intechopen.com/state.item.id.
97. Patil, S., Gao, Y.G., Lin, X., Li, Y., Dang, K., Tian, Y. *et al.*, The development of functional non-viral vectors for gene delivery. *Int. J. Mol. Sci.*, 20, 21, 5491, 2019, Available from: https://www.mdpi.com/1422-0067/20/21/5491/htm.
98. Putz, A.M., Savii, C., Ianăşi, C., Dudás, Z., Székely, K.N., Plocek, J. *et al.*, Pore ordering in mesoporous matrices induced by different directing agents. *J. Porous Mater.* 22, 2, 321–31, 2017, Available from: http://arxiv.org/abs/1702.00366.

99. Almásy, L., Putz, A.M., Tian, Q., Kopitsa, G.P., Khamova, T.V., Barabás, R. *et al.*, Hybrid mesoporous silica with controlled drug release. *J. Serb. Chem. Soc.*, 84, 9, 1027–39, 2019.
100. Kamegawa, R., Naito, M., Miyata, K., Functionalization of silica nanoparticles for nucleic acid delivery. *Nano Res.*, 11, 10, 5219–39, 2018, Available from: https://link.springer.com/article/10.1007/s12274-018-2116-7.
101. Castillo, R.R., Lozano, D., Vallet-Regí, M., Mesoporous silica nanoparticles as carriers for therapeutic biomolecules. *Pharmaceutics*, 12, 5, 2020, Available from: /pmc/articles/PMC7284475/.
102. Testa, C., Zammataro, A., Pappalardo, A., Trusso Sfrazzetto, G., Catalysis with carbon nanoparticles. *RSC Adv.*, 9, 47, 27659–64, 2019, Available from: https://pubs.rsc.org/en/content/articlehtml/2019/ra/c9ra05689k.
103. Mostafavi, E. and Zare, H., Carbon-based nanomaterials in gene therapy. *OpenNano*, 7, 100062, 2022.
104. Allegri, M., Perivoliotis, D.K., Bianchi, M.G., Chiu, M., Pagliaro, A., Koklioti, M.A. *et al.*, Toxicity determinants of multi-walled carbon nanotubes: The relationship between functionalization and agglomeration. *Toxicol. Rep.* 3, 230, 2016, Available from: /pmc/articles/PMC5615827/.
105. Francis, A.P. and Devasena, T., Toxicity of carbon nanotubes: A review. *Toxicol. Ind. Health*, 34, 3, 200–10, 2018, Available from: https://pubmed.ncbi.nlm.nih.gov/29506458/.
106. Anzar, N., Hasan, R., Tyagi, M., Yadav, N., Narang, J., Carbon nanotube - a review on synthesis, properties and plethora of applications in the field of biomedical science. *Sens. Int.*, 1, 100003, 2020.
107. Yusaf, T., Mahamude, A.S.F., Farhana, K., Harun, W.S.W., Kadirgama, K., Ramasamy, D. *et al.*, A comprehensive review on graphene nanoparticles: Preparation, properties, and applications. *Sustainability*, 14, 19, 12336, 2022.
108. Wu, X., Mu, F., Wang, Y., Zhao, H., Graphene and graphene-based nanomaterials for DNA detection: A review. *Molecules: A Journal of Synthetic Chemistry and Natural Product Chemistry*, 23, 8, 2018, Available from: /pmc/articles/PMC6222676/.
109. Weng, Y., Huang, Q., Li, C., Yang, Y., Wang, X., Yu, J. *et al.*, Improved nucleic acid therapy with advanced nanoscale biotechnology. *Mol. Ther. Nucleic Acids*, 19, 581–601, 2020.
110. Kaemmerer, W.F., How will the field of gene therapy survive its success? *Bioeng. Transl. Med.*, 3, 2, 166–77, 2018.
111. Suschak, J.J., Williams, J.A., Schmaljohn, C.S., Advancements in DNA vaccine vectors, non-mechanical delivery methods, and molecular adjuvants to increase immunogenicity. *Hum. Vaccin. Immunother.*, 13, 12, 2837–48, 2017, https://doi.org/101080/2164551520171330236, Available from: https://www.tandfonline.com/doi/abs/10.1080/21645515.2017.1330236.
112. La-Beck, N.M. and Gabizon, A.A., Nanoparticle interactions with the immune system: Clinical implications for liposome-based cancer chemotherapy. *Front. Immunol.*, 8, APR, 416, 2017.

113. Papini, E., Tavano, R., Mancin, F., Opsonins and dysopsonins of nanoparticles: Facts, concepts, and methodological guidelines. *Front. Immunol.* 11, 2020, Available from: /pmc/articles/PMC7587406/.
114. Nguyen, V.H. and Lee, B.J., Protein corona: A new approach for nanomedicine design. *Int. J. Nanomed.* 12, 3137, 2017, Available from: /pmc/articles/PMC5402904/.
115. Zhang, M., Gao, S., Yang, D., Fang, Y., Lin, X., Jin, X. *et al.*, Influencing factors and strategies of enhancing nanoparticles into tumors *in vivo*. *Acta Pharm. Sin. B*, 11, 8, 2265–85, 2021.
116. Sims, L.B., Curtis, L.T., Frieboes, H.B., Steinbach-Rankins, J.M., Enhanced uptake and transport of PLGA-modified nanoparticles in cervical cancer. *J. Nanobiotechnology*, 14, 1, 33, 2016, Available from: /pmc/articles/PMC4840861/.
117. Estapé Senti, M., de Jongh, C.A., Dijkxhoorn, K., Verhoef, J.J.F., Szebeni, J., Storm, G. *et al.*, Anti-PEG antibodies compromise the integrity of PEGylated lipid-based nanoparticles via complement. *J. Controlled Release*, 341, 475–86, 2022.
118. Hu, Y., Hou, Y., Wang, H., Lu, H., Polysarcosine as an alternative to PEG for therapeutic protein conjugation. *Bioconjug. Chem.*, 29, 7, 2232–8, 2018, Available from: https://pubs.acs.org/doi/abs/10.1021/acs.bioconjchem.8b00237.
119. Rosenblum, D., Joshi, N., Tao, W., Karp, J.M., Peer, D., Progress and challenges towards targeted delivery of cancer therapeutics. *Nat. Commun.*, 9, 1, 1–12, 2018, Available from: https://www.nature.com/articles/s41467-018-03705-y.
120. Manzari, M.T., Shamay, Y., Kiguchi, H., Rosen, N., Scaltriti, M., Heller, D.A., Targeted drug delivery strategies for precision medicines. *Nat. Rev. Mater.*, 6, 4, 351–70, 2021, Available from: https://www.nature.com/articles/s41578-020-00269-6.
121. Zhang, R., Qin, X., Kong, F., Chen, P., Pan, G., Improving cellular uptake of therapeutic entities through interaction with components of cell membrane. *Drug Deliv.*, 26, 1, 328, 2019, Available from: /pmc/articles/PMC6442206/.
122. Tasset, A., Bellamkonda, A., Wang, W., Pyatnitskiy, I., Ward, D., Peppas, N. *et al.*, Overcoming barriers in non-viral gene delivery for neurological applications. *Nanoscale*, 14, 10, 3698–719, 2022, Available from: https://pubs.rsc.org/en/content/articlehtml/2022/nr/d1nr06939j.
123. Jiang, C., Chen, J., Li, Z., Wang, Z., Zhang, W., Liu, J., Recent advances in the development of polyethylenimine-based gene vectors for safe and efficient gene delivery. *Expert. Opin. Drug Deliv.*, 16, 4, 363–76, 2019, https://doi.org/101080/1742524720191604681, Available from: https://www.tandfonline.com/doi/abs/10.1080/17425247.2019.1604681.
124. Behzadi, S., Serpooshan, V., Tao, W., Hamaly, M.A., Alkawareek, M.Y., Dreaden, E.C. *et al.*, Cellular uptake of nanoparticles: Journey inside the cell. *Chem. Soc. Rev.*, 46, 14, 4218, 2017, Available from: /pmc/articles/PMC5593313/.

125. Sousa De Almeida, M., Susnik, E., Drasler, B., Taladriz-Blanco, P., Petri-Fink, A., Rothen-Rutishauser, B., Understanding nanoparticle endocytosis to improve targeting strategies in nanomedicine. *Chem. Soc. Rev.*, 50, 9, 5397–434, 2021, Available from: https://pubs.rsc.org/en/content/articlehtml/2021/cs/d0cs01127d.
126. Foroozandeh, P. and Aziz, A.A., Insight into cellular uptake and intracellular trafficking of nanoparticles. *Nanoscale Res. Lett.*, 13, 1, 1–12, 2018, Available from: https://nanoscalereslett.springeropen.com/articles/10.1186/s11671-018-2728-6.
127. Cahill, K., Cell-penetrating peptides, electroporation and drug delivery. *IET Syst. Biol.* 4, 6, 367–78, 2010 Available from: https://pubmed.ncbi.nlm.nih.gov/21073236/.
128. Cevher, E., Sezer, A.D., Çağlar, E.S., Cevher, E., Sezer, A.D., Çağlar, E.S., Gene delivery systems: Recent progress in viral and non-viral therapy, in: *Recent Advances in Novel Drug Carrier Systems,* 2012, Available from: https://www.intechopen.com/state.item.id.
129. Ruseska, I. and Zimmer, A., Internalization mechanisms of cell-penetrating peptides. *Beilstein J. Nanotechnol.* 11, 101, 2020, Available from: /pmc/articles/PMC6964662/.
130. Pei, D. and Buyanova, M., Overcoming endosomal entrapment in drug delivery. *Bioconjug. Chem.*, 30, 2, 273, 2019, Available from: /pmc/articles/PMC6501178/.
131. Mazumdar, S., Chitkara, D., Mittal, A., Exploration and insights into the cellular internalization and intracellular fate of amphiphilic polymeric nanocarriers. *Acta Pharm. Sin. B*, 11, 4, 903–24, 2021.
132. Vykoukal, J.V., Fahrmann, J.F., Thompson, T.C., Caveolin and lipid domains–close companions in managing cellular pathways. *Cancer Metastasis Rev.*, 39, 2, 341–2, 2020, Available from: https://link.springer.com/article/10.1007/s10555-020-09891-w.
133. Dudău, M., Codrici, E., Tanase, C., Gherghiceanu, M., Enciu, A.M., Hinescu, M.E., Caveolae as potential hijackable gates in cell communication. *Front. Cell Dev. Biol.*, 8, 1228, 2020.
134. Liu, C.G., Han, Y.H., Kankala, R.K., Wang, S.B., Chen, A.Z., Subcellular performance of nanoparticles in cancer therapy. *Int. J. Nanomed.* 15, 675, 2020, Available from: /pmc/articles/PMC7008395/.
135. Hill, M.M., Bastiani, M., Luetterforst, R., Kirkham, M., Kirkham, A., Nixon, S.J. *et al.*, PTRF-Cavin, a conserved cytoplasmic protein required for caveola formation and function. *Cell,* 132, 1, 113–24, 2008, Available from: https://pubmed.ncbi.nlm.nih.gov/18191225/.
136. Recouvreux, M.V. and Commisso, C., Macropinocytosis: A metabolic adaptation to nutrient stress in cancer. *Front. Endocrinol. (Lausanne)*, 8, SEP, 261, 2017.
137. Stow, J.L., Hung, Y., Wall, A.A., Macropinocytosis: Insights from immunology and cancer. *Curr. Opin. Cell Biol.*, 65, 131–40, 2020.

138. Falcone, S., Cocucci, E., Podini, P., Kirchhausen, T., Clementi, E., Meldolesi, J., Macropinocytosis: Regulated coordination of endocytic and exocytic membrane traffic events. *J. Cell Sci.* 119, 4758–69, 2006, Available from: https://journals.biologists.com/jcs/article/119/22/4758/29438/Macropinocytosis-regulated-coordination-of.
139. Kerr, M.C. and Teasdale, R.D., Defining macropinocytosis. *Traffic*, 10, 4, 364–71, 2009, Available from: https://pubmed.ncbi.nlm.nih.gov/19192253/.
140. Sahay, G., Alakhova, D.Y., Kabanov, A.V., Endocytosis of nanomedicines. *J. Controlled Release* 145, 3, 182–95, 2010, Available from: https://pubmed.ncbi.nlm.nih.gov/20226220/.
141. Brock, R., The uptake of arginine-rich cell-penetrating peptides: Putting the puzzle together. *Bioconjug. Chem.* 25, 5, 863–8, 2014, Available from: https://pubmed.ncbi.nlm.nih.gov/24679171/.
142. Margus, H., Padari, K., Pooga, M., Cell-penetrating peptides as versatile vehicles for oligonucleotide delivery. *Mol. Ther.* 20, 3, 525, 2012, Available from: /pmc/articles/PMC3293609/.
143. Wang, C., Zhao, T., Li, Y., Huang, G., White, M.A., Gao, J., Investigation of endosome and lysosome biology by ultra pH-sensitive nanoprobes. *Adv. Drug Deliv. Rev.* 113, 87, 2017, Available from: /pmc/articles/PMC5339051/.
144. Kongkatigumjorn, N., Smith, S.A., Chen, M., Fang, K., Yang, S., Gillies, E.R. *et al.*, Controlling endosomal escape using pH-responsive nanoparticles with tunable disassembly. *ACS Appl. Nano Mater.*, 1, 7, 3164–73, 2018, Available from: https://pubs.acs.org/doi/abs/10.1021/acsanm.8b00338.
145. Gujrati, M., Malamas, A., Shin, T., Jin, E., Sun, Y., Lu, Z.R., Multifunctional cationic lipid-based nanoparticles facilitate endosomal escape and reduction-triggered cytosolic siRNA release. *Mol. Pharm.*, 11, 8, 2734–44, 2014, Available from: https://pubs.acs.org/doi/full/10.1021/mp400787s.
146. Hall, A., Lächelt, U., Bartek, J., Wagner, E., Moghimi, S.M., Polyplex evolution: Understanding biology, optimizing performance. *Mol. Ther.*, 25, 7, 1476, 2017, Available from: /pmc/articles/PMC5498806/.
147. Brock, D.J., Kondow-Mcconaghy, H.M., Hager, E.C., Pellois, J.P., Endosomal escape and cytosolic penetration of macromolecules mediated by synthetic delivery agents. *Bioconjug. Chem.*, 30, 2, 293, 2019, Available from: /pmc/articles/PMC6561124/.
148. Boussif, O., Lezoualc'h, F., Zanta, M.A., Mergny, M.D., Scherman, D., Demeneix, B. *et al.*, A versatile vector for gene and oligonucleotide transfer into cells in culture and *in vivo*: Polyethylenimine. *Proc. Natl. Acad. Sci.*, 92, 16, 7297–301, 1995, Available from: https://www.pnas.org/doi/abs/10.1073/pnas.92.16.7297.
149. Midoux, P., Pichon, C., Yaouanc, J.J., Jaffrès, P.A., Chemical vectors for gene delivery: A current review on polymers, peptides and lipids containing histidine or imidazole as nucleic acids carriers. *Br. J. Pharmacol.* 157, 2, 166–78, 2009, Available from: https://pubmed.ncbi.nlm.nih.gov/19459843/.

150. Bus, T., Traeger, A., Schubert, U.S., The great escape: How cationic polyplexes overcome the endosomal barrier. *J. Mater. Chem. B*, 6, 43, 6904–18, 2018, Available from: https://pubs.rsc.org/en/content/articlehtml/2018/tb/c8tb00967h.
151. Zhang, X.X., McIntosh, T.J., Grinstaff, M.W., Functional lipids and lipoplexes for improved gene delivery. *Biochimie*, 94, 1, 42, 2012, Available from: /pmc/articles/PMC3771368/.
152. Dean, D.A., Strong, D.D., Zimmer, W.E., Nuclear entry of nonviral vectors. *Gene Ther.*, 12, 11, 881–90, 2005, Available from: https://pubmed.ncbi.nlm.nih.gov/15908994/.
153. Kruppa, A.J. and Buss, F., Motor proteins at the mitochondria–cytoskeleton interface. *J. Cell Sci.*, 134, 7, 1073–87, 2021, Available from: /pmc/articles/PMC8077471/.
154. Cabukusta, B. and Neefjes, J., Mechanisms of lysosomal positioning and movement. *Traffic*, 19, 10, 761–9, 2018, Available from: https://onlinelibrary.wiley.com/doi/full/10.1111/tra.12587.
155. Lu, J., Wu, T., Zhang, B., Liu, S., Song, W., Qiao, J. *et al.*, Types of nuclear localization signals and mechanisms of protein import into the nucleus. *Cell Commun. Signal.*, 19, 1, 1–10, 2021, Available from: https://biosignaling.biomedcentral.com/articles/10.1186/s12964-021-00741-y.
156. Cartier, R. and Reszka, R., Utilization of synthetic peptides containing nuclear localization signals for nonviral gene transfer systems. *Gene Ther.*, 9, 3, 157–67, 2002, Available from: https://www.nature.com/articles/3301635.
157. Cao, S., Zhou, Q., Chen, J.L., Jiang, N., Wang, Y.J., Deng, Q. *et al.*, Enhanced effect of nuclear localization signal peptide during ultrasound-targeted microbubble destruction-mediated gene transfection. *Mol. Med. Rep.*, 16, 1, 565–72, 2017, Available from: http://www.spandidos-publications.com/10.3892/mmr.2017.6661/abstract.
158. Bitoque, D.B., Morais, J., Oliveira, A.V., Sequeira, R.L., Calado, S.M., Fortunato, T.M. *et al.*, Human-derived NLS enhance the gene transfer efficiency of chitosan. *Biosci. Rep.*, 41, 1, 2021, Available from: /pmc/articles/PMC7789810/.
159. Wang, H.Y., Chen, J.X., Sun, Y.X., Deng, J.Z., Li, C., Zhang, X.Z. *et al.*, Construction of cell penetrating peptide vectors with N-terminal stearylated nuclear localization signal for targeted delivery of DNA into the cell nuclei. *J. Controlled Release*, 155, 1, 26–33, 2011.
160. Rabbani, P.S., Zhou, A., Borab, Z.M., Frezzo, J.A., Srivastava, N., More, H.T. *et al.*, Novel lipoproteoplex delivers Keap1 siRNA based gene therapy to accelerate diabetic wound healing. *Biomaterials*, 132, 1–15, 2017.
161. Tian, Z., Liang, G., Cui, K., Liang, Y., Wang, Q., Lv, S. *et al.*, Insight into the prospects for RNAi therapy of cancer. *Front. Pharmacol.*, 12, 308, 2021.
162. Liu, Z., Wang, S., Tapeinos, C., Torrieri, G., Känkänen, V., El-Sayed, N. *et al.*, Non-viral nanoparticles for RNA interference: Principles of design and practical guidelines. *Adv. Drug Deliv. Rev.*, 174, 576–612, 2021.

163. Jere, D., Jiang, H.L., Kim, Y.K., Arote, R., Choi, Y.J., Yun, C.H. *et al.*, Chitosan-graft-polyethylenimine for Akt1 siRNA delivery to lung cancer cells. *Int. J. Pharm.*, 378, 1–2, 194–200, 2009.
164. Kanazawa, T., Sugawara, K., Tanaka, K., Horiuchi, S., Takashima, Y., Okada, H., Suppression of tumor growth by systemic delivery of anti-VEGF siRNA with cell-penetrating peptide-modified MPEG–PCL nanomicelles. *Eur. J. Pharm. Biopharm.*, 81, 3, 470–7, 2012.
165. Lammers, T. and Ferrari, M., The success of nanomedicine. *Nano Today*, 31, 100853, 2020.
166. Fletcher, N.L., Kempe, K., Thurecht, K.J., Next-generation polymeric nanomedicines for oncology: Perspectives and future directions. *Macromol. Rapid Commun.*, 41, 18, 2000319, 2020, Available from: https://onlinelibrary.wiley.com/doi/full/10.1002/marc.202000319.
167. Astruc, D., Introduction to nanomedicine. *Molecules*, 2015, 21, 1, 4, 2016, Available from: https://www.mdpi.com/1420-3049/21/1/4/htm.
168. Barz, M., Luxenhofer, R., Schillmeier, M., Quo vadis nanomedicine? *Nanomedicine (Lond.)* 10, 20, 3089–91, 2015, Available from: https://pubmed.ncbi.nlm.nih.gov/26465063/.
169. Lammers, T., SMART drug delivery systems: Back to the future vs. clinical reality. *Int. J. Pharm.*, 454, 1, 527, 2013, Available from: /pmc/articles/PMC3745778/.
170. Hua, S., de Matos, M.B.C., Metselaar, J.M., Storm, G., Current trends and challenges in the clinical translation of nanoparticulate nanomedicines: Pathways for translational development and commercialization. *Front. Pharmacol.*, 9, 2018.
171. Hare, J.I., Lammers, T., Ashford, M.B., Puri, S., Storm, G., Barry, S.T., Challenges and strategies in anti-cancer nanomedicine development: An industry perspective. *Adv. Drug Deliv. Rev.*, 108, 25–38, 2017, Available from: https://pubmed.ncbi.nlm.nih.gov/27137110/.
172. Izci, M., Maksoudian, C., Manshian, B.B., Soenen, S.J., The use of alternative strategies for enhanced nanoparticle delivery to solid tumors. *Chem. Rev.*, 121, 3, 1746–803, 2021, Available from: https://pubs.acs.org/doi/full/10.1021/acs.chemrev.0c00779.
173. Pathak, C., Vaidya, F.U., Pandey, S.M., Mechanism for development of nano-based drug delivery system, in: *Applications of Targeted Nano Drugs and Delivery Systems: Nanoscience and Nanotechnology in Drug Delivery*, pp. 35–67, 2019.
174. Ma, X. and Williams, R.O., Polymeric nanomedicines for poorly soluble drugs in oral delivery systems: An update. *J. Pharm. Investig.*, 48, 1, 61–75, 2017, Available from: https://link.springer.com/article/10.1007/s40005-017-0372-2.
175. Grabrucker, A.M., Ruozi, B., Belletti, D., Pederzoli, F., Forni, F., Vandelli, M.A. *et al.*, Nanoparticle transport across the blood brain barrier. *Tissue Barriers.*,

4, 1, 2016, https://doi.org/101080/21688370201611153568, Available from: https://www.tandfonline.com/doi/abs/10.1080/21688370.2016.1153568.

176. Tsou, Y.H., Zhang, X.Q., Zhu, H., Syed, S., Xu, X., Drug delivery to the brain across the blood–brain barrier using nanomaterials. *Small*, 13, 43, 1701921, 2017, Available from: https://onlinelibrary.wiley.com/doi/full/10.1002/smll.201701921.
177. Ceña, V. and Játiva, P., Nanoparticle crossing of blood–brain barrier: A road to new therapeutic approaches to central nervous system diseases. *Nanomedicine (Lond).*, 13, 13, 1513–6, 2018, https://doi.org/102217/nnm-2018-0139, Available from: https://www.futuremedicine.com/doi/10.2217/nnm-2018-0139.
178. Hu, Y. and Hammarlund-Udenaes, M., Perspectives on nanodelivery to the brain: Prerequisites for successful brain treatment. *Mol. Pharm.*, 17, 11, 4029–39, 2020, Available from: https://pubs.acs.org/doi/full/10.1021/acs.molpharmaceut.0c00881.
179. Moradi Kashkooli, F., Soltani, M., Souri, M., Controlled anti-cancer drug release through advanced nano-drug delivery systems: Static and dynamic targeting strategies. *J. Controlled Release*, 327, 316–49, 2020.
180. Miao, L., Lin, C.M., Huang, L., Stromal barriers and strategies for the delivery of nanomedicine to desmoplastic tumors. *J. Controlled Release*, 219, 192–204, 2015.
181. Feng, X., Dixon, H., Glen-Ravenhill, H., Karaosmanoglu, S., Li, Q., Yan, L. *et al.*, Smart nanotechnologies to target tumor with deep penetration depth for efficient cancer treatment and imaging. *Adv. Ther. (Weinh.)* 2, 10, 1900093, 2019, Available from: https://onlinelibrary.wiley.com/doi/full/10.1002/adtp.201900093.
182. Muntimadugu, E., Kommineni, N., Khan, W., Exploring the potential of nanotherapeutics in targeting tumor microenvironment for cancer therapy. *Pharmacol. Res.*, 126, 109–22, 2017.
183. Dewhirst, M.W. and Secomb, T.W., Transport of drugs from blood vessels to tumour tissue. *Nat. Rev. Cancer*, 17, 12, 738–50, 2017, Available from: https://www.nature.com/articles/nrc.2017.93.
184. Tinkle, S., Mcneil, S.E., Mühlebach, S., Bawa, R., Borchard, G., Barenholz, Y.C. *et al.*, Nanomedicines: Addressing the scientific and regulatory gap. *Ann. N. Y. Acad. Sci.*, 1313, 1, 35–56, 2014, Available from: https://onlinelibrary.wiley.com/doi/full/10.1111/nyas.12403.
185. Pautler, M. and Brenner, S., Nanomedicine: Promises and challenges for the future of public health. *Int. J. Nanomed.*, 5, 1, 803, 2010, Available from: /pmc/articles/PMC2962275/.
186. Bremer-Hoffmann, S., Halamoda-Kenzaoui, B., Borgos, S.E., Identification of regulatory needs for nanomedicines. *J. Interdiscip. Nanomed.*, 3, 1, 4–15, 2018.

187. Caster, J.M., Patel, A.N., Zhang, T., Wang, A., Investigational nanomedicines in 2016: A review of nanotherapeutics currently undergoing clinical trials. *Wiley Interdiscip. Rev. Nanomed. Nanobiotechnol.*, 9, 1, 2017.
188. Đorđević, S., Gonzalez, M.M., Conejos-Sánchez, I., Carreira, B., Pozzi, S., Acúrcio, R.C. *et al.*, Current hurdles to the translation of nanomedicines from bench to the clinic. *Drug Deliv. Transl. Res.*, 12, 3, 500–25, 2021, Available from: https://europepmc.org/articles/PMC8300981.
189. Tinkle, S., Mcneil, S.E., Mühlebach, S., Bawa, R., Borchard, G., Barenholz, Y.C. *et al.*, Nanomedicines: Addressing the scientific and regulatory gap. *Ann. N. Y. Acad. Sci.*, 1313, 1, 35–56, 2014, Available from: https://onlinelibrary.wiley.com/doi/full/10.1111/nyas.12403.
190. Khan, M.N., Translational nano-medicine lab to clinic, in: *Intelligent Nanomaterials for Drug Delivery Applications*, pp. 141–62, 2020.
191. Ouranidis, A., Vavilis, T., Mandala, E., Davidopoulou, C., Stamoula, E., Markopoulou, C.K. *et al.*, mRNA therapeutic modalities design, formulation and manufacturing under pharma 4.0 principles. *Biomedicines*, 10, 1, 50, 2022, 2021, Available from: https://www.mdpi.com/2227-9059/10/1/50/htm.
192. Colino, C.I., Lanao, J.M., Gutierrez-Millan, C., Targeting of hepatic macrophages by therapeutic nanoparticles. *Front. Immunol.* 11, 218, 2020 Available from: /pmc/articles/PMC7065596/.
193. Wani, T.U., Raza, S.N., Khan, N.A., Nanoparticle opsonization: Forces involved and protection by long chain polymers. *Polym. Bull.*, 77, 7, 3865–89, 2019, Available from: https://link.springer.com/article/10.1007/s00289-019-02924-7.
194. Pandey, A. and Mishra, A.K., Immunomodulation, toxicity, and therapeutic potential of nanoparticles. *Biotech. (Basel (Switzerland))* 11, 3, 2022, Available from: https://pubmed.ncbi.nlm.nih.gov/36134916/.
195. Hatakeyama, H., Akita, H., Harashima, H., The polyethyleneglycol dilemma: Advantage and disadvantage of PEGylation of liposomes for systemic genes and nucleic acids delivery to tumors. *Biol. Pharm., Bull.* 36, 6, 892–9, 2013, Available from: https://pubmed.ncbi.nlm.nih.gov/23727912/.
196. Nosova, A.S., Koloskova, O.O., Nikonova, A.A., Simonova, V.A., Smirnov, V.V., Kudlay, D. *et al.*, Diversity of PEGylation methods of liposomes and their influence on RNA delivery. *Medchemcomm.*, 10, 3, 369–77, 2019, Available from: https://europepmc.org/articles/PMC6457174.
197. Yoo, J., Park, C., Yi, G., Lee, D., Koo, H., Active targeting strategies using biological ligands for nanoparticle drug delivery systems. *Cancers (Basel)* 11, 5, 2019, Available from: /pmc/articles/PMC6562917/.
198. Fang, Y., Xue, J., Gao, S., Lu, A., Yang, D., Jiang, H. *et al.*, Cleavable PEGylation: A strategy for overcoming the "PEG dilemma" in efficient drug delivery. *Drug Deliv.*, 24, 22–32, 2017, https://doi.org/101080/1071754420171388451, Available from: https://www.tandfonline.com/doi/abs/10.1080/10717544.2017.1388451.

199. Qin, X., Zhang, H., Xing, X., Wang, P., Yan, J., Liu, D. *et al.*, Robust strategies in nuclear-targeted cancer therapy based on functional nanomaterials. *Mater. Des.*, 221, 110999, 2022.
200. Degors, I.M.S., Wang, C., Rehman, Z.U., Zuhorn, I.S., Carriers break barriers in drug delivery: Endocytosis and endosomal escape of gene delivery vectors. *Acc. Chem. Res.*, 52, 7, 1750–60, 2019, Available from: https://pubs.acs.org/doi/full/10.1021/acs.accounts.9b00177.
201. Pei, D. and Buyanova, M., Overcoming endosomal entrapment in drug delivery. *Bioconjug. Chem.*, 30, 2, 273, 2019, Available from: /pmc/articles/PMC6501178/.
202. Hare, J.I., Lammers, T., Ashford, M.B., Puri, S., Storm, G., Barry, S.T., Challenges and strategies in anti-cancer nanomedicine development: An industry perspective. *Adv. Drug Deliv. Rev.*, 108, 25–38, 2017, Available from: https://pubmed.ncbi.nlm.nih.gov/27137110/.
203. Agrahari, V. and Agrahari, V., Facilitating the translation of nanomedicines to a clinical product: Challenges and opportunities. *Drug Discovery Today*, 23, 5, 974–91, 2018.
204. Accomasso, L., Cristallini, C., Giachino, C., Risk assessment and risk minimization in nanomedicine: A need for predictive, alternative, and 3Rs strategies. *Front. Pharmacol.*, 9, MAR, 228, 2018, Available from: /pmc/articles/PMC5890110/.
205. Siegrist, S., Cörek, E., Detampel, P., Sandström, J., Wick, P., Huwyler, J., Preclinical hazard evaluation strategy for nanomedicines. *Nanotoxicology.*, 13, 1, 73–99, 2018, https://doi.org/101080/17435390201815O5000, Available from: https://www.tandfonline.com/doi/abs/10.1080/17435390.2018.1505000.
206. Gagliardi, A., Giuliano, E., Venkateswararao, E., Fresta, M., Bulotta, S., Awasthi, V. *et al.*, Biodegradable polymeric nanoparticles for drug delivery to solid tumors. *Front. Pharmacol.*, 12, 17, 2021.
207. Witika, B.A., Makoni, P.A., Matafwali, S.K., Chabalenge, B., Mwila, C., Kalungia, A.C. *et al.*, Biocompatibility of biomaterials for nanoencapsulation: Current approaches. *Nanomaterials,* 10, 9, 1–40, 2020, Available from: /pmc/articles/PMC7557593/.
208. Jackman, J.A., Mészáros, T., Fülöp, T., Urbanics, R., Szebeni, J., Cho, N.J., Comparison of complement activation-related pseudoallergy in miniature and domestic pigs: Foundation of a validatable immune toxicity model. *Nanomedicine*, 12, 4, 933–43, 2016.
209. Tirumala, M.G., Anchi, P., Raja, S., Rachamalla, M., Godugu, C., Novel methods and approaches for safety evaluation of nanoparticle formulations: A focus towards *in vitro* models and adverse outcome pathways. *Front. Pharmacol.,* 12, 2021, Available from: /pmc/articles/PMC8458898/.
210. Zielińska, A., Costa, B., Ferreira, M.V., Miguéis, D., Louros, J.M.S., Durazzo, A. *et al.*, Nanotoxicology and nanosafety: Safety-by-design and testing at a

glance. *Int. J. Environ. Res. Public Health,* 17, 13, 1–22, 2020, Available from: /pmc/articles/PMC7369733/.

211. Li, Y., Wang, J., Zhao, F., Bai, B., Nie, G., Nel, A.E. *et al.*, Nanomaterial libraries and model organisms for rapid high-content analysis of nanosafety. *Natl. Sci. Rev.,* 5, 3, 365–88, 2018, Available from: https://academic.oup.com/nsr/article/5/3/365/4597983.
212. Mittal, D. and Kaul, G., Towards nano-risk assessment with high throughput screening and high content analysis: An intelligent testing strategy, in: *Comprehensive Nanoscience and Nanotechnology,* vol. 1–5, pp. 343–60, 2019, Available from: https://www.researchgate.net/publication/324733567_Towards_Nano-Risk_Assessment_With_High_Throughput_Screening_and_High_Content_Analysis_An_Intelligent_Testing_Strategy.
213. Kumar, V., Sharma, N., Maitra, S.S., *In vitro* and *in vivo* toxicity assessment of nanoparticles. *Int. Nano Lett.*, 7, 4, 243–56, 2017, Available from: https://link.springer.com/article/10.1007/s40089-017-0221-3.
214. Soares, S., Sousa, J., Pais, A., Vitorino, C., Nanomedicine: Principles, properties, and regulatory issues. *Front. Chem.*, 6, 360, 2018.
215. Farjadian, F., Ghasemi, A., Gohari, O., Roointan, A., Karimi, M., Hamblin, M.R., Nanopharmaceuticals and nanomedicines currently on the market: Challenges and opportunities. *Nanomedicine,* 14, 1, 93, 2019, Available from: /pmc/articles/PMC6391637/.
216. Murday, J.S., Siegel, R.W., Stein, J., Wright, J.F., Translational nanomedicine: Status assessment and opportunities. *Nanomedicine*, 5, 3, 251–73, 2009.
217. Ramos, T.I., Villacis-Aguirre, C.A., López-Aguilar, K.V., Padilla, L.S., Altamirano, C., Toledo, J.R. *et al.*, The Hitchhiker's guide to human therapeutic nanoparticle development. *Pharmaceutics,* 14, 2, 2022, Available from: https://pubmed.ncbi.nlm.nih.gov/35213980/.
218. Hertig, J.B., Shah, V.P., Flühmann, B., Mühlebach, S., Stemer, G., Surugue, J. *et al.*, Tackling the challenges of nanomedicines: Are we ready? *Am. J. Health Syst. Pharm.,* 78, 12, 1047–56, 2021, Available from: https://pubmed.ncbi.nlm.nih.gov/33599767/.
219. Foulkes, R., Man, E., Thind, J., Yeung, S., Joy, A., Hoskins, C., The regulation of nanomaterials and nanomedicines for clinical application: Current and future perspectives. *Biomater. Sci.,* 8, 17, 4653–64, 2020 Available from: https://pubs.rsc.org/en/content/articlehtml/2020/bm/d0bm00558d.
220. Sarwal, A., Singh, G., Singh, K., Garg, S., Recent interventions for nanotechnology based drug products: Insights into the regulatory aspects. *Curr. Pharm. Des.,* 24, 43, 5219–28, 2018, Available from: https://pubmed.ncbi.nlm.nih.gov/30652639/.
221. Allan, J., Belz, S., Hoeveler, A., Hugas, M., Okuda, H., Patri, A. *et al.*, Regulatory landscape of nanotechnology and nanoplastics from a global perspective. *Regul. Toxicol. Pharm.*, 122, 104885, 2021.

222. Bhatt, S., Punetha, V. D., Pathak, R., Punetha, M., Graphene in nanomedicine: A review on nano-bio factors and antibacterial activity. *Colloids Surf. B.*, 226, 1–22, 2023. https://doi.org/10.1016/j.colsurfb.2023.113323
223. Bhatt, S., Pathak, R., Punetha, V.D. *et al.*, Recent advances and mechanism of antimicrobial efficacy of graphene-based materials: A review. *J. Mater. Sci.*, 58, 7839–7867, 2023. https://doi.org/10.1007/s10853-023-08534-z
224. Bhatt, S., Punetha, V.D., Pathak, R., Punetha, M., Two-dimensional carbon nanomaterial-based biosensors: Micromachines for advancing the medical diagnosis, in: *Recent Advances in Graphene Nanophotonics. Advanced Structured Materials*, Patel, S.K., Taya, S.A., Das, S., Vasu Babu, K. (eds.), vol. 190, pp. 181–224, Springer, Cham., 2023. https://doi.org/10.1007/978-3-031-28942-2_9
225. Pathak, R., Punetha, V. D., Bhatt, S., Punetha, M., Multifunctional role of carbon dot-based polymer nanocomposites in biomedical applications: A review. *J. Mater. Sci.*, 58, 15, 6419–6443, 2023. https://doi.org/10.1007/s10853-023-08408-4

14

Biomaterials-Based Vaccination in Cancer Therapy

Rishav Sharma, Rishabha Malviya* and Sonali Sundram

Department of Pharmacy, School of Medical and Allied Sciences, Galgotias University, Greater Noida, Uttar Pradesh, India

Abstract

According to the WHO's Cancer Report, one in eight deaths around the globe in the year 2000 were brought on by cancer. Unfortunately, cancer has a high mortality rate (about 25%) in the developed world. This highlights the critical need for therapies that effectively combat cancer. The creation of immunotherapies to combat cancer is at the forefront of current cancer therapy strategies. Tumor-specific RNA is being studied as a potential component of vaccinations. When employed alone or in conjunction with dendritic cell transfection, RNA has many of the same benefits as a vaccine as other types of DNA-based vaccines, including the capacity to produce protective immune responses while being practically, practically, safely, and effectively given. In order to provide a comprehensive overview of RNA vaccination, this study focuses on data from *in vitro* research and current clinical trials.

Keywords: Vaccine, cancer, biomaterials, RNA vaccine, DNA vaccine, immunotherapy

14.1 Introduction

Individualized treatment plans are necessary to provide the most effective care for patients due to the wide variety of cancer malignancies' root causes. The objective is to control cells and ongoing processes to combat illness, and molecular intervention is a key strategy. For some time now, the notion of using one's own immune system in the treatment of cancer has piqued

**Corresponding author*: rishabhamalviya19@gmail.com

Rishabha Malviya, Sonali Sundram and Neelam Jain (eds.) Integration of Biomaterials for Gene Therapy, (417–434) © 2023 Scrivener Publishing LLC

interest due to its apparent simplicity and potential efficacy. As a result, research on the subject has been going on since the 1970s [1–3]. During this time period, the concept of immunotherapy was born with the first attempts to use RNA to provoke immune responses. Animals inoculated with human tumor cells had their lymphoid organs harvested for their so-called immunological RNA, which was then injected into human patients as an adjuvant therapy [1, 2]. Despite the apparent simplicity of the concept, researchers have encountered significant obstacles when attempting to generate highly precise and efficient immune responses against chosen targets inside the immune system. The fact that the cause of cancer can be traced back to a transformed self is a blessing and a curse. It's less of a challenge since cancer cells look and behave differently than regular body cells. Complex because, first, these cells are still self-cells and, second, they adopt a wide variety of strategies to avoid being recognized by the immune system. However, many different strategies have been developed to educate the immune system to detect and destroy malignant tumor cells by exploiting their distinctions from healthy cells. The generation of immunotherapeutic vaccines is aimed at triggering and sustaining an immune response that protects against malignancies. Therefore, the immune system must be able to prevent the tumor from returning by eliminating any remaining tumor cells. Multiple types of leukocytes, such as antigen-presenting cells, CD4+ and CD8+ T cells, and B cells, must be activated for an effective immune response. It will cause the body to generate memory cells, which could defend against tumor progression after the original cells have been eliminated. The numerous types of messenger RNA are now being studied for their vaccine potential. These entire tumor-associated amyloid mRNA and entire tumor cell RNA are utilized in these methods to elicit targeted CTL responses [4, 5].

14.2 Tumor-Associated Antigens

Tumor-associated antigens (TAAs) have been, and continue to be, a critical target for vaccine development [6]. In order to stimulate the immune system, it is required to identify immunogenic chemicals that are highly expressed in malignant tissues but very weakly expressed in normal adult tissues. Many TAAs are expressed at higher levels in cancer cells while being present in both tumor and normal tissues. Antigens that are extensively distributed should be used with caution since they induce tolerance via the negative selection processes of T cell development. The induction of autoimmune illnesses may be possible if the immune system is stimulated to produce antibodies or other reactions to certain antigens, provided that tolerance can be overcome. Differentiation antigens make up a subset of conserved TAAs. They are expressed only in select malignancies and the tissues from which those cancers originated, rather than everywhere. Melan-A/MART-1 [7] is a differentiation antigen that is nearly exclusively expressed in melanoma cells and melanocytes. Differentiation antigens are only found in certain organs, yet it is still difficult to induce tolerance to them. Furthermore, autoimmune effects may be induced by tampering with these TAAs. Antigen-induced immune responses are those that are specifically produced in tumor tissues would be a more secure strategy to use the immune system. The tumor-associated antigens (TAAs) are of two sorts, and both are tumor-specific. Somatic tissues do not express cancer/testis (CT) antigens [8] but these antigens are detected in a wide variety of tumor tissues and the germline. Due to a lack of expression of MHC class I and class II molecules, CT antigens are not presented in the germline. This implies that the tumor itself is the sole place where the antigens will be identified. However, in the absence of common tumor antigens, the induction of immune responses will not result in autoimmunity. One CT antigen that has been exploited as a therapeutic target is NY-ESO-1 [9, 10]. Tumors are unique in their expression of mutation antigens. Multiple mutations are common in cancer cells, and these results in the translation of proteins with different properties. Malignant tumors become more aggressive if mutations are oncogenic or cancer suppressor gene mutations. Because the immune system does not recognize mutation antigens as self, and because no resistance has been built against them, they are an excellent target for immunotherapy. CDK4 [10], Kras [11], caspase8 [12], and the transforming growth factor beta receptor [13–15] are all examples of mutant antigens. Tumor-associated antigens (TAAs) of viral origin are another subset of TAAs. The development of tumor cells may be triggered by exposure to a number of viruses, including HTLV-1, HIV, and EBV. In particular, prior viral an infection is associated with the onset of NHLs and MMs. The modified cells now display viral antigens on their surface, providing a potential new platform for vaccine development. The immunoglobulin type may also be targeted in diseases like NHL and MM.

14.3 Vaccine Delivery

The next step, after distinguishing between tumor and healthy tissue, is to train the immune system to recognize these distinctions. The TAAs are given to the patient in order to achieve this goal. It remains unclear is the optimal approach. Decisions on the appropriate antigen and the best

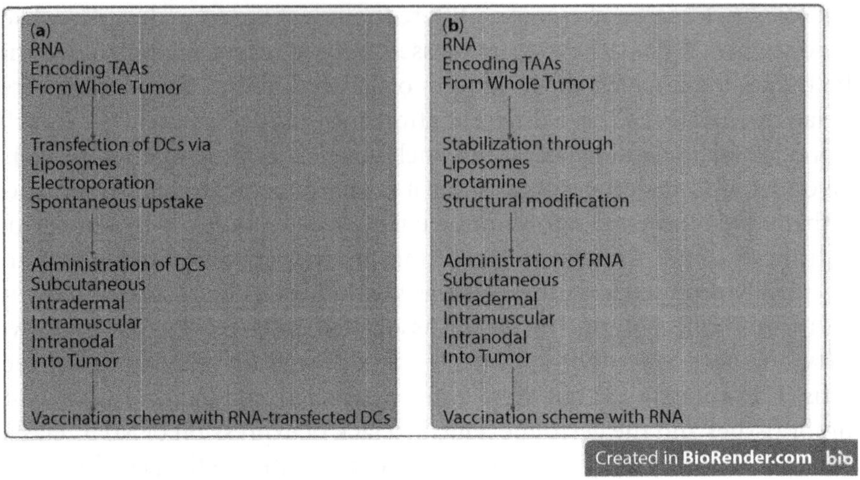

Figure 14.1 Overview of RNA-vaccination strategies employing DCs (a) and pure/stabilized RNA (b).

route for administering it are crucial. Numerous strategies have been used, depending on the antigen used [16] TAA peptides or proteins in their whole have been tested, as have irradiation tumor lysates and apoptotic tumor cell remnants, as well as TAA-encoding plasmids and TAA/whole tumor cell RNA. Optimizing antigen delivery technique is just as important as selecting the right antigen. In addition to GM-CSF, IFN-, or IL-12, the aforementioned tumor antigens might be employed as vaccinations on their own. Dendritic cells may also be used to introduce the antigens into the body, in addition to direct vaccination. To that purpose, this article will examine the potential of RNA vaccines, both standalone and in tandem with dendritic cells, in the fight against the disease. Figure 14.1 offers an overview of the two methods.

14.4 Dendritic Cells

Dendritic cells are being widely examined in the search for a successful cancer vaccine (DCs). Antigen presentation cells (APCs) including DCs are one of the most potent populations in this task, alongside B cells and macrophages. When they are still their early stages, DCs are widely distributed throughout the body via the circulatory and lymphatic systems, they are most often found in tissues. MHC class II antigens typically produced, but hardly are costimulatory molecules such as CD80 (B7.1) or CD86

(B7.2). DCs are susceptible to antigens while moving about the body's periphery. Receptor-mediated endocytosis, phagocytosis, and macropinocytosis all seem to be ways in which cells take up these antigens. After consuming antigen, DCs migrate to afferent lymphoid organs, in which they undergo maturation. Developing DCs display a wide range of functional and morphological diversity [17, 18]. DCs and resident T cells interact through adhesion molecules that they share, establishing immunological synapses. T cells are activated in response to the processed antigens presented by DCs, and antigen-specific responses are created. MHC class I or class II molecules are involved in antigen presentation depending on the kind of antigen. DCs, like almost every other cell in the body, may deliver antigens by way of MHC class I molecules. When antigens have an internal source, this pathway is used. In order to be evacuated from the cell, MHC class I molecules should be loaded with peptides that have been generated in the cytosol, degraded by proteasomes, and then transferred to the endoplasmic reticulum (ER). CD8+ cytotoxic T cells interact with MHC class I molecules through their T cell receptors in such areas (CTLs). In contrast to MHC class II molecules, DCs are able to present antigens from the outside. Antigens incorporated into a cell are destroyed by cathepsins in the endosomal and lysosomal compartments. Synthesized peptides are presented to CD4+ T lymphocytes through MHC class II molecules. Foreign antigens could also be presented by DCs via MHC class I molecules, in addition to the "conventional" mechanisms. After entering the cell, antigens are transferred through intracellular divisions into the cytosol, whereby they initiate processing and presentation through the MHC class I pathway. Researchers have proven that cross priming is beneficial for generating effective CTL responses [19–21].

14.5 *In Vitro* Generation of Dendritic Cells

Vaccine techniques that use DCs need substantial numbers of these cells. There are already a number of recognized techniques for producing large amounts *in vitro*. Precursor CD34+ cells obtained from bone marrow or blood have the potential to transform into DCs after being exposed to cytokine mixtures including granulocyte-macrophage colony-stimulating factor (GM-CSF), interleukin-4 (IL-4) or tumors necrosis factor-alpha (TNF-) [44, 46]. Myeloid DCs may also be generated using a different strategy including CD14+ monocytes isolated from blood and the cytokines GM-CSF and IL-4 or CD40 ligand [22–26]. Even though these cells are young and express just a small number of costimulatory molecules on their surface, they may develop into more mature cells in response to various inflammatory stimuli, such as Toll-like receptor ligands.

14.6 Usage of RNA

Ribonucleic acids (RNAs) are monophosphate (mono-) ribonucleotide (nucleotide) strings that contain four nucleotides (adenine, guanine, cytidine, and uridine). The ribose sugars› terminal 3›- and 5›-phosphate residues serve as links between the two. Messenger RNA (mRNA), which encodes the instructions for making proteins, is employed in immunotherapy. Initial vaccination strategies included obtaining mRNA from patients› own tumor tissue. The main drawback of this method was that it required a substantial volume of tumor cells. Preparing large quantities of RNA *in vitro* is now feasible. It is possible to get high-quality RNA encoding many patient-specific TAAs from a small number of tumor cells by extracting RNA and creating tumor cDNA libraries [27, 28]. Using mRNA as a vaccine has a number of benefits, including the fact that it can be obtained in big amounts with no trouble. Vaccines made from RNA molecules are thought to be risk-free since they decay quickly. As contrast to plasmid DNA, they are rapidly eliminated from the body and are unable to permanently alter the gene expression of cells by integrating into the genome [29]. The aforementioned details also suggest that adverse consequences such as the development of autoimmune diseases or anti-DNA antibodies are very improbable after RNA immunization. Although immunizations based on peptides, mRNA applications are not limited by the MHC. CD8+ CTLs and CD4+ T helper cells may be induced, and an immune response can be mounted against a wide range of antigens. The likelihood of tumor escape is decreased by a polyclonal immune response. Because RNA is so unstable, it is impossible to control how much of a gene is expressed at any one time. RNA transfection only requires cytoplasmic insertion, while transfection with DNA also has to occur in the nucleus. All of these things argue in favor of using vaccines based on RNA.

14.7 RNA-Pulsed DCs as Vaccines

DCs produced *in vitro* are used as tumor antigen-presenting cells, which is a fundamental strategy in RNA vaccination. Transfection of one or many TAA-encoding RNAs, or complete tumor cell RNA, into DCs after they have been generated is possible. The transfected DCs will next convert the foreign DNA into proteins. Inside the cell, these proteins catalyze breakdown processes that yield peptides that are displayed on major histocompatibility complex class I and class II molecules. There have been several *in vitro* studies showing that pulsing DCs with either whole tumor RNA or RNA encoding specific TAAs resulted in the generation of tumor-specific CTLs. Gilboa's group first outlined this method, which has since served as the foundation for further research [30, 31]. Instead of using cell lines, researchers in several studies induced CTL responses using RNA extracted from individuals with tumors. Renal tumor cell RNA was used to transfect DCs in a study by Heiser *et al.* [32]. Polyclonal CTL responses were produced, and both original and metastasized tumor cells were lysed. Compared to CTL responses elicited against a single antigen, polyclonal T cell activity seemed to be more robust. However, the strength of immune responses triggered by individual TAAs has been studied in different contexts. Using CEA mRNA transfection, Nair *et al.* [24] induced specific CTL responses *in vitro* using DCs. Grunebach *et al.* compared the effects of Her-2/neu RNA electroporated into DCs and cotransfection with two distinct TAAs. When compared to solo TAA transfection, they observed that co-transfection elevated costimulatory molecules and enhanced immune responses. Responses from CD8 and CD4 T cells were stimulated [33]. This maturation process is kicked off early when RNA is transfected into DCs. Whether or not these stimuli are sufficient, and whether or not additional signals result in more competent cells, remains an open subject. Multiple other maturation stimuli have been studied for their impact on DC phenotype and their ability to promote T cell proliferation and, by extension, the induction of targeted immunological responses. Transfecting DCs with Mart-1 mRNA and partially maturing them with CD40 ligand, Onaitis *et al.* [34]. CD40L-stimulated samples demonstrated enhanced T-cell proliferation because of increased IL-12 secretion by the treated DCs. After successive activation with a cytokine cocktail, the same group reported augmented Mart-1-specific T lymphocyte development [35], which has been found to be mediated by increased IL-12 production (TNF-, IL-1, IL-6, and PGE2). Additionally, they noted that several ligands for the Toll-like receptors (TLRs) may effectively mature DCs [36, 37]. T cell response induction and improvement have been studied in mice trials. It was shown by Boczkowski *et al.* [38] that mice might be protected from tumors that expressed OVA by inducing OVA-specific CTLs. To develop a vaccination against tumors, Naka *et al.* [39] cotransfected mouse DCs with tumor RNA and GMCSF RNA. Since the animals have previously undergone immunotherapeutic therapy, any tumor regrowth should be prevented. Tumor development was restrained and enhanced CTL responses were produced by the co-transfected DCs. Kim *et al.* looked at whether altering the TAA may improve immune responses against the CEA antigen. The researchers pumped DCs with RNA that encoded a fusion gene composed of shortened CEA (CEA) lacking its signal sequences, calreticulin (CRT), and the HIV TAT protein transduction domain (TAT-PTD). Mice that were given the vaccination had improved CD4 and CD8 responses to CEA, which in turn reduced tumor development and increased the animals' chances of survival [40]. The majority of breast and non-small cell lung malignancies, as well as tumors of the pancreas and stomach, express CEA, making it a useful tumor marker. Recent phase I/II studies have examined the TAA to determine its potential use in vaccine treatment. Three patients with resected pancreatic cancer received immunizations with CEA mRNA transfected DCs [41]. Patients received injections on a monthly basis for six months, and after more than 2.5 years, there was no evidence of illness recurrence. DCs supplied with increasing amounts of CEA mRNA have been given to patients with tumors expressing CEA (phase I trial) or excised hepatic metastases of colon cancer (phase II study). Of the patients who participated in the phase I study, only one had a complete recovery, two exhibited partial improvements, and three showed stable disease. Even in the second phase of the research, the clinical response was dismal. Disease returned in 9 out of 13 individuals. These illustrations may stand in for a number of similar clinical phase I/II investigations that have been conducted. Although substantial immune responses may be found *in vitro* after vaccination and production of particular CTLs is effective, minimal impact is seen when analyzing clinical responses. There was no evidence of tumor regression or protection against recurrence in the vast majority of treated individuals who received vaccines. Tolerance induction, the effects of medicinal preparation, and the presence of myeloid suppressive cells and regulatory T cells may account for the discrepancy between *in vitro* and *in vivo* results. Another crucial consideration is the antigen which must be transfected into DCs. They may be "charged" with either a single TAA, a mixture of TAAs, or entire cell tumor RNA. The latter has the potential to stimulate immunological responses against several antigens. Whether or if this method is superior than using a number of other TAAs remains an open question that needs to be researched further. Whole RNA may be isolated from tumor tissues and utilized in vaccine production if there is enough of the tumor material available. A cDNA library may be constructed, cloned, amplified in E. coli, and then RNA can be synthesized *in vitro* if this is not the case [42]. Although *in vitro* transcription has a positive effect in terms of reducing the potential patient population, altering the RNA's composition mixes due to the prevalence of shorter fragments [42].

14.8 RNA Vaccines

Dendritic cell antigen presentation may be facilitated *in vitro* using RNA vaccines, but there is also an indirect method that uses RNA vaccines. Transfection of RNA into DCs does not need any ancillary methods, such as liposomes or electroporation. Vaccination using unmodified RNA is therefore a natural choice. The efficient introduction of mRNA into patients has been the subject of much research and experimentation. It is generally accepted that RNA molecules are unstable and decay quickly. This is why there was once some debate about whether or not injecting the RNA molecules themselves was a good idea. Naked RNA injection into mouse skeletal muscle, however, demonstrated *in vivo* expression of the encoded protein by Wolff *et al.* [43]. Additionally, it was shown that the administration of unprotected RNA was sufficient to elicit targeted cytotoxic T lymphocyte responses in patients. Even Nevertheless, work has been done to stabilize RNA molecules to make transport more efficient. Coding the nucleic acid on gold particles and then "gene-gun delivery" [44] is a potential method of RNA administration. The particles act as shuttles, transporting the RNA molecules over the skin. Expression and presentation to T cells of the encoded proteins occurs after incorporation by DCs. Liposomal RNA packing is another strategy that has been studied. Cationic lipids in liposomes form stable compounds with negatively charged nucleic acid units via electrostatic interactions [45]. Liposomes are useful as adjuvants because they stabilize RNA and activate immune cells on their own [46, 47]. Condensation of RNA to protamine [48], an arginine-rich protein necessary for DNA condensation during spermatogenesis, is another means to preserve RNA. Immunostimulatory complexes of proline and RNA are formed spontaneously *in vitro* and remain stable for many hours. Human dendritic cells (DCs), monocytes (M), B cells (B), natural killer and granulocytes are stimulated by protamine through a MyD88-dependent mechanism, and it also protects RNA. The stability and translation efficiency of RNA can be altered by extending the poly-A tail at the 3′-end and changing the Cap structure at the 5'-end. Transcriptional efficiency is greatly enhanced when an Antireverse Cap Analog (ARCA) is used in place of the endogenous 7-methylguanosine triphosphate. In mice, DCs are activated by phosphothioate RNA through the threat receptor MyD88 [49]. The best injection location for immunostimulatory RNA has been studied, and this topic has been considered in relation to RNA-transfected DCs. Subcutaneous, intradermal, intramuscular, and intranasal injection methods are all viable options for administering RNA on its own. The strongest immune responses seem to be generated by injections near or into the lymph nodes.

14.9 Optimization of Immunotherapy

Self-tolerance relies on CD4+CD25+ regulatory T cells, which also act as immunological suppressors. They regulate the immune system and lessen the likelihood of damaging T cell responses. Patients with tumors who have a high number of Tregs tend to have a shorter survival rate. Therefore, eliminating Tregs might improve survival rates and enhance immunological responses in patients. Dannull *et al.* [50] significantly reduced the number of regulatory T cells in renal cell cancer patients by using RNA transfected DCs as a vaccination. To eliminate Tregs, they used DAB389IL-2 (ONTAK), a recombinant IL-2 diphtheria toxin conjugate that targets and eliminates only CD25+ regulatory T cells. Combination therapy significantly enhanced tumor-specific CD8 and CD4 T cell responses, in compared to DC vaccine injections alone. Targeted CTLA-4 is another method for disabling the immune system. Cytotoxic T lymphocyte antigen (CTLA)-4 is a protein that inhibits T cell proliferation by mimicking the activity of costimulatory B7 molecules. The chemokine receptor CD28 (CD28) is bound by both B7 and CTLA-4. CTLA-4 inhibits T cell proliferation through an interaction with CD28, which has a lower affinity for B7. Following T cell activation, an uptick in CTLA-4 expression increases the likelihood of an interaction with CD28, creating a negative feedback process. A number of autoimmune disorders are caused by CTLA-4 polymorphisms [51], demonstrating the crucial role this molecule plays. Mice with lympho-proliferative disorders that do not express CTLA-4 [52, 53] provide further evidence for the role of CTLA-4 in regulating T cell responses. Inhibiting CTLA-4 may result in the production of more robust T cell responses. The humanized anti-CTLA-4 inhibitory antibodies Tremelimumab and Ipilimumab work in similar ways. Improvements in clinical responses were seen in individuals treated for melanoma after their use in clinical studies [87]. There were serious adverse effects, such as diarrhea, panhypopituitarism, and autoimmune thyroiditis, along with these encouraging outcomes. These immune-related adverse events (IRAEs) most likely accompany beneficial clinical outcomes. CTLA-4 inhibition has the potential to cause autoimmune disorders if used for an extended period of time, hence its safety should be a primary focus of future research. Currently, researchers are looking at using CTLA-4 suppression in conjunction with other immune modulatory mechanisms [54]. Combining RNA vaccination with tyrosine kinase inhibitors is another strategy to boost anticancer effects (TKIs). Sorafenib and sunitinib are two TKIs that block the signaling pathways within cells that promote cell growth and blood vessel formation, respectively. Renal cell carcinoma (RCC) and hepatocellular carcinoma are two cancers that sorafenib is used to treat (HCC). Sunitinib is used to treat RCC as well as gastrointestinal (GI) tumors. Pretreatment with sorafenib inhibited the induction of antigen-specific T lymphocytes in mice, but sunitinib had no such impact [55]. The maturation of DCs and their ability to stimulate T cells were both unaffected by sunitinib in human monocyte-derived DCs, whereas both were suppressed by sorafenib. These results suggest that sunitinib might be an effective option for use in conjunction with RNA vaccines.

14.10 Cancer Treatment Through RNA Interference

The use of information theory to cancer treatment is novel and fruitful [56]. The clinical responses seen with siRNA-based drugs were short-lived and often led to tumor recurrence or progression [57]. Gene targeting is challenging for most solid tumors because of tumor heterogeneity, many signaling pathways, cross-links, and both vertical and horizontal feedback loops. Self-medication is another source of drug resistance [58]. At least three treatments are recommended to reduce the likelihood of medication resistance. By focusing on these three characteristics, we may develop more effective therapies and lower cancer incidence [59]. In order to determine which signaling pathways are most important and which genes are implicated, scientists are now engaged in a collaborative effort to create a centralized database based on the notion of targeting these three variables. Individualized treatments are the basis of the next generation of RNAi-based cancer therapeutics (Figure 14.2). The use of RNA interference (RNAi) to interrupt signaling pathways and create a triple RNAi-based vaccination against cancer will be the strategy used. By lowering local and systemic immunosuppression associated with immunotherapy, damaging the ratio of tumor/T-cell activation, inducing apoptosis to boost antigen presentation, and affecting tumor development are all ways in which this strategy functions. These medications' effects are not time-bound, and they may induce memory T-cell responses that persist for a long period [60].

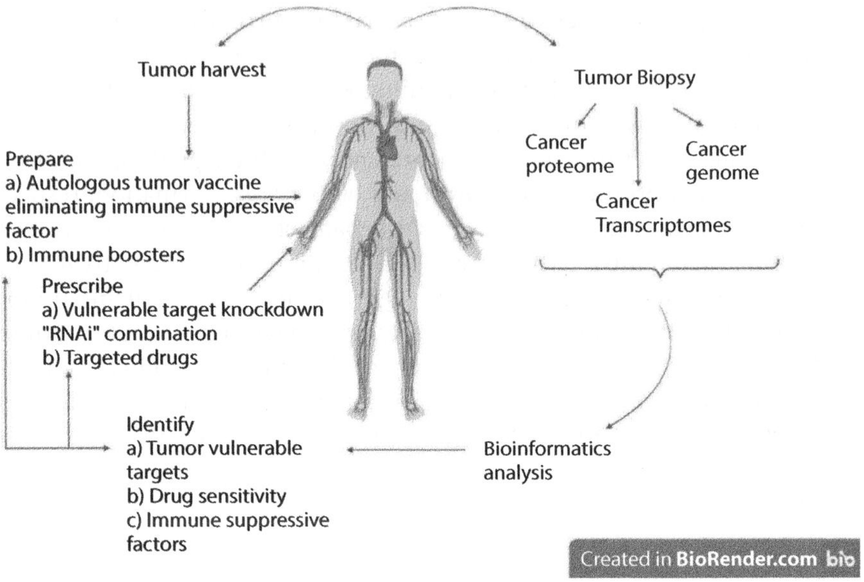

Figure 14.2 The technique for adopting cancer-specific customized medications is depicted schematically.

14.11 Conclusion

Over the last two decades, scientists have worked tirelessly on immunotherapies to combat cancer. The fundamental idea is to make use of the differences between the tumor and the surrounding healthy tissue. Tumor-associated antigens, which seem to be indication of the molecular differentiation of altered tissues, have been an important discovery in support of ongoing research. Several *in vitro* research and *in vivo* studies have used TAAs as the starting point for eliciting targeted immune responses against tumor cells. Vaccination using mRNA encoding TAAs or entire tumor RNA is a practical and seemingly safe option. Dendritic cells, given their role as antigen-presenting cells, may be transfected with RNA to introduce TAAs to T cells and elicit an immunological response. More recently, research has examined the efficacy of immunization experiments that include the use of naked RNA in addition to potential enhancers. There is a wide range of options for tweaking both approaches during vaccination delivery. As a result of this variety, current studies almost never repeat a method again, instead altering specific variables with each new investigation. It's possible that further tests are required to determine which vaccine strategies work best. However, RNA vaccination might be the answer to curing many forms of cancer. Too far, clinical results have been uncommon *in vivo* studies, despite the fact that they all have the commonality of inducing the creation of particular T cells following patient immunization. Instead of tumor disappearing, it is more common to see the illness progressing or returning. Thus, modern research is focused on making vaccinations more efficient. Combining RNA vaccination with the activation of the immune system through cytokines and TLR ligands, as well as the suppression of cell populations that contribute to immune response suppression, is one such option. When the best vaccination mechanisms have been identified, patients will be given individualized therapy to combat their condition. An effective immune response may be generated and sustained if the patient's immune system is stimulated. RNA immunization, it is hoped, will make it possible to rein in cancer.

References

1. Pilch, Y.H., de Kernion, J.B., Skinner, D.G., Ramming, K.P., Schick, P.M., Fritze, D., Brower, P., Kern, D.H., Immunotherapy of cancer with "Immune" RNA: A preliminary report. *Am. J. Surg.*, 132, 5, 631–7, 1976.
2. Pilch, Y.H., Ramming, K.P., Dekernion, J.B., Preliminary studies of specific immunotherapy of cancer with immune RNA. *Cancer*, 40, S5, 2747–57, 1977.
3. Steele, G., Wang, B.S., Richie, J., Wilson, R.E., Ervin, T., Yankee, R., Fallon, M., Mannick, J.A., *In vivo* effect and parallel *in vitro* lymphocyte-mediated tumor cytolysis after phase I xenogeneic immune RNA treatment of patients with widespread melanoma or metastatic renal cell carcinoma. *Cancer Res.*, 40, 7, 2377–82, 1980.
4. Hoerr, I., Obst, R., Rammensee, H.G., Jung, G., *In vivo* application of RNA leads to induction of specific cytotoxic T lymphocytes and antibodies. *Eur. J. Immunol.*, 30, 1, 1–7, 2000.
5. Weide, B., Carralot, J.P., Reese, A., Scheel, B., Eigentler, T.K., Hoerr, I., Rammensee, H.G., Garbe, C., Pascolo, S., Results of the first phase I/II clinical vaccination trial with direct injection of mRNA. *J. Immunother.*, 31, 2, 180–8, 2008.
6. van der Bruggen, P., Traversari, C., Chomez, P., Lurquin, C., De Plaen, E., Van den Eynde, B., Knuth, A., Boon, T., A gene encoding an antigen recognized by cytolytic T lymphocytes on a human melanoma. *Science*, 254, 5038, 1643–7, 1991.
7. Coulie, P.G., Brichard, V., Van Pel, A., Wölfel, T., Schneider, J., Traversari, C., Mattei, S., De Plaen, E., Lurquin, C., Szikora, J.P., Renauld, J.C., A new gene coding for a differentiation antigen recognized by autologous cytolytic T lymphocytes on HLA-A2 melanomas. *J. Exp. Med.*, 180, 1, 35–42, 1994.

8. Caballero, O.L. and Chen, Y.T., Cancer/testis antigens: Potential targets for immunotherapy, in: *Innate Immune Regulation and Cancer Immunotherapy*, pp. 347–69, 2012.
9. Chen, Y.T., Scanlan, M.J., Sahin, U., Türeci, Ö., Gure, A.O., Tsang, S., Williamson, B., Stockert, E., Pfreundschuh, M., Old, L.J., A testicular antigen aberrantly expressed in human cancers detected by autologous antibody screening. *Proc. Natl. Acad. Sci.*, 94, 5, 1914–8, 1997.
10. Wolfel, T., Hauer, M., Schneider, J., Serrano, M., A p16INK4a-insensitive CDK4 mutant targeted by cytolytic T lymphocytes in a human melanoma. *Science*, 269, 5228, 1281, 1995.
11. Fossum, B., Gedde-Dahl, I.I.I.T., Hansen, T., Eriksen, J.A., And, E.T., Gaudernack, G., Overlapping epitopes encompassing a point mutation (12 Gly→ Arg) in p21 ras can be recognized by HLA-DR,-DP and-DQ restricted T cells. *Eur. J. Immunol.*, 23, 10, 2687–91, 1993.
12. Mandruzzato, S., Brasseur, F., Andry, G., Boon, T., Bruggen, P.V., A CASP-8 mutation recognized by cytolytic T lymphocytes on a human head and neck carcinoma. *J. Exp. Med.*, 186, 5, 785–93, 1997.
13. Markowitz, S., Wang, J., Myeroff, L., Parsons, R., Sun, L., Lutterbaugh, J., Fan, R.S., Zborowska, E., Kinzler, K.W., Vogelstein, B., Brattain, M., Inactivation of the type II TGF-β receptor in colon cancer cells with microsatellite instability. *Science*, 268, 5215, 1336–8, 1995.
14. Parsons, R., Myeroff, L.L., Liu, B., Willson, J.K., Markowitz, S.D., Kinzler, K.W., Vogelstein, B., Microsatellite instability and mutations of the transforming growth factor β type II receptor gene in colorectal cancer. *Cancer Res.*, 55, 23, 5548–50, 1995.
15. Sæterdal, I., Bjørheim, J., Lislerud, K., Gjertsen, M.K., Bukholm, I.K., Olsen, O.C., Nesland, J.M., Eriksen, J.A., Møller, M., Lindblom, A., Gaudernack, G., Frameshift-mutation-derived peptides as tumor-specific antigens in inherited and spontaneous colorectal cancer. *Proc. Natl. Acad. Sci.*, 98, 23, 13255–60, 2001.
16. Bakker, A.B., Schreurs, M.W., De Boer, A.J., Kawakami, Y., Rosenberg, S.A., Adema, G.J., Figdor, C.G., Melanocyte lineage-specific antigen gp100 is recognized by melanoma-derived tumor-infiltrating lymphocytes. *J. Exp. Med.*, 179, 3, 1005–9, 1994.
17. Brichard, V., Van Pel, A., Wölfel, T., Wölfel, C., De Plaen, E., Lethé, B., Coulie, P., Boon, T., The tyrosinase gene codes for an antigen recognized by autologous cytolytic T lymphocytes on HLA-A2 melanomas. *J. Exp. Med.*, 178, 2, 489–95, 1993.
18. Wölfel, T., Van Pel, A., Brichard, V., Schneider, J., Seliger, B., Zum Büschenfelde, K.H., Boon, T., Two tyrosinase nonapeptides recognized on HLA-A2 melanomas by autologous cytolytic T lymphocytes. *Eur. J. Immunol.*, 24, 3, 759–64, 1994.

19. Bevan, M.J., Cross-priming for a secondary cytotoxic response to minor H antigens with H-2 congenic cells which do not cross-react in the cytotoxic assay. *J. Exp. Med.*, 143, 5, 1283–8, 1976.
20. Bevan, M.J., Priming for a cytotoxic response to minor histocompatibility antigens: Antigen specificity and failure to demonstrate a carrier effect. *J. Immunol.*, 118, 4, 1370–4, 1977.
21. Brossart, P. and Bevan, M.J., Presentation of exogenous protein antigens on major histocompatability complex class I molecules by dendritic cells: Pathway of presentation and regulation by cytokines. *Blood, The Journal of the American Society of Hematology*, 90, 4, 1594–9, 1997.
22. Caux, C., Dezutter-Dambuyant, C., Schmitt, D., Banchereau, J., GM-CSF and TNF-α cooperate in the generation of dendritic Langerhans cells. *Nature*, 360, 6401, 258–61, 1992.
23. Caux, C., Vanbervliet, B., Massacrier, C., Dezutter-Dambuyant, C., de Saint-Vis, B., Jacquet, C., Yoneda, K., Imamura, S., Schmitt, D., Banchereau, J., CD34+ hematopoietic progenitors from human cord blood differentiate along two independent dendritic cell pathways in response to GM-CSF+ TNF alpha. *J. Exp. Med.*, 184, 2, 695–706, 1996.
24. Inaba, K., Inaba, M., Romani, N., Aya, H., Deguchi, M., Ikehara, S., Muramatsu, S., Steinman, R.M., Generation of large numbers of dendritic cells from mouse bone marrow cultures supplemented with granulocyte/ macrophage colony-stimulating factor. *J. Exp. Med.*, 176, 6, 1693–702, 1992.
25. Peters, J.H., Xu, H., Ostermeier, D., Friedrichs, D., Gieseler, R.K., Ruppert, J., Signals required for differentiating dendritic cells from human monocytes *in vitro*, in: *Dendritic Cells in Fundamental and Clinical Immunology*, pp. 275–80, 1993.
26. Romani, N., Gruner, S., Brang, D., Kämpgen, E., Lenz, A., Trockenbacher, B., Konwalinka, G., Fritsch, P.O., Steinman, R.M., Schuler, G., Proliferating dendritic cell progenitors in human blood. *J. Exp. Med.*, 180, 1, 83–93, 1994.
27. Sallusto, F. and Lanzavecchia, A., Efficient presentation of soluble antigen by cultured human dendritic cells is maintained by granulocyte/macrophage colony-stimulating factor plus interleukin 4 and downregulated by tumor necrosis factor alpha. *J. Exp. Med.*, 179, 4, 1109–18, 1994.
28. Lee, A.W., Truong, T., Bickham, K., Fonteneau, J.F., Larsson, M., Da Silva, I., Somersan, S., Thomas, E.K., Bhardwaj, N., A clinical grade cocktail of cytokines and PGE2 results in uniform maturation of human monocyte-derived dendritic cells: Implications for immunotherapy. *Vaccine*, 20, A8–22, 2002.
29. Carralot, J.P., Weide, B., Schoor, O., Probst, J., Scheel, B., Teufel, R., Hoerr, I., Garbe, C., Rammensee, H.G., Pascolo, S., Production and characterization of amplified tumor-derived cRNA libraries to be used as vaccines against metastatic melanomas. *Genet. Vaccines Ther.*, 3, 1–0, 2005.

30. Scheel, B., Teufel, R., Probst, J., Carralot, J.P., Geginat, J., Radsak, M., Jarrossay, D., Wagner, H., Jung, G., Rammensee, H.G., Hoerr, I., Toll-like receptor-dependent activation of several human blood cell types by protamine-condensed mRNA. *Eur. J. Immunol.*, 35, 5, 1557–66, 2005.
31. Scheel, B., Aulwurm, S., Probst, J., Stitz, L., Hoerr, I., Rammensee, H.G., Weller, M., Pascolo, S., Therapeutic anti-tumor immunity triggered by injections of immunostimulating single-stranded RNA. *Eur. J. Immunol.*, 36, 10, 2807–16, 2006.
32. Heiser, A., Maurice, M.A., Yancey, D.R., Coleman, D.M., Dahm, P., Vieweg, J., Human dendritic cells transfected with renal tumor RNA stimulate polyclonal T-cell responses against antigens expressed by primaryand metastatic tumors. *Cancer Res.*, 61, 8, 3388–93, 2001.
33. Milazzo, C., Reichardt, V.L., Müller, M.R., Grünebach, F., Brossart, P., Induction of myeloma-specific cytotoxic T cells using dendritic cells transfected with tumor-derived RNA. *Blood, The Journal of the American Society of Hematology*, 101, 3, 977–82, 2003.
34. Oka, Y., Elisseeva, O.A., Tsuboi, A., Ogawa, H., Tamaki, H., Li, H., Oji, Y., Kim, E.H., Soma, T., Asada, M., Ueda, K., Human cytotoxic T-lymphocyte responses specific for peptides of the wild-type Wilms' tumor gene (WT1) product. *Immunogenetics*, 51, 99–107, 2000.
35. Caux, C., Massacrier, C., Vanbervliet, B., Dubois, B., Van Kooten, C., Durand, I., Banchereau, J., Activation of human dendritic cells through CD40 cross-linking. *J. Exp. Med.*, 180, 4, 1263–72, 1994.
36. Hart, D.N. and Prickett, T.C., Intercellular adhesion molecule-2 (ICAM-2) expression on human dendritic cells. *Cell. Immunol.*, 148, 2, 447–54, 1993.
37. Prickett, T.C., Mckenzie, J.L., Hart, D.N., Adhesion molecules on human tonsil dendritic cells 1. *Transplantation.*, 53, 2, 483–90, 1992.
38. Bevan, M.J., Cross-priming for a secondary cytotoxic response to minor H antigens with H-2 congenic cells which do not cross-react in the cytotoxic assay. *J. Exp. Med.*, 143, 5, 1283–8, 1976.
39. Boczkowski, D., Nair, S.K., Snyder, D., Gilboa, E., Dendritic cells pulsed with RNA are potent antigen-presenting cells *in vitro* and *in vivo*. *J. Exp. Med.*, 184, 2, 465–72, 1996.
40. Caux, C., Dezutter-Dambuyant, C., Schmitt, D., Banchereau, J., GM-CSF and TNF-α cooperate in the generation of dendritic Langerhans cells. *Nature*, 360, 6401, 258–61, 1992.
41. Milazzo, C., Reichardt, V.L., Müller, M.R., Grünebach, F., Brossart, P., Induction of myeloma-specific cytotoxic T cells using dendritic cells transfected with tumor-derived RNA. *Blood, The Journal of the American Society of Hematology*, 101, 3, 977–82, 2003.
42. Heiser, A., Maurice, M.A., Yancey, D.R., Coleman, D.M., Dahm, P., Vieweg, J., Human dendritic cells transfected with renal tumor RNA stimulate polyclonal T-cell responses against antigens expressed by primaryand metastatic tumors. *Cancer Res.*, 61, 8, 3388–93, 2001.

43. Grünebach, F., Kayser, K., Weck, M.M., Müller, M.R., Appel, S., Brossart, P., Cotransfection of dendritic cells with RNA coding for HER-2/neu and 4-1BBL increases the induction of tumor antigen specific cytotoxic T lymphocytes. *Cancer Gene Ther.*, 12, 9, 749–56, 2005.
44. Heiser, A., Coleman, D., Dannull, J., Yancey, D., Maurice, M.A., Lallas, C.D., Dahm, P., Niedzwiecki, D., Gilboa, E., Vieweg, J., Autologous dendritic cells transfected with prostate-specific antigen RNA stimulate CTL responses against metastatic prostate tumors. *J. Clin. Invest.*, 109, 3, 409–17, 2002.
45. Bedrosian, I., Mick, R., Xu, S., Nisenbaum, H., Faries, M., Zhang, P., Cohen, P.A., Koski, G., Czerniecki, B.J., Intranodal administration of peptide-pulsed mature dendritic cell vaccines results in superior CD8+ T-cell function in melanoma patients. *J. Clin. Oncol.*, 21, 20, 3826–35, 2003.
46. Fong, L., Brockstedt, D., Benike, C., Wu, L., Engleman, E.G., Dendritic cells injected via different routes induce immunity in cancer patients. *J. Immunol.*, 166, 6, 4254–9, 2001.
47. Kyte, J.A., Mu, L., Aamdal, S., Kvalheim, G., Dueland, S., Hauser, M., Gullestad, H.P., Ryder, T., Lislerud, K., Hammerstad, H., Gaudernack, G., Phase I/II trial of melanoma therapy with dendritic cells transfected with autologous tumor-mRNA. *Cancer Gene Ther.*, 13, 10, 905–18, 2006.
48. Dueland, S., Mu, L.J., Kvalheim, G., Hauser, M., Waehre, H., Aamdal, S., Gaudernack, G., Dendritic cells transfected with allo-tumor mRNA as cancer vaccine in treatment of hormone resistant prostate cancer patients. *J. Clin. Oncol.*, 23, 16_suppl, 2541, 2005.
49. De Vries, I.J., Lesterhuis, W.J., Barentsz, J.O., Verdijk, P., Van Krieken, J.H., Boerman, O.C., Oyen, W.J., Bonenkamp, J.J., Boezeman, J.B., Adema, G.J., Bulte, J.W., Magnetic resonance tracking of dendritic cells in melanoma patients for monitoring of cellular therapy. *Nat. Biotechnol.*, 23, 11, 1407–13, 2005.
50. Dannull, J., Su, Z., Rizzieri, D., Yang, B.K., Coleman, D., Yancey, D., Zhang, A., Dahm, P., Chao, N., Gilboa, E., Vieweg, J., Enhancement of vaccine-mediated antitumor immunity in cancer patients after depletion of regulatory T cells. *J. Clin. Invest.*, 115, 12, 3623–33, 2005.
51. Kristiansen, O.P., Larsen, Z.M., Pociot, F., CTLA-4 in autoimmune diseases–a general susceptibility gene to autoimmunity? *Genes Immun.*, 1, 3, 170–84, 2000.
52. Tivol, E.A., Borriello, F., Schweitzer, A.N., Lynch, W.P., Bluestone, J.A., Sharpe, A.H., Loss of CTLA-4 leads to massive lymphoproliferation and fatal multiorgan tissue destruction, revealing a critical negative regulatory role of CTLA-4. *Immunity*, 3, 5, 541–7, 1995.
53. Ribas, A., Camacho, L.H., Lopez-Berestein, G., Pavlov, D., Bulanhagui, C.A., Millham, R., Comin-Anduix, B., Reuben, J.M., Seja, E., Parker, C.A., Sharma, A., Antitumor activity in melanoma and anti-self responses in a phase I trial with the anti-cytotoxic T lymphocyte-associated antigen 4 monoclonal antibody CP-675,206. *J. Clin. Oncol.*, 23, 8968–77, 25, 2005.

54. Robert, C. and Ghiringhelli, F., What is the role of cytotoxic T lymphocyte–associated antigen 4 blockade in patients with metastatic melanoma?. *The Oncologist.*, 14, 8, 848–61, 2009.
55. Burnett, J.C. and Rossi, J.J., RNA-based therapeutics: Current progress and future prospects. *Chem. Biol.*, 19, 1, 60–71, 2012.
56. Burnett, J.C., Rossi, J.J., Tiemann, K., Current progress of siRNA/shRNA therapeutics in clinical trials. *Biotechnol. J.*, 6, 9, 1130–46, 2011.
57. Mansoori, B., Shotorbani, S.S., Baradaran, B., RNA interference and its role in cancer therapy. *Adv. Pharm. Bull.*, 4, 4, 313, 2014.
58. Davis, M.E., Zuckerman, J.E., Choi, C.H., Seligson, D., Tolcher, A., Alabi, C.A., Yen, Y., Heidel, J.D., Ribas, A., Evidence of RNAi in humans from systemically administered siRNA via targeted nanoparticles. *Nature*, 464, 7291, 1067–70, 2010.
59. Damase, T.R., Sukhovershin, R., Boada, C., Taraballi, F., Pettigrew, R.I., Cooke, J.P., The limitless future of RNA therapeutics. *Front. Bioeng. Biotechnol.*, 9, 161, 2021.
60. Halloy, F., Biscans, A., Bujold, K.E., Debacker, A., Hill, A.C., Lacroix, A., Luige, O., Strömberg, R., Sundstrom, L., Vogel, J., Ghidini, A., Innovative developments and emerging technologies in RNA therapeutics. *RNA Biol.*, 19, 1, 313–32, 2022.

Index

Printed and bound by CPI Group (UK) Ltd, Croydon, CR0 4YY

17/09/2026

14964279-0004